Beginning iOS 5 Development

Exploring the iOS SDK

Dave Mark
Jack Nutting
Jeff LaMarche

Apress®

Beginning iOS 5 Development: Exploring the iOS SDK

ISBN-13 (pbk): 978-1-4302-3605-4

ISBN-13 (electronic): 978-1-4302-3606-1

President and Publisher: Paul Manning
Lead Editor: Tom Welsh
Technical Reviewer: Mark Dalrymple
Editorial Board: Steve Anglin, Mark Beckner, Ewan Buckingham, Gary Cornell, Morgan Ertel,
    Jonathan Gennick, Jonathan Hassell, Robert Hutchinson, Michelle Lowman,
    James Markham, Matthew Moodie, Jeff Olson, Jeffrey Pepper, Douglas Pundick, Ben
    Renow-Clarke, Dominic Shakeshaft, Gwenan Spearing, Matt Wade, Tom Welsh
Coordinating Editor: Kelly Moritz
Copy Editor: Marilyn Smith
Compositor: MacPS, LLC
Indexer: BIM Indexing & Proofreading Services
Artist: SPi Global
Cover Designer: Anna Ishchenko

Distributed to the book trade worldwide by Springer Science+Business Media New York, 233 Spring Street, 6th Floor, New York, NY 10013. Phone 1-800-SPRINGER, fax (201) 348-4505, e-mail orders-ny@springer-sbm.com, or visit www.springeronline.com.

For information on translations, please e-mail rights@apress.com, or visit www.apress.com.

Apress and friends of ED books may be purchased in bulk for academic, corporate, or promotional use. eBook versions and licenses are also available for most titles. For more information, reference our Special Bulk Sales–eBook Licensing web page at www.apress.com/bulk-sales.

Any source code or other supplementary materials referenced by the author in this text is available to readers at www.apress.com. For detailed information about how to locate your book's source code, go to www.apress.com/source-code/.

This book is dedicated to the memory of Steve Jobs.
We continue to be inspired by his spirit and his vision.

Contents at a Glance

Contents

About the Authors

 **Dave Mark** is a longtime Mac developer and author, who has written a number of books on Mac and iOS development, including *Beginning iPhone 4 Development* (Apress, 2011), *More iPhone 3 Development* (Apress, 2010), *Learn C on the Mac* (Apress, 2008), *Ultimate Mac Programming* (Wiley, 1995), and *The Macintosh Programming Primer* series (Addison-Wesley, 1992). Dave was one of the founders of MartianCraft, an iOS and Android development house. Dave loves the water and spends as much time as possible on it, in it, or near it. He lives with his wife and three children in Virginia.

 **Jack Nutting** has been using Cocoa since the olden days, long before it was even called Cocoa. He has used Cocoa and its predecessors to develop software for a wide range of industries and applications, including gaming, graphic design, online digital distribution, telecommunications, finance, publishing, and travel. When he is not working on Mac or iOS projects, he is developing web applications with Ruby on Rails. Jack is a passionate proponent of Objective-C and the Cocoa frameworks. At the drop of a hat, he will speak at length on the virtues of dynamic dispatch and runtime class manipulations to anyone who will listen (and even to some who won't). Jack has written several books on iOS and Mac development, including *Beginning iPhone 4 Development* (Apress, 2011), *Learn Cocoa on the Mac* (Apress, 2010), and *Beginning iPad Development for iPhone Developers* (Apress, 2010). He blogs from time to time at www.nuthole.com.

 **Jeff LaMarche** is a Mac and iOS developer with more than 20 years of programming experience. Jeff has written a number of iOS and Mac development books, including *Beginning iPhone 4 Development* (Apress, 2011), *More iPhone 3 Development* (Apress, 2010), and *Learn Cocoa on the Mac* (Apress, 2010). Jeff is a principal at MartianCraft, an iOS and Android development house. He has written about Cocoa and Objective-C for *MacTech Magazine*, as well as articles for Apple's developer web site. Jeff also writes about iOS development for his widely read blog at www.iphonedevelopment.blogspot.com.

About the Technical Reviewer

 **Mark Dalrymple** is a longtime Mac and Unix programmer, working on cross-platform tool kits, Internet publishing tools, high-performance web servers, and end-user desktop applications. He is also the principal author of *Learn Objective-C on the Mac* (Apress, 2009) and *Advanced Mac OS X Programming* (Big Nerd Ranch, 2005). In his spare time, Mark plays trombone and bassoon, and makes balloon animals.

Acknowledgments

This book could not have been written without our mighty, kind, and clever families, friends, and cohorts. First and foremost, eternal thanks to Terry, Weronica, and Deneen for putting up with us, and for keeping the rest of the universe at bay while we toiled away on this book. This project saw us tucked away in our writers' cubby for many long hours, and somehow, you didn't complain once. We are lucky men.

This book could not have been written without the fine folks at Apress. Clay Andres brought us to Apress in the first place and carried the first few iterations of this book on his back. Dominic Shakeshaft and Steve Anglin were the gracious masterminds who dealt with all of our complaints with a smile on their faces, and somehow found solutions that made sense and made this book better. Kelly Moritz, our wonderful and gracious coordinating editor, was the irresistible force to our slowly movable object. Tom Welsh, our developmental editor, helped us with some terrific feedback along the way. They kept the book on the right track and always pointed in the right direction. Marilyn Smith, copy editor *extraordinaire*, you were such a pleasure to work with! Jeffrey Pepper, Frank McGuckin, Brigid Duffy, and the Apress production team took all these pieces and somehow made them whole. Dylan Wooters assembled the marketing message and got it out to the world. To all the folks at Apress, thank you, thank you, thank you!

A very special shout-out to our incredibly talented technical reviewer, Mark Dalrymple. In addition to providing insightful feedback, Mark tested all the code in this book and helped keep us on the straight and narrow. Thanks, Mark!

Finally, thanks to our children for their patience while their dads were working so hard. This book is for you, Maddie, Gwynnie, Ian, Kai, Henrietta, Dorotea, Daniel, Kelley, and Ryan.

Preface

Hard as it is for us to believe, you now hold in your hands (or see on your screen) the fourth edition of this book. In the years since we set out on this journey, we've poured more blood, sweat, and tears than we ever imagined into this book, in an attempt to give developers the best introduction to the fantastic and sometimes surprising world of Cocoa Touch development. We've also had a lot of fun along the way, and we hope that you will, too.

This edition of the book has been rebuilt from the ground up to cover the exciting new changes Xcode 4 brings to the table. Apple reengineered huge portions of Xcode when transitioning from Xcode 3 to Xcode 4, and again as it moved to the current version (as of this writing), Xcode 4.2. We've followed suit. Every project in the book has been written from scratch using the amazing technology built into Xcode 4.2.

And, of course, as the title of this new edition implies, each and every project was designed to work properly under iOS 5. The iOS SDK has evolved significantly with this latest iOS release. As you might expect, there are many new changes to the project templates and a lot of new ways to do the things you've always done. And, of course, there's a lot of new technology to master. We've written entirely new chapters on using both storyboards and iCloud, we've covered new strategies for dealing with table views, and we've re-created every example project using the Automatic Reference Counting (ARC) feature to simplify memory management.

In short, we've made this latest edition the biggest, most substantial version of the book so far. Whether you're new to iOS development or have been working with it for a while, we think you'll like the new material covered by this volume. If you haven't made it through a previous edition of this book yet, if you feel a bit fuzzy still, or if you just want to help us out as authors, by all means, pick up this fourth edition. We do appreciate your support. Be sure to check out the book's official community forum at http://iphonedevbook.com, and drop us a line to let us know about your amazing new apps. We look forward to seeing you on the forum. Happy coding!

Dave, Jack, and Jeff

Chapter 1

Welcome to the Jungle

So, you want to write iPhone, iPod touch, and iPad applications? Well, we can't say that we blame you. iOS, the core software of all of these devices, is an exciting platform that has been seeing explosive growth since it first came out in 2007. The rise of the mobile software platform means that people are using software everywhere they go. With the release of iOS 5, and the latest incarnation of the iOS software development kit (SDK), things have only gotten better and more interesting.

What This Book Is

This book is a guide to help you get started down the path to creating your own iOS applications. Our goal is to get you past the initial learning curve, to help you understand the way iOS applications work and how they are built.

As you work your way through this book, you will create a number of small applications, each designed to highlight specific iOS features and show you how to control or interact with those features. If you combine the foundation you'll gain through this book with your own creativity and determination, and then add in the extensive and well-written documentation provided by Apple, you'll have everything you need to build your own professional iPhone and iPad applications.

> **TIP:** Dave, Jack, and Jeff have a forum set up for this book. It's a great place to meet like-minded folks, get your questions answered, and even answer other people's questions. The forum is at `http://iphonedevbook.com`. Be sure to check it out!

What You Need

Before you can begin writing software for iOS, you'll need a few items. For starters, you'll need an Intel-based Macintosh running Lion (OS X 10.7) or later. Any recent Intel-based Macintosh computer—laptop or desktop—should work just fine.

You'll also need to sign up to become a registered iOS developer. Apple requires this step before you're allowed to download the iOS SDK.

To sign up as a developer, just navigate to `http://developer.apple.com/ios/`. That will bring you to a page similar to the one shown in Figure 1–1.

Figure 1–1. *Apple's iOS Dev Center web site*

First, click the button labeled *Log in*. You'll be prompted for your *Apple ID*. If you don't have an Apple ID, click the *Create Apple ID* button, create one, and then log in. Once you are logged in, you'll be taken to the main iOS development page. Not only will you see a link to the SDK download, but you'll also find links to a wealth of documentation, videos, sample code, and the like—all dedicated to teaching you the finer points of iOS application development.

The most important tool you'll be using to develop iOS applications is called Xcode. Xcode is Apple's integrated development environment (IDE). Xcode includes tools for creating and debugging source code, compiling applications, and performance tuning the applications you've written.

You can find a download link for Xcode on `http://developer.apple.com/ios/` once you've signed up. You can also download Xcode from the Macintosh App Store, which you can access from your Mac's Apple menu.

SDK VERSIONS AND SOURCE CODE FOR THE EXAMPLES

As the versions of the SDK and Xcode evolve, the mechanism for downloading them will also change. Sometimes the SDK and Xcode are featured as separate downloads; other times, they will be merged as a single download. Bottom line: you want to download the latest released (non-beta) version of Xcode and the iOS SDK.

This book has been written to work with the latest version of the SDK. In some places, we have chosen to use new functions or methods introduced with iOS 5 that may prove incompatible with earlier versions of the SDK. We'll be sure to point those situations out as they arise in this book.

Be sure to download the latest and greatest source code archives from the book's web site at `http://iphonedevbook.com` or from the book's page on `http://apress.com`. We'll update the code as new versions of the SDK are released, so be sure to check the site periodically.

Developer Options

The free SDK download option includes a simulator that will allow you to build and run iPhone and iPad apps on your Mac. This is perfect for learning how to program for iOS. However, the simulator does *not* support many hardware-dependent features, such as the accelerometer and camera. Also, the free option will not allow you to download your applications onto your actual iPhone or other device, and it does not give you the ability to distribute your applications on Apple's App Store. For those capabilities, you'll need to sign up for one of the other options, which aren't free:

▨ The Standard program costs $99/year. It provides a host of development tools and resources, technical support, distribution of your application via Apple's App Store, and, most important, the ability to test and debug your code on an iOS device, rather than just in the simulator.

▤   The Enterprise program costs $299/year. It is designed for companies developing proprietary, in-house iOS applications and for those developing applications for the Apple's App Store with more than one developer working on the project.

For more details on these programs, visit `http://developer.apple.com/programs/ios` and `http://developer.apple.com/programs/ios/enterprise` to compare the two.

Because iOS supports an always-connected mobile device that uses other companies' wireless infrastructure, Apple has needed to place far more restrictions on iOS developers than it ever has on Mac developers (who are able—at least as of this writing—to write and distribute programs with absolutely no oversight or approval from Apple). Even though the iPod touch and the Wi-Fi–only versions of the iPad don't use anyone else's infrastructure, they're still subject to these same restrictions.

Apple has not added restrictions to be mean, but rather as an attempt to minimize the chances of malicious or poorly written programs being distributed that could degrade performance on the shared network. Developing for iOS may seem like it presents a lot of hoops to jump through, but Apple has expended quite an effort to make the process as painless as possible. And also consider that $99 is still considerably less than buying, for example, Visual Studio, which is Microsoft's software development IDE.

This may seem obvious, but you'll also need an iPhone, iPod touch, or iPad. While much of your code can be tested using the iOS simulator, not all programs can be. And even those that can run on the simulator really need to be thoroughly tested on an actual device before you ever consider releasing your application to the public.

> **NOTE:** If you are going to sign up for the Standard or Enterprise program, you should do it right now. The approval process can take a while, and you'll need that approval to be able to run your applications on an actual device. Don't worry, though, because all the projects in the first several chapters and the majority of the applications in this book will run just fine on the iOS simulator.

What You Need to Know

This book assumes that you already have some programming knowledge. It assumes that you understand the fundamentals of object-oriented programming (you know what objects, loops, and variables are, for example). It also assumes that you are familiar with the Objective-C programming language. Cocoa Touch, the part of the SDK that you will be working with through most of this book, uses the latest version of Objective-C, which contains several new features not present in earlier versions. But don't worry if you're not familiar with the more recent additions to the Objective-C language. We highlight any of the new language features we take advantage of, and explain how they work and why we are using them.

You should also be familiar with iOS itself, as a user. Just as you would with any platform for which you wanted to write an application, get to know the nuances and quirks of the iPhone, iPad, or iPod touch. Take the time to get familiar with the iOS interface and with the way Apple's iPhone and/or iPad applications look and feel.

NEW TO OBJECTIVE-C?

If you have not programmed in Objective-C before, here are a few resources to help you get started:

- Check out *Learn Objective-C on the Mac*, an excellent and approachable introduction to Objective-C by Mac programming experts Mark Dalrymple and Scott Knaster (Apress, 2009):

http://www.apress.com/book/view/9781430218159

- See Apple's introduction to the language, *Learning Objective-C: A Primer*.

http://developer.apple.com/library/ios/#referencelibrary/↵
     GettingStarted/Learning_Objective-C_A_Primer

- Take a look at *The Objective-C Programming Language*, a very detailed and extensive description of the language and a great reference guide:

http://developer.apple.com/library/ios/#documentation/Cocoa/Conceptual/Ob
jectiveC

That last one is also available as a free download from iBooks on your iPhone, iPod touch, or iPad. It's perfect for reading on the go! Apple has released several developer titles in this format, and we hope that more are on the way. Search for "Apple developer publications" in iBooks to find them.

What's Different About Coding for iOS?

If you have never programmed in Cocoa or its predecessors NeXTSTEP or OpenStep, you may find Cocoa Touch—the application framework you'll be using to write iOS applications—a little alien. It has some fundamental differences from other common application frameworks, such as those used when building .NET or Java applications. Don't worry too much if you feel a little lost at first. Just keep plugging away at the exercises, and it will all start to fall into place after a while.

If you have written programs using Cocoa or NeXTSTEP, a lot in the iOS SDK will be familiar to you. A great many classes are unchanged from the versions that are used to develop for Mac OS X. Even those that are different tend to follow the same basic principles and similar design patterns. However, several differences exist between Cocoa and Cocoa Touch.

Regardless of your background, you need to keep in mind some key differences between iOS development and desktop application development. These differences are discussed in the following sections.

Only One Active Application

On iOS, only one application can be active and displayed on the screen at any given time. Since iOS 4, applications have been able to run in the background after the user presses the home button, but even that is limited to a narrow set of situations, and you must code for it specifically.

When your application isn't active or running in the background, it doesn't receive any attention from the CPU whatsoever, which will wreak havoc with open network connections and the like. iOS 5 makes great strides forward in allowing background processing, but making your apps play nicely in this situation will require some effort on your part.

Only One Window

Desktop and laptop operating systems allow many running programs to coexist, each with the ability to create and control multiple windows. However, iOS gives your application just one "window" to work with. All of your application's interaction with the user takes place inside this one window, and its size is fixed at the size of the screen.

Limited Access

Programs on a computer pretty much have access to everything the user who launched them does. However, iOS seriously restricts what your application can access.

You can read and write files only from the part of iOS's file system that was created for your application. This area is called your application's **sandbox**. Your sandbox is where your application will store documents, preferences, and every other kind of data it may need to retain.

Your application is also constrained in some other ways. You will not be able to access low-number network ports on iOS, for example, or do anything else that would typically require root or administrative access on a desktop computer.

Limited Response Time

Because of the way it is used, iOS needs to be snappy, and it expects the same of your application. When your program is launched, you need to get your application open, preferences and data loaded, and the main view shown on the screen as fast as possible—in no more than a few seconds.

At any time when your program is running, it may have the rug pulled out from under it. If the user presses the home button, iOS goes home, and you must quickly save everything and quit. If you take longer than five seconds to save and give up control, your application process will be killed, regardless of whether you are finished saving.

Note that in iOS 5, this situation is ameliorated somewhat by the existence of new API that allows your app to ask for additional time to work when it's about to go dark.

Limited Screen Size

The iPhone's screen is really nice. When introduced, it was the highest resolution screen available on a consumer device, by far.

But the iPhone display just isn't all that big, and as a result, you have a lot less room to work with than on modern computers. The screen is just 640 × 960 on the latest retina display devices (iPhone 4 and fourth-generation iPod touch) and 320 × 480 pixels on older devices. And that 640 × 960 retina display is crammed into the same old form factor, so you can't count on fitting more controls or anything like that; they will all just be higher resolution than before.

The iPad increases the available space a bit by offering a 1024 × 768 display, but even today, that's not so terribly large. To give an interesting contrast, at the time of this writing, Apple's least expensive iMac supports 1920 × 1080 pixels, and its least expensive notebook computer, the MacBook, supports 1280 × 800 pixels. On the other end of the spectrum, Apple's largest current monitor, the 27-inch LED Cinema Display, offers a whopping 2560 × 1440 pixels.

Limited System Resources

Any old-time programmers who are reading this are likely laughing at the idea of a machine with at least 256MB of RAM and 8GB of storage being in any way resource-constrained, but it is true. Developing for iOS is not, perhaps, in exactly the same league as trying to write a complex spreadsheet application on a machine with 48KB of memory. But given the graphical nature of iOS and all it is capable of doing, running out of memory is very easy.

The iOS devices available right now have either 256MB or 512MB of physical RAM, though that will likely increase over time. Some of that memory is used for the screen buffer and by other system processes. Usually, no more than half of that memory is left for your application to use, and the amount can be considerably less, especially now that apps can run in the background.

Although that may sound like it leaves a pretty decent amount of memory for such a small computer, there is another factor to consider when it comes to memory on iOS. Modern computer operating systems like Mac OS X will take chunks of memory that aren't being used and write them out to disk in something called a **swap file**. The swap file allows applications to keep running, even when they have requested more memory than is actually available on the computer. iOS, however, will not write volatile memory, such as application data, out to a swap file. As a result, the amount of memory available to your application is constrained by the amount of unused physical memory in the iOS device.

Cocoa Touch has built-in mechanisms for letting your application know that memory is getting low. When that happens, your application must free up unneeded memory or risk being forced to quit.

No Garbage Collection, but...

We mentioned earlier that Cocoa Touch uses Objective-C, but one of the key new features of that language is not available with iOS: Cocoa Touch does not support garbage collection. The need to do manual memory management when programming for iOS has been a bit of a stumbling block for many programmers new to the platform, especially those coming from languages that offer garbage collection.

With the version of Objective-C supported by iOS 5, however, this particular stumbling block is basically gone. iOS 5 introduces a feature called Automatic Reference Counting (ARC), which gets rid of the need to manually manage memory for Objective-C objects. We'll talk about ARC in Chapter 3.

Some New Stuff

Since we've mentioned that Cocoa Touch is missing some features that Cocoa has, it seems only fair to mention that the iOS SDK contains some functionality that is not currently present in Cocoa or, at least, is not available on every Mac:

- The iOS SDK provides a way for your application to determine the iOS device's current geographic coordinates using Core Location.

- Most iOS devices have built-in cameras and photo libraries, and the SDK provides mechanisms that allow your application to access both.

- iOS devices have a built-in accelerometer (and, in the latest iPhone and iPod touch, a gyroscope) that lets you detect how your device is being held and moved.

A Different Approach

Two things iOS devices don't have are a physical keyboard and a mouse, which means you have a fundamentally different way of interacting with the user than you do when programming for a general-purpose computer. Fortunately, most of that interaction is handled for you. For example, if you add a text field to your application, iOS knows to bring up a keyboard when the user clicks in that field, without you needing to write any extra code.

> **NOTE:** Current devices do allow you to connect an external keyboard via Bluetooth, which gives you a nice keyboard experience and saves some screen real estate, but this is still a fairly rare usage. Connecting a mouse is still not an option.

What's in This Book

Here is a brief overview of the remaining chapters in this book:

- In Chapter 2, you'll learn how to use Xcode's partner in crime, Interface Builder, to create a simple interface, placing some text on the screen.

- In Chapter 3, you'll start interacting with the user, building a simple application that dynamically updates displayed text at runtime based on buttons the user presses.

- Chapter 4 will build on Chapter 3 by introducing you to several more of iOS's standard user interface controls. We'll also demonstrate how to use alerts and action sheets to prompt users to make a decision or to inform them that something out of the ordinary has occurred.

- In Chapter 5, we'll look at handling autorotation and autosize attributes, the mechanisms that allow iOS applications to be used in both portrait and landscape modes.

- In Chapter 6, we'll move into more advanced user interfaces and explore creating applications that support multiple views. We'll show you how to change which view is being shown to the user at runtime, which will greatly enhance the potential of your apps.

- Tab bars and pickers are part of the standard iOS user interface. In Chapter 7, we'll look at how to implement these interface elements.

- In Chapter 8, we'll cover table views, the primary way of providing lists of data to the user and the foundation of hierarchical navigation-based applications. You'll also see how to let the user search in your application data.

- One of the most common iOS application interfaces is the hierarchical list that lets you drill down to see more data or more details. In Chapter 9, you'll learn what's involved in implementing this standard type of interface.

- iOS 5 brings a new way to design your apps called **storyboards**. Chapter 10 covers this great new feature.

- The iPad, with its different form factor from the other iOS devices, requires a different approach to displaying a GUI and provides some components to help make that happen. In Chapter 11, we'll show you how to use the iPad-specific parts of the SDK.

- In Chapter 12, we'll look at implementing application settings, which is iOS's mechanism for letting users set their application-level preferences.

▨ Chapter 13 covers data management on iOS. We'll talk about creating objects to hold application data and see how that data can be persisted to iOS's file system. We'll also discuss the basics of using Core Data, which allows you to save and retrieve data easily.

▨ Another new feature of iOS 5 is iCloud, which allows your document to store data online and sync it between different instances of the application. Chapter 14 shows you how to get started with iCloud.

▨ Since iOS 4, developers have access to a new approach to multithreaded development using Grand Central Dispatch, and also have the ability to make their apps run in the background in certain circumstances. In Chapter 15, we'll show you how that's done.

▨ Everyone loves to draw, so we'll look at doing some custom drawing in Chapter 16. We'll use basic drawing functions in Quartz 2D and OpenGL ES.

▨ The multitouch screen common to all iOS devices can accept a wide variety of gestural inputs from the user. In Chapter 17, you'll learn all about detecting basic gestures, such as the pinch and swipe. We'll also look at the process of defining new gestures and talk about when new gestures are appropriate.

▨ iOS is capable of determining its latitude and longitude thanks to Core Location. In Chapter 18, we'll build some code that makes use of Core Location to figure out where in the world your device is and use that information in our quest for world dominance.

▨ In Chapter 19, we'll look at interfacing with iOS's accelerometer and gyroscope, which is how your device knows which way it's being held and the speed and direction in which it is moving. We'll explore some of the fun things your application can do with that information.

▨ Nearly every iOS device has a camera and a library of pictures, both of which are available to your application, if you ask nicely! In Chapter 20, we'll show you how to ask nicely.

▨ iOS devices are currently available in more than 90 countries. In Chapter 21, we'll show you how to write your applications in such a way that all parts can be easily translated into other languages. This helps expand the potential audience for your applications.

▨ By the end of this book, you'll have mastered the fundamental building blocks for creating iPhone and iPad applications. But where do you go from here? In Chapter 22, we'll explore the logical next steps for you to take on your journey to master the iOS SDK.

What's New in This Update?

Since the first edition of this book hit the bookstores, the growth of the iOS development community has been phenomenal. The SDK has continually evolved, with Apple releasing a steady stream of SDK updates.

Well, we've been busy, too! The second we found out about iOS SDK 5, we immediately went to work, updating every single project to ensure not only that the code compiles using the latest version of Xcode and the SDK, but also that each one takes advantage of the latest and greatest features offered by Cocoa Touch. We made a ton of subtle changes throughout the book, and added a good amount of substantive changes as well, including two brand-new chapters: one on storyboarding and another on iCloud. And, of course, we reshot every screen shown in the book.

Are You Ready?

iOS is an incredible computing platform and an exciting new frontier for your development pleasure. Programming for iOS is going to be a new experience—different from working on any other platform. For everything that looks familiar, there will be something alien, but as you work through the book's code, the concepts should all come together and start to make sense.

Keep in mind that the exercises in this book are not simply a checklist that, when completed, magically grants you iOS developer guru status. Make sure you understand what you did and why before moving on to the next project. Don't be afraid to make changes to the code. Observing the results of your experimentation is one of the best ways you can wrap your head around the complexities of coding in an environment like Cocoa Touch.

That said, if you have your iOS SDK installed, turn the page. If not, get to it! Got it? Good. Then let's go!

Chapter 2

Appeasing the Tiki Gods

As you're probably well aware, it has become something of a tradition to call the first project in any book on programming "Hello, World." We considered breaking this tradition, but were scared that the tiki gods would inflict some painful retribution on us for such a gross breach of etiquette. So, let's do it by the book, shall we?

In this chapter, we're going to use Xcode to create a small iOS application that will display the text "Hello, World!" We'll look at what's involved in creating an iOS application project in Xcode, work through the specifics of using Xcode's Interface Builder to design our application's user interface, and then run our application on the iOS simulator. After that, we'll give our application an icon to make it feel more like a real iOS application.

We have a lot to do here, so let's get going.

Setting Up Your Project in Xcode

By now, you should have Xcode and the iOS SDK installed on your machine. You should also download the book project archive from the book web site (http://www.iphonedevbook.com/forum/forum.php). The book forums are a great place to download the latest book source code, get your questions answered, and meet up with like-minded people. Of course, you can also find the source code on the Apress web site.

> **NOTE:** Even though you have the complete set of project files at your disposal in this book's project archive, you'll get more out of the book if you create each project by hand, rather than simply running the version you downloaded. By doing that, you'll gain familiarity and expertise working with the various application development tools.
>
> There's no substitute for actually creating applications; software development is not a spectator sport.

The project we're going to build in this chapter is contained in the *02 Hello World* folder of the project archive.

Before we can start, we need to launch Xcode. Xcode is the tool that we'll use to do most of what we do in this book, but it's not installed in the */Applications* folder as with most Mac applications. If you've already installed the developer tools as outlined in the previous chapter, you'll find Xcode located in */Developer/Applications*. You'll be using Xcode a lot, so you might want to consider dragging it to your dock so you'll have ready access to it.

If this is your first time using Xcode, don't worry; we'll walk you through every step involved in creating a new project. Apple recently released a new, completely rewritten version of Xcode that's quite a bit different than the previous version. If you're already an old hand but haven't worked with Xcode 4, you will find that quite a bit has changed.

When you first launch Xcode, you'll be presented with a welcome window like the one shown in Figure 2–1. From here, you can choose to create a new project, connect to a version-control system to check out an existing project, or select from a list of recently opened projects. The welcome window also contains links to iOS and Mac OS X technical documentation, tutorial videos, news, sample code, and other useful items. All of this functionality can be accessed from the Xcode menu as well, but this window gives you a nice starting point, covering some of the most common tasks you're likely to want to do after launching Xcode. If you feel like poking through the information here for a few minutes, by all means, go right ahead. When you're finished, close the window, and we'll proceed. If you would rather not see this window in the future, just uncheck the *Show this window when Xcode launches* checkbox before closing it.

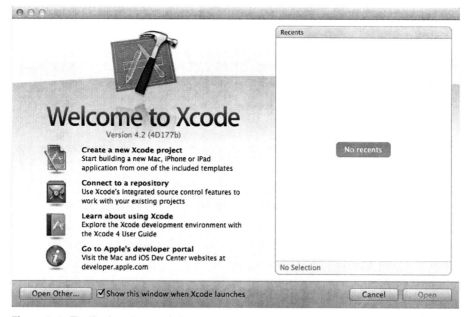

Figure 2–1. *The Xcode welcome window*

NOTE: If you have an iPhone, iPad, or iPod touch connected to your machine, you might see a message when you first launch Xcode asking whether you want to use that device for development. For now, click the *Ignore* button. Alternatively, the *Organizer* window, which shows (among other things) the devices that have been synchronized with your computer, might appear. In that case, just close the *Organizer* window. If you choose to join the paid iOS Developer Program, you will gain access to a program portal that will tell you how to use your iOS device for development and testing.

Create a new project by selecting **New ➤ New Project...** from the **File** menu (or by pressing ⇧⌘N). A new project window will open, and will show you the project template selection sheet (see Figure 2–2). From this sheet, you'll choose a project template to use as a starting point for building your application. The pane on the left side of the sheet is divided into two main sections: *iOS* and *Mac OS X*. Since we're building an iOS application, select *Application* in the *iOS* section to reveal the iOS application templates.

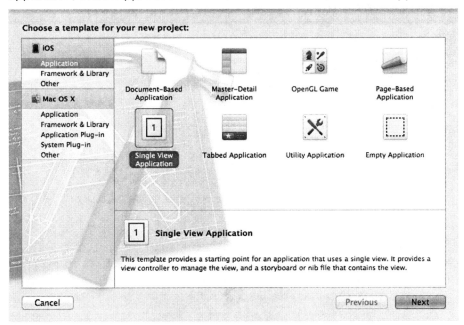

Figure 2–2. *The project template selection sheet lets you select from various templates when creating a new project.*

Each of the icons shown in the upper-right pane in Figure 2–2 represents a separate project template that can be used as a starting point for your iOS applications. The icon labeled *Single View Application* is the simplest template and the one we'll be using for the first several chapters. The other templates provide additional code and/or resources needed to create common iPhone and iPad application interfaces, as you'll see in later chapters.

Click the *Single View Application* icon (as in Figure 2–2), and then click the *Next* button. You'll see the project options sheet, which should look like Figure 2–3. On this sheet, you need to specify the *Product Name* and *Company Identifier* for your project. Xcode will combine the two of those to generate a unique *Bundle Identifier* for your app. Name your product *Hello World*, and then enter *com.apress* in the *Company Identifier* field, as shown in Figure 2–3. Later, after you've signed up for the developer program and learned about provisioning profiles, you'll want to use your own company identifier. We'll talk more about the bundle identifier later in the chapter.

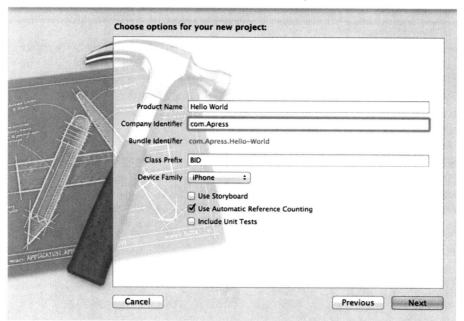

Figure 2–3. *Selecting a product name and company identifier for your project. Use these settings for now.*

The next text box is labeled *Class Prefix*, and we should populate this with a sequence of at least three capital letters. These characters will be added to the beginning of the name of all classes that Xcode creates for us. This is done to avoid naming conflicts with Apple (who reserves the use of all two-letter prefixes) and other developers whose code we might use. In Objective-C, having more than one class with the same name will prevent your application from being built.

For the projects in the book, we're going to use the prefix *BID*, which stands for **B**eginning **i**Phone **D**evelopment. While there are likely to be many classes named, for example, `ViewController`, far fewer classes are likely to be named `BIDMyViewController`, which means a lot less chance of conflicts.

We also need to specify the *Device Family*. In other words, Xcode wants to know if we're building an app for the iPhone and iPod touch, if we're building an app for the iPad, or if we're building a universal application that will run on all iOS devices. Select *iPhone* for the *Device Family* if it's not already selected. This tells Xcode that we'll be targeting this particular app at the iPhone and iPod touch, which have the same screen size. For the first part of the book, we'll be using the iPhone device family, but don't worry—we'll cover the iPad also.

There are three checkboxes on this sheet. You should check the middle option, *Use Automatic Reference Counting*, but uncheck the other two. Automatic Reference Counting (ARC) is a new feature of the Objective-C language, introduced with iOS 5, that makes your life much easier. We'll talk briefly about ARC in the next chapter.

The *Use Storyboard* option will be covered starting in Chapter 10. The other option— *Include Unit Tests*—will set up your project in such a way that you can add special pieces of code to your project, called **unit tests**, which are not part of your application, but run every time you create your application to test certain functionality. Unit tests allow you to identify when a change made to your code breaks something that was previously working. Although it can be a valuable tool, we won't be using automated unit testing in this book, so you can leave its box unchecked.

Click *Next* again, and you'll be asked where to save your new project using a standard save sheet, as shown in Figure 2–4. If you haven't already done so, jump over to the Finder and create a new master directory for these book projects, and then return to Xcode and navigate into that directory. Before you click the *Create* button, be sure to uncheck the *Create local git repository for this project* checkbox. With the *Source Control* checkbox unchecked, create the new project by clicking the *Create* button.

> **NOTE:** A source control repository is a tool used to keep track of changes made to an application's source code and resources while it's being built. It also facilitates multiple developers working on the same application at the same time by providing tools to resolve conflicts when they arise. We won't be using source control in this book, so you can leave its box unchecked.

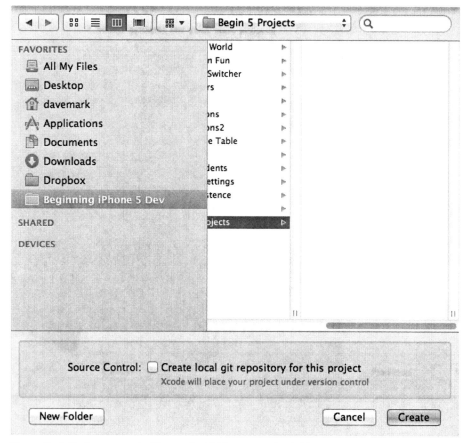

Figure 2–4. *Saving your project in a project folder on your hard drive*

The Xcode Workspace Window

After you dismiss the save sheet, Xcode will create and then open your project. You will see a new **workspace window**, as shown in Figure 2–5. There's a lot of information crammed into this window, and it's where you will be spending a lot of your iOS development time.

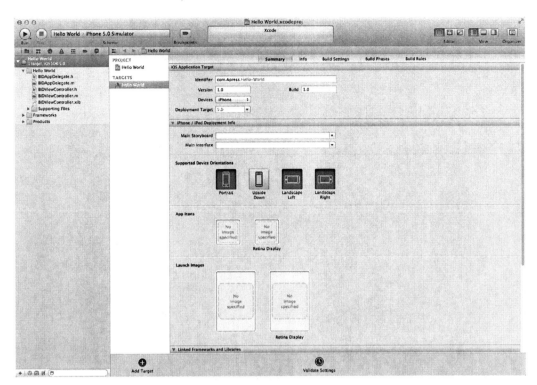

Figure 2–5. *The Hello World project in Xcode*

Even if you are an old hand with earlier versions of Xcode, you'll still benefit from reading through this section, as a *lot* has changed since the last release of Xcode 3.*x*. Let's take a quick tour.

The Toolbar

The top of the Xcode workspace window is called the **toolbar** (see Figure 2–6). On the left side of the toolbar are controls to start and stop running your project, a popup menu to select the scheme you want to run, and a button to toggle breakpoints on and off. A **scheme** brings together target and build settings, and the toolbar popup menu lets you select a specific setup with just one click.

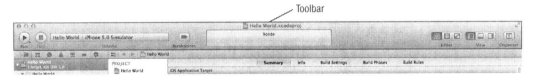

Figure 2–6. *The Xcode toolbar*

The big box in the middle of the toolbar is the **activity view**. As its name implies, the activity view displays any actions or processes that are currently happening. For example, when you run your project, the activity view gives you a running commentary on the various steps it's taking to build your application. If you encounter any errors or warnings, that information is displayed here as well. If you click the warning or error, you'll go directly to the issues navigator, which provides more information about the warning or error, as described in the next section.

On the right side of the toolbar are three sets of buttons. The left set, labeled *Editor*, lets you switch between three different editor configurations:

- The **standard view** gives you a single pane dedicated to editing a file or project-specific configuration values.

- The incredibly powerful **assistant view** splits the editor pane into two panes, left and right. The pane on the right is generally used to display a file that relates to the file on the left, or that you might need to refer to while editing the file on the left. You can manually specify what goes into each pane, or you can let Xcode decide what's most appropriate for the task at hand. For example, if you're editing the implementation of an Objective-C class (the *.m* file), Xcode will automatically show you that class's header file (the *.h* file) in the right pane. If you're designing your user interface on the left, Xcode will show you the code that user interface is able to interact with on the right. You'll see the assistant view at work throughout the book.

- The versions button converts the editor pane into a time-machine-like comparison view that works with source code management systems such as Subversion and Git. You can compare the current version of a source file with a previously committed version or compare any two earlier versions with each other.

To the right of the editor button set is another set of buttons that show and hide the navigator pane and the utility pane, on the left and right side of the editor pane. Click those buttons to see these panes in action.

Finally, the rightmost button brings up the *Organizer* window, which is where you'll find the bulk of nonproject-specific functionality. It's used as the documentation viewer for Apple's API documentation, shows you all the source code repositories that Xcode knows about, keeps a list of all projects you've opened, and maintains a list of all devices that you've synchronized with this computer.

The Navigator View

Just below the toolbar, on the left side of the workspace window, is the **navigator view**. The navigator view offers seven configurations that give you different views into your project. Click one of the icons at the top of the navigator view to switch among the following navigators, going from left to right:

- **Project navigator**: This view contains a list of files that are used by your project (see Figure 2–7). You can store references to everything you expect—from source code files to artwork, data models, property list (or **plist**) files (discussed in the "A Closer Look at Our Project" section later in this chapter), and even other project files. By storing multiple projects in a single workspace, multiple projects can easily share resources. If you click any file in the navigator view, that file will display in the editor pane. In addition to viewing the file, you can also edit the file (if it's a file that Xcode knows how to edit).

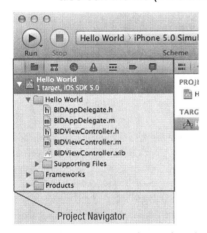

Figure 2–7. *The Xcode navigator view showing the project navigator. Click one of the seven icons at the top of the view to switch navigators.*

- **Symbol navigator**: As its name implies, this navigator focuses on the **symbols** defined in the workspace (see Figure 2–8). Symbols are basically the items that the compiler recognizes, such as Objective-C classes, enumerations, structs, and global variables.

Figure 2–8. *The Xcode navigator view showing the symbol navigator. Open the disclosure triangle to explore the files and symbols defined within each group.*

- **Search navigator**: You'll use this navigator to perform searches on all the files in your workspace (see Figure 2–9). You can select *Replace* from the *Find* popup menu, and do a search and replace on all or just selected portions of the search results. For richer searches, select *Show Find Options* from the popup menu tied to the magnifying glass in the search field.

Figure 2–9. *The Xcode navigator view showing the search navigator. Be sure to check out the popup menus hidden under the word Find and under the magnifying glass in the search field.*

▓ **Issues navigator**: When you build your project, any errors or warnings will appear in this navigator, and a message detailing the number of errors will appear in the activity view at the top of the window (see Figure 2–10). When you click an error in the issues navigator, you'll jump to the appropriate line of code in the editor pane.

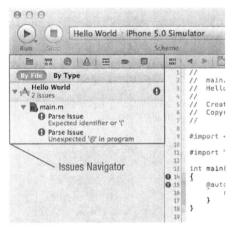

Figure 2–10. *The Xcode navigator view showing the issues navigator. This is where you'll find your compiler errors and warnings.*

▓ **Debug navigator**: This navigator is your main view into the debugging process (see Figure 2–11). If you are new to debugging, you might check out this part of the *Xcode 4 User Guide*:

```
http://developer.apple.com/library/mac/#documentation/
ToolsLanguages/Conceptual/Xcode4UserGuide/Debugging/Debugging.html
```

The debug navigator lists the stack frame for each active thread. A **stack frame** is a list of the functions or methods that have been called previously, in the order they were called. Click a method, and the associated code appears in the editor pane. In the editor, there will be a second frame, where you can control the debugging process, display and modify data values, and access the low-level debugger. A slider at the bottom of the debug navigator allows you to control the level of detail it tracks. Slide to the extreme right to see everything, including all the system calls. Slide to the extreme left to see only your calls. The default setting of right in the middle is a good place to start.

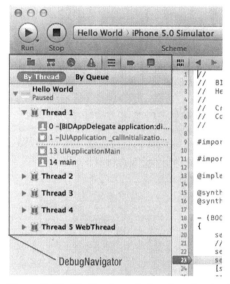

Figure 2–11. *The Xcode navigator view showing the debug navigator. Be sure to try out the detail slider at the bottom of the window, which allows you to specify the level of debug detail you want to see.*

- **Breakpoint navigator**: The breakpoint navigator lets you see all the breakpoints that you've set (see Figure 2–12). Breakpoints are, as the name suggests, points in your code where the application will stop running (or **break**), so that you can look at the values in variables and do other tasks needed to debug your application. The list of breakpoints in this navigator is organized by file. Click a breakpoint in the list, and that line will appear in the editor pane. Be sure to check out the popup at the lower-left corner of the workspace window when in the breakpoint navigator. The plus popup lets you add an exception or symbolic breakpoint, and the minus popup deletes any selected breakpoints.

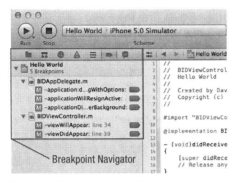

Figure 2-12. *The Xcode navigator view showing the breakpoint navigator. The list of breakpoints is organized by file.*

■ **Log navigator**: This navigator keeps a history of your recent build results and run logs (see Figure 2-13). Click a specific log, and the build command and any build issues are displayed in the edit pane.

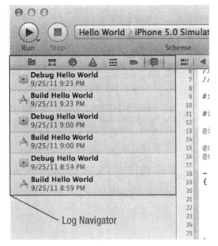

Figure 2-13. *The Xcode navigator view showing the log navigator. The log navigator displays a list of builds, with the details associated with a selected view displayed in the edit pane.*

The Jump Bar

With a single click, the **jump bar** allows you to jump to a specific element in the hierarchy you are currently navigating. For example, Figure 2-14 shows a source file being edited in the edit pane. The jump bar is just above the source code. Here's how it breaks down:

■ The funky looking icon at the left end of the jump bar is actually a popup menu that displays submenus listing recent files, unsaved files, counterparts, superclasses and subclasses, siblings, categories, includes, and files that include the current file.

- To the right of the über menu are left and right arrows that take you back to the previous file and return to the next file, respectively.

- The jump bar includes a segmented popup that displays the files for the current project that can be displayed for the current editor. In Figure 2–14, we're in the source code editor, so we see all the source files in our project. At the tail end of the jump bar is a popup that shows the methods and other symbols contained by the currently selected file. The jump bar in Figure 2–14 shows the file *BIDAppDelegate.m*, with a submenu listing the symbols defined in that file.

Figure 2–14. *The Xcode editor pane showing the jump bar, with a source code file selected. The submenu shows the list of methods in the selected file.*

The jump bar is incredibly powerful. Look for it as you make your way through the various interface elements that make up Xcode 4.

> **TIP:** If you're running Xcode under Lion (Mac OS X 10.7), there's full support for full-screen mode. Just click the full-screen button in the upper right of the project window to try out distraction-free, full-screen coding!

XCODE KEYBOARD SHORTCUTS

If you prefer navigating with keyboard shortcuts instead of mousing to on-screen controls, you'll like what Xcode has to offer. Most actions that you will do regularly in Xcode have keyboard shortcuts assigned to them, such as ⌘B to build your application or ⌘N to create a new file.

You can change all of Xcode's keyboard shortcuts, as well as assign shortcuts to commands that don't already have one using Xcode's preferences, under the *Key Bindings* tab.

A really handy keyboard shortcut is ⇧⌘O, which is Xcode's Open Quickly feature. After pressing it, start typing the name of a file, setting, or symbol, and Xcode will present you with a list of options. When you narrow down the list to the file you want, hitting return will open it in the editing pane, allowing you to switch files in just a few keystrokes.

The Utility Pane

As we mentioned earlier, the second-to-last button on the right side of the Xcode toolbar opens and closes the utility pane. Like an inspector, the utility pane is context-sensitive, with contents that change depending on what is being displayed in the editor pane. You'll see examples throughout the book.

Interface Builder

Earlier versions of Xcode included an interface design tool called Interface Builder, which allowed you to build and customize your project's user interface. One of the major changes introduced in Xcode 4 is the integration of Interface Builder into the workspace itself. Interface Builder is no longer a separate stand-alone application, which means you don't need to jump back and forth between Xcode and Interface Builder as your code and interface evolve. Huzzah!

We'll be working extensively with Xcode's interface-building functionality throughout the book, digging into all its nooks and crannies. In fact, we'll do our first bit of interface building a bit later in this chapter.

New Compiler and Debugger

One of the most important changes brought in by Xcode 4 lies under the hood: a brand-new compiler and low-level debugger. Both are significantly faster and smarter than their predecessors.

The new compiler, LLVM 3, generates code that is faster by far than that generated by GCC, which was the default compiler in previous versions of Xcode. In addition to creating faster code, LLVM also knows more about your code, so it can generate smarter, more precise error messages and warnings.

LLVM can also offer more precise code completion, and it can make educated guesses as to the actual intent of a piece of code when it produces a warning, offering a popup menu of likely fixes. This makes errors like misspelled symbol names, mismatched parentheses, and missing semicolons a breeze to find and fix.

LLVM brings to the table a sophisticated **static analyzer** that can scan your code for a wide variety of potential problems, including problems with Objective-C memory management. In fact, LLVM is so smart about this that it can handle most memory management tasks for you, as long as you abide by a few simple rules when writing your code. We'll look at the wonderful new ARC feature, which we mentioned earlier, starting in the next chapter.

A Closer Look at Our Project

Now that we've explored the Xcode workspace window, let's take a look at the files that make up our new *Hello World* project. Switch to the project navigator by clicking the leftmost of the seven navigator icons on the left side of your workspace (as discussed in the "The Navigator View" section earlier in the chapter) or by pressing ⌘1.

> **TIP:** The seven navigator configurations can be accessed using the keyboard shortcuts ⌘1 to ⌘7. The numbers correspond to the icons starting on the left, so ⌘1 is the project navigator, ⌘2 is the symbol navigator, and so on up to ⌘7, which takes you to the log navigator.

The first item in the project navigator list bears the same name as your project—in this case, *Hello World*. This item represents your entire project, and it's also where project-specific configuration can be done. If you single-click it, you'll be able to edit a number of project configuration settings in Xcode's editor. You don't need to worry about those project-specific settings now, however. At the moment, the defaults will work fine.

Flip back to Figure 2–7. Notice that the disclosure triangle to the left of *Hello World* is open, showing a number of subfolders (which are called **groups** in Xcode):

- *Hello World*: The first folder, which is always named after your project, is where you will spend the bulk of your time. This is where most of the code that you write will generally go, as will the files that make up your application's user interface. You are free to create subfolders under the *Hello World* folder to help organize your code, and you're even allowed to use other groups if you prefer a different organizational approach. While we won't touch most of the files in this folder until next chapter, there is one file we will explore when we make use of Interface Builder in the next section:

    - *BIDViewController.xib* contains the user interface elements specific to your project's main view controller.

- *Supporting Files*: This folder contains source code files and resources that aren't Objective-C classes but that are necessary to your project. Typically, you won't spend a lot of time in the *Other Sources* folder. When you create a new iPhone application project, this folder contains four files:

    - *Hello_World-Info.plist* is a property list that contains information about the application. We'll look briefly at this file in the "Some iPhone Polish—Finishing Touches" section later in this chapter.

- *InfoPlist.strings* is a text file that contains human-readable strings that may be referenced in the info property list. Unlike the info property list itself, this file can be localized, allowing you to include multiple language translations in your application (a topic we'll cover in Chapter 21).

- *main.m* contains your application's `main()` method. You normally won't need to edit or change this file. In fact, if you don't know what you're doing, it's really a good idea not to touch it.

- *Hello_World_Prefix.pch* is a list of header files from external frameworks that are used by your project (the extension *.pch* stands for **prec**ompiled **h**eader). The headers referenced in this file are typically ones that aren't part of your project and aren't likely to change very often. Xcode will precompile these headers and then continue to use that precompiled version in future builds, which will reduce the amount of time it takes to compile your project whenever you select **Build** or **Run**. It will be a while before you need to worry about this file, because the most commonly used header files are already included for you.

- *Frameworks*: This folder is a special kind of library that can contain code as well as resources, such as image and sound files. Any framework or library that you add to this folder will be linked into your application, and your code will be able to use any objects, functions, and resources contained in that framework or library. The most commonly needed frameworks and libraries are linked into your project by default, so most of the time, you will not need to add anything to this folder. If you do need less commonly used libraries and frameworks, it's easy to add them to the *Frameworks* folder. We'll show you how to add frameworks in Chapter 7.

- *Products*: This folder contains the application that this project produces when it is built. If you expand *Products*, you'll see an item called *Hello World.app*, which is the application that this particular project creates. *Hello World.app* is this project's only product. Because we have never built it, *Hello World.app* is red, which is Xcode's way of telling you that a file reference points to something that is not there.

> **NOTE:** The "folders" in the navigator area do not necessarily correspond to folders in your Mac's file system. These are logical groupings within Xcode to help you keep everything organized, and to make it faster and easier to find what you're looking for while working on your application. Often, the items contained in those two project folders are stored directly in the project's directory, but you can store them anywhere—even outside your project folder if you want. The hierarchy inside Xcode is completely independent of the file system hierarchy, so moving a file out of the *Classes* folder in Xcode, for example, will not change the file's location on your hard drive.
>
> It is possible to configure a group to use a specific file system directory using the utility pane. However, by default, new groups added to your project are completely independent of the file system, and their contents can be contained anywhere.

Introducing Xcode's Interface Builder

In your workspace window's project navigator, expand the *Hello World* group, if it's not already open, and then select the file *BIDViewController.xib*. As soon as you do, the file will open in the editor pane, as shown in Figure 2–15. You should see a graph paper background, which makes a nice backdrop for editing interfaces. This is Xcode's Interface Builder (sometimes referred to as IB), which is where you'll design your application's user interface.

Interface Builder has a long history. It has been around since 1988 and has been used to develop applications for NeXTSTEP, OpenStep, Mac OS X, and now iOS devices such as iPhone and iPad. As we noted earlier, before Xcode 4, Interface Builder was a separate application that was installed along with Xcode and worked in tandem with it. Now, Interface Builder is fully integrated into Xcode.

Interface Builder supports two file types: an older format that uses the extension *.nib* and a newer format that uses the extension *.xib*. The iOS project templates all use *.xib* files by default, but for the first 20 years it existed, all Interface Builder files had the extension *.nib*, and as a result, most developers took to calling Interface Builder files "nib files." Interface Builder files are often called nib files regardless of whether the extension actually used for the file is *.xib* or *.nib*. In fact, Apple still uses the terms *nib* and *nib file* throughout its documentation.

The gray vertical bar on the left edge of the graph paper is known as the **dock**. The dock contains an icon for each top-level object in the nib file. If you click the triangle-in-a-circle icon just to the right of the bottom of the dock, you'll see a list view representation of those objects. Click the icon again to return to the icon view.

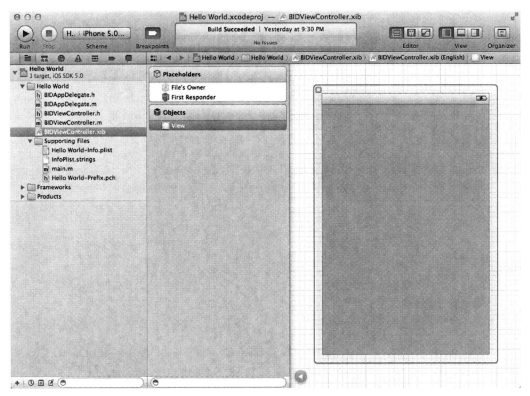

Figure 2–15. *We selected BIDViewController.xib in the project navigator. This opened the file in Interface Builder. Note the graph paper background in the editor pane. The gray vertical bar to the left of the graph paper is called the dock.*

The top two icons in the nib file are called *File's Owner* and *First Responder*, which are special items that every nib file has and which we'll talk about more in a moment. Each of the remaining icons represents a single instance of an Objective-C class that will be created automatically for you when this nib file is loaded. Our nib file has one additional icon beyond the required *File's Owner* and *First Responder*. That third icon—the one below the horizontal line—represents a view object. This is the view that will be shown when our application launches, and it was created for us when we selected the *Single View Application* template.

Now, let's say that you want to create an instance of a button. You could create that button by writing code, but creating an interface object by dragging a button out of a library and specifying its attributes is so much simpler, and it results in exactly the same thing happening at runtime.

The *BIDViewController.xib* file we are looking at right now is loaded automatically when your application launches—for the moment, don't worry about how—so it is the right place to add the objects that make up your application's user interface. When you create objects in Interface Builder, they'll be instantiated in your program when that nib file is loaded. You'll see many examples of this process throughout this book.

What's in the Nib File?

As we mentioned earlier, the contents of the nib file are represented by icons or a list in the dock immediately to the left of the editor pane (see Figure 2–15). Every nib file starts off with the same two icons: *File's Owner* and *First Responder*. These two are created automatically and cannot be deleted. Furthermore, they are visually separated from the objects you add to the nib file by a divider. From that, you can probably guess that they are important.

- *File's Owner* represents the object that loaded the nib file from disk. In other words, *File's Owner* is the object that "owns" this copy of the nib file.

- *First Responder* is, in very basic terms, the object with which the user is currently interacting. If, for example, the user is currently entering data into a text field, that field is the current first responder. The first responder changes as the user interacts with the user interface, and the *First Responder* icon gives you a convenient way to communicate with whatever control or other object is the current first responder, without needing to write code to determine which control or view that might be.

We'll talk more about these objects starting in the next chapter, so don't worry if you're a bit fuzzy right now on when you would use *First Responder* or what constitutes the "owner" of a nib.

Every other icon in this window, other than these first two special cases, represents an object instance that will be created when the nib file loads, exactly as if you had written code to alloc and init a new Objective-C object. In our case, there is a third icon called *View* (see Figure 2–15).

The *View* icon represents an instance of the UIView class. A UIView object is an area that a user can see and interact with. In this application, we will have only one view, so this icon represents everything that the user can see in our application. Later, we'll build more complex applications that have more than one view. For now, just think of this as what the user can see when using your application.

> **NOTE:** Technically speaking, our application will actually have more than one view. All user interface elements that can be displayed on the screen—including buttons, text fields, and labels—are descendents of UIView. When you see the term *view* used in this book, however, we will generally be referring to only actual instances of UIView, and this application has only one of those.

If you click the *View* icon, a depiction of an iPhone-sized screen will open (if it's not already displayed). This is where you can design your user interface graphically.

The Library

As shown in Figure 2–16, the utility view, which makes up the right side of the workspace, is divided into two sections. If you're not currently seeing the utility view, click the rightmost of the three *View* buttons in the toolbar, select View ➤ Utilities ➤ Show Utilities, or press ⌥⌘0 (option-command-zero).

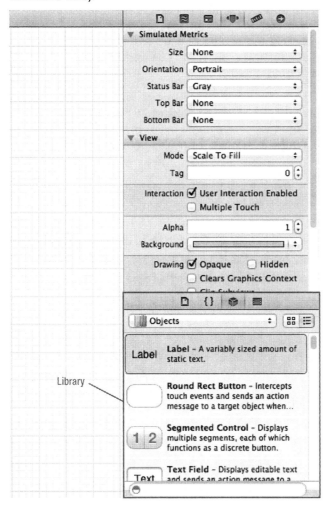

Figure 2–16. *The library is where you'll find stock objects from the UIKit that are available for use in Interface Builder. Everything above the library but below the toolbar is known collectively as the inspector.*

The bottom half of the utility view is called the **library pane**, or just plain **library**. The library is a collection of reusable items you can use in your own programs. The four icons in the bar at the top of the library pane divide the library into four sections:

- **File template library**: This section contains a collection of file templates you can use when you need to add a new file to your project. For example, if you want to add a new Objective-C class to your project, drag an Objective-C class file from the file template library.

- **Code snippet library**: This section features a collection of code snippets you can drag into your source code files. Can't remember the syntax for Objective-C fast enumeration? That's fine—just drag that particular snippet out of the library, and you don't need to look it up. Have you written something you think you'll want to use again later? Select it in your text editor and drag it to the code snippet library.

- **Object library**: This section is filled with reusable objects, such as text fields, labels, sliders, buttons, and just about any object you would ever need to design your iOS interface. We'll use the object library extensively in this book to build the interfaces for our sample programs.

- **Media library**: As its name implies, this section is for all your media, including pictures, sounds, and movies.

> **NOTE:** The items in the object library are primarily from the iOS UIKit, which is a framework of objects used to create an app's user interface. UIKit fulfills the same role in Cocoa Touch as AppKit does in Cocoa. The two frameworks are similar conceptually, but because of differences in the platforms, there are obviously many differences between them. On the other hand, the Foundation framework classes, such as NSString and NSArray, are shared between Cocoa and Cocoa Touch.

Note the search field at the bottom of the library. Do you want to find a button? Type *button* in the search field, and the current library will show only items with *button* in the name. Don't forget to clear the search field when you are finished searching.

Adding a Label to the View

Let's give Interface Builder a try. Click the object library icon (it looks like a cube) at the top of the library to bring up the object library. Now scroll through the library to find a *Table View*. That's it—keep scrolling, and you'll find it. Or wait! There's a better way: just type the words *Table View* in the search field. Isn't that so much easier?

> **TIP:** Here's a nifty shortcut: press ^⌥⌘3 to jump to the search field and highlight its contents.

Now find a *Label* in the library. It is likely on or near the top of the list. Next, drag the label onto the view we saw earlier. (If you don't see the view in your editor pane, click the *View* icon in the Interface Builder dock.) As your cursor appears over the view, it will turn into the standard, "I'm making a copy of something" green plus sign you know from the Finder. Drag the label to the center of the view. A pair of blue guidelines—one vertical and one horizontal—will appear when your label is centered. It's not vital that the label be centered, but it's good to know those guidelines are there. Figure 2–17 shows what our workspace looked like just before we released our drag.

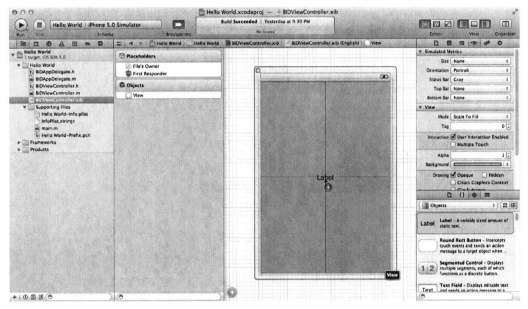

Figure 2–17. *We've found a label in our library and dragged it onto our view. Note that we typed label into the library search field to limit our object list to those containing the word label.*

User interface items are stored in a hierarchy. Most views can contain **subviews**, though there are some, like buttons and most other controls, that can't. Interface Builder is smart. If an object does not accept subviews, you will not be able to drag other objects onto it.

We'll add our label as a subview of our main view (the view named *View*), which will cause it to show up automatically when that view is displayed to the user. Dragging a *Label* from the library to the view called *View* adds an instance of UILabel as a subview of our application's main view.

Let's edit the label so it says something profound. Double-click the label you just created, and type the text *Hello, World!* Next, click off the label, and then reselect it and drag the label to recenter it, or position it wherever you want it to appear on the screen.

Guess what? Once we save, we're finished. Select **File ➤ Save**, or press ⌘S. Then click the popup menu at the upper left of the Xcode workspace window and choose **iPhone Simulator** (the popup might also include a version number—choose the latest and greatest) so that our app will run in the simulator. If you are a member of Apple's paid iOS Developer Program, you can try running your app on your phone. In this book, we'll stick with the simulator as much as possible, since running in the simulator doesn't require any paid membership.

Ready to run? Select **Product ➤ Run** or press ⌘**R**. Xcode will compile your app and launch it in the iPhone simulator, as shown in Figure 2–18.

> **NOTE:** If your iOS device is connected to your Mac when you build and run, things might not go quite as planned. In a nutshell, in order to be able to build and run your applications on your iPhone, iPad, or iPod touch, you must sign up and pay for one of Apple's iOS Developer Programs, and then go through the process of configuring Xcode appropriately. When you join the program, Apple will send you the information you'll need to get this done. In the meantime, most of the programs in this book will run just fine using the iPhone or iPad simulator.

Figure 2–18. *Here's the Hello, World program in its full iPhone glory!*

When you are finished admiring your handiwork, you can head back over to Xcode. Xcode and the simulator are separate applications.

TIP: You are welcome to quit the simulator once you finish examining your app, but you'll just be restarting it in a moment. If you leave the simulator running and ask Xcode to run your application again, Xcode will ask you if you want to stop your existing app first or run the app as a second instance, leaving the first instance running as well. If this seems confusing, feel free to quit the simulator each time you finish testing your app. No one will know!

Wait a second! That's it? But we didn't write any code. That's right.

Pretty neat, huh?

Well, how about if we wanted to change some of the properties of the label, like the text size or color? We would need to write code to do that, right? Nope. Let's see just how easy it is to make changes.

Changing Attributes

Head back to Xcode and single-click the *Hello World* label so that it is selected. Now turn your attention to the area above the library pane. This part of the utility pane is called the **inspector**. Like the library, the inspector pane is topped by a series of icons, each of which changes the inspector to view a specific type of data. To change the attributes of the label, we'll need the fourth icon from the left, which brings up the object attributes inspector, as shown in Figure 2–19.

TIP: The inspector, like the project navigator, has keyboard shortcuts corresponding to each of its icons. The inspector's keyboard shortcuts start with ⌥⌘1 for the leftmost icon, ⌥⌘2 for the next icon, and so on. Unlike the project navigator, the number of icons in the inspector is context-sensitive and changes depending on which object is selected in the navigator and/or editor.

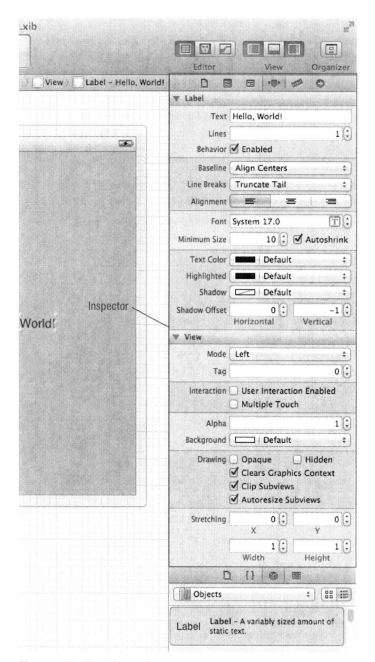

Figure 2–19. *The object attributes inspector showing our label's attributes*

Go ahead and change the label's appearance to your heart's delight. Feel free to play around with the font, size, and color of the text. Note that if you increase the font size, you may need to resize the label itself to make room for larger text. Once you're finished playing, save the file and select **Run** again. The changes you made should show up in your application, once again without writing any code.

> **NOTE:** Don't worry too much about what all of the fields in the object attributes inspector mean, or fret if you can't get one of your changes to show up. As you make your way through the book, you'll learn a lot about the object attributes inspector and what each of the fields does.

By letting you design your interface graphically, Interface Builder frees you to spend time writing the code that is specific to your application, instead of writing tedious code to construct your user interface.

Most modern application development environments have some tool that lets you build your user interface graphically. One distinction between Interface Builder and many of these other tools is that Interface Builder does not generate any code that must be maintained. Instead, Interface Builder creates Objective-C objects, just as you would do in your own code, and then serializes those objects into the nib file so that they can be loaded directly into memory at runtime. This avoids many of the problems associated with code generation and is, overall, a more powerful approach.

Some iPhone Polish—Finishing Touches

Now let's put a last bit of spit and polish on our application to make it feel a little more like an authentic iPhone application. First, run your project. When the simulator window appears, click the iPhone's home button (the black button with the white square at the very bottom of the window). That will bring you back to the iPhone home screen, as shown in Figure 2–20. Notice anything a bit, well, boring?

Take a look at the Hello World icon at the top of the screen. Yeah, that icon will never do, will it? To fix it, you need to create an icon and save it as a portable network graphic (*.png*) file. Actually, you should create two icons. One needs to be 114 × 114 pixels in size, and the other needs to be 57 × 57 pixels. Why two icons? Well, the iPhone 4 introduced the **Retina display**, which was exactly double the resolution of earlier iPhone models. The smaller icon will be used on non-Retina devices, and the larger one will be used on devices with a Retina display.

Do not try to match the style of the buttons that are already on the phone when you create the icons; your iPhone will automatically round the edges and give it that nice, glassy appearance. Just create normal flat, square images. We have provided two icon images in the project archive *02 - Hello World* folder, called *icon.png* and *icon@2x.png*, which you can use if you don't want to create your own. The @2x in the name of the larger file is a special naming convention that identifies the file as the Retina version of the same file with the same name minus the @2x.

Figure 2–20. *Our Hello, World icon is just plain boring. It needs a real icon!*

> **NOTE:** For your application's icon, you must use *.png* images, but you should actually use that format for all images in your iOS projects. Xcode automatically optimizes *.png* images at build time, which makes them the fastest and most efficient image type for use in iOS apps. Even though most common image formats will display correctly, you should use *.png* files unless you have a compelling reason to use another format.

After you've designed your app icon, press ⌘1 to open the project navigator, and then click the topmost row in the navigator—the one with the blue icon and the name *Hello World*. Now, turn your attention to the editing pane.

On the left side of the editing pane, you'll see a white column with list entries labeled *PROJECT* and *TARGETS*. Make sure that the *Hello World* target is selected. To the right of that column, you'll see a big, gray settings pane. At the top of that pane is a series of five tabs. Select the *Summary* tab. In the *Summary* tab, scroll down, looking for a section labeled *App Icons* (see Figure 2–21). This is where we'll drag our newly added icons.

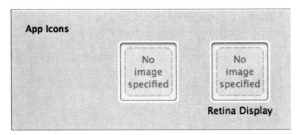

Figure 2–21. *The App Icon boxes on your project's Summary tab. This is where you can set your application's icon.*

From the Finder, drag *icon.png* to the left rectangle. This will copy *icon.png* into your project and set it as your application's icon. Next, drag *icon@2x.png* from the Finder to the right rectangle, which will set that as you application's Retina display icon.

If you look back in the project navigator, you'll notice that the two images were added to your project, but not inside a folder (see Figure 2–22). To keep our project organized, select *icon.png* and *icon@2x.png*, and drag them to the *Supporting Files* group.

Figure 2–22. *When the icons are added to your project, they're not placed in a subfolder. If you want to keep your project organized, you'll need to move them yourself.*

Let's take a look at what Xcode did with those icons, behind the scenes. In Xcode's project navigator, look in the *Supporting Files* folder again, and then single-click the *Hello_World-Info.plist* file. This is a **property list** file that contains some general information about our application, including specifics on our project icon files.

When you select *Hello_World-Info.plist*, the property list will appear in the editor pane. Within the property list, find a row with the label *Icon files* in the left column. The corresponding right column in that same row should say *(2 items)*. This row holds an array, which means it can hold multiple values. In this case, there's one row for each icon that was specified. Single-click the disclosure triangle immediately to the left of the name *Icon Files*, and you'll see the two items in the array, as shown in Figure 2–23.

Key	Type	Value
Localization native development region	String	en
Bundle display name	String	${PRODUCT_NAME}
Executable file	String	${EXECUTABLE_NAME}
▼ Icon files	Array	(2 items)
Item 0	String	icon.png
Item 1	String	icon@2x.png
▶ Icon files (iOS 5)	Diction...	(2 items)
Bundle identifier	String	com.Apress.${PRODUCT_NAME:rfc1034identifier}
InfoDictionary version	String	6.0
Bundle name	String	${PRODUCT_NAME}
Bundle OS Type code	String	APPL
Bundle versions string, short	String	1.0
Bundle creator OS Type code	String	????
Bundle version	String	1.0
Application requires iPhone environmer	Boolean	YES
▶ Supported interface orientations	Array	(3 items)

Figure 2–23. *Expanding the disclosure triangle shows the contents of the Icon Files array. Inside the array, you'll find a row for each of our two icon files.*

Looking at the plist contents in Figure 2–23, you might notice another row with the key *Icon Files (iOS 5)*. If you click the disclosure triangle next to that entry, you'll see that it contains two named entries: one called *Primary Icon* and another called *Newsstand Icon*. If you expand *Primary Icon*, you'll see the same thing you saw under *Icon Files*. Don't be too concerned about this. If you set your icons using Xcode the way we just did, Xcode will always configure the property list correctly.

The reason that the same icon information is represented twice is that prior to iOS 5, there was only one icon for an app, so a single array (*Icon Files*) was sufficient for holding the information about the icons. With iOS 5, Apple introduced a way to specify other types of icons for your application, including one to be used within Apple's Newsstand app. We won't be covering Newsstand in this book, so you don't need to worry about when or why you would specify an icon for that. Just be aware that iOS 5 introduced a new way to specify icons, and for the time being, apps will be supporting both the old and new methods.

> **NOTE:** If you were to just copy the two icon image files into your Xcode project and do nothing else, your icon would actually show up anyway. Huh? Why's that? By default, if no icon file name is provided, the SDK looks for a resource named *icon.png* and uses that. You also don't need to tell it about the *@2x* version of the icon. iOS knows to look for that on a device with a Retina display. To be safe, however, you should future-proof your app and always specify your application's icons in the info property list.

Now, take a look at the other rows in *Hello_World-Info.plist*. While most of these settings are fine as they are, one in particular deserves a moment of our attention: *Bundle identifier*. This is that unique identifier we entered when we created our project. This value should always be set. The standard naming convention for bundle identifiers is to use one of the top-level Internet domains, such as *com* or *org* followed by a period, then the name of your company or organization followed by another period, and finally, the name of your application.

When we created this project, we were prompted for a bundle identifier, and we entered *com.apress*. The value at the end of the string is a special code that will be replaced with your application's name when your application is built. This allows you to tie your application's bundle identifier to its name. If you need to change your application's unique identifier after creating the project, this is where you would do it.

Now compile and run your app. When the simulator has finished launching, press the button with the white square to go home, and check out your snazzy new icon. Ours is shown in Figure 2–24.

Figure 2–24. *Your application now has a snazzy icon!*

NOTE: If you want to clear out old applications from the iPhone simulator's home screen, you can choose **iPhone Simulator ➤ Reset Content and Settings….** from the iPhone simulator's application menu.

Bring It on Home

Pat yourself on the back. Although it may not seem like you accomplished all that much in this chapter, we actually covered a lot of ground. You learned about the iOS project templates, created an application, learned a ton about Xcode 4, started using Interface Builder, and learned how to set your application icon and bundle identifier.

The Hello, World program, however, is a strictly one-way application. We show some information to the users, but we never get any input from them. When you're ready to see how to go about getting input from the user of an iOS device and taking actions based on that input, take a deep breath and turn the page.

Chapter **3**

Handling Basic Interaction

Our Hello, World application was a good introduction to iOS development using Cocoa Touch, but it was missing a crucial capability: the ability to interact with the user. Without that, our application is severely limited in terms of what it can accomplish.

In this chapter, we're going to write a slightly more complex application—one that will feature two buttons as well as a label, as shown in Figure 3–1. When the user taps either of the buttons, the label's text will change. This may seem like a rather simplistic example, but it demonstrates the key concepts involved in creating interactive iOS apps.

Figure 3–1. *The simple two-button application we will build in this chapter*

The Model-View-Controller Paradigm

Before diving in, a bit of theory is in order. The designers of Cocoa Touch were guided by a concept called **Model-View-Controller** (MVC), which is a very logical way of dividing the code that makes up a GUI-based application. These days, almost all object-oriented frameworks pay a certain amount of homage to MVC, but few are as true to the MVC model as Cocoa Touch.

The MVC pattern divides all functionality into three distinct categories:

- **Model**: The classes that hold your application's data.

- **View**: Made up of the windows, controls, and other elements that the user can see and interact with.

- **Controller**: The code that binds together the model and view. It contains the application logic that decides how to handle the user's inputs.

The goal in MVC is to make the objects that implement these three types of code as distinct from one another as possible. Any object you create should be readily identifiable as belonging in one of the three categories, with little or no functionality that could be classified as being either of the other two. An object that implements a button, for example, shouldn't contain code to process data when that button is tapped, and an implementation of a bank account shouldn't contain code to draw a table to display its transactions.

MVC helps ensure maximum reusability. A class that implements a generic button can be used in any application. A class that implements a button that does some particular calculation when it is clicked can be used only in the application for which it was originally written.

When you write Cocoa Touch applications, you will primarily create your view components using a visual editor within Xcode called Interface Builder, although you will also modify, and sometimes even create, your user interfaces from code.

Your model will be created by writing Objective-C classes to hold your application's data or by building a data model using something called Core Data, which you'll learn about in Chapter 13. We won't be creating any model objects in this chapter's application, because we do not need to store or preserve data, but we will introduce model objects as our applications get more complex in future chapters.

Your controller component will typically be composed of classes that you create and that are specific to your application. Controllers can be completely custom classes (NSObject subclasses), but more often, they will be subclasses of one of several existing generic controller classes from the UIKit framework, such as UIViewController, which you'll see shortly. By subclassing one of these existing classes, you will get a lot of functionality for free and won't need to spend time recoding the wheel, so to speak.

As we get deeper into Cocoa Touch, you will quickly start to see how the classes of the UIKit framework follow the principles of MVC. If you keep this concept in the back of your mind as you develop, you will end up creating cleaner, more easily maintained code.

Creating Our Project

It's time to create our next Xcode project. We're going to use the same template that we used in the previous chapter: *Single View Application*. By starting with this simple template again, it will be easier for you to see how the view and controller objects work together in an iOS application. We'll use some of the other templates in later chapters.

Launch Xcode and select **File ➤ New ➤ New Project...** or press ⇧⌘N. Select the *Single View Application* template, and then click *Next*.

You'll be presented with the same options sheet as you saw in the previous chapter. In the *Product Name* field, type the name of our new application, *Button Fun*. The *Company Identifier* field should still have the value you used in the previous chapter, so you can leave that alone. In the *Class Prefix* field, use the same value as you did in the previous chapter: *BID*.

Just as we did with Hello, World, we're going to write an iPhone application, so select *iPhone* for *Device Family*. We're not going to use storyboards or unit tests, so you can leave both of those options unchecked. However, we do want to use ARC, so check the *Use Automatic Reference Counting* box. We'll explain ARC later in the chapter. Figure 3–2 shows the completed options sheet.

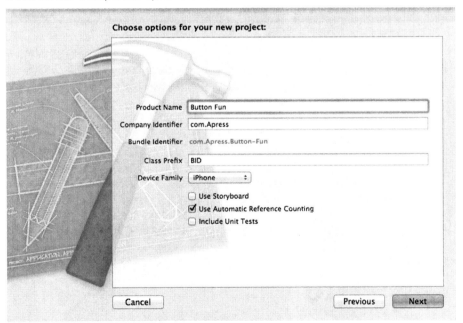

Figure 3–2. *Naming your project and selecting options*

Hit *Next*, and you'll be prompted for a location for your project. You can leave the *Create local git repository* checkbox unchecked. Save the project with the rest of your book projects.

Looking at the View Controller

A little later in this chapter, we'll design a view (or user interface) for our application using Interface Builder, just as we did in the previous chapter. Before we do that, we're going to look at and make some changes to the source code files that were created for us. Yes, Virginia, we're actually going to write some code in this chapter.

Before we make any changes, let's look at the files that were created for us. In the project navigator, the *Button Fun* group should already be expanded, but if it's not, click the disclosure triangle next to it (see Figure 3–3).

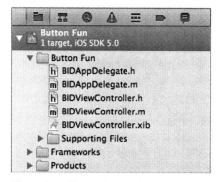

Figure 3–3. *The project navigator showing the class files that were created for us by the project template. Note that our class prefix was automatically incorporated into the class file names.*

The *Button Fun* folder should contain four source code files (the ones that end in *.h* or *.m*) and a single nib file. These four source code files implement two classes that our application needs: our application delegate and the view controller for our application's only view. Notice that Xcode automatically added the prefix we specified to all of our class names.

We'll look at the application delegate a little later in the chapter. First, we'll work with the view controller class that was created for us.

The controller class called BIDViewController is responsible for managing our application's view. The BID part of the name is derived automatically from the class prefix we specified, and the ViewController part of the name identifies that this class is, well, a view controller. Click *BIDViewController.h* in the *Groups & Files* pane, and take a look at the contents of the class's header file:

```
#import <UIKit/UIKit.h>

@interface BIDViewController : UIViewController

@end
```

Not much to it, is there? BIDViewController is a subclass of UIViewController, which is one of those generic controller classes we mentioned earlier. It is part of the UIKit and, by subclassing it, we get a bunch of functionality for free. Xcode doesn't know what our application-specific functionality is going to be, but it does know that we're going to have some, so it has created this class for us to write that application-specific functionality.

Understanding Outlets and Actions

In Chapter 2, you used Xcode's Interface Builder to design a user interface. A moment ago, you saw the shell of a view controller class. There must be some way for the code in this view controller class to interact with the objects in the nib file, right?

Absolutely! A controller class can refer to objects in a nib file by using a special kind of property called an **outlet**. Think of an outlet as a pointer that points to an object within the nib. For example, suppose you created a text label in Interface Builder (as we did in Chapter 2) and wanted to change the label's text from within your code. By declaring an outlet and connecting that outlet to the label object, you would then be able to use the outlet from within your code to change the text displayed by the label. You'll see how to do just that in this chapter.

Going in the opposite direction, interface objects in our nib file can be set up to trigger special methods in our controller class. These special methods are known as **action methods** (or just **actions**). For example, you can tell Interface Builder that when the user taps a button, a specific action method within your code should be called. You could even tell Interface Builder that when the user first touches a button, it should call one action method, and then later when the finger is lifted off the button, it should call a different action method.

Prior to Xcode 4, we would have needed to create our outlets and actions here in the view controller's header file before we could go to Interface Builder and start connecting outlets and actions. Xcode 4's assistant view gives us a much faster and more intuitive approach that lets us create and connect outlets and actions simultaneously, a process we're going to look at shortly. But before we start making connections, let's talk about outlets and actions in a little more detail. Outlets and actions are two of the most basic building blocks you'll use to create iOS apps, so it's important that you understand what they are and how they work.

Outlets

Outlets are special Objective-C class properties that are declared using the keyword IBOutlet. Declaring an outlet is done in your controller class header file, and might look something like this:

```
@property (nonatomic, retain) IBOutlet UIButton *myButton;
```

This example is an outlet called *myButton*, which can be set to point to any button in Interface Builder.

The IBOutlet keyword is defined like this:

```
#ifndef IBOutlet
#define IBOutlet
#endif
```

Confused? IBOutlet does absolutely nothing as far as the compiler is concerned. Its sole purpose is to act as a hint to tell Xcode that this is a property that we're going to want to connect to an object in a nib file. Any property that you create and want to connect to an object in a nib file must be preceded by the IBOutlet keyword. Fortunately, Xcode will now create outlets for us automatically.

OUTLET CHANGES

Over time, Apple has changed the way that outlets are declared and used. Since you are likely to run across older code at some point, let's look at how outlets have changed.

In the first version of this book, we declared both a property and its underlying instance variable for our outlets. At that time, properties were a new construct in the Objective-C language, and they required you to declare a corresponding instance variable, like this:

```
@interface MyViewController : UIViewController
{
    UIButton *myButton;
}
@property (nonatomic, retain) UIButton *myButton;
@end
```

Back then, we placed the IBOutlet keyword before the instance variable declaration, like this:

```
IBOutlet UIButton *myButton;
```

This was how Apple's sample code was written at the time, and also how the IBOutlet keyword had traditionally been used in Cocoa and NeXTSTEP.

By the time we wrote the second edition of the book, Apple had moved away from placing the IBOutlet keyword in front of the instance variable, and it became standard to place it within the property declaration, like this:

```
@property (nonatomic, retain) IBOutlet UIButton *myButton;
```

Even though both approaches continued to work (and still do), we followed Apple's lead and changed the book code so that the IBOutlet keyword was in the property declaration rather than in the instance variable declaration.

When Apple switched the default compiler from GCC to LLVM recently, it stopped being necessary to declare instance variables for properties. If LLVM finds a property without a matching instance variable, it will create one automatically. As a result, in this edition of the book, we've stopped declaring instance variables for our outlets altogether.

All of these approaches do exactly the same thing, which is to tell Interface Builder about the existence of an outlet. Placing the IBOutlet keyword on the property declaration is Apple's current recommendation, so that's what we're going to use. But we wanted to make you aware of the history in case you come across older code that has the IBOutlet keyword on the instance variable.

You can read more about Objective-C properties in the book *Learn Objective-C on the Mac* by Mark Dalrymple and Scott Knaster (Apress, 2009) and in the document called *Introduction to the Objective-C Programming Language*, available from Apple's Developer web site at http://developer.apple.com/documentation/Cocoa/Conceptual/ObjectiveC.

Actions

In a nutshell, actions are methods that are declared with a special return type, `IBAction`, which tells Interface Builder that this method can be triggered by a control in a nib file. The declaration for an action method will usually look like this:

```
- (IBAction)doSomething:(id)sender;
```

or like this:

```
- (IBAction)doSomething;
```

The actual name of the method can be anything you want, but it must have a return type of `IBAction`, which is the same as declaring a return type of `void`. A void return type is how you specify that a method does not return a value. Also, the method must either take no arguments or take a single argument, usually called `sender`. When the action method is called, `sender` will contain a pointer to the object that called it. For example, if this action method was triggered when the user taps a button, `sender` would point to the button that was tapped. The `sender` argument exists so that you can respond to multiple controls using a single action method. It gives you a way to identify which control called the action method.

> **TIP** There's actually a third, lesser-used type of `IBAction` declaration that looks like this:
>
> ```
> - (IBAction)doSomething:(id)sender
>                forEvent:(UIEvent *)event;
> ```
>
> We'll talk about control events starting in the next chapter.

It won't hurt anything if you declare an action method with a `sender` argument and then ignore it. You will likely see a lot of code that does just that. Action methods in Cocoa and NeXTSTEP needed to accept `sender` whether they used it or not, so a lot of iOS code, especially early iOS code, was written that way.

Now that you understand what actions and outlets are, you'll see how they work as we design our user interface. Before we start doing that, however, we have one quick piece of housekeeping to do to keep everything neat and orderly.

Cleaning Up the View Controller

Single-click *BIDViewController.m* in the project navigator to open the implementation file. As you can see, there's a fair bit of boilerplate code that was provided for us by the project template we chose. These methods are ones that are commonly used in `UIViewController` subclasses, so Xcode gave us stub implementations of them, and we can just add our code there. However, we don't need most of these stub implementations for this project, so all they're doing is taking up space and making our code harder to read. We're going to do our future selves a favor and delete what we don't need.

Delete all the methods except for `viewDidUnload`. When you're finished, your implementation should look like this:

```
#import "BIDViewController.h"
@implementation BIDViewController

- (void)viewDidUnload
{
    [super viewDidUnload];
    // Release any retained subviews of the main view.
    // e.g. self.myOutlet = nil;
}

@end
```

That's much simpler, huh? Don't worry about all those methods you just deleted. You'll be introduced to most of them throughout the course of the book.

The method we've left in is one that every view controller with outlets should implement. When a view is unloaded, which can happen when the system needs to make additional memory available, it's important to `nil` out your outlets. If you don't, the memory used by those outlets will not be released. Fortunately, all we need to do is leave this empty implementation in place, and Xcode will take care of releasing any outlets we create, as you'll see in this chapter.

Designing the User Interface

Make sure you save the changes you just made, and then single-click *BIDViewController.xib* to open your application's view in Xcode's Interface Builder (see Figure 3–4). As you'll remember from the previous chapter, the gray window that shows up in the editor represents your application's one and only view. If you look back at Figure 3–1, you can see that we need to add two buttons and a label to this view.

Let's take a second to think about our application. We're going to add two buttons and a label to our user interface, and that process is very similar to what we did in the previous chapter. However, we're also going to need outlets and actions to make our application interactive.

The buttons will need to each trigger an action method on our controller. We could choose to make each button call a different action method, but since they're going to do essentially the same task (update the label's text), we will need to call the same action method. We'll differentiate between the two buttons using that `sender` argument we discussed earlier in the section on actions. In addition to the action method, we'll also need an outlet connected to the label so that we can change the text that the label displays.

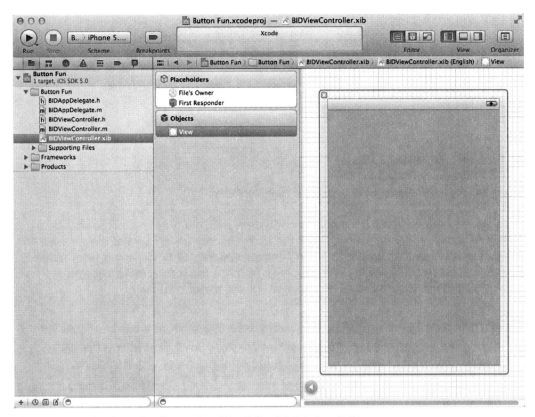

Figure 3–4. *BIDViewController.xib open for editing in Xcode's Interface Builder*

Let's add the buttons first, and then place the label. We'll create the corresponding actions and outlets as we design our interface. We could also manually declare our actions and outlets, and then connect our user interface items to them, but why do extra work when Xcode will do it for us?

Adding the Buttons and Action Method

Our first order of business is to add two buttons to our user interface. We'll then have Xcode create an empty action method for us and connect both buttons to that action method. This will cause the buttons, when tapped by the user, to call that action method. Any code we place in that action method will be executed when the user taps the button.

Select **View ➤ Utilities ➤ Show Object Library** or press ^⌥⌘3 to open the object library. Type *UIButton* into the object library's search box (you actually need to type only the first four characters, *UIBu*, to narrow down the list). Once you're finished typing, only one item should appear in the object library: *Round Rect Button* (see Figure 3–5).

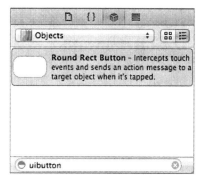

Figure 3–5. *The Round Rect Button as it appears in the object library*

Drag *Round Rect Button* from the library and drop it on the gray view. This will add a button to your application's view. Place the button along the left side of the view, using the blue guidelines that appear to place it the appropriate distance from the left edge. For vertical placement, use the blue guideline to place the button halfway down in the view. You can use Figure 3–1 as a placement guide, if that helps.

> **NOTE:** The little, blue guidelines that appear as you move objects around in Interface Builder are there to help you stick to the *iOS Human Interface Guidelines* (usually referred to as "the HIG"). Apple provides the HIG for people designing iPhone and iPad applications. The HIG tells you how you should—and shouldn't—design your user interface. You really should read it, because it contains valuable information that every iOS developer needs to know. You'll find it at `http://developer.apple.com/iphone/library/documentation/UserExperience/Conceptual/MobileHIG/`.

Double-click the newly added button. This will allow you to edit the button's title. Give this button a title of *Left*.

Now, it's time for some Xcode 4 magic. Select **View ➤ Assistant Editor ➤ Show Assistant Editor** or press ⌥⌘↩ to open the assistant editor. You can also show and hide the assistant editor by clicking the middle editor button in the collection of seven buttons on the upper-right side of the project window (see Figure 3–6).

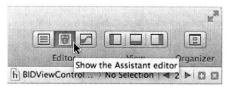

Figure 3–6. *The Show the Assistant editor toggle button*

Unless you specifically request otherwise (see the options in the **Assistant Editor** menu), the assistant editor will appear to the right of the editing pane. The left side will continue to show Interface Builder, but the right will display *BIDViewController.h*, which is the header file for the view controller that "owns" this nib.

> **TIP:** After opening the assistant editor, you may need to resize your window to have enough room to work. If you're on a smaller screen, like the one on a MacBook Air, you might need to close the utility view and/or project navigator to give yourself enough room to use the assistant editor effectively. You can do this easily using the three view buttons in the upper-right side of the project window (see Figure 3–6).

Remember the *File's Owner* icon we discussed in the previous chapter? The object that loads a nib is considered its **owner**, and with the case of nibs like this one that define the user interface for one of an application's views, the owner of the nib is the corresponding view controller class. Because our view controller class is the file's owner, the assistant editor knows to show us the header of the view controller class, which is the most likely place for us to connect actions and outlets.

As you saw earlier, there's really not much in *BIDViewController.h*. It's just an empty UIViewController subclass. But it won't be an empty subclass for long!

We're now going to ask Xcode to automatically create a new action method for us and associate that action with the button we just created.

To do this, begin by clicking your new button so it is selected. Now, hold down the control key on your keyboard, and then click and drag from the button over to the source code in the assistant editor. You should see a blue line running from the button to your cursor (see Figure 3–7). This blue line is how we connect objects in nibs to code or other objects.

> **Tip.** You can drag that blue line to anything you want to connect to your button: to the header file in the assistant editor, to the *File's Owner* icon, to any of the other icons on the left side of the editing pane, or even to other objects in the nib.

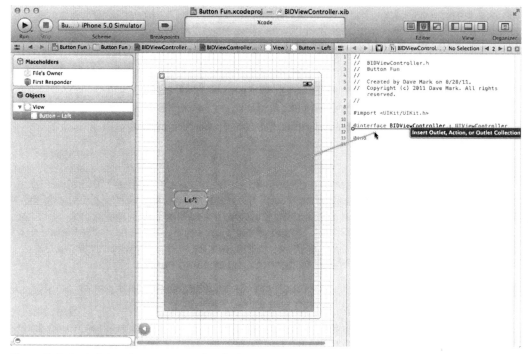

Figure 3–7. *Control-dragging to source code will give you the option to create an outlet, action, or outlet collection.*

If you move your cursor so it's between the @interface and @end keywords (as shown in Figure 3–7), a gray box will appear, letting you know that releasing the mouse button will insert an outlet, an action, or an outlet collection for you.

> **NOTE:** We make use of actions and outlets in this book, but we do not use outlet collections. Outlet collections allow you to connect multiple objects of the same kind to a single NSArray property, rather than creating a separate property for each object.

To finish this connection, release your mouse button, and a floating popup will appear, like the one shown in Figure 3–8. This window lets you customize your new action. In the window, click the popup menu labeled *Connection*, and change the selection from *Outlet* to *Action*. This tells Xcode that we want to create an action instead of an outlet.

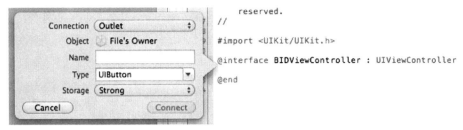

Figure 3–8. *The floating popup that appears after you control-drag to source code*

The popup will change to look like Figure 3–9. In the *Name* field, type *buttonPressed*. When you're finished, do *not* hit return. Pressing return would finalize our outlet, and we're not quite ready to do that. Instead, press tab to move to the *Type* field and type in *UIButton*, replacing the default value of *id*.

NOTE: As you probably remember, an *id* is a generic pointer that can point to any Objective-C object. We could leave this as *id*, and it would work fine, but if we change it to the class we expect to call the method, the compiler can warn us if we try to do this from the wrong type of object. There are times when you'll want the flexibility to be able to call the same action method from different types of controls, and in those cases, you would want to leave this set to *id*. In our case, we're only going to call this method from buttons, so we're letting the Xcode and LLVM know that. Now, it can warn us if we unintentionally try to connect something else to it.

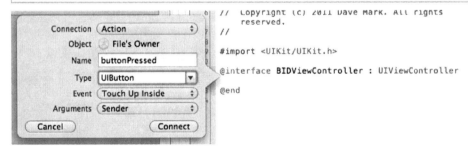

Figure 3–9. *Changing the connection type to Action changes the appearance of the popup.*

There are two fields below *Type*, which we will leave at their default values. The *Event* field lets you specify when the method is called. The default value of *Touch Up Inside* fires when the user lifts a finger off the screen if, and only if, the finger is still on the button. This is the standard event to use for buttons. This gives the user a chance to reconsider. If the user moves a finger off the button before lifting it off the screen, the method won't fire.

The *Arguments* field lets you choose between the three different method signatures that can be used for action methods. We want the `sender` argument so that we can tell which button called the method. That's the default, so we just leave it as is.

Hit the return key or click the *Connect* button, and Xcode will insert the action method for you. Your *BIDViewController.h* file should now look like this:

```
#import <UIKit/UIKit.h>

@interface BIDViewController : UIViewController
- (IBAction)buttonPressed:(id)sender;

@end
```

> **NOTE:** Over time, Apple will tweak both Xcode and the code templates we've been using. When that happens, you may need to make some adjustments to our step-by-step instructions. In the current example, we would expect to see `UIButton` instead of `id` in the declaration of the `buttonPressed` parameter. Likely, this will eventually be tweaked, and you'll need to make a change or two to this approach. But this is no big deal; that's the nature of the beast.

Xcode has now added a method declaration to your class's header file for you. Single-click *BIDViewController.m* to look at the implementation file, and you'll see that it has also added a method stub for you.

```
- (IBAction)buttonPressed:(id)sender {
}
```

In a few moments, we'll come back here to write the code that needs to run when the user taps either button. In addition to creating the method declaration and implementation, Xcode has also connected that button to this action method and stored that information in the nib file. That means we don't need to do anything else to make that button call this method when our application runs.

Go back to *BIDViewController.xib* and drag out another button, this time placing the button on the right side of the screen. After placing it, double-click it and change its name to *Right*. The blue lines will pop up to help you align it with the right margin, as you saw before, and they will also help you align the button vertically with the other button.

> **TIP:** Instead of dragging a new object out from the library, you could hold down the option key and drag the original object (the *Left* button in this example) over. Holding down the option key tells Interface Builder to make a copy of the object you drag.

This time, we don't want to create a new action method. Instead, we want to connect this button to the existing one that Xcode created for us a moment ago. How do we do that? We do it pretty much the same way as we did for the first button.

After changing the name of the button, control-click the new button and drag toward your header file again. This time, as your cursor gets near the declaration of `buttonPressed:`, that method should highlight, and you'll get a gray popup saying *Connect Action* (see Figure 3–10). When you see that popup, release the mouse button.

and Xcode will connect this button to the existing action method. That will cause this button, when tapped, to trigger the same action method as the other button.

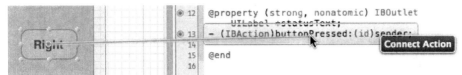

Figure 3–10. *Dragging to an existing action to whitespace will connect the button to an existing action.*

Note that this will work even if you control-drag to connect your button to a method in your implementation file. In other words, you can control-drag from your new button to the buttonPressed declaration in *BIDViewController.h* or to the buttonPressed method implementation in *BIDViewController.m*. Xcode 4 sure am smart!

Adding the Label and Outlet

In the object library, type *Label* into the search field to find the Label user interface item (see Figure 3–11). Drag the *Label* to your user interface, somewhere above the two buttons you placed earlier. After placing it, use the resize handles the stretch the label from the left margin to the right margin. That should give it plenty of room for the text we'll be displaying to the user.

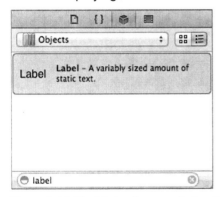

Figure 3–11. *The label as it appears in the object library*

Labels, by default, are left-justified, but we want this one to be centered. Select View ➤ Utilities ➤ Show Attributes Inspector (or press ⌥⌘4) to bring up the attributes inspector (see Figure 3–12). Make sure the label is selected, and then look in the attributes inspector for the *Alignment* buttons. Select the middle *Alignment* button to center the label's text.

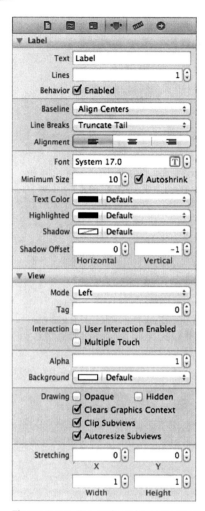

Figure 3–12. *The attribute inspector for the label*

Before the user taps a button, we don't want the label to say anything, so double-click the label (so the text is selected) and press the delete button on your keyboard. That will delete the text currently assigned to the label. Hit return to commit your changes. Even though you won't be able to see the label when it's not selected, don't worry—it's still there.

> **TIP:** If you have invisible user interface elements, like empty labels, and want to be able to see where they are, select **Canvas** from the **Assistant Editor** menu, and then from the submenu that pops up, turn on **Show Bounds Rectangles**.

All that's left is to create an outlet for the label. We do this exactly the way we created and connected actions earlier. Make sure the assistant editor is open and displaying *BIDViewController.h*. If you need to switch files, use the popup above the assistant editor.

Next, select the label in Interface Builder and control-drag from the label to the header file. Drag until your cursor is right above the existing action method. When you see something like Figure 3–13, let go of the mouse button, and you'll see the popup window again (shown earlier in Figure 3–8).

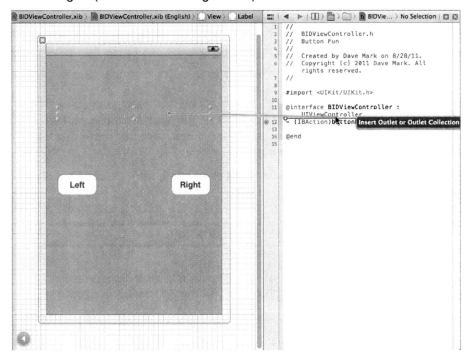

Figure 3–13. *Control-dragging to create an outlet*

We want to create an outlet, so leave the *Connection* at the default type of *Outlet*. We want to choose a descriptive name for this outlet so we'll remember what it is used for when we're working on our code. Type *statusText* into the *Name* field. Leave the *Type* field set to *UILabel*. The final field, labeled *Storage*, can be left at the default value.

Hit return to commit your changes, and Xcode will insert the outlet property into your code. Your controller class's header file should now look like this:

```
#import <UIKit/UIKit.h>

@interface BIDViewController : UIViewController
@property (strong, nonatomic) IBOutlet UILabel *statusText;
- (IBAction)buttonPressed:(id)sender;
@end
```

Now, we have an outlet, and Xcode has automagically connected the label to our outlet. This means that if we make any changes to statusText in code, those changes will affect the label on our user interface. If we set the text property on statusText, for example, it will change what text is displayed to the user.

Single-click *BIDViewController.m* in the project navigator to look at the implementation of our controller. There, you'll see that Xcode has inserted a @synthesize statement for us for the property that it created. It also did something else. Remember that method that we left in our code when we deleted the boilerplate methods? Look at it now:

```
- (void)viewDidUnload {
    [self setStatusText:nil];
    [super viewDidUnload];
    // Release any retained subviews of the main view.
    // e.g. self.myOutlet = nil;
}
```

See that line of code above the call to super? Xcode added that automatically as well. When our view is unloaded, we need to "let go" of all of our outlets; otherwise, their memory can't be freed. Assigning a value of nil to the outlet does just that—it allows the previous value to be released from memory.

Essentially, control-dragging to create the outlet did everything we needed to set up the outlet for use.

AUTOMATIC REFERENCE COUNTING

If you're already familiar with Objective-C, or if you've read earlier versions of this book, you might have noticed that we don't have a dealloc method. We're not releasing our instance variables!

Warning! Warning! Danger, Will Robinson!
Actually, Will, you can relax. We're quite OK. There's no danger at all—really.
It's no longer necessary to release objects. Well, that's not entirely true. It is necessary, but the LLVM 3.0 compiler that Apple started shipping with iOS 5 is so smart that it will release objects for us, using a new feature called Automatic Reference Counting, or ARC, to do the heavy lifting. That means no more dealloc methods, and no more worrying about calling release or autorelease.
ARC applies to only Objective-C objects, not to Core Foundation objects or to memory allocated with malloc() and the like, and there are some caveats and gotchas that can trip you up, but for the most part, worrying about memory management is a thing of the past.
To learn more about ARC, check out the ARC release notes at this URL:

```
https://developer.apple.com/library/ios/#releasenotes/ObjectiveC/RN
-TransitioningToARC/
```

ARC is very cool, but it's not magic. You should still understand the basic rules of memory management in Objective-C to avoid getting in trouble with ARC. To brush up on the Objective-C memory management contract, read Apple's *Memory Management Programming Guide* at this URL:

```
https://developer.apple.com/library/ios/#documentation/Cocoa/Concep
tual/MemoryMgmt/
```

Writing the Action Method

So far, we've designed our user interface, and wired up both outlets and actions to our user interface. All that's left to do is to use those actions and outlets to set the text of the label when a button is pressed. You should still be in *BIDViewController.m*, but if you're not, single-click that file in the project navigator to open it in the editor. Find the empty `buttonPressed:` method that Xcode created for us earlier.

In order to differentiate between the two buttons, we're going to use the `sender` parameter. We'll retrieve the title of the button that was pressed using `sender`, and then create a new string based on that title, and assign that as the label's text. Add the bold code below to your empty method:

```
- (IBAction)buttonPressed:(UIButton *)sender {
    NSString *title = [sender titleForState:UIControlStateNormal];
    statusText.text = [NSString stringWithFormat:@"%@ button pressed.", title];
}
```

This is pretty straightforward. The first line retrieves the tapped button's title using `sender`. Since buttons can have different titles depending on their current state, we use the `UIControlStateNormal` parameter to specify that we want the title when the button is in its normal, untapped state. This is usually the state you want to specify when asking a control (a button is a type of control) for its title. We'll look at control states in more detail in Chapter 4.

The next line creates a new string by appending the text "button pressed." to the title we retrieved in the previous line. So, if the left button, which has a title of *Left*, is tapped, this line will create a string that says "Left button pressed." This new string is assigned to the label's text property, which is how we change the text that the label is displaying.

MESSAGE NESTING

Objective-C messages are often nested by some developers. You may come across code like this in your travels:

```
statusText.text = [NSString stringWithFormat:@"%@ button pressed.",
    [sender titleForState:UIControlStateNormal]];
```

This one line of code will function exactly the same as the two lines of code that make up our `buttonPressed:` method. This is because Objective-C methods can be nested, which essentially substitutes the return value from the nested method call.

For the sake of clarity, we won't generally nest Objective-C messages in the code examples in this book, with the exception of calls to `alloc` and `init`, which, by long-standing convention, are almost always nested.

Trying It Out

Guess what? We're basically finished. Are you ready to try out our app? Let's do it!

Select **Product ➤ Run**. If you run into any compile or link errors, go back and compare your code changes to those shown in this chapter. Once your code builds properly, Xcode will launch the iPhone simulator and run your application. When you tap the right button, the text "Right button pressed." should appear (as in Figure 3–1). If you then tap the left button, the label will change to say "Left button pressed."

Looking at the Application Delegate

Well, cool, your application works! Before we move on to our next topic, let's take a minute to look through the two source code files we have not yet examined, *BIDAppDelegate.h* and *BIDAppDelegate.m*. These files implement our **application delegate**.

Cocoa Touch makes extensive use of **delegates**, which are classes that take responsibility for doing certain tasks on behalf of another object. The application delegate lets us do things at certain predefined times on behalf of the UIApplication class. Every iOS application has one and only one instance of UIApplication, which is responsible for the application's run loop and handles application-level functionality such as routing input to the appropriate controller class. UIApplication is a standard part of the UIKit, and it does its job mostly behind the scenes, so you generally don't need to worry about it.

At certain well-defined times during an application's execution, UIApplication will call specific methods on its delegate, if there is a delegate and that delegate implements the method. For example, if you have code that needs to fire just before your program quits, you would implement the method applicationWillTerminate: in your application delegate and put your termination code there. This type of delegation allows your application to implement common application-wide behavior without needing to subclass UIApplication or, indeed, without needing to know anything about the inner workings of UIApplication.

Click *BIDAppDelegate.h* in the project navigator to see the application delegate's header file. It should look similar to this:

```
#import <UIKit/UIKit.h>

@class BIDViewController;

@interface BIDAppDelegate : UIResponder <UIApplicationDelegate>

@property (strong, nonatomic) UIWindow *window;

@property (strong, nonatomic) BIDViewController *viewController;

@end
```

One thing worth pointing out is this line of code:

```
@interface BIDAppDelegate : UIResponder <UIApplicationDelegate>
```

Do you see that value between the angle brackets? This indicates that this class conforms to a protocol called UIApplicationDelegate. Hold down the option key. Your cursor should turn into crosshairs. Move your cursor so that it is over the word UIApplicationDelegate. Your cursor should turn into a pointing hand with a question mark in the center, and the word UIApplicationDelegate should be highlighted, as if it were a link in a browser (see Figure 3–14).

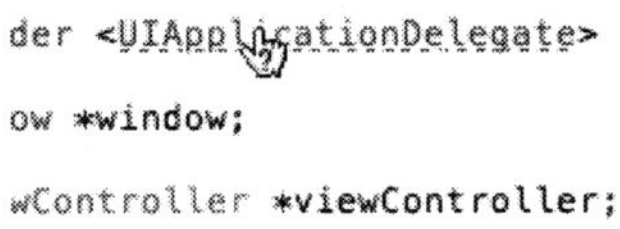

```
der <UIApplicationDelegate>

ow *window;

wController *viewController;
```

Figure 3–14. *When you hold down the option key in Xcode and point at a symbol in your code, the symbol is highlighted and your cursor changes into a pointing hand with a question mark.*

With the option key still held down, click this link. This will open a small popup window showing a brief overview of the UIApplicationDelegate protocol, as shown in Figure 3–15.

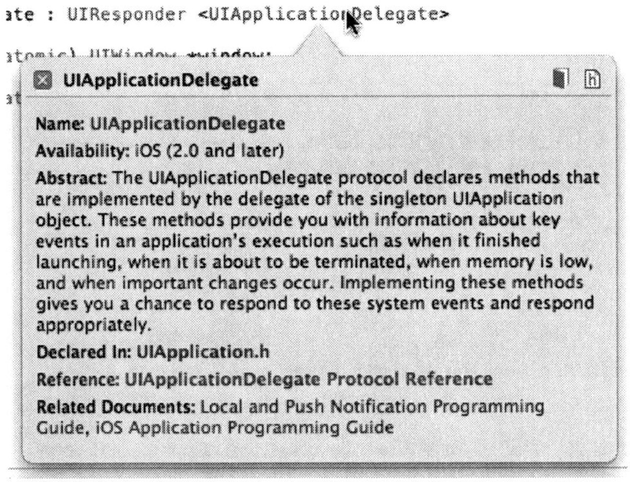

Figure 3–15. *When we option-clicked <UIApplicationDelegate> from within our source code, Xcode popped up this window, called the Quick Help panel, which describes the protocol.*

Notice the two icons in the upper-right corner of this new popup documentation window (see Figure 3–15). Click the left icon to view the full documentation for this symbol, or click the right icon to view the symbol's definition in a header file. This same trick works with class, protocol, and category names, as well as method names displayed in the editor pane. Just option-click a word, and Xcode will search for that word in the documentation browser.

Knowing how to quickly look up things in the documentation is definitely worthwhile, but looking at the definition of this protocol is perhaps more important. Here's where you'll find which methods the application delegate can implement and when those methods will be called. It's probably worth your time to read over the descriptions of these methods.

> **NOTE:** If you've worked with Objective-C before but not with Objective-C 2.0, you should be aware that protocols can now specify optional methods. UIApplicationDelegate contains many optional methods. However, you do not need to implement any of the optional methods in your application delegate unless you have a reason to do so.

Back in the project navigator, click *BIDAppDelegate.m* to see the implementation of the application delegate. It should look something like this:

```objc
#import "BIDAppDelegate.h"

#import "BIDViewController.h"

@implementation BIDAppDelegate

@synthesize window = _window;
@synthesize viewController = _viewController;

- (BOOL)application:(UIApplication *)application
didFinishLaunchingWithOptions:(NSDictionary *)launchOptions
{
    self.window = [[UIWindow alloc] initWithFrame:[[UIScreen mainScreen] bounds]];
    // Override point for customization after application launch.
    self.viewController = [[BIDViewController alloc]
initWithNibName:@"BIDViewController" bundle:nil];
    self.window.rootViewController = self.viewController;
    [self.window makeKeyAndVisible];
    return YES;
}

- (void)applicationWillResignActive:(UIApplication *)application
{
    /*
    Sent when the application is about to move from active to inactive state. This can
occur for certain types of temporary interruptions (such as an incoming phone call or
SMS message) or when the user quits the application and it begins the transition to the
background state.
    Use this method to pause ongoing tasks, disable timers, and throttle down OpenGL ES
frame rates. Games should use this method to pause the game.
    */
}

- (void)applicationDidEnterBackground:(UIApplication *)application
{
    /*
    Use this method to release shared resources, save user data, invalidate timers, and
store enough application state information to restore your application to its current
state in case it is terminated later.
```

```
    If your application supports background execution, this method is called instead of
applicationWillTerminate: when the user quits.
    */
}

- (void)applicationWillEnterForeground:(UIApplication *)application
{
    /*
    Called as part of the transition from the background to the inactive state; here you
can undo many of the changes made on entering the background.
    */
}

- (void)applicationDidBecomeActive:(UIApplication *)application
{
    /*
    Restart any tasks that were paused (or not yet started) while the application was
inactive. If the application was previously in the background, optionally refresh the
user interface.
    */
}

- (void)applicationWillTerminate:(UIApplication *)application
{
    /*
    Called when the application is about to terminate.
    Save data if appropriate.
    See also applicationDidEnterBackground:.
    */
}

@end
```

At the top of the file, you can see that our application delegate has implemented one of those protocol methods covered in the documentation, called `application:didFinishLaunchingWithOptions:`. As you can probably guess, this method fires as soon as the application has finished all the setup work and is ready to start interacting with the user.

Our delegate version of `application:didFinishLaunchingWithOptions:` creates a window, and then it creates an instance of our controller class by loading the nib file that contains our view. It then adds that controller's view as a subview to the application's window, which makes the view visible. This is how the view we designed is shown to the user. You don't need to do anything to make this happen; it's all part of the code generated by the template we used to build this project, but it's good to know that it happens here.

We just wanted to give you a bit of background on application delegates and show how this all ties together before closing this chapter.

Bring It on Home

This chapter's simple application introduced you to MVC, creating and connecting outlets and actions, implementing view controllers, and using application delegates. You learned how to trigger action methods when a button is tapped and saw how to change the text of a label at runtime. Although we built a simple application, the basic concepts we used are the same as those that underlie the use of all controls under iOS, not just buttons. In fact, the way we used buttons and labels in this chapter is pretty much the way that we will implement and interact with most of the standard controls under iOS.

It's critical that you understand everything we did in this chapter and why we did it. If you don't, go back and redo the parts that you don't fully understand. This is important stuff! If you don't make sure you understand everything now, you will only get more confused as we get into creating more complex interfaces later in this book.

In the next chapter, we'll take a look at some of the other standard iOS controls. You'll also learn how to use alerts to notify the user of important happenings and how to use action sheets to indicate that the user needs to make a choice before proceeding. When you feel you're ready to proceed, give yourself a pat on the back for being such an awesome student, and head on over to the next chapter.

Chapter 4

More User Interface Fun

In Chapter 3, we discussed MVC and built an application using it. You learned about outlets and actions, and used them to tie a button control to a text label. In this chapter, we're going to build an application that will take your knowledge of controls to a whole new level.

We'll implement an image view, a slider, two different text fields, a segmented control, a couple of switches, and an iOS button that looks more like, well, an iOS button. You'll see how to set and retrieve the values of various controls. You'll learn how to use action sheets to force the user to make a choice, and how to use alerts to give the user important feedback. You'll also learn about control states and the use of stretchable images to make buttons look the way they should.

Because this chapter's application uses so many different user interface items, we're going to work a little differently than we did in the previous two chapters. We'll break our application into pieces, implementing one piece at a time, and bouncing back and forth between Xcode and the iPhone simulator, testing each piece before we move on to the next. Dividing the process of building a complex interface into smaller chunks makes it much less intimidating, as well as more like the actual process you'll go through when building your own applications. This code-compile-debug cycle makes up a large part of a software developer's typical day.

A Screen Full of Controls

As we mentioned, the application we're going to build in this chapter is a bit more complex than the one we created in Chapter 3. We'll still use only a single view and controller, but as you can see in Figure 4–1, there's a lot more going on in this one view.

Figure 4–1. *The Control Fun application, featuring text fields, labels, a slider, and several other stock iPhone controls*

The logo at the top of the iPhone screen is an **image view**, and in this application, it does nothing more than display a static image. Below the logo are two **text fields**: one that allows the entry of alphanumeric text and one that allows only numbers. Below the text fields is a **slider**. As the user moves the slider, the value of the label next to it will change so that it always reflects the slider's current value.

Below the slider is a **segmented control** and two **switches**. The segmented control will toggle between two different types of controls in the space below it. When the application first launches, two switches will appear below the segmented control. Changing the value of either switch will cause the other one to change its value to match. Now, this isn't something you would likely do in a real application, but it does demonstrate how to change the value of a control programmatically and how Cocoa Touch animates certain actions without you needing to do any work.

Figure 4–2 shows what happens when the user taps the segmented control. The switches disappear and are replaced by a button. When the *Do Something* button is pressed, an action sheet pops up, asking if the user really meant to tap the button (see Figure 4–3). This is the standard way of responding to input that is potentially dangerous or that could have significant repercussions, since it gives the user a chance to stop potential badness from happening. If *Yes, I'm Sure!* is selected, the application will put up an alert, letting the user know that everything is OK (see Figure 4–4).

Figure 4–2. *Tapping the segmented controller on the left side causes a pair of switches to be displayed. Tapping the right side causes a button to be displayed.*

Figure 4–3. *Our application uses an action sheet to solicit a response from the user.*

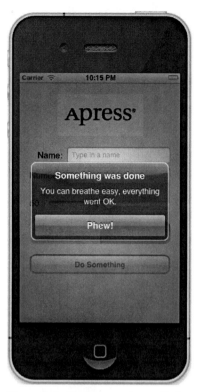

Figure 4–4. *Alerts are used to notify the user when important things happen. We use one here to confirm that everything went OK.*

Active, Static, and Passive Controls

Interface controls are in used in three basic modes: active, static (or inactive), and passive. The buttons that we used in the previous chapter are classic examples of active controls. You push them, and something happens—usually, a piece of code that you wrote fires.

Although many of the controls that you will use will directly trigger action methods, not all controls will. The image view that we'll be implementing in this chapter is a good example of a control being used statically. Even though a UIImageView can be configured to trigger action methods, in our application, the image view is passive—the user cannot do anything with it. Text fields and image controls are often used in this manner.

Some controls can work in a passive mode, simply holding on to a value that the user has entered until you're ready for it. These controls don't trigger action methods, but the user can interact with them and change their values. A classic example of a passive control is a text field on a web page. Although it's possible to create validation code that fires when the user tabs out of a field, the vast majority of web page text fields are simply containers for data that's submitted to the server when you click the submit button. The text fields themselves usually don't cause any code to fire, but when the submit button is clicked, the text field's data goes along for the ride.

On an iOS device, most of the available controls can be used in all three modes, and nearly all of them can function in more than one mode, depending on your needs. All iOS controls are subclasses of UIControl and, because of that, are capable of triggering action methods. Many controls can be used passively, and all of them can be made inactive or invisible. For example, using one control might trigger another inactive control to become active. However, some controls, such as buttons, really don't serve much purpose unless they are used in an active manner to trigger code.

There are some behavioral differences between controls on iOS and those on your Mac. Here are a few examples:

- Because of the multitouch interface, all iOS controls can trigger multiple actions depending on how they are touched. The user might trigger a different action with a finger swipe across the control than with just a tap.

- You could have one action fire when the user presses down on a button and a separate action fire when the finger is lifted off the button.

- You could have a single control call multiple action methods on a single event. You could have two different action methods fire on the touch up inside event, meaning that both methods would be called when the user's finger is lifted after touching that button.

> **NOTE:** Although controls can trigger multiple methods on iOS, the vast majority of the time, you're probably better off implementing a single action method that does what you need for a particular use of a control. Though you won't usually need this capability, it's good to keep it in mind when working in Interface Builder. Connecting an event to an action in Interface Builder does *not* disconnect a previously connected action from the same control! This can lead to surprising misbehaviors in your app, where a control will trigger multiple action methods. Keep an eye open when retargeting an event in Interface Builder, and make sure to remove old actions before connecting to new ones.

Another major difference between iOS and the Mac stems from the fact that, normally, iOS devices do not have a physical keyboard. The standard iOS software keyboard is actually just a view filled with a series of button controls that are managed for you by the system. Your code will likely never directly interact with the iOS keyboard.

Creating the Application

Let's get started. Fire up Xcode if it's not already open, and create a new project called *Control Fun*. We're going to use the Single View Application template again, so create your project just as you did in the previous two chapters.

Now that you've created your project, let's get the image we'll use in our image view. The image must be imported into Xcode before it will be available for use inside Interface Builder, so we'll import it now. You can use the image named *apress_logo.png* in the project archives in the *04 - Control Fun* folder, or you can use an image of your own choosing. If you use your own image, make sure that it is a *.png* image sized correctly for the space available. It should be fewer than 100 pixels tall and a maximum of 300 pixels wide so that it can fit comfortably at the top of the view without being resized.

Add the image to the *Supporting Files* folder of your project by dragging the image from the Finder to the *Supporting Files* folder in the project navigator. When prompted, check the checkbox that says *Copy items into destination group's folder (if needed)*, and then click *Finish*.

Implementing the Image View and Text Fields

With the image added to your project, your next step is to implement the five interface elements at the top of the application's screen: the image view, the two text fields, and the two labels (see Figure 4–5).

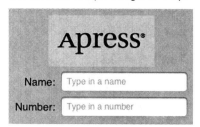

Figure 4–5. *The image view, labels, and text fields we will implement first*

Adding the Image View

In the project navigator, click *BIDViewController.xib* to open the file in Interface Builder, Xcode's nib editor. You'll see the familiar graph paper background and single gray view where you can lay out your application's interface.

If the object library is not open, select **View ➤ Utilities ➤ Show Object Library** to open it. Scroll about one-fourth of the way through the list until you find *Image View* (see Figure 4–6), or just type *image view* in the search field. Remember that the object library is the third icon on top of the library pane. You won't find *Image View* under any of the other icons.

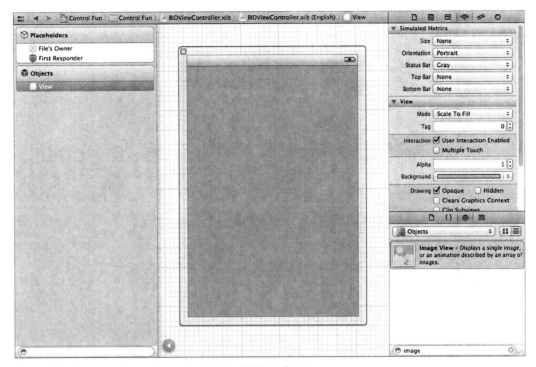

Figure 4–6. *The Image View element in Interface Builder's library*

Drag an image view onto the view in the nib editor. Notice that as you drag your image view out of the library, it changes size twice. As the drag makes its way out of the library pane, it takes the shape of a horizontal rectangle. Then, when your drag enters the frame of the view, the image view resizes to be the size of the view, minus the status bar at the top. This behavior is normal and, in many cases, exactly what you want, as often the first image you place in a view is a background image. Release the drag inside the view, taking care to be sure that the new UIImageView snaps to the sides and bottom of the surrounding view. In this particular case, we actually don't want our image view to take the entire space, so use the drag handles to resize the image view to the approximate size of the image you imported into Xcode. Don't worry about getting it exactly right yet; we'll take care of that in the next section. Figure 4–7 shows our resized UIImageView.

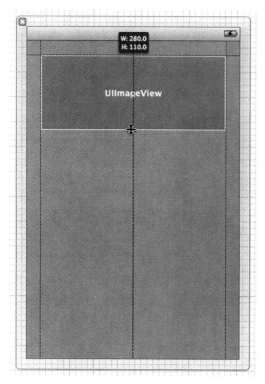

Figure 4–7. *Our resized UIImageView, sized to accommodate the image we will place here*

Remember that if you ever encounter difficulty selecting an item in the nib editor, you can switch the nib editor's dock to list view by clicking the small triangle icon below the dock. Now, click the item you want selected in the list and, sure enough, that item will be selected in the nib editor.

To get at an object that is nested inside another object, click the disclosure triangle to the left of the enclosing object to reveal the nested object. In our case, to select the image view, first click the disclosure triangle to the left of the view. Then, when the image view appears in the dock, click it, and the corresponding image view in the nib editor will be selected.

With the image view selected, bring up the object attributes inspector by pressing ⌥⌘4, and you should see the editable options of the UIImageView class, as shown in Figure 4–8.

Figure 4–8. *The image view attributes inspector. We selected our image from the Image popup at the top of the inspector, and this populated the image view with our image.*

The most important setting for our image view is the topmost item in the inspector, labeled *Image*. Click the little arrow to the right of the field to see a popup menu listing the available images, which should include any images you added to your Xcode project. Select the image you added earlier. Your image should now appear in your image view.

Resizing the Image View

As it turns out, the image we used is a fair amount smaller than the image view in which it was placed. If you take another look at Figure 4–8, you'll notice that the image we used was scaled to completely fill the image view. A big clue that this is so is the *Mode* setting in the attributes inspector, which is set to *Scale To Fill*.

Though we could keep our app this way, it's generally a good idea to do any image scaling before runtime, as image scaling takes time and processor cycles. Let's resize our image view to the exact size of our image.

Make sure the image view is selected and that you can see the resize handles. Now select the image view one more time. You should see the outline of the image view replaced by a thick, gray border. Finally, press ⌘= or select **Editor ➤ Size to Fit Content**. This will resize the image view to match the size of its contents.

Now that the image view is resized, move it into its final position. You'll need to click off it, and then click it again to reselect it. Now drag the image view so the top hits the blue guideline toward the top of your view and it is centered according to the centering blue guideline (see Figure 4–9). Note that you can also center an item in its containing view by choosing **Editor ➤ Alignment ➤ Align Horizontal Center in Container**.

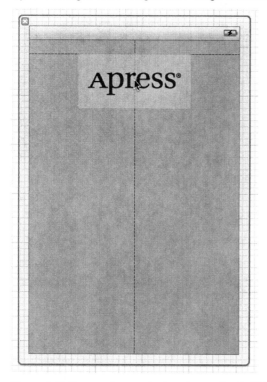

Figure 4–9. *Once we have resized our image view to fit the size of its image, we drag it into position using the view's blue guidelines.*

TIP: Dragging and resizing views in Interface Builder can be tricky. Don't forget about the hierarchical list mode, activated by clicking the small triangle icon at the bottom of the nib editor's dock. When it comes to resizing, hold down the option key, and Interface Builder will draw some helpful red lines on the screen that make it much easier to get a sense of the image view's size. This trick won't work with dragging, since the option key will prompt Interface Builder to make a copy of the dragged object. However, if you select **Editor ➤ Canvas ➤ Show Bounds Rectangles**, Interface Builder will draw a line around all of your interface items, making them easier to see. You can turn off those lines by selecting **Show Bounds Rectangles** a second time.

Setting View Attributes

Select your image view, and then switch your attention back over to the attributes inspector. Below the *Image View* section of the inspector is the *View* section. As you may have deduced, the pattern here is that the attributes that are specific to the selected object are shown at the top, followed by more general attributes that apply to the selected object's parent class. In this case, the parent class of UIImageView is UIView, so the next section is simply labeled *View*, and it contains attributes that any view class will have.

The Mode Attribute

The first option in the view inspector is a popup menu labeled **Mode**. The **Mode** menu defines how the view will display its content. This determines how the image will be aligned inside the view and whether it will be scaled to fit. Feel free to play with the various options, but the default value of *Scale To Fill* will work fine for now.

Keep in mind that choosing any option that causes the image to scale will potentially add processing overhead, so it's best to avoid those and size your images correctly before you import them. If you want to display the same image at multiple sizes, generally it's better to have multiple copies of the image at different sizes in your project, rather than force your iOS device to do scaling at runtime. Of course, there are times when scaling at runtime is appropriate; this is a guideline, not a rule.

Tag

The next item, *Tag*, is worth mentioning, though we won't be using it in this chapter. All subclasses of UIView, including all views and controls, have a property called tag, which is just a numeric value that you can set here or in code. The tag is designed for your use; the system will never set or change its value. If you assign a tag value to a control or view, you can be sure that the tag will always have that value unless you change it.

Tags provide an easy, language-independent way of identifying objects on your interface. Let's say you have five different buttons, each with a different label, and you want to use a single action method to handle all five buttons. In that case, you probably need some way to differentiate among the buttons when your action method is called. Sure, you could look at the button's title, but code that does that probably won't work when your application is translated into Swahili or Sanskrit. Unlike labels, tags will never change, so if you set a tag value here in Interface Builder, you can then use that as a fast and reliable way to check which control was passed into an action method in the sender argument.

Interaction Checkboxes

The two checkboxes in the *Interaction* section have to do with user interaction. The first checkbox, *User Interaction Enabled*, specifies whether the user can do anything at all with this object. For most controls, this box will be checked, because if it's not, the control will never be able to trigger action methods. However, image views default to unchecked because they are often used just for the display of static information. Since all we're doing here is displaying a picture on the screen, there is no need to turn this on.

The second checkbox is *Multiple Touch*, and it determines whether this control is capable of receiving multitouch events. Multitouch events allow complex gestures like the pinch gesture used to zoom in in many iOS applications. We'll talk more about gestures and multitouch events in Chapter 13. Since this image view doesn't accept user interaction at all, there's no reason to turn on multitouch events, so leave the checkbox unchecked.

The Alpha Value

The next item in the inspector is *Alpha*. Be careful with this one. Alpha defines how transparent your image is—how much of what's beneath it shows through. It's defined as a floating-point number between 0.0 and 1.0, where 0.0 is fully transparent and 1.0 is completely opaque. If you have any value less than 1.0, your iOS device will draw this view with some amount of transparency so that any objects underneath it show through. With a value of less than 1.0, even if there's nothing actually underneath your image, you will cause your application to spend processor cycles calculating transparency, so don't set *Alpha* to anything other than 1.0 unless you have a very good reason for doing so.

Background

The next item down, *Background*, is a property inherited from UIView, and determines the color of the background for the view. For image views, this matters only when an image doesn't fill its view and is letterboxed or when parts of the image are transparent. Since we've sized our view to perfectly match our image, this setting will have no visible effect, so we can leave it alone.

Drawing Checkboxes

Below *Background* are a series of *Drawing* checkboxes. The first one is labeled *Opaque*. That should be checked by default; if not, click to check that checkbox. This tells iOS that nothing behind your view needs to be drawn and allows iOS's drawing methods to do some optimizations that speed up drawing.

You might be wondering why we need to select the *Opaque* checkbox when we've already set the value of *Alpha* to 1.0 to indicate no transparency. The alpha value applies to the parts of the image to be drawn, but if an image doesn't completely fill the image view, or there are holes in the image thanks to an alpha channel, the objects below will still show through, regardless of the value set in *Alpha*. By selecting *Opaque*, we are telling iOS that nothing below this view ever needs to be drawn no matter what, so it does not need to waste processing time with anything below our object. We can safely select the *Opaque* checkbox, because we earlier selected *Size To Fit*, which caused the image view to match the size of the image it contains.

The *Hidden* checkbox does exactly what you think it does. If it's checked, the user can't see this object. Hiding an object can be useful at times, as you'll see later in this chapter when we hide our switches and button, but the vast majority of the time—including now—you want this to remain unchecked.

The next checkbox, *Clears Graphics Context*, will rarely need to be checked. When it is checked, iOS will draw the entire area covered by the object in transparent black before it actually draws the object. Again, it should be turned off for the sake of performance and because it's rarely needed. Make sure this checkbox is unchecked (it is likely checked by default).

Clip Subviews is an interesting option. If your view contains subviews, and those subviews are not completely contained within the bounds of its parent view, this checkbox determines how the subviews will be drawn. If *Clip Subviews* is checked, only the portions of subviews that lie within the bounds of the parent will be drawn. If *Clip Subviews* is unchecked, subviews will be drawn completely, even if they lie outside the bounds of the parent.

It might seem that the default behavior should be the opposite of what it actually is— *Clip Subviews* should be unchecked by default. Calculating the clipping area and displaying only part of the subviews is a somewhat costly operation, mathematically speaking, and most of the time, a subview won't lay outside the bounds of its superview. You can turn on *Clip Subviews* if you really need it for some reason, but it is off by default for the sake of performance.

The last checkbox in this section, *Autoresize Subviews*, tells iOS to resize any subviews if this view is resized. Leave this checked (since we don't allow our view to be resized, it really does not matter whether it's checked or not).

Stretching

Next up is a section simply labeled *Stretching*. You can leave your yoga mat in the closet though, because the only stretching going on here is in the form of rectangular views being redrawn as they're resized on the screen. The idea is that rather than the entire content of a view being stretched uniformly, you can keep the outer edges of a view, such as the bezeled edge of a button, looking the same even as the center portion stretches.

The four floating-point values set here let you declare which portion of the rectangle is stretchable by specifying a point at the upper-left corner of the view and the size of the stretchable area, all in the form of a number between 0.0 and 1.0, representing a portion of the overall view size. For example, if you wanted to keep 10% of each edge not stretchy, you would specify 0.1 for both *X* and *Y*, and 0.8 for both *Width* and *Height*. In this case, we're going to leave the default values of 0.0 for X and Y, and 1.0 for Width and Height. Most of the time, you will not change these values.

Adding the Text Fields

With your image view finished, it's time to bring on the text fields. Grab a text field from the library, and drag it into the *View*, underneath the image view. Use the blue guidelines to align it with the right margin and snug it just under your image view (see Figure 4–10).

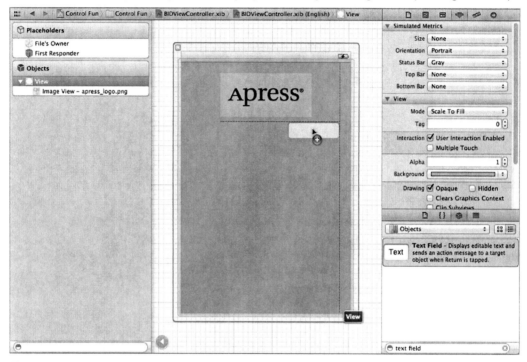

Figure 4–10. *We dragged a text field out of the library and dropped it onto the view, just below our image view and touching the right blue guideline.*

A horizontal blue guideline will appear just above the text field when you move it very close to the bottom of your image view. That guideline tells you when the object you are dragging is the minimum reasonable distance from an adjacent object. You can leave your text field there for now, but to give it a balanced appearance, consider moving the text field just a little farther down. Remember that you can always edit your nib file again in order to change the position and size of interface elements without needing to change code or reestablish connections.

After you drop the text field, grab a label from the library, and drag that over so it is aligned with the left margin of the view and vertically with the text field you placed earlier. Notice that multiple blue guidelines will pop up as you move the label around, making it easy to align the label to the text field using the top, bottom, or middle of the label. We're going to align the label and the text field using the middle of those guidelines (see Figure 4–11).

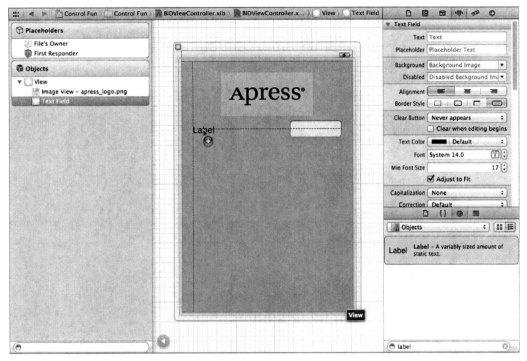

Figure 4–11. *Aligning the label and text field using the baseline guide*

Double-click the label you just dropped, change it to read *Name*: instead of *Label* (note the colon character at the end of the label), and press the return key to commit your changes.

Next, drag another text field from the library to the view, and use the guidelines to place it below the first text field (see Figure 4–12).

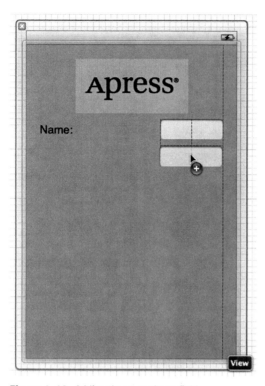

Figure 4–12. *Adding the second text field*

Once you've added the second text field, grab another label from the library, and place it on the left side, below the existing label. Again, use the middle blue guideline to align your new label with the second text field. Double-click the new label, and change it to read *Number*: (don't forget the colon).

Now, let's expand the size of the bottom text field to the left, so it snugs up against the right side of the label. Why start with the bottom text field? We want the two text fields to be the same size, and the bottom label is longer.

Single-click the bottom text field, and drag the left resize dot to the left until a blue guideline appears to tell you that you are as close as you should ever be to the label (see Figure 4–13). This particular guideline is somewhat subtle—it's only as tall as the text field itself, so keep your eyes peeled.

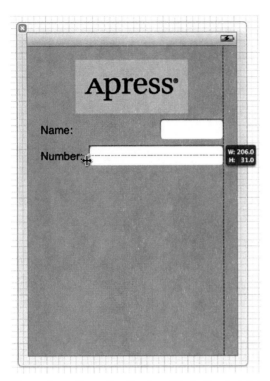

Figure 4–13. *Expanding the size of the bottom text field*

Now, expand the top text field in the same way so that it matches the bottom one in size. Once again, a blue guideline provides some help, and this one is easier to spot.

We're basically finished with the text fields except for one small detail. Look back at Figure 4–5. Do you see how the *Name*: and *Number*: are right-aligned? Right now, ours are both against the left margin. To align the right sides of the two labels, click the *Name*: label, hold down the shift key, and click the *Number*: label so both labels are selected. Now select **Editor ➤ Align ➤ Right Edges**.

When you are finished, the interface should look very much like the one shown in Figure 4–5. The only difference is the light-gray text in each text field. We'll add that now.

Select the top text field and press ⌥⌘4 to bring up the attributes inspector (see Figure 4–14). The text field is one of the most complex iOS controls, as well as one of the most commonly used. Let's take a walk through the settings, beginning from the top of the inspector.

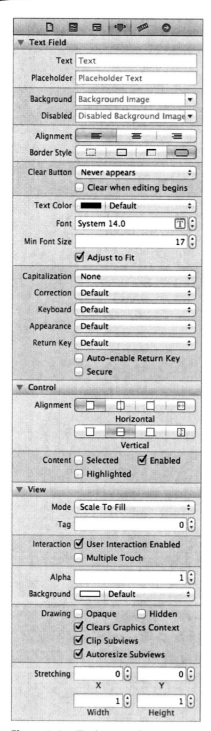

Figure 4–14. *The inspector for a text field showing the default values*

Text Field Inspector Settings

In the first field, *Text*, you can set a default value for the text field. Whatever you type here will show up in the text field when your application launches.

The second field, *Placeholder*, allows you to specify a bit of text that will be displayed in gray inside the text field, but only when the field does not have a value. You can use a placeholder instead of a label if space is tight, or you can use it to clarify what the user should type into this text field. Type in the text *Type in a name* as the placeholder for our currently selected text field, and then hit return to commit the change.

The next two fields, *Background* and *Disabled*, are used only if you need to customize the appearance of your text field, which is completely unnecessary and actually ill-advised the vast majority of the time. Users expect text fields to look a certain way. We're going to skip over these fields, leaving them set to their defaults.

Below these fields are three buttons for controlling the alignment of the text displayed in the field. We'll leave this setting at the default value of left-aligned (the leftmost button).

Next are four buttons labeled *Border Style*. These allow you to change the way the text field's edge will be drawn. The default value (the rightmost button) creates the text field style that users are most accustomed to seeing for normal text fields in an iOS application. Feel free to try all four different styles. When you're finished experimenting, set this setting back to the rightmost button.

Below the border setting is a *Clear Button* popup button, which lets you choose when the clear button should appear. The clear button is the small *X* that can appear at the right end of a text field. Clear buttons are typically used with search fields and other fields where you would be likely to change the value frequently. They are not typically included on text fields used to persist data, so leave this at the default value of *Never appears*.

The *Clear when editing begins* checkbox specifies what happens when the user touches this field. If this box is checked, any value that was previously in this field will be deleted, and the user will start with an empty field. If this box is unchecked, the previous value will remain in the field, and the user will be able to edit it. Leave this checkbox unchecked.

After that is a series of fields that let you set the font, font color, and minimum font size. We'll leave the *Text Color* at the default value of black. Note that the *Text Color* popup is divided into two parts. The right side allows you to select from a set of preselected colors, and the left side gives you access to a color well to more precisely specify your color.

The *Font* setting is divided into three parts. On the right is a control that lets you increment or decrement the text size, one point at a time. The left side allows you to manually edit the font name and size. Finally, click the *T*-in-a-box icon to bring up a popup window that lets you set the various font attributes. We'll leave the *Font* at its default setting of *System 14.0*.

Following the *Font* setting is a control that lets you set the minimum font size that the text field will use for displaying its text. Leave that at its default value for now.

The *Adjust to Fit* checkbox specifies whether the size of the text should shrink if the text field is reduced in size. Adjusting to fit will keep the entire text visible in the view, even if the text would normally be too big to fit in the allotted space. This checkbox works in conjunction with the minimum font size setting. No matter the size of the field, the text will not be resized below that minimum size. Specifying a minimum size allows you to make sure that the text doesn't get too small to be readable.

The next section defines how the keyboard will look and behave when this text field is being used. Since we're expecting a name, let's change the *Capitalization* popup to *Words*. This causes every word to be automatically capitalized, which is what you typically want with names.

The next three popups—*Correction*, *Keyboard*, and *Appearance*—can be left at their default values. Take a minute to look at each to get a sense of what these settings do.

Next is the *Return Key* popup. The return key is the key on the lower right of the keyboard, and its label changes based on what you're doing. If you are entering text into Safari's search field, for example, then it says *Search*. In an application like ours, where the text fields share the screen with other controls, *Done* is the right choice. Make that change here.

If the *Auto-enable Return Key* checkbox is checked, the return key is disabled until at least one character is typed into the text field. Leave this unchecked, because we want to allow the text field to remain empty if the user prefers not to enter anything.

The *Secure* checkbox specifies whether the characters being typed are displayed in the text field. You would check this checkbox if the text field were being used as a password field. Leave it unchecked for our app.

The next section allows you to set control attributes inherited from UIControl, but these generally don't apply to text fields and, with the exception of the *Enabled* checkbox, won't affect the field's appearance. We want to leave these text fields enabled so that the user can interact with them. Leave the default settings in this section.

The last section on the inspector, *View*, should look familiar. It's identical to the section of the same name on the image view inspector we looked at earlier. These are attributes inherited from the UIView class, and since all controls are subclasses of UIView, they all share this section of attributes. As you did earlier for the image view, check the *Opaque* checkbox, and uncheck *Clears Graphics Context* and *Clip Subviews*, for the reasons we discussed earlier.

Setting the Attributes for the Second Text Field

Next, single-click the second text field in the *View* window, and return to the inspector. In the *Placeholder* field, type *Type in a number*, and make sure *Clear When Editing Begins* is unchecked. A little farther down, click the *Keyboard* popup menu. Since we want the user to enter only numbers, not letters, select *Number Pad*. This ensures that the users will be presented with a keyboard containing only numbers, meaning they won't be able to enter alphabetical characters, symbols, or anything other than numbers. We don't need to set the *Return Key* value for the numeric keypad, because that style of keyboard doesn't have a return key, so all of the other inspector settings can stay at the default values. As you did earlier, check the *Opaque* checkbox, and uncheck *Clears Graphics Context* and *Clip Subviews*.

Creating and Connecting Outlets

We are almost ready to take our app for its first test drive. For this first part of the interface, all that's left is creating and connecting our outlets. The image view and labels on our interface do not need outlets because we don't need to change them at runtime. The two text fields, however, are passive controls that hold data we'll need to use in our code, so we need outlets pointing to each of them.

As you probably remember from the previous chapter, Xcode 4 allows us to create and connect outlets at the same time using the assistant editor. Go into the assistant editor now by selecting the middle toolbar button labeled *Editor* or by selecting **View ➤ Assistant Editor ➤ Show Assistant Editor**.

Make sure your nib file is selected in the project navigator. If you don't have a large amount of screen real estate, you might also want to select **View ➤ Utilities ➤ Hide Utilities** to hide the utility pane during this step. When you bring up the assistant editor, the nib editing pane will be split in two, with Interface Builder in one half and *BIDViewController.h* in the other (see Figure 4–15). This new editing area—the one showing *BIDViewController.h*—is the assistant.

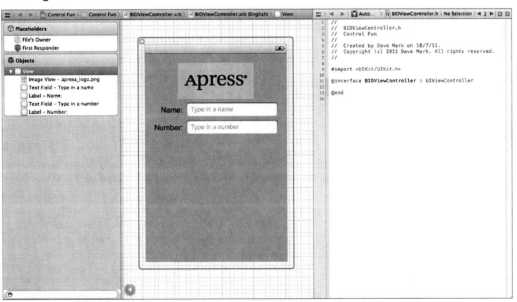

Figure 4–15. *The nib editing area with the assistant turned on. You can see the assistant area on the right, showing the code from BIDViewController.h.*

You'll see that the upper boundary of the assistant includes a jump bar, much like the normal editor pane. One important addition to the assistant's jump bar is a new set of "smart" selections, which let you switch between a variety of files that Xcode believes are relevant, based on what appears in the main view. By default, it shows a group of files labeled *Top Level Objects*, including your own source code for the controller class (since it's one of the top-level objects in the nib), as well as headers for UIResponder and UIView, since those are also represented at the top level of the nib. Take a few minutes to click around the jump bar at the top of the assistant, just to get a feel for what's what. Once you have a sense of the jump bar and files represented there, move on.

Now comes the fun part. Make sure *BIDViewController.h* is still showing in the assistant (use the jump bar to return there if necessary). Now control-drag from the top text field in the view over to the *BIDViewController.h* source code, right below the @interface line. You should see a gray popup that reads *Insert Outlet, Action, or Outlet Collection* (see Figure 4–16). Release the mouse button, and you'll get the same popup you saw in the previous chapter. We want to create an outlet called nameField, so type *nameField* into the *Name* field (say that five times fast!), and then hit return.

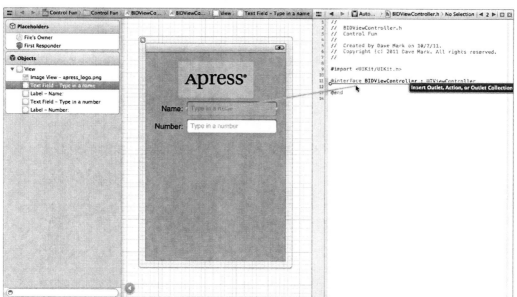

Figure 4–16. *With the assistant turned on, we control-drag over to the declaration of nameField to connect that outlet.*

You now have a property called nameField in BIDViewController, and it has been connected to the top text field. Do the same for the second text field, creating and connecting it to a property called *numberField*.

Closing the Keyboard

Let's see how our app works, shall we? Select **Product** ➤ **Run**. Your application should come up in the iPhone simulator. Click the *Name* text field. The traditional keyboard should appear. Type in a name. Now, tap the *Number* field. The numeric keypad should appear (see Figure 4–17). Cocoa Touch gives us all this functionality for free just by adding text fields to our interface.

Figure 4–17. *The keyboard comes up automatically when you touch either the text field or the number field.*

Woo-hoo! But there's a little problem. How do you get the keyboard to go away? Go ahead and try. We'll wait right here while you do.

Closing the Keyboard When Done Is Tapped

Because the keyboard is software-based, rather than a physical keyboard, we need to take a few extra steps to make sure the keyboard goes away when the user is finished with it. When the user taps the *Done* button on the text keyboard, a *Did End On Exit* event will be generated, and at that time, we need to tell the text field to give up control so that the keyboard will go away. In order to do that, we need to add an action method to our controller class.

Select *BIDViewController.h* in the project navigator, and add the following line of code, shown in bold:

```
#import <UIKit/UIKit.h>


@interface BIDViewController : UIViewController
@property (strong, nonatomic) IBOutlet UITextField *nameField;
@property (strong, nonatomic) IBOutlet UITextField *numberField;

- (IBAction)textFieldDoneEditing:(id)sender;

@end
```

When you selected the header file in the project navigator, you probably noticed that the assistant we opened earlier has adapted to having a source code file selected in the main editor pane, and now automatically shows the selected file's counterpart. If you select a *.h* file, the assistant will automatically show the matching *.m* file, and vice versa. This is a remarkably handy addition to Xcode 4! As a result of this behavior, *BIDViewController.m* is now shown in the assistant view, ready for us to implement this method.

Add this action method at the bottom of *BIDViewController.m*, just before the @end:

```
- (IBAction)textFieldDoneEditing:(id)sender {
    [sender resignFirstResponder];
}
```

As you learned in Chapter 2, the first responder is the control with which the user is currently interacting. In our new method, we tell our control to resign as a first responder, giving up that role to the previous control the user worked with. When a text field yields first responder status, the keyboard associated with it goes away.

Save both of the files you just edited. Let's hop back to the nib file and trigger this action from both of our text fields.

Select *BIDViewController.xib* in the project navigator, single-click the *Name* text field, and press ⌥⌘6 to bring up the connections inspector. This time, we don't want the *Touch Up Inside* event that we used in the previous chapter. Instead, we want *Did End On Exit*, since that event will fire when the user taps the *Done* button on the text keyboard.

Drag from the circle next to *Did End On Exit* to the *File's Owner* icon, and connect it to the *textFieldDoneEditing:* action. You can also do this by dragging to the textFieldDoneEditing: method in the assistant view. Repeat this procedure with the other text field, save your changes, and then press ⌘R to run the app again.

When the simulator appears, click the *Name* field, type in something, and then tap the *Done* button. Sure enough, the keyboard drops away, just as you expected. All right! What about the *Number* field, though? Um, where's the *Done* button on that one (see Figure 4–17)?

Well, crud! Not all keyboard layouts feature a *Done* button. We could force the user to tap the *Name* field and then tap *Done*, but that's not very user-friendly, is it? And we most definitely want our application to be user-friendly. Let's see how to handle this situation.

Touching the Background to Close the Keyboard

Can you recall what Apple's iPhone applications do in this situation? Well, in most places where there are text fields, tapping anywhere in the view where there's no active control will cause the keyboard to go away. How do we implement that?

The answer is probably going to surprise you because of its simplicity. Our view controller has a property called view that it inherited from UIViewController. This view property corresponds to the *View* in the nib file. The view property points to an instance of UIView in the nib that acts as a container for all the items in our user interface. It has no appearance in the user interface, but it covers the entire iPhone window, sitting "below" all of the other user interface objects. It is sometimes referred to as a nib's **container view** because its main purpose is to simply hold other views and controls. For all intents and purposes, the container view is the background of our user interface.

Using Interface Builder, we can change the class of the object that view points to so that its underlying class is UIControl instead of UIView. Because UIControl is a subclass of UIView, it is perfectly appropriate for us to connect our view property to an instance of UIControl. Remember that when a class subclasses another object, it is just a more specific version of that class, so a UIControl *is* a UIView. If we simply change the instance that is created from UIView to UIControl, we gain the ability to trigger action methods. Before we do that, though, we need to create an action method that will be called when the background is tapped.

We need to add one more action to our controller class. Add the following line to your *BIDViewController.h* file:

```
#import <UIKit/UIKit.h>

@interface BIDViewController : UIViewController
@property (strong, nonatomic) IBOutlet UITextField *nameField;
@property (strong, nonatomic) IBOutlet UITextField *numberField;

- (IBAction)textFieldDoneEditing:(id)sender;

- (IBAction)backgroundTap:(id)sender;
@end
```

Save the header file.

Now, switch over to the implementation file and add the following method at the end of the file, just before @end:

```
- (IBAction)backgroundTap:(id)sender {
    [nameField resignFirstResponder];
    [numberField resignFirstResponder];
}
```

This method simply tells both text fields to yield first responder status if they have it. It is perfectly safe to call `resignFirstResponder` on a control that is not the first responder, so we can call it on both text fields without needing to check whether either is the first responder.

> **TIP:** You'll be switching between header and implementation files a lot as you code. Fortunately, in addition to the convenience provided by the assistant, Xcode also has a key combination that will switch between counterparts quickly. The default key combination is ⌃⌘⇧, although you can change it to anything you want using Xcode's preferences.

Save this file. Now, select the nib file again. Make sure your dock is in list mode (click the triangle icon to the bottom right of the dock to switch to list view). Single-click *View* so it is selected. Do *not* select one of your view's subitems. We want the container view itself.

Next, press ⌥⌘3 to bring up the **identity inspector** (see Figure 4–18). This is where you can change the underlying class of any object instance in your nib file.

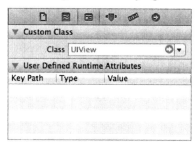

Figure 4–18. *We switched Interface Builder to list view, and then selected our view. We then switched to the identity inspector, which allows us to change the underlying class of any object instance in our nib.*

The field labeled *Class* should currently say *UIView*. If not, you likely don't have the container view selected. Now, change that setting to *UIControl*. Press return to commit the change. All controls that are capable of triggering action methods are subclasses of `UIControl`, so by changing the underlying class, we have just given this view the ability to trigger action methods. You can verify this by pressing ⌥⌘6 to bring up the connections inspector. You should now see all the events that you saw when you were connecting buttons to actions in the previous chapter.

Drag from the *Touch Down* event to the *File's Owner* icon (see Figure 4–19), and choose the *backgroundTap:* action. Now, touches anywhere in the view without an active control will trigger our new action method, which will cause the keyboard to retract. Connecting to *File's Owner* like this is exactly the same as connecting to the method in the code. For a view controller nib file, the *File's Owner* is the view controller class, so that was just a slightly different way of achieving the exact same result.

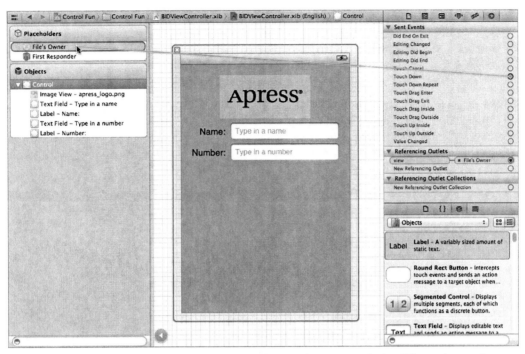

Figure 4–19. *By changing the class of our view from UIView to UIControl, we gain the ability to trigger action methods on any of the standard events. We'll connect the view's Touch Down event to the backgroundTap: action.*

> **NOTE:** You might be wondering why we selected *Touch Down* instead of *Touch Up Inside*, as we did in the previous chapter. The answer is that the background isn't a button. It's not a control in the eyes of the user, so it wouldn't occur to most users to try to drag their finger somewhere to cancel the action.

Save the nib file, and then compile and run your application again. This time, the keyboard should disappear not only when the *Done* button is tapped, but also when you tap anywhere that's not an active control, which is the behavior that your users will expect.

Excellent! Now that we have this section all squared away, are you ready to move onto the next group of controls?

Adding the Slider and Label

Now it's time to add the slider and accompanying label. Remember that the value in the label will change as the slider is used. Select *BIDViewController.xib* in the project navigator so we can add more items to our application's user interface.

Before we place the slider, let's add a bit of breathing room to our design. The blue guidelines we used to determine the spacing between the top text field and the image above it are really suggestions for minimum proximity. In other words, the blue guidelines tell you, "Don't get any closer than this." Drag the two text fields and their labels down a bit, using Figure 4–1 as a guide. Now let's add the slider.

From the object library, bring over a slider and arrange it below the *Number* text field, using the right-side blue guideline as a stopping point, and leaving a little breathing room below the bottom text field. Our slider ended up about halfway down the view. Single-click the newly added slider to select it, and then press ⌥⌘4 to go back to the object attributes inspector if it's not already visible (see Figure 4–20).

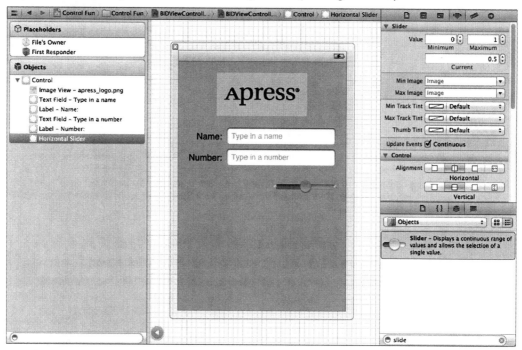

Figure 4–20. *The inspector showing default attributes for a slider*

A slider lets you choose a number in a given range. Use the inspector to set the *Minimum* value to *1.00*, the *Maximum* value to *100.00*, and the *Current* value to *50.00*. Leave the *Update Events, Continuous* checkbox checked. This ensures a continuous flow of events as the slider's value changes. That's all we need to worry about for now.

Bring over a label and place it next to the slider, using the blue guidelines to align it vertically with the slider and to align its left edge with the left margin of the view (see Figure 4–21).

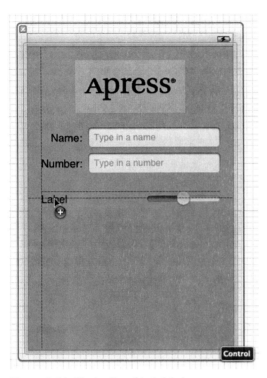

Figure 4–21. *Placing the slider's label*

Double-click the newly placed label, and change its text from *Label* to *100*. This is the largest value that the slider can hold, and we can use that to determine the correct width of the slider. Since "100" is shorter than "Label," we can make the label shorter. Resize the label by grabbing the right-middle resize dot and dragging to the left. Make sure you stop resizing before the text starts to get smaller. If it does start to get smaller, bring the resize dot back to the right until it returns to its original size. You can also automatically size the label to fit the text, as we discussed earlier, by pressing ⌘= or by selecting **Editor ➤ Size to Fit Content**.

Next, resize the slider by single-clicking the slider to select it and dragging the left resize dot to the left until the blue guidelines indicate that you should stop.

Now, double-click the label again, and change its value to *50*. That is the starting value of the slider, and we need to change it back to make sure that the interface looks correct at launch time. Once the slider is used, the code we just wrote will make sure the label continues to show the correct value.

Creating and Connecting the Actions and Outlets

All that's left to do with these two controls is to connect the outlet and action. We need an outlet that points to the label so that we can update the label's value when the slider is used, and we're going to need an action method for the slider to call as it's changed.

Make sure you're using the assistant editor and editing *BIDViewController.h*, and then control-drag from the slider to just above the @end declaration in the assistant editor. When the popup window appears, change the *Connection* popup menu to *Action*, and then type *sliderChanged* in the *name* field. Hit return to create and connect the action.

Next, control-drag from the newly added label over to the assistant editor. This time, drag to just below the last property and above the first action method. When the popup comes up, type *sliderLabel* into the *Name* text field, and then hit return to create and connect the outlet.

Implementing the Action Method

Though Xcode has created and connected our action method, it's still up to us to actually write the code that makes up the action method so it does what it's supposed to do. Save the nib, then in the project navigator, single-click *BIDViewController.m* and look for the `sliderChanged:` method, which should be empty. Add the following code to that method:

```
- (IBAction)sliderChanged:(id)sender {
  UISlider *slider = (UISlider *)sender;
  int progressAsInt = (int)roundf(slider.value);
  sliderLabel.text = [NSString stringWithFormat:@"%d", progressAsInt];
}
```

The first line in the method assigns sender to a `UISlider` pointer so that the compiler will let us use `UISlider` methods and properties without warnings. We then retrieve the current value of the slider, round it down to the nearest integer, and assign it to an integer variable. The last line of code creates a string containing that number and assigns it to the label.

Save the file. Next, press ⌘R to build and launch your app in the iPhone simulator, and try out the slider. As you move it, you should see the label's text change in real time. Another piece falls into place. Now, let's look at implementing the switches.

Implementing the Switches, Button, and Segmented Control

Back to Xcode we go once again. Getting dizzy yet? This back and forth may seem a bit strange, but it's fairly common to bounce around between source code and nib files in Xcode, and testing your app in the iOS simulator while you're developing.

Our application will have two switches, which are small controls that can have only two states: on and off. We'll also add a segmented control to hide and show the switches. Along with that control, we'll add a button that is revealed when the segmented control's right side is tapped. Let's implement those next.

Back in the nib file, drag a segmented control from the object library (see Figure 4–22) and place it on the *View* window, a little below the slider.

TIP: To give you a sense of the spacing we're going for, take a look at the image view with the Apress logo. We tried to leave about the same amount of space above the image view as below the image view. We did the same thing with the slider: we tried to leave about the same amount of space above the slide as below. Just a suggestion from the boyeez.

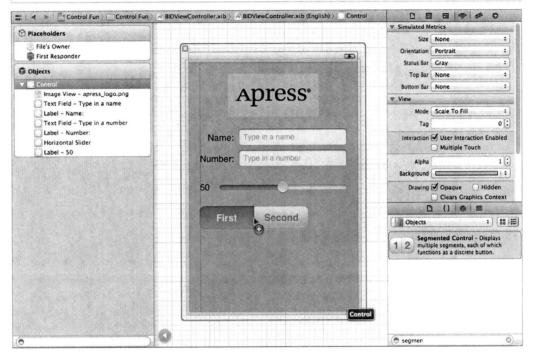

Figure 4–22. *Dragging a segmented control from the library to the left side of the parent view. Next, we'll resize the segmented control so it stretches to the right side of the view.*

Expand the width of the segmented control so that it stretches from the view's left margin to its right margin. Double-click the word *First* on the segmented control and change the title from *First* to *Switches*. After doing that, repeat the process with the *Second* segment, renaming it *Button* (see Figure 4–23).

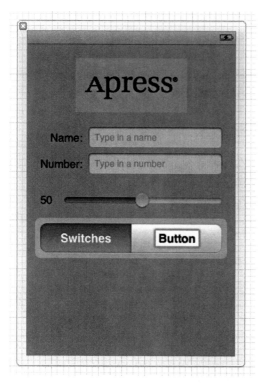

Figure 4–23. *Renaming the segments in the segmented control*

Adding Two Labeled Switches

Next, grab a switch from the library, and place it on the view, below the segmented control and against the left margin. Drag a second switch and place it against the right margin, aligned vertically with the first switch (see Figure 4–24).

> **TIP:** Holding down the option key and dragging an object in Interface Builder will create a copy of that item. When you have many instances of the same object to create, it can be faster to drag only one object from the library and then option-drag as many copies as you need.

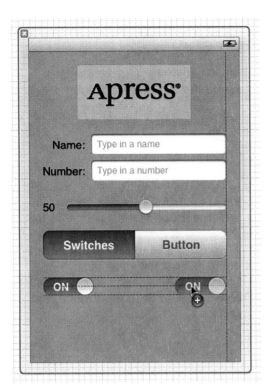

Figure 4–24. *Adding the switches to the view*

Connecting and Creating Outlets and Actions

Before we add the button, we'll create outlets for the two switches and connect them. The button that we'll be adding next will actually sit on top of the switches, making it harder to control-drag to and from them, so we want to take care of the switch connections before we add the button. Since the button and the switches will never be visible at the same time, having them in the same physical location won't be a problem.

Using the assistant editor, control-drag from the switch on the left to just below the last outlet in your header file. When the popup appears, name the outlet *leftSwitch* and hit return. Repeat with the other switch, naming its outlet *rightSwitch*.

Now, select the left switch again by single-clicking it. Control-drag once more to the assistant editor. This time, drag to right above the @end declaration before letting go. When the popup appears, change the *Connection* popup to *Action*, give it a name of *switchChanged:*, and hit return to create the new action. Repeat with the right switch, but instead of creating a new action, drag to the *switchChanged:* action that was just created and connect to it instead. Just as we did in the previous chapter, we're going to use a single method to handle both switches.

Finally, control-drag from the segmented control to the assistant editor, right above the @end declaration. Insert a new action method called *toggleControls:*.

Implementing the Switch Actions

Save the nib file and single-click *BIDViewController.m*. Look for the `switchChanged:` method that was added for you automatically, and add the following code to it:

```
- (IBAction)switchChanged:(id)sender {
  UISwitch *whichSwitch = (UISwitch *)sender;
  BOOL setting = whichSwitch.isOn;
  [leftSwitch setOn:setting animated:YES];
  [rightSwitch setOn:setting animated:YES];
}
```

The `switchChanged:` method is called whenever one of the two switches is tapped. In this method, we simply grab the value of `sender`, which represents the switch that was pressed, and use that value to set both switches. Now, `sender` is always going to be either `leftSwitch` or `rightSwitch`, so you might be wondering why we're setting them both. The reason is one of practicality. It's less work to just set the value of both switches every time than to determine which switch made the call and set only the other one. Whichever switch called this method will already be set to the correct value, and setting it again to that same value won't have any effect.

Adding the Button

Next, go back to Interface Builder and drag a *Round Rect Button* from the library to your view. Add this button directly on top of the leftmost button, aligning it with the left margin and vertically aligning its center with the two switches (see Figure 4–25).

Figure 4–25. *Adding a round rect button on top of the existing switches*

Now, grab the right-center resize handle and drag all the way to the right until you reach the blue guideline that indicates the right margin. The button should completely cover the two switches (see Figure 4–26).

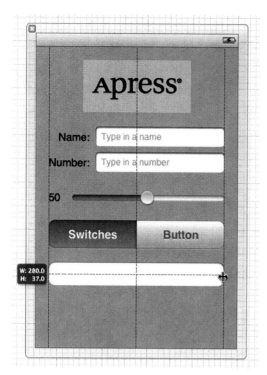

Figure 4–26. *The round rect button, once placed and resized, will completely obscure the two switches.*

Double-click the newly added button and give it a title of *Do Something.*

Connecting and Creating the Button Outlets and Actions

Control-drag from the new button to the assistant editor, just below the last outlet already in the header. When the popup appears, create a new outlet called *doSomethingButton.* After you've done that, control-drag from the button a second time to just above the @end declaration. This time, instead of creating an outlet, create an action called *buttonPressed:.*

If you save your work and take the application for a test drive, you'll see that the segmented control will be live, but it doesn't do anything particularly useful yet. We need to add some logic to make the button and switches hide and unhide.

We also need to mark our button as hidden from the start. We didn't want to do that before because it would have made it harder to connect the outlets and actions. Now that we've done that, let's hide the button. We'll show the button when the user taps the right side of the segmented control, but when the application starts, we want the button hidden. Press ⌥⌘4 to bring up the attributes inspector, scroll down to the *View* section, and click the *Hidden* checkbox. The button will disappear.

Implementing the Segmented Control Action

Save the nib file and single-click *BIDViewController.m*. Look for the `toggleControls:` method that Xcode created for us and add the following code to it:

```
- (IBAction)toggleControls:(id)sender {
    // 0 == switches index
    if ([sender selectedSegmentIndex] == 0) {
        leftSwitch.hidden = NO;
        rightSwitch.hidden = NO;
        doSomethingButton.hidden = YES;
    }
    else {
        leftSwitch.hidden = YES;
        rightSwitch.hidden = YES;
        doSomethingButton.hidden = NO;
    }
}
```

This code looks at the `selectedSegmentIndex` property of sender, which tells us which of the sections is currently selected. The first section, called `switches`, has an index of 0, a fact that we've written down in a comment so that when we later revisit the code, we know what's going on. Depending on which segment is selected, we hide or show the appropriate controls.

At this point, save and try running the application in the iOS simulator. If you've typed everything correctly, you should be able to switch between the button and the pair of switches using the segmented control, and if you tap either switch, the other one will change its value as well. The button, however, still doesn't do anything. Before we implement it, we need to talk about action sheets and alerts.

Implementing the Action Sheet and Alert

Action sheets and **alerts** are both used to provide the user with feedback. as follows:

- Action sheets are used to force the user to make a choice between two or more items. The action sheet comes up from the bottom of the screen and displays a series of buttons (see Figure 4–3). Users are unable to continue using the application until they have tapped one of the buttons. Action sheets are often used to confirm a potentially dangerous or irreversible action such as deleting an object.

- Alerts appear as a blue, rounded rectangle in the middle of the screen (see Figure 4–4). Just like action sheets, alerts force users to respond before they are allowed to continue using the application. Alerts are usually used to inform the user that something important or out of the ordinary has occurred. Unlike action sheets, alerts may be presented with only a single button, although you have the option of presenting multiple buttons if more than one response is appropriate.

> **NOTE:** A view that forces users to make a choice before they are allowed to continue using their application is known as a **modal view**.

Conforming to the Action Sheet Delegate Method

Remember back in Chapter 3 when we talked about the application delegate? Well, UIApplication is not the only class in Cocoa Touch that uses delegates. In fact, delegation is a common design pattern in Cocoa Touch. Action sheets and alerts both use delegates so that they know which object to notify when they're dismissed. In our application, we'll need to be notified when the action sheet is dismissed. We don't need to know when the alert is dismissed, because we're just using it to notify the user of something, not to actually solicit a choice.

In order for our controller class to act as the delegate for an action sheet, it needs to conform to a protocol called UIActionSheetDelegate. We do that by adding the name of the protocol in angle backets after the superclass in our class declaration. Add the following protocol declaration to *BIDViewController.h*:

```
#import <UIKit/UIKit.h>

@interface BIDViewController : UIViewController <UIActionSheetDelegate>
@property (strong, nonatomic) IBOutlet UITextField *nameField;
@property (strong, nonatomic) IBOutlet UITextField *numberField;
. . .
```

Showing the Action Sheet

Let's switch over to *BIDViewController.m* and implement the button's action method. We actually need to implement another method in addition to our existing action method: the UIActionSheetDelegate method that the action sheet will use to notify us that it has been dismissed.

First, look for the empty buttonPressed: method that Xcode created for you. Add the following code to that method to create and show the action sheet:

```
- (IBAction)buttonPressed:(id)sender {

    UIActionSheet *actionSheet = [[UIActionSheet alloc]
        initWithTitle:@"Are you sure?"
        delegate:self
        cancelButtonTitle:@"No Way!"
        destructiveButtonTitle:@"Yes, I'm Sure!"
        otherButtonTitles:nil];
    [actionSheet showInView:self.view];

}
```

Next, add a new method just after the existing buttonPressed: method:

```
- (void)actionSheet:(UIActionSheet *)actionSheet
    didDismissWithButtonIndex:(NSInteger)buttonIndex
{
    if (buttonIndex != [actionSheet cancelButtonIndex])
    {
        NSString *msg = nil;

        if (nameField.text.length > 0)
        msg = [[NSString alloc] initWithFormat:
            @"You can breathe easy, %@, everything went OK.",
            nameField.text];
        else
            msg = @"You can breathe easy, everything went OK.";

        UIAlertView *alert = [[UIAlertView alloc]
                        initWithTitle:@"Something was done"
                        message:msg
                        delegate:self
                        cancelButtonTitle:@"Phew!"
                        otherButtonTitles:nil];
        [alert show];
    }
}
```

What exactly did we do there? Well, first, in the doSomething: action method, we allocated and initialized a UIActionSheet object, which is the object that represents an action sheet (in case you couldn't puzzle that one out for yourself):

```
UIActionSheet *actionSheet = [[UIActionSheet alloc]
            initWithTitle:@"Are you sure?"
            delegate:self
            cancelButtonTitle:@"No Way!"
            destructiveButtonTitle:@"Yes, I'm Sure!"
            otherButtonTitles:nil];
```

The initializer method takes a number of parameters. Let's look at each of them in turn.

The first parameter is the title to be displayed. Refer back to Figure 4–3 to see how the title we're supplying will be displayed at the top of the action sheet.

The next argument is the delegate for the action sheet. The action sheet's delegate will be notified when a button on that sheet has been tapped. More specifically, the delegate's actionSheet:didDismissWithButtonIndex: method will be called. By passing self as the delegate parameter, we ensure that our version of actionSheet:didDismissWithButtonIndex: will be called.

Next, we pass in the title for the button that users will tap to indicate they do not want to proceed. All action sheets should have a cancel button, though you can give it any title that is appropriate to your situation. You do not want to use an action sheet if there is no choice to be made. In situations where you want to notify the user without giving a choice of options, an alert view is more appropriate.

The next parameter is the destructive button, and you can think of this as the "yes, please go ahead" button, though once again, you can assign it any title.

The last parameter allows you to specify any number of other buttons that you may want shown on the sheet. This final argument can take a variable number of values, which is one of the nice features of the Objective-C language. If we had wanted two more buttons on our action sheet, we could have done it like this:

```
UIActionSheet *actionSheet = [[UIActionSheet alloc]
    initWithTitle:@"Are you sure?"
    delegate:self
    cancelButtonTitle:@"No Way!"
    destructiveButtonTitle:@"Yes, I'm Sure!"
    otherButtonTitles:@"Foo", @"Bar", nil];
```

This code would have resulted in an action sheet with four buttons. You can pass as many arguments as you want in the otherButtonTitles parameter, as long as you pass nil as the last one. Of course, there is a practical limitation on how many buttons you can have, based on the amount of screen space available.

After we create the action sheet, we tell it to show itself:

```
[actionSheet showInView:self.view];
```

Action sheets always have a parent, which must be a view that is currently visible to the user. In our case, we want the view that we designed in Interface Builder to be the parent, so we use self.view. Note the use of Objective-C dot notation. self.view is equivalent to saying [self view], using the accessor to return the value of our view property.

Why didn't we just use view, instead of self.view? view is a private instance variable of our parent class UIViewController, which means we can't access it directly, but instead must use an accessor method.

Well, that wasn't so hard, was it? In just a few lines of code, we showed an action sheet and required the user to make a decision. iOS will even animate the sheet for us without requiring us to do any additional work. Now, we just need to find out which button the user tapped. The other method that we just implemented, actionSheet:didDismissWithButtonIndex, is one of the UIActionSheetDelegate methods, and since we specified self as our action sheet's delegate, this method will automatically be called by the action sheet when a button is tapped.

The argument buttonIndex will tell us which button was actually tapped. But how do we know which button index refers to the cancel button and which one refers to the destructive button? Fortunately, the delegate method receives a pointer to the UIActionSheet object that represents the sheet, and that action sheet object knows which button is the cancel button. We just need look at one of its properties, cancelButtonIndex:

```
if (buttonIndex != [actionSheet cancelButtonIndex])
```

This line of code makes sure the user didn't tap the cancel button. Since we gave the user only two options, we know that if the cancel button wasn't tapped, the destructive button must have been tapped, and it's OK to proceed. Once we know the user didn't cancel, the first thing we do is create a new string that will be displayed to the user. In a real application, here you would do whatever processing the user requested. We're just going to pretend we did something, and notify the user using an alert.

If the user has entered a name in the top text field, we'll grab that, and we'll use it in the message that we'll display in the alert. Otherwise, we'll just craft a generic message to show.

```
NSString *msg = nil;

if (nameField.text.length > 0)
    msg = [[NSString alloc] initWithFormat:
        @"You can breathe easy, %@, everything went OK.",
        nameField.text];
else
    msg = @"You can breathe easy, everything went OK.";
```

The next lines of code are going to look kind of familiar. Alert views and action sheets are created and used in a very similar manner.

```
UIAlertView *alert = [[UIAlertView alloc]
    initWithTitle:@"Something was done"
    message:msg
    delegate:nil
    cancelButtonTitle:@"Phew!"
    otherButtonTitles:nil];
```

Again, we pass a title to be displayed. We also pass a more detailed message, which is that string we just created. Alert views have delegates, too, and if we needed to know when the user had dismissed the alert view or which button was tapped, we could specify self as the delegate here, just as we did with the action sheet. If we had done that, we would now need to conform our class to the UIAlertViewDelegate protocol also, and implement one or more of the methods from that protocol. In this case, we're just informing the user of something and giving the user only one button. We don't really care when the button is tapped, and we already know which button will be tapped, so we just specify nil here to indicate that we don't need to be pinged when the user is finished with the alert view.

Alert views, unlike action sheets, are not tied to a particular view, so we just tell the alert view to show itself without specifying a parent view. After that, it's just a matter of some memory cleanup, and we're finished. Save the file. Then build, run, and try out the completed application.

Spiffing Up the Button

If you compare your running application to Figure 4–2, you might notice an interesting difference. Your *Do Something* button doesn't look like the one in the figure. And it doesn't look like the button on the action sheet or those in other iPhone applications, does it? The default round rect button doesn't really look that spiffy, so let's take care of that before we finish up the app.

Most of the buttons you see on your iOS device are drawn using images. Don't worry; you don't need to create images in an image editor for every button. All you need to do is specify a kind of template image that iOS will use when drawing your buttons.

It's important to keep in mind that your application is sandboxed. You can't get to the template images that are used in other applications on your iOS device or the ones used by iOS itself, so you must make sure that any images you need are in your application's bundle. So, where can you get these image templates?

Fortunately, Apple has provided a bunch for you. You can get them from the iPhone sample application called UICatalog, available from the iOS Developer Library:

`http://developer.apple.com/library/ios/#samplecode/UICatalog/index.html`

Alternatively, you can simply copy the images from the *04 - Control Fun* folder from this book's project archive. Yes, it is OK to use these images in your own applications, because Apple's sample code license specifically allows you to use and distribute them.

So, from either the *04 - Control Fun* folder or the *Images* subfolder of the UICatalog project's folder, add the two images named *blueButton.png* and *whiteButton.png* to your Xcode project.

If you tap one of the buttons in the project navigator, you'll see that there's not much to them. There's a trick to using them for your buttons.

Go back to the nib file you've been working on and single-click the *Do Something* button. Yeah, we know, the button is now invisible because we marked it as hidden, but you should have no problem seeing the ghost image. In addition, you can also click the button in the dock's list.

With the button selected, press ⌥⌘4 to open the attributes inspector. In the inspector, use the first popup menu to change the type from *Rounded Rect to Custom*. You'll see in the inspector that you can specify an image for your button, but we're not going to do that, because these image templates need to be handled a little differently.

Using the viewDidLoad Method

UIViewController, our controller's superclass, has a method called viewDidLoad that we can override if we need to modify any of the objects that were created from our nib. Because we can't do what we want completely in Interface Builder, we're going to take advantage of viewDidLoad.

Save your nib. Then switch over to *BIDViewController.m* and look for the viewDidLoad method. The Xcode project template created an empty version of this method for you. Find it, and add the following code to it. When you're finished, we'll talk about what the method does.

```
- (void)viewDidLoad {
    [super viewDidLoad];
    // Do any additional setup after loading the view, typically from a nib.
    UIImage *buttonImageNormal = [UIImage imageNamed:@"whiteButton.png"];
    UIImage *stretchableButtonImageNormal = [buttonImageNormal
```

```
                              stretchableImageWithLeftCapWidth:12 topCapHeight:0];
    [doSomethingButton setBackgroundImage:stretchableButtonImageNormal
                              forState:UIControlStateNormal];

    UIImage *buttonImagePressed = [UIImage imageNamed:@"blueButton.png"];
    UIImage *stretchableButtonImagePressed = [buttonImagePressed
                              stretchableImageWithLeftCapWidth:12 topCapHeight:0];
    [doSomethingButton setBackgroundImage:stretchableButtonImagePressed
                              forState:UIControlStateHighlighted];
}
```

This code sets the background image for the button based on those template images we added to our project. It specifies that, while being touched, the button should change from using the white image to the blue image. This short method introduces two new concepts: **control states** and **stretchable images**. Let's look at each of them in turn.

Control States

Every iOS control has four possible control states and is always in one, and only one, of these states at any given moment:

- **Normal**: The most common state is the normal control state, which is the default state. It's the state that controls are in when not in any of the other states.

- **Highlighted**: The highlighted state is the state a control is in when it's currently being used. For a button, this would be while the user has a finger on the button.

- **Disabled**: Controls are in the disabled state when they have been turned off, which can be done by unchecking the *Enabled* checkbox in Interface Builder or setting the control's enabled property to NO.

- **Selected**: Only some controls support the selected state. It is usually used to indicate that the control is turned on or selected. Selected is similar to highlighted, but a control can continue to be selected when the user is no longer directly using that control.

Certain iOS controls have attributes that can take on different values depending on their state. For example, by specifying one image for UIControlStateNormal and a different image for UIControlStateHighlighted, we are telling iOS to use one image when the user has a finger on the button and a different image the rest of the time.

Stretchable Images

Stretchable images are an interesting concept. A stretchable image is a resizable image that knows how to resize itself intelligently so that it maintains the correct appearance. For these button templates, we don't want the edges to stretch evenly with the rest of the image. **End caps** are the parts of an image, measured in pixels, that should not be resized. We want the bevel around the edges to stay the same, no matter what size we make the button, so we specify a left end cap size of 12.

Because we pass in the new stretchable image to our button, rather than the image template, iOS knows how to draw the button properly at any size. We could now go in and change the size of the button in the nib file, and it would still be drawn correctly. If we had specified the button image directly in the nib file, it would resize the entire image evenly, and our button would look weird at most sizes.

> **TIP:** How did we know what value to use for the end caps? It's simple really: we copied from Apple's sample code.

Why don't you save the file and try out our app? The *Do Something* button should now look a little more iPhone-ish, but everything should work the same.

Crossing the Finish Line

This was a big chapter. Conceptually, we didn't hit you with too much new stuff, but we took you through the use of a good number of controls and showed you many different implementation details. You got a lot more practice with outlets and actions, and saw how to use the hierarchical nature of views to your advantage. You learned about control states and stretchable images, and you also learned how to use both action sheets and alerts.

There's a lot going on in this little application. Feel free to go back and play with it. Change values, experiment by adding and modifying code, and see what different settings in Interface Builder do. There's no way we could take you through every permutation of every control available in iOS, but the application you just put together is a good starting point and covers a lot of the basics.

In the next chapter, we're going to look at what happens when the user rotates an iOS device from portrait to landscape orientation or vice versa. You're probably well aware that many apps change their displays based on the way the user is holding the device, and we're going to show you how to do that in your own applications.

Chapter 5

Autorotation and Autosizing

The iPhone and iPad are amazing pieces of engineering. Apple engineers found all kinds of ways to squeeze maximum functionality into a pretty darn small package. One example of this is how these devices can be used in either portrait (tall and skinny) or landscape (short and wide) mode, and how that can be changed at runtime simply by rotating the device. You can see an example of this behavior, which is called **autorotation**, in iOS's web browser, Mobile Safari (see Figure 5–1).

In this chapter, we'll cover autorotation in detail. We'll start with an overview of the ins and outs of autorotation, and then move on to different ways of implementing that functionality in your apps.

Figure 5–1. *Like many iOS applications, Mobile Safari changes its display based on how it is held, making the most of the available screen space.*

The Mechanics of Autorotation

Autorotation might not be right for every application. Several of Apple's iPhone applications support only a single orientation. Contacts can be edited only in portrait mode, for example. However, iPad applications are different. Apple recommends that all applications (with the exception of immersive apps like games that are inherently designed around a particular layout) should support every orientation.

In fact, all of Apple's own iPad apps work fine in both orientations. Many of them use the orientations to show different views of your data. For example, the Mail and Notes apps use landscape orientation to display a list of items (folders, messages, or notes) on the left and the selected item on the right, and portrait orientation to let you focus on the details of just the selected item.

For iPhone apps, the base rule is that if autorotation enhances the user experience, you should add it to your application. For iPad apps, the rule is you should add autorotation unless you have a compelling reason not to. Fortunately, Apple did a great job of hiding the complexities of autorotation in iOS and in the UIKit, so implementing this behavior in your own iOS applications is actually quite easy.

Autorotation is specified in the view controller. If the user rotates the device, the active view controller will be asked if it's OK to rotate to the new orientation (which you'll see how to do in this chapter). If the view controller responds in the affirmative, the application's window and views will be rotated, and the window and view will be resized to fit the new orientation.

On the iPhone and iPod touch, a view that starts in portrait mode will be 320 points wide and 480 points tall. On the iPad, portrait mode means 768 points wide and 1024 points tall. The amount of screen real estate available for your app will be decreased by 20 points vertically if your app is showing the **status bar**. The status bar is the 20-point strip at the top of the screen (see Figure 5–1) that shows information like signal strength, time, and battery charge.

When the device is switched to landscape mode, the view rotates, along with the application's window, and is resized to fit the new orientation, so that it is 480 points wide by 320 points tall (iPhone and iPod touch) or 1024 points wide by 768 points tall (iPad). As before, the vertical space actually available to your app is reduced by 20 points if you're showing the status bar, which most apps do.

Points, Pixels, and the Retina Display

You might be wondering why we're talking about "points" instead of pixels. Earlier versions of this book did, in fact, refer to screen sizes in pixels rather than points. The reason for this change is Apple's introduction of the **retina display**.

The retina display is Apple's marketing term for the high-resolution screen on the iPhone 4, iPhone 4s, and later-generation iPod touches. It doubles the screen resolution from the original 320 × 480 pixels to 640 × 960 pixels.

Fortunately, you don't need to do a thing in most situations to account for this. When we work with on-screen elements, we specify dimensions and distances in *points*, not in pixels. For older iPhones and all iPads, points and pixels are equivalent. One point is one pixel. On more recent model iPhones and iPod touches, however, a point equates to 4 pixels and the screen is still 320 × 480 points, even though there are actually 640 × 960 pixels. Think of it as a "virtual resolution," with iOS automatically mapping points to the physical pixels of your screen. We'll talk more about this in Chapter 16.

In typical applications, most of the work in actually moving the pixels around the screen is managed by iOS. Your application's main job in all this is making sure everything fits nicely and looks proper in the resized window.

Autorotation Approaches

Your application can take three general approaches when managing rotation. Which one you use depends on the complexity of your interface. We'll look at all three approaches in this chapter.

With simpler interfaces, you can specify the correct **autosize attributes** for all of the objects that make up your interface. Autosize attributes tell the iOS device how your controls should behave when their enclosing view is resized. If you've worked with Cocoa on Mac OS X, you're already familiar with the basic process, because it is the same one used to specify how Cocoa controls behave when the user resizes the window in which they are contained.

Autosize attributes are quick and easy to use, but they aren't appropriate for all applications. More complex interfaces must handle autorotation in a different manner. For more complex views, you have two additional approaches:

- Manually reposition the objects in your view in code when notified that your view is rotating.

- Actually design two different versions of your view in Xcode's Interface Builder: one view for portrait mode and a separate view for landscape mode.

In both cases, you will need to override methods from `UIViewController` in your view's controller class.

Let's get started, shall we? We'll look at autosizing first.

Handling Rotation Using Autosize Attributes

We'll create a simple app to demonstrate using autosize attributes. Start a new *Single View Application* project in Xcode, and call it *Autosize*. Choose *iPhone* as the *Device Family* and make sure to use ARC. Before we lay out our GUI in a nib file, we need to tell iOS that our view supports autorotation. We do that by modifying the view controller class.

Configuring Supported Orientations

First, we need to specify which orientations our application supports. When your window appeared, it should have opened to your project settings. If not, click the top line in the project navigator (the one named after your project), and then make sure you're on the *Summary* tab. Among the options available in the summary, you should see a section called *iPhone / iPod Deployment Info* and, within that, a section called *Supported Device Orientations* (see Figure 5–2).

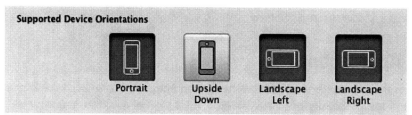

Figure 5–2. *The Summary tab for our project shows, among other things, the supported device orientations.*

This is how you identify which orientations your application supports. It doesn't necessarily mean that every view in your application will use all of the selected orientations, but if you're going to support an orientation in any of your application's views, that orientation must be selected here.

> **NOTE:** The four buttons shown in Figure 5–2 are actually just a shortcut to adding and deleting entries in your application's *Info.plist* file. If you single-click *Autosize-Info.plist* in the *Supporting Files* folder in the project navigator, you should see an entry called either *UISupportedInterfaceOrientations* or *Supported interface orientations*, with three subentries for the three orientations currently selected. Selecting and deselecting those buttons in the project summary simply adds and removes items from this array. Using the buttons is easier and less prone to error, so we definitely recommend using the buttons, but we thought you should know what they do.

Have you noticed that the *Upside Down* orientation is off by default? That's because if the phone rings while it is being held upside down, the phone is likely to remain upside down when you answer it. iPad app projects default to all four orientations being supported because the iPad is meant to be used in any orientation. Since our project is an iPhone project, we can leave the buttons as they are set.

We've identified the orientations our app will support, but that's not all we need to do. We also must specify for each view controller which orientations are supported, and that must be a subset of the orientations selected here.

Specifying Rotation Support

Single-click *BIDViewController.m*. In the code that's already there, you'll see a method called `shouldAutorotateToInterfaceOrientation:` provided for you, courtesy of the template:

```
- (BOOL)shouldAutorotateToInterfaceOrientation:
                        (UIInterfaceOrientation)interfaceOrientation {

    // Return YES for supported orientations
    return (interfaceOrientation != UIInterfaceOrientationPortraitUpsideDown);
}
```

This method is iOS's way of asking a view controller if it's OK to rotate to a specific orientation. Four defined orientations correspond to the four general ways that an iOS device can be held:

- `UIInterfaceOrientationPortrait`
- `UIInterfaceOrientationPortraitUpsideDown`
- `UIInterfaceOrientationLandscapeLeft`
- `UIInterfaceOrientationLandscapeRight`

In the case of the iPhone, the template defaults to supporting all orientations except upside down, just as you saw in the supported orientations. If we had instead created an iPad project, the default version of the `shouldAutorotateToInterfaceOrientation:` method created by the template would just return YES.

When the iOS device is changed to a new orientation, the `shouldAutorotateToInterfaceOrientation:` method is called on the active view controller. The parameter `interfaceOrientation` will contain one of the four values in the preceding list, and this method needs to return either YES or NO to signify whether the application's window should be rotated to match the new orientation. Because every view controller subclass can implement this differently, it is possible for one application to support autorotation with some of its views but not with others, or for one view controller to support certain orientations under certain conditions.

Code Sense in Action

Have you noticed that the defined system constants on the iPhone are always designed so that values that work together start with the same letters? One reason why `UIInterfaceOrientationPortrait`, `UIInterfaceOrientationPortraitUpsideDown`, `UIInterfaceOrientationLandscapeLeft`, and `UIInterfaceOrientationLandscapeRight` all begin with `UIInterfaceOrientation` is to let you take advantage of Xcode's **Code Sense** feature.

You've probably noticed that as you type, Xcode frequently tries to complete the word you are typing. That's Code Sense in action.

Developers cannot possibly remember all the various defined constants in the system, but you can remember the common beginning for the groups you use frequently. When you need to specify an orientation, simply type *UIInterfaceOrientation* (or even *UIInterf*), and then press the escape key to bring up a list of all matches. (In Xcode's preferences, you can change that matching key from escape to something else.) You can use the arrow keys to navigate the list that appears and make a selection by pressing the tab or return key. This is much faster than needing to look up the values in the documentation or header files.

Once again, the template has predicted what we would need, so we can leave this code untouched for now. However, feel free to play around with this method by returning YES or NO for different orientations.

> **NOTE:** iOS actually has two different types of orientations. The one we're discussing here is the **interface orientation**. There's also a separate but related concept of **device orientation**. Device orientation specifies how the device is currently being held. Interface orientation is which way the stuff on the screen is rotated. If you turn a standard iPhone app upside down, the device orientation will be upside down, but the interface orientation will be one of the other three, since iPhone apps typically don't support portrait upside down.

Designing an Interface with Autosize Attributes

In Xcode, select *BIDViewController.xib* to edit the file in Interface Builder. One nice thing about using autosize attributes is that they require very little code. We do need to specify which orientations we support in code, but the rest of the autoresize implementation can be done right here in Interface Builder.

To see how this works, drag six *Round Rect Buttons* from the library over to your view, and place them as shown in Figure 5–3. Double-click each button, and assign a title to each one so you can tell them apart later. We've used *UL* for the upper-left button, *UR* for the upper-right button, *L* for the middle-left button, *R* for the middle-right button, *LL* for the lower-left button, and *LR* for the lower-right button.

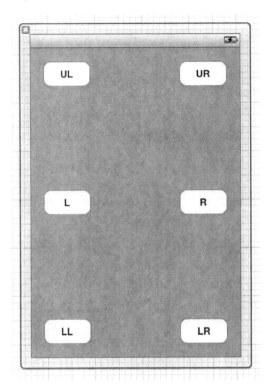

Figure 5–3. *Adding six labeled buttons to the interface*

Let's see what happens now that we've specified that we support autorotation but haven't set any autosize attributes. Build and run the app. Once the iPhone simulator comes up, select **Hardware ➤ Rotate Left**, which will simulate turning the iPhone to landscape mode. Take a look at Figure 5–4. Oh, dear.

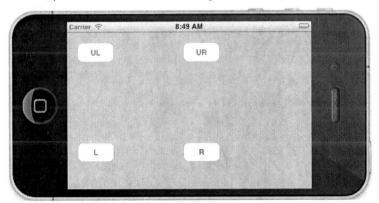

Figure 5–4. *Well, that's not very useful, is it? Where are buttons LL and LR?*

Most controls default to a setting that has them stay where they are in relation to the left side and top of the screen. There are some exceptions to that rule, but it's usually true.

For some controls, this is perfectly appropriate. The upper-left button (*UL*), for example, is probably right where we want it to appear. The rest of them, however, do not fare as well.

Quit the simulator, and let's get to work fixing the GUI so that it adapts to the screen size in a sensible way.

Using the Size Inspector's Autosize Attributes

Single-click the upper-left button on your view, and then press   ⌘5 to bring up the **size inspector**, which should look like Figure 5–5.

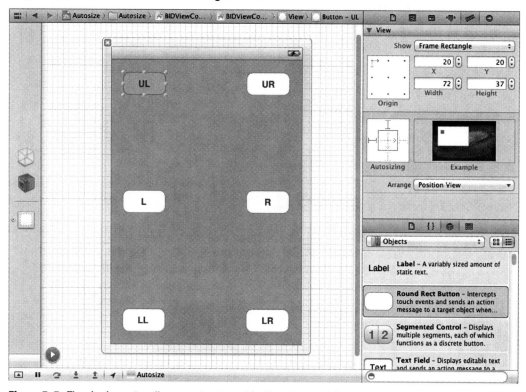

Figure 5–5. *The size inspector allows you to set an object's autosize attributes.*

The size inspector allows you to set, among other things, an object's autosize attributes. The box on the left in Figure 5–6 is where you actually set the attributes. The box on the right runs a little animation (move your cursor over the box to bring the animation to life) that will show you how the object will behave during a resize. In the box on the left, the inner square represents the current object. If a button is selected, the inner square represents that button.

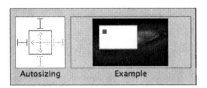

Figure 5-6. *The Autosizing section of the size inspector*

The red arrows inside the inner square represent the horizontal and vertical space inside the selected object. Clicking either arrow will change it from dashed to solid or from solid back to dashed. If the horizontal arrow is solid, the width of the object is free to change as the window resizes; if the horizontal arrow is dashed, iOS will try to keep the width of the object at its original value if possible. The same is true for the height of the object and the vertical arrow.

The four red *I* shapes outside the inner box represent the distance between the edge of the selected object and the same edge of the view that contains it. If the *I* is dashed, the space is flexible; if it's solid red, the amount of space should be kept constant if possible.

Huh?

Perhaps this concept will make a little more sense if you actually see it in action. Figure 5-6 represents the default autosize settings, which specify that the object's size will remain constant as its superview is resized, and that the distance from the left and top edges should also stay constant. If you look at the animation next to the autosize control, you can see how the object will behave during a resize. Notice that the inner box stays in the same place relative to the left and top edges of the parent view as the parent view changes in size.

Try this experiment. With your upper-left (*UL*) button selected, click both of the solid red *I* shapes (to the top and left of the inner box) so they become dashed and look like the ones shown in Figure 5-7. With all possible lines set to dashed, the size of the object will be kept the same, and it will float in the middle of the superview as the superview is resized.

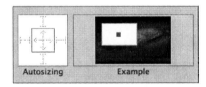

Figure 5-7. *With all dashed lines, your control floats in the parent and keeps its size.*

Now, click the vertical arrow inside the box and the *I* shape both above and below the box so that your autosize attributes look like the ones shown in Figure 5-8.

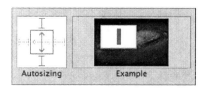

Figure 5–8. *This configuration allows the vertical size of the object to change.*

Here, we are indicating that the vertical size of our object can change, and that the distance from the top of our object to the top of the window and the distance from the bottom of our object to the bottom of the window should stay constant. With this configuration, the width of the object won't change, but its height will.

Change the autosize attributes a few more times, and watch the animation until you grok how different settings will impact the behavior when the view is rotated and resized.

Setting the Buttons' Autosize Attributes

Now, let's set the autosize attributes for our six buttons. Go ahead and see if you can figure them out. If you get stumped, take a look at Figure 5–9, which shows the autosize attributes needed for each button in order to keep all of the buttons on the screen when the phone is rotated.

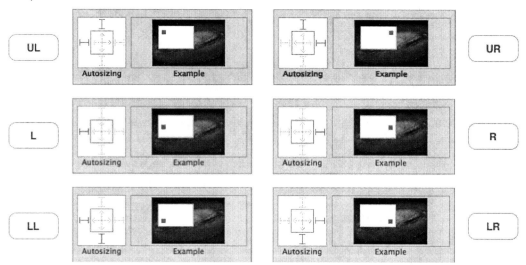

Figure 5–9. *Autosize attributes for all six buttons*

Once you have the attributes set as shown Figure 5–9, save the nib file, and then build and run the app. This time, when the iPhone simulator comes up, you should be able to select **Hardware ➤ Rotate Left** or **Rotate Right** and have all the buttons stay on the screen (see Figure 5–10). If you rotate back, they should return to their original positions. This technique will work for a great many applications.

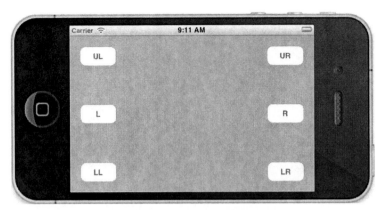

Figure 5–10. *The buttons in their new positions after rotating*

In this example, we kept our buttons the same size, so now all of our buttons are visible and usable, but there is a lot of unused space on the screen. Perhaps it would be better if we allowed the width or height of our buttons to change so that there will be less empty space on the interface? Feel free to experiment with the autosize attributes of these six buttons, and perhaps even add some other buttons. Play around until you feel comfortable with the way autosize works.

In the course of your experimentation, you're bound to notice that sometimes no combination of autosize attributes will give you exactly what you want. In some cases, you'll need to rearrange your interface more drastically than can be handled with this technique. For those situations, a little more code is in order. Let's take a look at that next.

Restructuring a View When Rotated

Back in your nib file, single-click each of the buttons, and use the size inspector to change the *Width* and *Height* fields to *125*, which will set the width and height of the buttons to 125 points. If you like, you can select all six buttons and use the size inspector to change them all at once. When you are finished, rearrange your buttons using the blue guidelines so that your view looks like Figure 5–11.

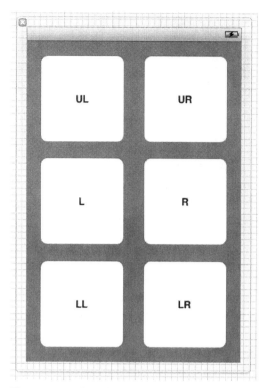

Figure 5–11. *View after resizing all the buttons*

Can you guess what's going to happen now when we rotate the screen? Well, assuming that you kept the buttons' autosize attributes to the settings shown in Figure 5–9, you probably won't be pleased. The buttons will overlap and look like Figure 5–12, because there simply isn't enough height on the screen in landscape mode to accommodate three buttons that are 125 points tall.

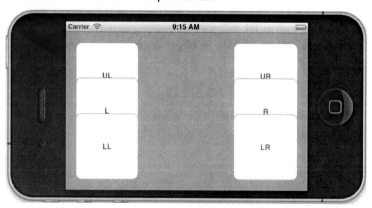

Figure 5–12. *Not exactly what we want. Too much overlap. We need another solution.*

We could accommodate this scenario using the autosize attributes by allowing the height of the buttons to change, but that's wouldn't make the best use of our screen real estate, because it would leave a large gap in the middle of the screen. If there was room for six square buttons in portrait mode, there should still be room for six square buttons in landscape mode—we just need to shuffle them around a bit. One way we can handle this is to specify new positions for each of the buttons when the view is rotated.

Creating and Connecting Outlets

Edit *BIDViewController.xib* and bring up the assistant editor (as you did in the previous chapter). Make sure you can see *BIDViewController.h* in addition to the GUI layout area, and then control-drag from each of the six buttons to the header file on the right to create six outlets called buttonUL, buttonUR, buttonL, buttonR, buttonLL, and buttonLR. Be sure each of the new outlets is specified as *weak*.

Once you've connected all six buttons to new outlets, save the nib. Your header file should look like this:

```
# import <UIKit/UIKit.h>

@interface BIDViewController : UIViewController
@property (weak, nonatomic) IBOutlet UIButton *buttonUL;
@property (weak, nonatomic) IBOutlet UIButton *buttonUR;
@property (weak, nonatomic) IBOutlet UIButton *buttonL;
@property (weak, nonatomic) IBOutlet UIButton *buttonR;
@property (weak, nonatomic) IBOutlet UIButton *buttonLL;
@property (weak, nonatomic) IBOutlet UIButton *buttonLR;

@end
```

Moving the Buttons on Rotation

To move these buttons to make the best use of space, we need to override the method willAnimateRotationToInterfaceOrientation:duration: in *BIDViewController.m*. This method is called automatically after a rotation has happened but before the final rotation animations have occurred.

Add the following method at the bottom of *BIDViewController.m*, just above the @end:

```
- (void)willAnimateRotationToInterfaceOrientation:(UIInterfaceOrientation)
                    interfaceOrientation duration:(NSTimeInterval)duration {

    if (UIInterfaceOrientationIsPortrait(interfaceOrientation)) {
        buttonUL.frame = CGRectMake(20, 20, 125, 125);
        buttonUR.frame = CGRectMake(175, 20, 125, 125);
        buttonL.frame = CGRectMake(20, 168, 125, 125);
        buttonR.frame = CGRectMake(175, 168, 125, 125);
        buttonLL.frame = CGRectMake(20, 315, 125, 125);
        buttonLR.frame = CGRectMake(175, 315, 125, 125);
    } else {
        buttonUL.frame = CGRectMake(20, 20, 125, 125);
```

```
        buttonUR.frame = CGRectMake(20, 155, 125, 125);
        buttonL.frame = CGRectMake(177, 20, 125, 125);
        buttonR.frame = CGRectMake(177, 155, 125, 125);
        buttonLL.frame = CGRectMake(328, 20, 125, 125);
        buttonLR.frame = CGRectMake(328, 155, 125, 125);
    }
}
```

The size and position of all views, including controls such as buttons, are specified in a property called `frame`, which is a `struct` of type `CGRect`. `CGRectMake` is a function provided by Apple that lets you easily create a `CGRect` by specifying the x and y positions along with the width and height.

Save this code. Now build and run the application to see it in action. Try rotating, and watch how the buttons end up in their new positions.

Swapping Views

Moving controls to different locations, as we did in the previous section, can be a very tedious process, especially with a complex interface. Wouldn't it be nice if we could just design the landscape and portrait views separately, and then swap them out when the phone is rotated?

Well, we can. But it's a moderately complicated option, which you'll likely use only in the case of very complex interfaces.

While controls on both views can trigger the same actions, we need to be able to keep track of the fact that multiple outlets will be pointing to objects performing the same function. For example, if we had a button called *foo*, we would actually have two copies of that button—one in the landscape layout and one in the portrait layout—and any change we make to one needs to be made to the other. So, if we wanted to disable or hide that button, we would need to disable or hide both of the foo buttons.

We could handle this by using multiple outlets, perhaps `fooPortrait` and `fooLandscape`, one pointing to each button. In fact, in previous editions of the book, that's exactly what we did. Life is better now. There's a relatively new feature of iOS called an **outlet collection** that we can use to make our code a little simpler and easier to manage. An outlet collection is exactly like an outlet in every way except for one. Whereas an outlet can point to only a single element, an outlet collection is actually an array and can point to any number of objects. This will allow us to have a single property that points to both versions of the same button.

To demonstrate how this works, we'll build an app with separate views for portrait and landscape orientation. Although the interface we'll build is not complex enough to justify the technique we're using, keeping the interface simple will help clarify the process.

Create a new project in Xcode using the *Single View Application* template again (we'll start working with other templates in the next chapter). Call this project *Swap*. The application will start up in portrait mode, with two buttons, one on top of the other (see Figure 5–13).

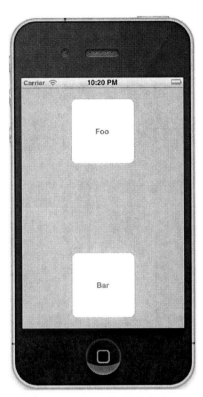

Figure 5–13. *The Swap application at launch. This is the portrait view and its two buttons.*

Rotating the phone swaps in a completely different view, specifically designed for landscape orientation. The landscape view will also feature two buttons with the exact same labels (see Figure 5–14), so the users won't know they're looking at two different views.

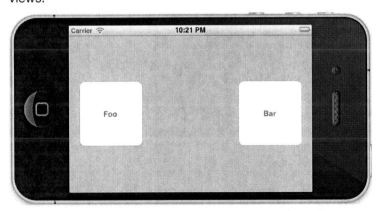

Figure 5–14. *Similar but not the same. This is the landscape view, with two different buttons.*

When a button is tapped, it will show an alert identifying which button was tapped. We won't pull the button's name from sender the way we did in Chapter 3, however. We'll use the outlet collection to determine which button was tapped.

Designing the Two Views

We'll need two views in our nib. We can use the existing view that Xcode created for us as one of them, but we'll need to add a second view. The easiest way to get that second view is to duplicate the existing view, and then make the necessary changes.

Select *BIDViewController.xib* to edit the file in Interface Builder. In the nib editor dock, there should be three icons. The bottom one represents the view that Xcode created for us. Hold down the option key on your keyboard, and then click and drag that icon downward. When you see the green plus on your icon, that's your indication that you've moved far enough for a copy. Release the mouse button to duplicate the view.

Single-click the newly added view and bring up the attributes inspector by pressing ⌥z4. Under the heading *Simulated Metrics*, look for a popup menu called *Orientation*. Change it from *Portrait* to *Landscape*.

We'll need access to both of these views in our code so that we can swap between them, so we need a pair of outlets. Make sure the assistant editor is turned on and displaying *BIDViewController.h*. Control-drag from the portrait view over to *BIDViewController.h*, and when prompted, create an outlet called *portrait*. Make sure you specify *Strong* in the *Storage* popup menu. Do the same thing from the landscape view, creating an outlet called *landscape*.

The next step is to drag in our buttons. Go to the object library and drag out a pair of *Rounded Rect Buttons* onto each of our views. Use Figure 5–15 as a guide. Click each button and use the size inspector (**View ➤ Utilities ➤ Size**) to change the *Width* and *Height* attributes to *125*. Center each button and drag it to the blue guideline near one edge of the view. Double-click each button and change its label to either *Foo* or *Bar*. Again, Figure 5–15 will set you straight.

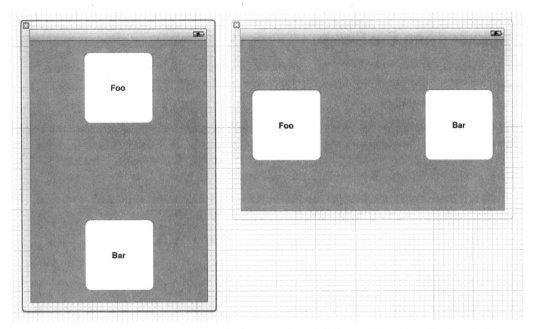

Figure 5–15. We dragged two buttons onto each of our two views, labeling one Foo and one Bar in each view.

Now, let's create and connect the button outlets. Again, make sure the assistant is turned on and displaying *BIDViewController.h*. Control-drag from the *Foo* button on the landscape view to the header file on the right. When prompted, change the *Connection* popup menu from *Outlet* to *Outlet Collection*, and give it a name of *foos*. Next, drag from the *Foo* button on the portrait view and connect it to the existing *foos* outlet connection. To reiterate, you first control-dragged from the *Foo* button on the landscape view to create the outlet collection, and then control-dragged from the other view's *Foo* button to connect it to that same outlet collection.

Repeat those steps with the *Bar* buttons. Control-drag from one of them to create a new outlet collection named `bars`, and then control-drag from the other *Bar* button to connect it to the same collection.

Lastly, we need to create an action method and connect all four buttons to it. Control-drag from the *Foo* button on the landscape view to *BIDViewController.h*, and when prompted, change the connection type from *Outlet* to *Action*. Give the action a name of *buttonTapped:*. Then connect the other three buttons to that action and save the nib.

Implementing the Swap

Single-click *BIDViewController.m* to open your view controller's implementation file for editing. First, at the top of the file, add the following C macro:

```
#define degreesToRadians(x) (M_PI * (x) / 180.0)
```

This macro just allows us to convert between degrees and radians, which we'll need to do in our code to handle swapping in rotated views. Scroll down a little, and add the following method after the last @synthesize call. It's a little scary looking, but don't worry; we'll explain what's going on after you've finished typing.

```
- (void)willAnimateRotationToInterfaceOrientation:(UIInterfaceOrientation)
interfaceOrientation duration:(NSTimeInterval)duration {

    if (interfaceOrientation == UIInterfaceOrientationPortrait) {
        self.view = self.portrait;
        self.view.transform = CGAffineTransformIdentity;
        self.view.transform =
        CGAffineTransformMakeRotation(degreesToRadians(0));
        self.view.bounds = CGRectMake(0.0, 0.0, 320.0, 460.0);
    }
    else if (interfaceOrientation == UIInterfaceOrientationLandscapeLeft) {
        self.view = self.landscape;
        self.view.transform = CGAffineTransformIdentity;
        self.view.transform =
        CGAffineTransformMakeRotation(degreesToRadians(-90));
        self.view.bounds = CGRectMake(0.0, 0.0, 480.0, 300.0);
    }
    else if (interfaceOrientation ==
            UIInterfaceOrientationLandscapeRight) {
        self.view = self.landscape;
        self.view.transform = CGAffineTransformIdentity;
        self.view.transform =
        CGAffineTransformMakeRotation(degreesToRadians(90));
        self.view.bounds = CGRectMake(0.0, 0.0, 480.0, 300.0);
    }
}
```

The method willAnimateRotationToInterfaceOrientation:duration: is actually a method from our superclass that we've overridden. This method is called as the rotation begins but before the rotation actually happens. Actions that we take in this method will be animated as part of the rotation animation.

In this method, we look at the orientation that we're rotating to and set the view property to either landscape or portrait, as appropriate for the new orientation, which makes sure the appropriate view is being shown. We then call CGAffineTransformMakeRotation, part of the Core Graphics framework, to create a **rotation transformation**.

A **transformation** is a mathematical description of changes to an object's size, position, or angle. Ordinarily, iOS takes care of setting the transform value automatically when the device is rotated. However, it handles this only for views that are in the view hierarchy, which means only the view already being shown is updated properly.

When we swap in our new view here, the view we're swapping in hasn't been adjusted by the system, so we need to make sure that we give it the correct transform for it to display correctly. That's what `willAnimateRotationToInterfaceOrientation:duration:` is doing each time it sets the view's `transform` property. Once the view has been rotated, we adjust its frame so that it fits snugly into the window at the current orientation.

Next, we need to implement our `buttonTapped:` method. Xcode has already created a stub implementation of this method for you. Add the following bold code to that existing method:

```
- (IBAction)buttonTapped:(id)sender {
    NSString *message = nil;

    if ([self.foos containsObject:sender])
        message = @"Foo button pressed";
    else
        message = @"Bar button pressed";

    UIAlertView *alert = [[UIAlertView alloc] initWithTitle:message
                                                    message:nil
                                                   delegate:nil
                                          cancelButtonTitle:@"Ok"
                                          otherButtonTitles:nil];
    [alert show];
}
```

There's nothing too surprising here. The outlet collections we created to point to the buttons are standard `NSArray` objects. To determine if `sender` is one of the *Foo* buttons, we simply check to see if `foos` contains it. If `foos` doesn't, then we know it's a *Bar* button.

Now, compile the app and give it a run.

Changing Outlet Collections

Our view-swapping app is obviously a rather simple example. In more complex user interfaces, you might need to make changes to user interface elements. In those cases, make sure that you make the same change to both the portrait and landscape versions.

Let's see how that works. Let's change the `buttonTapped:` method so that when a button is tapped, it disappears. We can't just use `sender` for that because we need to also hide the corresponding button in the other orientation.

Replace your existing implementation of buttonTapped: with this one:

```
- (IBAction)buttonTapped:(id)sender {
    if ([self.foos containsObject:sender]) {
        for (UIButton *oneFoo in foos) {
            oneFoo.hidden = YES;
```

```
        }
    }
    else {
        for (UIButton *oneBar in bars) {
            oneBar.hidden = YES;
        }
    }
}
```

Build and run the app again, and try it out. Tap one of the buttons, and then rotate to the other orientation. If you tapped the *Foo* button, you shouldn't see a *Foo* button on either the landscape or the portrait orientation. This is because we looped through the elements in the outlet collection and hid them all.

> **NOTE:** If you accidentally click both buttons, the only way to bring them back is to quit the simulator and rerun the project. Don't use this approach in your own applications.

Rotating Out of Here

In this chapter, you tried out three completely different approaches to supporting autorotation in your applications. You learned about autosize attributes and how to restructure your views, in code, when the iOS device rotates. You saw how to swap between two completely different views when the device rotates.

You also got your first taste of using multiple views in an application by swapping between two views from the same nib. In the next chapter, we're going to start looking at true multiview applications.

Every application we've written so far has used a single view controller, and all except the last one in this chapter used a single content view. A lot of complex iOS applications, such as Mail and Contacts, are made possible only by the use of multiple views and view controllers, and we're going to look at exactly how that works in Chapter 6.

Chapter **6**

Multiview Applications

Up until this point, we've written applications with a single view controller. While there certainly is a lot you can do with a single view, the real power of the iOS platform emerges when you can switch out views based on user input. Multiview applications come in several different flavors, but the underlying mechanism is the same, regardless of how the app may appear on the screen.

In this chapter, we're going to focus on the structure of multiview applications and the basics of swapping content views by building our own multiview application from scratch. We will write our own custom controller class that switches between two different content views, which will give you a strong foundation for taking advantage of the various multiview controllers that Apple provides.

But before we start building our application, let's see how multiple-view applications can be useful.

Common Types of Multiview Apps

Strictly speaking, we have worked with multiple views in our previous applications, since buttons, labels, and other controls are all subclasses of UIView, and they can all go into the view hierarchy. But when Apple uses the term *view* in documentation, it is generally referring to a UIView or one of its subclasses that has a corresponding view controller. These types of views are also sometimes referred to as **content views**, because they are the primary container for the content of your application.

The simplest example of a multiview application is a utility application. A utility application focuses primarily on a single view but offers a second view that can be used to configure the application or to provide more detail than the primary view. The Stocks application that ships with iPhone is a good example (see Figure 6–1). If you click the little *i* icon in the lower-right corner, the view flips over to let you configure the list of stocks tracked by the application.

Figure 6–1. *The Stocks application that ships with iPhone has two views: one to display the data and another to configure the stock list.*

There are also several tab bar applications that ship with the iPhone, such as the Phone application (see Figure 6–2) and the Clock application. A tab bar application is a multiview application that displays a row of buttons, called the **tab bar**, at the bottom of the screen. Tapping one of the buttons causes a new view controller to become active and a new view to be shown. In the Phone application, for example, tapping *Contacts* shows a different view than the one shown when you tap *Keypad*.

Figure 6–2. *The Phone application is an example of a multiview application using a tab bar*

Another common kind of multiview iPhone application is the navigation-based application, which features a navigation controller that uses a **navigation bar** to control a hierarchical series of views. The Settings application is a good example. In Settings, the first view you get is a series of rows, each row corresponding to a cluster of settings or a specific app. Touching one of those rows takes you to a new view where you can customize one particular set of settings. Some views present a list that allows you to dive even deeper. The navigation controller keeps track of how deep you go and gives you a control to let you make your way back to the previous view.

For example, if you select the *Sounds* preference, you'll be presented a view with a list of sound-related options. At the top of that view is a navigation bar with a left arrow that takes you back to the previous view if you tap it. Within the sound options is a row labeled *Ringtone*. Tap *Ringtone*, and you're taken to a new view featuring a list of ringtones and a navigation bar that takes you back to the main *Sounds* preference view (see Figure 6–3). A navigation-based application is useful when you want to present a hierarchy of views.

Figure 6–3. *The iPhone Settings application is an example of a multiview application using a navigation bar.*

On the iPad, most navigation-based applications, such as Mail, are implemented using a **split view**, where the navigation elements appear on the left side of the screen, and the item you select to view or edit appears on the right. You'll learn more about split views and other iPad-specific GUI elements in Chapter 10.

Because views are themselves hierarchical in nature, it's even possible to combine different mechanisms for swapping views within a single application. For example, the iPhone's iPod application uses a tab bar to switch between different methods of organizing your music, and a navigation controller and its associated navigation bar to allow you to browse your music based on that selection. In Figure 6–4, the tab bar is at the bottom of the screen, and the navigation bar is at the top of the screen.

Figure 6–4. *The iPod application uses both a navigation bar and a tab bar.*

Some applications make use of a **toolbar**, which is often confused with a tab bar. A tab bar is used for selecting one and only one option from among two or more. A toolbar can hold buttons and certain other controls, but those items are not mutually exclusive. A perfect example of a toolbar is at the bottom of the main Safari view (see Figure 6–5). If you compare the toolbar at the bottom of the Safari view with the tab bar at the bottom of the Phone or iPod application, you'll find the two pretty easy to tell apart. The tab bar is divided into clearly defined segments, while typically, the toolbar is not divided this way.

Figure 6–5. *Mobile Safari features a toolbar at the bottom. The toolbar is like a free-form bar that allows you to include a variety of controls.*

Each of these multiview application types uses a specific controller class from the UIKit. Tab bar interfaces are implemented using the class UITabBarController, and navigation interfaces are implemented using UINavigationController.

The Architecture of a Multiview Application

The application we're going to build in this chapter, View Switcher, is fairly simple in appearance, but in terms of the code we're going to write, it's by far the most complex application we've yet tackled. View Switcher will consist of three different controllers, three nibs, and an application delegate.

When first launched, View Switcher will look like Figure 6–6, with a toolbar at the bottom containing a single button. The rest of the view will contain a blue background and a button yearning to be pressed.

Figure 6–6. *When you first launch the View Switcher application, you'll see a blue view with a button and a toolbar with its own button.*

When the *Switch Views* button is pressed, the background will turn yellow, and the button's title will change (see Figure 6–7).

Figure 6–7. *When you press the Switch Views button, the blue view flips over to reveal the yellow view.*

If either the *Press Me* or *Press Me, Too* button is pressed, an alert will pop up indicating which view's button was pressed (see Figure 6–8).

Figure 6–8. *When the Press Me or Press Me, Too button is pressed, an alert is displayed.*

Although we could achieve this same functionality by writing a single-view application, we're taking this more complex approach to demonstrate the mechanics of a multiview application. There are actually three view controllers interacting in this simple application: one that controls the blue view, one that controls the yellow view, and a third special controller that swaps the other two in and out when the *Switch Views* button is pressed.

Before we start building our application, let's talk about the way iPhone multiview applications are put together. Most multiview applications use the same basic pattern.

The Root Controller

The nib file is a key player here. For our View Switcher application, you'll find the file *MainWindow.xib* in your project window's *Resources* folder. That file contains the application delegate and the application's main window, along with the *File's Owner* and *First Responder* icons. We'll add an instance of a controller class that is responsible for managing which other view is currently being shown to the user. We call this controller the **root controller** (as in "the root of the tree" or "the root of all evil") because it is the first controller the user sees and the controller that is loaded when the application loads.

This root controller is often an instance of UINavigationController or UITabBarController, although it can also be a custom subclass of UIViewController.

In a multiview application, the job of the root controller is to take two or more other views and present them to the user as appropriate, based on the user's input. A tab bar controller, for example, will swap in different views and view controllers based on which tab bar item was last tapped. A navigation controller will do the same thing as the user drills down and backs up through hierarchical data.

> **NOTE:** The root controller is the primary view controller for the application and, as such, is the view that specifies whether it is OK to automatically rotate to a new orientation. However, the root controller can pass responsibility for tasks like that to the currently active controller.

In multiview applications, most of the screen will be taken up by a content view, and each content view will have its own controller with its own outlets and actions. In a tab bar application, for example, taps on the tab bar will go to the tab bar controller, but taps anywhere else on the screen will go to the controller that corresponds to the content view currently being displayed.

Anatomy of a Content View

In a multiview application, each view controller controls a content view, and these content views are where the bulk of your application's user interface is built. Each content view generally consists of up to three pieces: the view controller, the nib, and a subclass of UIView. Unless you are doing something really unusual, your content view will always have an associated view controller, will usually have a nib, and will sometimes subclass UIView. Although you can create your interface in code rather than using a nib file, few people choose that route because it is more time-consuming and the code is difficult to maintain. In this chapter, we'll be creating only a nib and a controller class for each content view.

In the *View Switcher* project, our root controller controls a content view that consists of a toolbar that occupies the bottom of the screen. The root controller then loads a blue view controller, placing the blue content view as a subview to the root controller view. When the root controller's *Switch Views* button (the button is in the toolbar) is pressed, the root controller swaps out the blue view controller and swaps in a yellow view controller, instantiating that controller if it needs to do so. Confused? Don't worry, because this will become clearer as we walk through the code.

Building View Switcher

Enough theory! Let's go ahead and build our project. Select File ➤ New ➤ New Project... or press ⇧⌘N. When the template selection sheet opens, select *Empty Application* (see Figure 6–9), and then click *Next*. On the next page of the assistant, enter *View Switcher* as the *Product Name*, leave *BID* as the *Class Prefix*, and set the *Device Family* popup button to *iPhone*. Also make sure the checkboxes labeled *Use Core Data and Include Unit Tests* are unchecked, and *Use Automatic Reference Counting* is checked. Click *Next* to continue. On the next screen, navigate to wherever you're saving your projects on disk, and click the *Create* button to create a new project directory.

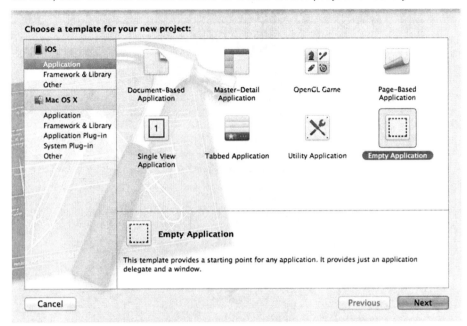

Figure 6–9. *Creating a new project using the Empty Application project template*

The template we just selected is actually even simpler than the *Single View Application* template we've been using up to now. This template will give us a window, an application delegate, and nothing else—no views, no controllers, no nothing.

> **NOTE:** The **window** is the most basic container in iOS. Each app has exactly one window that belongs to it, though it is possible to see more than one window on the screen at a time. For example, if your app is running and a Short Message Service (SMS) message comes in, you'll see the SMS message displayed in its window. Your app can't access that overlaid window because it belongs to the SMS app.

You won't use the *Empty Application* template very often when you're creating applications, but by starting from nothing for our example, you'll really get a feel for the way multiview applications are put together.

If they're not expanded already, take a second to expand the *View Switcher* folder in the project navigator, as well as the *Supporting Files* folder it contains. Inside the *View Switcher* folder, you'll find the two files that implement the application delegate. Within the *Supporting Files* folder, you'll find the *View Switcher-Info.plist* file, the *InfoPlist.strings* file (which contains the localized versions of your *Info.plist* file), the standard *main.m*, and the precompiled header file (*View Switcher-Prefix.pch*). Everything else we need for our application, we must create.

Creating Our View Controller and Nib Files

One of the more daunting aspects of building a multiview application from scratch is that we need to create several interconnected objects. We're going to create all the files that will make up our application before we do anything in Interface Builder and before we write any code. By creating all the files first, we'll be able to use Xcode's Code Sense feature to write our code faster. If a class hasn't been declared, Code Sense has no way to know about it, so we would need to type its name in full every time, which takes longer and is more error-prone.

Fortunately, in addition to project templates, Xcode also provides file templates for many standard file types, which helps simplify the process of creating the basic skeleton of our application.

Single-click the *View Switcher* folder in the project navigator, and then press ⌘N or select **File ➤ New ➤ New File…**. Take a look at the window that opens (see Figure 6–10).

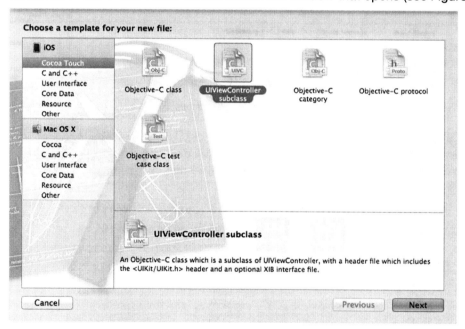

Figure 6–10. *The template we'll use to create a new view controller subclass*

If you select *Cocoa Touch* from the left pane, you will be given templates for a number of common Cocoa Touch classes. Select *UIViewController subclass* and click *Next*. On the next page of the assistant, you'll see a text field where you can enter the name of the new class. Type in *BIDSwitchViewController*, and then direct your attention to the three other controls that let you configure the subclass:

- The first is a combo box labeled *Subclass of*, with possible values of *UIViewController* and *UITableViewController*. If we wanted to create a table-based layout, for example, we might change this to *UITableViewController*. For our purposes, *UIViewController* will do what we need.

- The second is a checkbox labeled *Targeted for iPad*. If it's checked by default, you should uncheck it now (since we're not making an iPad GUI).

- The third is another checkbox, labeled *With XIB for user interface*. If that box is checked, uncheck it as well. If you left that checkbox checked, Xcode would create a nib file that corresponds to this controller class. We will start using that option in the next chapter, but for now, we want you to see how the different parts of the puzzle fit together by creating each individually.

Click *Next*. A window appears that lets you choose a particular directory in which to save the files and pick a group and target for your files. By default, this window will show the directory most relevant to the folder you selected in the project navigator. For the sake of consistency, you'll want to save the new class into the *View Switcher* folder, which Xcode set up when you created this project; it should already contain the BIDAppDelegate class. That's where Xcode puts all of the Objective-C classes that are created as part of the project, and it's as good a place as any for you to put your own classes.

About halfway down the window, you'll find the *Group* popup list. You'll want to add the new files to the *View Switcher* group. Finally, make sure the *View Switcher* target is selected in the *Targets* list before clicking the *Save* button.

Xcode should add two files to your *View Switcher* folder: *BIDSwitchViewController.h* and *BIDSwitchViewController.m*. BIDSwitchViewController will be your root controller—the controller that swaps the other views in and out. Now, we need to create the controllers for the two content views that will be swapped in and out. Repeat the same steps two more times to create *BIDBlueViewController.m*, *BIDYellowViewController.m*, and their *.h* counterparts, adding them to the same spot in the project hierarchy.

> **CAUTION:** Make sure you check your spelling, as a typo here will create classes that don't match the source code later in the chapter.

Our next step is to create a nib file for each of the two content views we just created. Single-click the *View Switcher* folder in the project navigator, and then press ⌘N or select File ➤ New ➤ New File... again. This time, select *User Interface* under the *iOS* heading in the left pane (see Figure 6–11). Next, select the icon for the *View* template, which will create a nib with a content view. Then click *Next*. On the next screen, select *iPhone* from the *Device Family* popup, and then click the *Next* button.

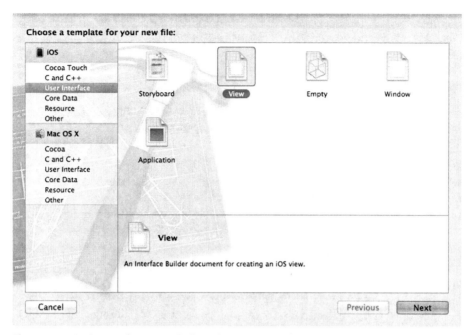

Figure 6–11. We're creating a new nib file, using the View template in the User Interface section.

When prompted for a file name, type *SwitchView.xib*. Just as you did earlier, you should choose the *View Switcher* folder as the save location. With *View Switcher* selected, ensure that *View Switcher* is selected from the *Group* popup menu and that the *View Switcher* target is checked, and then click *Save*. You'll know you succeeded when the file *SwitchView.xib* appears in the *View Switcher* group in the project navigator.

Now repeat the steps to create a second nib file called *BlueView.xib*, and once more to create *YellowView.xib*. After you've done that, you have all the files you need. It's time to start hooking everything together.

Modifying the App Delegate

Our first stop on the multiview express is the application delegate. Single-click the file *BIDAppDelegate.h* in the project navigator (make sure it's the app delegate and not *SwitchViewController.h*), and make the following changes to that file:

```
#import <UIKit/UIKit.h>
@class BIDSwitchViewController;
@interface BIDAppDelegate : UIResponder <UIApplicationDelegate>

@property (strong, nonatomic) UIWindow *window;
@property (strong, nonatomic) BIDSwitchViewController
    *switchViewController;
@end
```

The BIDSwitchViewController declaration you just added is a property that will point to our application's root controller. We need this because we are about to write code that will add the root controller's view to our application's main window when the application launches.

Now, we need to add the root controller's view to our application's main window. Click *BIDAppDelegate.m*, and add the following code:

```
#import "BIDAppDelegate.h"
#import "BIDSwitchViewController.h"
@implementation BIDAppDelegate

@synthesize window = _window;
@synthesize switchViewController;

- (BOOL)application:(UIApplication *)application
        didFinishLaunchingWithOptions:(NSDictionary *)launchOptions
{
    self.window = [[UIWindow alloc] initWithFrame:[[UIScreen mainScreen] bounds]];
    // Override point for customization after application launch
    self.switchViewController = [[BIDSwitchViewController alloc]
        initWithNibName:@"SwitchView" bundle:nil];
    UIView *switchView = self.switchViewController.view;
    CGRect switchViewFrame = switchView.frame;
    switchViewFrame.origin.y += [UIApplication
        sharedApplication].statusBarFrame.size.height;
    switchView.frame = switchViewFrame;
    [self.window addSubview:switchView];
    self.window.backgroundColor = [UIColor whiteColor];
    [self.window makeKeyAndVisible];
    return YES;
}

.
.
.

@end
```

Besides synthesizing the switchViewController property, we also create an instance of it and load its corresponding view from *SwitchView.xib*. Next, we change the view's geometry so it does not get hidden behind the status bar. If we were working with a nib that contained a view inside a window, we wouldn't need to make this change, but since we're creating this entire view hierarchy from scratch, we must manually adjust the view's frame so it snugs up against the bottom of the status bar.

After adjusting that view, we add it to the window, effectively making our switchViewController the root controller. Remember that the window is the only gateway to the user, so anything that needs to be displayed to the user must be added as a subview of the application's window.

If you go back to Chapter 5's *Swap* project and examine the code in *SwapAppDelegate.m*, you'll see that the template added the view controller's view to the application window for you. Since we're using a much simpler template for this project, we need to take care of that wiring together business ourselves.

Modifying BIDSwitchViewController.h

Because we're going to be setting up an instance of BIDSwitchViewController in *SwitchView.xib*, now is the time to add any needed outlets or actions to the *BIDSwitchViewController.h* header file.

We'll need one action method to toggle between the blue and yellow views. We won't create any outlets, but we will need two other pointers: one to each of the view controllers that we'll be swapping in and out. These don't need to be outlets, because we're going to create them in code rather than in a nib. Add the following code to *BIDSwitchViewController.h:*

```
#import <UIKit/UIKit.h>

@class BIDYellowViewController;
@class BIDBlueViewController;

@interface BIDSwitchViewController : UIViewController

@property (strong, nonatomic) BIDYellowViewController *yellowViewController;
@property (strong, nonatomic) BIDBlueViewController *blueViewController;

- (IBAction)switchViews:(id)sender;

@end
```

Now that we've declared the action we need, we can set this controller up in *SwitchView.xib*.

Adding a View Controller

Save your source code, and click *SwitchView.xib* to edit the central GUI for this app. Three icons appear in the nib's dock: *File's Owner, First Responder,* and *View* (see Figure 6–12).

Figure 6–12. *SwitchView.xib, showing three default icons in the dock, representing File's Owner, First Responder, and View*

By default, the *File's Owner* is configured to be an instance of NSObject. We'll need to change that to BIDSwitchViewController so that Interface Builder allows us to build connections to the BIDSwitchViewController outlets and actions. Single-click the *File's Owner* icon in the nib's dock, and press ⌥⌘3 to open the identity inspector (see Figure 6–13).

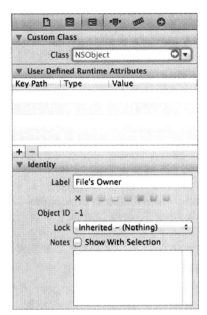

Figure 6–13. *Notice that the File's Owner Class field is currently set to NSObject in the identity inspector. We're about to change that to BIDSwitchViewController.*

The identity inspector allows you to specify the class of the currently selected object. Our *File's Owner* is currently specified as an NSObject, and it has no actions defined. Click inside the combo box labeled *Class*, which is at the top of the inspector and currently reads *NSObject*. Change the *Class* to *BIDSwitchViewController*.

Once you make that change, press ⌥⌘6 to switch to the connections inspector, where you will see that the switchViews: action method now appears in the section labeled *Received Actions* (see Figure 6–14). The connection inspector's *Received Actions* section shows all the actions defined for the current class. When we changed our *File's Owner* to a BIDSwitchViewController, the BIDSwitchViewController action switchViews: became available for connection. You'll see how we make use of this action in the next section.

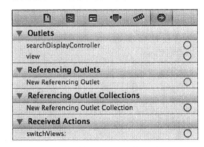

Figure 6–14. *The connections inspector showing that the switchViews: action has been added to the Received Actions section*

> **CAUTION:** If you don't see the switchViews: action as shown in Figure 6–14, check the spelling of your class file names. If you don't get the name exactly right, things won't match up. Watch your spelling!

Save your nib file and move to the next step.

Building a View with a Toolbar

We now need to build a view to add to BIDSwitchViewController. As a reminder, this new view controller will be our root view controller—the controller that is in play when our application is launched. BIDSwitchViewController's content view will consist of a toolbar that occupies the bottom of the screen. Its job is to switch between the blue view and the yellow view, so it will need a way for the user to change the views. For that, we're going to use a toolbar with a button. Let's build the toolbar view now.

Still in *SwitchView.xib*, in the nib's dock, click the *View* icon to make the view appear in the editing window (if it wasn't already there). This will also select the view. The view is an instance of UIView, and as you can see in Figure 6–15, it's currently empty and quite dull. This is where we'll start building our GUI.

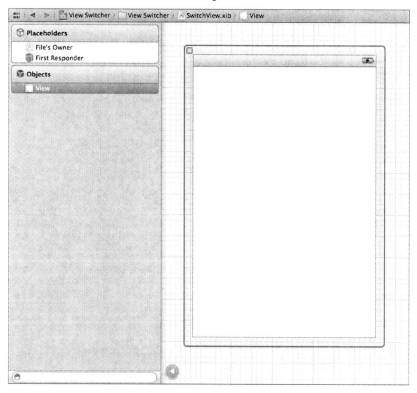

Figure 6–15. *The default view contained within our nib file, just waiting to be filled with interesting stuff!*

Now, let's add a toolbar to the bottom of the view. Grab a *Toolbar* from the library, drag it onto your view, and place it at the bottom, so that it looks like Figure 6–16.

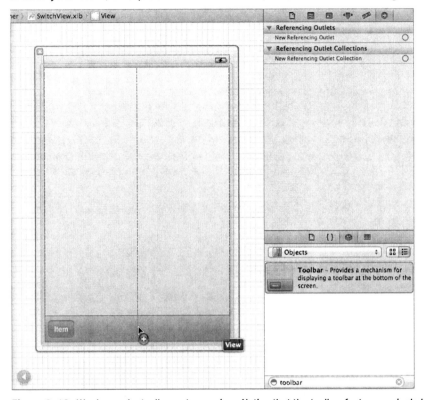

Figure 6–16. *We dragged a toolbar onto our view. Notice that the toolbar features a single button, labeled Item.*

The toolbar features a single button. We'll use that button to let the user switch between the different content views. Double-click the button, and change its title to *Switch Views*. Press the return key to commit your change.

Now, we can link the toolbar button to our action method. Before doing that, though, we should warn you: toolbar buttons aren't like other iOS controls. They support only a single target action, and they trigger that action only at one well-defined moment—the equivalent of a touch up inside event on other iOS controls.

Selecting a toolbar button in Interface Builder can be tricky. Click the view so we are all starting in the same place. Now, single-click the toolbar button. Notice that this selects the toolbar, not the button. Click the button a second time. This should select the button itself. You can confirm you have the button selected by switching to the object attributes inspector (⌥⌘4) and making sure the top group name is *Bar Button Item*.

Once you have the *Switch Views* button selected, control-drag from it over to the *File's Owner* icon, and select the *switchViews:* action. If the *switchViews:* action doesn't pop up, and instead you see an outlet called *delegate*, you've most likely control-dragged from the toolbar rather than the button. To fix it, just make sure you have the button rather than the toolbar selected, and then redo your control-drag.

> **TIP:** Remember that you can always view the nib's dock in list mode and use the disclosure triangles to drill down through the hierarchy to get to any element in the view hierarchy.

We have one more thing to do in this nib, which is to connect BIDSwitchViewController's view outlet to the view in the nib. The view outlet is inherited from the parent class, UIViewController, and gives the controller access to the view it controls. When we changed the underlying class of the file's owner, the existing outlet connections were broken. So, we need to reestablish the connection from the controller to its view. Control-drag from the *File's Owner* icon to the *View icon*, and select the *view* outlet to do that.

That's all we need to do here, so save your nib file. Next, let's get started implementing BIDSwitchViewController.

Writing the Root View Controller

It's time to write our root view controller. Its job is to switch between the blue view and the yellow view whenever the user clicks the *Switch Views* button.

In *BIDSwitchViewController.m*, first remove the comments around the viewDidLoad method. We'll be replacing that method in a moment. You can delete the remaining commented-out methods provided by the template if you want to shorten the code.

Start by adding this code to the top of the file:

```
#import "BIDSwitchViewController.h"
#import "BIDYellowViewController.h"
#import "BIDBlueViewController.h"

@implementation BIDSwitchViewController
@synthesize yellowViewController;
@synthesize blueViewController;
.
.
.
```

Next, replace viewDidLoad with this version:

```
- (void)viewDidLoad
{
    self.blueViewController = [[BIDBlueViewController alloc]
            initWithNibName:@"BlueView" bundle:nil];
    [self.view insertSubview:self.blueViewController.view atIndex:0];
    [super viewDidLoad];
}
```

Now, add in the switchViews: method:

```
- (IBAction)switchViews:(id)sender {
    if (self.yellowViewController.view.superview == nil) {
        if (self.yellowViewController == nil) {
            self.yellowViewController =
            [[BIDYellowViewController alloc] initWithNibName:@"YellowView"
                                                bundle:nil];
        }
        [blueViewController.view removeFromSuperview];
        [self.view insertSubview:self.yellowViewController.view atIndex:0];
    } else {
        if (self.blueViewController == nil) {
            self.blueViewController =
            [[BIDBlueViewController alloc] initWithNibName:@"BlueView"
                                                bundle:nil];
        }
        [yellowViewController.view removeFromSuperview];
        [self.view insertSubview:self.blueViewController.view atIndex:0];
    }
}
  .
  .
  .
```

Also, add the following code to the existing didReceiveMemoryWarning method:

```
- (void)didReceiveMemoryWarning {
    // Releases the view if it doesn't have a superview
    [super didReceiveMemoryWarning];

    // Release any cached data, images, etc, that aren't in use
    if (self.blueViewController.view.superview == nil) {
        self.blueViewController = nil;
    } else {
        self.yellowViewController = nil;
    }
}
```

The first method we modified, viewDidLoad, overrides a UIViewController method that is called when the nib is loaded. How could we tell? Hold down the option key and single-click the method name viewDidLoad. A documentation popup window will appear (see Figure 6–17). Alternatively, you can select **View ➤ Utilities ➤ Show Quick Help Inspector** to view similar information in the Quick Help panel. viewDidLoad is defined in our superclass, UIViewController, and is intended to be overridden by classes that need to be notified when the view has finished loading.

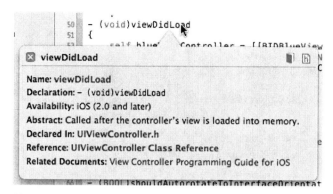

Figure 6–17. *This documentation window appears when you option-click the viewDidLoad method name.*

This version of viewDidLoad creates an instance of BIDBlueViewController. We use the initWithNibName:bundle: method to load the BIDBlueViewController instance from the nib file *BlueView.xib*. Note that the file name provided to initWithNibName:bundle: does not include the *.xib* extension. Once the BIDBlueViewController is created, we assign this new instance to our blueViewController property.

```
self.blueViewController = [[BIDBlueViewController alloc]
    initWithNibName:@"BlueView" bundle:nil];
```

Next, we insert the blue view as a subview of the root view. We insert it at index 0, which tells iOS to put this view behind everything else. Sending the view to the back ensures that the toolbar we created in Interface Builder a moment ago will always be visible on the screen, since we're inserting the content views behind it.

```
[self.view insertSubview:self.blueViewController.view atIndex:0];
```

Now, why didn't we load the yellow view here also? We're going to need to load it at some point, so why not do it now? Good question. The answer is that the user may never tap the *Switch Views* button. The user might just use the view that's visible when the application launches, and then quit. In that case, why use resources to load the yellow view and its controller?

Instead, we'll load the yellow view the first time we actually need it. This is called **lazy loading**, and it's a standard way of keeping memory overhead down. The actual loading of the yellow view happens in the switchViews: method, so let's take a look at that.

switchViews: first checks which view is being swapped in by seeing whether yellowViewController's view's superview is nil. This will return true if one of two things is true:

- If yellowViewController exists but its view is not being shown to the user, that view will not have a superview because it's not presently in the view hierarchy, and the expression will evaluate to true.

- If yellowViewController doesn't exist because it hasn't been created yet or was flushed from memory, it will also return true.

We then check to see whether `yellowViewController` is `nil`.

```
if (self.yellowViewController.view.superview == nil) {
```

If it is `nil`, that means there is no instance of `yellowViewController`, and we need to create one. This could happen because it's the first time the button has been pressed or because the system ran low on memory and it was flushed. In this case, we need to create an instance of `BIDYellowViewController` as we did for the `BIDBlueViewController` in the `viewDidLoad` method:

```
    if (self.yellowViewController == nil) {
        self.yellowViewController =
        [[BIDYellowViewController alloc] initWithNibName:@"YellowView"
                                                  bundle:nil];
    }
```

At this point, we know that we have a `yellowViewController` instance, because either we already had one or we just created it. We then remove `blueViewController`'s view from the view hierarchy and add the `yellowViewController`'s view:

```
    [blueViewController.view removeFromSuperview];
    [self.view insertSubview:self.yellowViewController.view atIndex:0];
```

If `self.yellowViewController.view.superview` is not `nil`, then we need to do the same thing, but for `blueViewController`. Although we create an instance of `BIDBlueViewController` in `viewDidLoad`, it is still possible that the instance has been flushed because memory got low. Now, in this application, the chances of memory running out are slim, but we're still going to be good memory citizens and make sure we have an instance before proceeding:

```
    } else {
        if (self.blueViewController == nil) {
            self.blueViewController =
            [[BIDBlueViewController alloc] initWithNibName:@"BlueView"
                                                    bundle:nil];
        }
        [yellowViewController.view removeFromSuperview];
        [self.view insertSubview:self.blueViewController.view atIndex:0];
    }
```

In addition to not using resources for the yellow view and controller if the *Switch Views* button is never tapped, lazy loading also gives us the ability to release whichever view is not being shown to free up its memory. iOS will call the `UIViewController` method `didReceiveMemoryWarning`, which is inherited by every view controller, when memory drops below a system-determined level.

Since we know that either view will be reloaded the next time it is shown to the user, we can safely release either controller. We do this by adding a few lines to the existing `didReceiveMemoryWarning` method:

```
- (void)didReceiveMemoryWarning {
    [super didReceiveMemoryWarning]; // Releases the view if it
                                     // doesn't have a superview
    // Release anything that's not essential, such as cached data
    if (self.blueViewController.view.superview == nil)
        self.blueViewController = nil;
```

```
        else
            self.yellowViewController = nil;
}
```

This newly added code checks to see which view is currently being shown to the user and releases the controller for the other view by assigning nil to its property. This will cause the controller, along with the view it controls, to be deallocated, freeing up its memory.

TIP: Lazy loading is a key component of resource management on iOS, and you should implement it anywhere you can. In a complex, multiview application, being responsible and flushing unused objects from memory can be the difference between an application that works well and one that crashes periodically because it runs out of memory.

Implementing the Content Views

The two content views that we are creating in this application are extremely simple. They each have one action method that is triggered by a button, and neither one needs any outlets. The two views are also nearly identical. In fact, they are so similar that they could have been represented by the same class. We chose to make them two separate classes because that's how most multiview applications are constructed.

Let's declare an action method in each of the header files. First, in *BIDBlueViewController.h*, add the following declaration:

```
#import <UIKit/UIKit.h>
@interface BIDBlueViewController : UIViewController
- (IBAction)blueButtonPressed;
@end
```

Save the file. Then add the following line to *BIDYellowViewController.h*:

```
#import <UIKit/UIKit.h>

@interface BIDYellowViewController : UIViewController
- (IBAction)yellowButtonPressed;
@end
```

Save this file as well.

Next, select *BlueView.xib* to open it in Interface Builder so we can make a few changes. First, we need to specify that the class that will load this nib from the file system is BIDBlueViewController. Single-click the *File's Owner* icon and press ⌥⌘3 to bring up the identity inspector. *File's Owner* defaults to *NSObject*; change it to *BIDBlueViewController.*

Single-click the *View* icon in the dock, and then press ⌥⌘4 to bring up the object attributes inspector. In the inspector's *View* section, click the color well that's labeled *Background*, and use the popup color picker to change the background color of this view to a nice shade of blue. Once you are happy with your blue, close the color picker.

Next, we'll change the size of the view in the nib. In the object attributes inspector, the top section is labeled *Simulated Metrics* (see Figure 6–18). If we set these drop-down menus to reflect which top and bottom elements are used in our application, Interface Builder will automatically calculate the size of the remaining space.

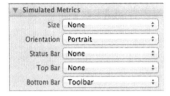

Figure 6–18. *The Simulated Metrics section of the view's attributes inspector*

The status bar is already specified, but here's a tricky spot: since this view is going to be contained inside the view we created in *SwitchView.xib*, we shouldn't actually specify a status bar, since doing so will shift our content a bit inside the containing view. So, click the *Status Bar* popup button, and then click *None*. Next, select the *Bottom Bar* popup and choose *Toolbar* to indicate that the enclosing view has a toolbar.

These settings will cause Interface Builder to calculate the correct size for our view automatically, so that we know how much space we have to work with. You can press ⌥⌘5 to bring up the size inspector to confirm this. After making the change, the height of the window should be 436 pixels, and the width should still be 320 pixels.

Drag a *Round Rect Button* from the library over to the view, using the guidelines to center the button in the view, both vertically and horizontally. Double-click the button, and change its title to *Press Me*. Next, with the button still selected, switch to the connections inspector (by pressing ⌥⌘6), drag from the *Touch Up Inside* event to the *File's Owner* icon, and connect to the *blueButtonPressed* action method.

We have one more thing to do in this nib, which is to connect BIDBlueViewController's view outlet to the view in the nib, just as we did earlier in *SwitchView.xib*. Control-drag from the *File's Owner* icon to the *View icon*, and select the *view* outlet.

Save the nib, and then go the project navigator and click *YellowView.xib*. We're going to make almost exactly the same changes to this nib file.

First, click the *File's Owner* icon in the dock and use the identity inspector to change its class to *BIDYellowViewController*.

Next, select the view and switch to the object attributes inspector. There, click the *Background* color well and select a bright yellow, and then close the color picker. Also, in the *Simulated Metrics* section, select *Toolbar* from the *Bottom Bar* popup, and switch the *Status Bar* popup to *None*.

Next, drag out a *Round Rect Button* from the library and use the guidelines to center it in the view. Then change its title to *Press Me, Too*. With the button still selected, use the connections inspector to drag from the *Touch Up Inside* event to the *File's Owner* icon, and connect to the *yellowButtonPressed* action method.

Finally, control-drag from the *File's Owner* icon to the *View icon*, and select the *view* outlet.

When you're finished, save the nib, and get ready to enter some more code.

The two action methods we're going to implement do nothing more than show an alert (as we did in Chapter 4's Control Fun application), so go ahead and add the following code to *BIDBlueViewController.m:*

```
#import "BIDBlueViewController.h"

@implementation BIDBlueViewController

- (IBAction)blueButtonPressed {
    UIAlertView *alert = [[UIAlertView alloc]
        initWithTitle:@"Blue View Button Pressed"
              message:@"You pressed the button on the blue view"
             delegate:nil
    cancelButtonTitle:@"Yep, I did."
    otherButtonTitles:nil];
    [alert show];
}
...
```

Save the file. Next, switch over to *BIDYellowViewController.m*, and add this very similar code to that file:

```
#import "BIDYellowViewController.h"

@implementation BIDYellowViewController

- (IBAction)yellowButtonPressed {
    UIAlertView *alert = [[UIAlertView alloc]
        initWithTitle:@"Yellow View Button Pressed"
              message:@"You pressed the button on the yellow view"
             delegate:nil
    cancelButtonTitle:@"Yep, I did."
    otherButtonTitles:nil];
    [alert show];
}
...
```

Save your code, and let's take this bad boy for a spin. If your app crashes on launch or when you switch views, go back and make sure you connected all three `view` outlets.

When our application launches, it shows the view we built in *BlueView.xib*. When you tap the *Switch Views* button, it will change to show the view that we built in *YellowView.xib*. Tap it again, and it goes back to the view in *BlueView.xib*. If you tap the button centered on the blue or yellow view, you'll get an alert view with a message indicating which button was pressed. This alert shows that the correct controller class is being called for the view that is being shown.

The transition between the two views is kind of abrupt, though. Gosh, if only there were some way to make the transition look nicer.

Of course, there is a way to make the transition look nicer! We can animate the transition in order to give the user visual feedback of the change.

Animating the Transition

UIView has several class methods we can call to indicate that the transition between views should be animated, to indicate the type of transition that should be used, and to specify how long the transition should take.

Go back to *BIDSwitchViewController.m*, and replace your switchViews: method with this new version:

```
- (IBAction)switchViews:(id)sender {
    [UIView beginAnimations:@"View Flip" context:nil];
    [UIView setAnimationDuration:1.25];
    [UIView setAnimationCurve:UIViewAnimationCurveEaseInOut];

    if (self.yellowViewController.view.superview == nil) {
        if (self.yellowViewController == nil) {
            self.yellowViewController =
            [[BIDYellowViewController alloc] initWithNibName:@"YellowView"
                                                     bundle:nil];
        }
        [UIView setAnimationTransition:
         UIViewAnimationTransitionFlipFromRight
                               forView:self.view cache:YES];

        [self.blueViewController.view removeFromSuperview];
        [self.view insertSubview:self.yellowViewController.view atIndex:0];
    } else {
        if (self.blueViewController == nil) {
            self.blueViewController =
            [[BIDBlueViewController alloc] initWithNibName:@"BlueView"
                                                   bundle:nil];
        }
        [UIView setAnimationTransition:
         UIViewAnimationTransitionFlipFromLeft
                               forView:self.view cache:YES];

        [self.yellowViewController.view removeFromSuperview];
        [self.view insertSubview:self.blueViewController.view atIndex:0];
    }
    [UIView commitAnimations];
}
```

Compile this new version, and run your application. When you tap the *Switch Views* button, instead of the new view just snapping into place, the old view will flip over to reveal the new view, as shown in Figure 6–19.

Figure 6–19. *One view transitioning to another, using the flip style of animation*

In order to tell iOS that we want a change animated, we need to declare an **animation block** and specify how long the animation should take. Animation blocks are declared by using the UIView class method beginAnimations:context:, like so:

```
[UIView beginAnimations:@"View Flip" context:NULL];
[UIView setAnimationDuration:1.25];
```

beginAnimations:context: takes two parameters. The first is an animation block title. This title comes into play only if you take more direct advantage of Core Animation, the framework behind this animation. For our purposes, we could have used nil. The second parameter is a (void *) that allows you to specify an object (or any other C data type) whose pointer you would like associated with this animation block. We used NULL here, since we don't need to do that.

After that, we set the **animation curve**, which determines the timing of the animation. The default, which is a linear curve, causes the animation to happen at a constant speed. The option we set here, UIViewAnimationCurveEaseInOut, specifies that the animation should start slow but speed up in the middle, and then slow down again at the end. This gives the animation a more natural, less mechanical appearance.

```
[UIView setAnimationCurve:UIViewAnimationCurveEaseInOut];
```

Next, we need to specify the transition to use. At the time of this writing, four iOS view transitions are available:

- `UIViewAnimationTransitionFlipFromLeft`
- `UIViewAnimationTransitionFlipFromRight`
- `UIViewAnimationTransitionCurlUp`
- `UIViewAnimationTransitionCurlDown`

We chose to use two different effects, depending on which view was being swapped in. Using a left flip for one transition and a right flip for the other makes the view seem to flip back and forth.

The `cache` option speeds up drawing by taking a snapshot of the view when the animation begins and using that image, rather than redrawing the view at each step of the animation. You should always cache the animation unless the appearance of the view may need to change during the animation.

```
[UIView setAnimationTransition:UIViewAnimationTransitionFlipFromRight
                 forView:self.view cache:YES];
```

Then we remove the currently shown view from our controller's view, and instead add the other view.

When we're finished specifying the changes to be animated, we call `commitAnimations` on `UIView`. Everything between the start of the animation block and the call to `commitAnimations` will be animated together.

Thanks to Cocoa Touch's use of Core Animation under the hood, we're able to do fairly sophisticated animation with only a handful of code.

Switching Off

Whoo-boy! Creating our own multiview controller was a lot of work, wasn't it? You should have a very good grasp on how multiview applications are put together now that you've built one from scratch.

Although Xcode contains project templates for the most common types of multiview applications, you need to understand the overall structure of these types of applications so you can build them yourself from the ground up. The delivered templates are incredible time-savers, but at times, they simply won't meet your needs.

In the next few chapters, we're going to continue building multiview applications to reinforce the concepts from this chapter and to give you a feel for how more complex applications are put together. In Chapter 7, we'll construct a tab bar application. Let's get going!

Tab Bars and Pickers

In the previous chapter, you built your first multiview application. In this chapter, you're going to build a full tab bar application with five different tabs and five different content views. Building this application will reinforce a lot of what you learned in Chapter 6. Now, you're too smart to spend a whole chapter doing stuff you already sort of know how to do, so we're going to use those five content views to demonstrate a type of iOS control that we have not yet covered. The control is called a **picker view**, or just a **picker.**

You may not be familiar with the name, but you've almost certainly used a picker if you've owned an iPhone or iPod touch for more than, say, 10 minutes. Pickers are the controls with dials that spin. You use them to input dates in the Calendar application or to set a timer in the Clock application (see Figure 7–1). On the iPad, the picker view isn't quite as common, since the larger display lets you present other ways of choosing among multiple items, but even there, it's used in the Calendar application.

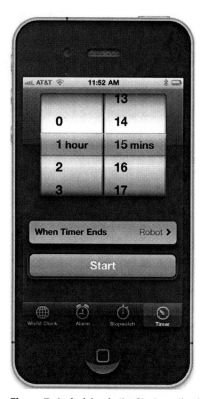

Figure 7–1. *A picker in the Clock application*

Pickers are a bit more complex than the iOS controls you've seen so far, and as such, they deserve a little more attention. Pickers can be configured to display one dial or many. By default, pickers display lists of text, but they can also be made to display images.

The Pickers Application

This chapter's application, Pickers, will feature a tab bar. As you build Pickers, you'll change the default tab bar so it has five tabs, add an icon to each of the tab bar items, and then create a series of content views and connect each view to a tab.

The application's content views will feature five different pickers:

- **Date picker**: The first content view we'll build will have a date picker, which is the easiest type of picker to implement (see Figure 7–2). The view will also have a button that, when tapped, will display an alert that shows the date that was picked.

Figure 7–2. *The first tab will show a date picker.*

■ **Single-component picker**: The second tab will feature a picker with a single list of values (see Figure 7–3). This picker is a little more work to implement than a date picker. You'll learn how to specify the values to be displayed in the picker by using a delegate and a data source.

Figure 7–3. *A picker displaying a single list of values*

■ **Multicomponent picker**: In the third tab, we're going to create a
picker with two separate wheels. The technical term for each of these
wheels is a **picker component**, so here we are creating a picker with
two components. You'll see how to use the data source and delegate
to provide two independent lists of data to the picker (see Figure 7–4).
Each of this picker's components can be changed without impacting
the other one.

Figure 7–4. *A two-component picker, showing an alert that reflects our selection*

■ **Picker with dependent components**: In the fourth content view, we'll build another picker with two components. But this time, the values displayed in the component on the right will change based on the value selected in the component on the left. In our example, we're going to display a list of states in the left component and a list of that state's ZIP codes in the right component (see Figure 7–5).

Figure 7–5. *In this picker, one component is dependent on the other. As you select a state in the left component, the right component changes to a list of ZIP codes in that state.*

■ **Custom picker with images**: Last, but most certainly not least, we're going to have some fun with the fifth content view. We'll demonstrate how to add image data to a picker, and we're going to do it by writing a little game that uses a picker with five components. In several places in Apple's documentation, the picker's appearance is described as looking a bit like a slot machine. Well, then, what could be more fitting than writing a little slot machine game (see Figure 7–6)? For this picker, the user won't be able to manually change the values of the components, but will be able to select the *Spin* button to make the five wheels spin to a new, randomly selected value. If three copies of the same image appear in a row, the user wins.

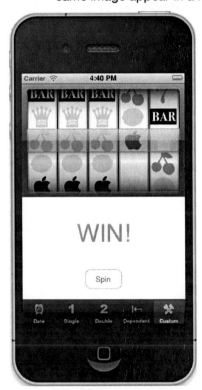

Figure 7–6. *Our fifth component picker. Note that we do not condone using your iPhone as a tiny casino.*

Delegates and Data Sources

Before we dive in and start building our application, let's look at what makes pickers more complex than the other controls you've used so far. With the exception of the date picker, you can't use a picker by just grabbing one in the object library, dropping it on your content view, and configuring it. You also need to provide each picker with both a picker **delegate** and a picker **data source**.

By this point, you should be comfortable using delegates. We've already used application delegates and action sheet delegates, and the basic idea is the same here. The picker defers several jobs to its delegate. The most important of these is the task of determining what to actually draw for each of the rows in each of its components. The picker asks the delegate for either a string or a view that will be drawn at a given spot on a given component. The picker gets its data from the delegate.

In addition to the delegate, pickers need to have a data source. In this instance, the name *data source* is a bit of a misnomer. The data source tells the picker how many components it will be working with and how many rows make up each component. The data source works like the delegate, in that its methods are called at certain, prespecified times. Without a data source and a delegate, pickers cannot do their job; in fact, they won't even be drawn.

It's very common for the data source and the delegate to be the same object, and just as common for that object to be the view controller for the picker's enclosing view, which is the approach we'll be using in this application. The view controllers for each of our application's content panes will be the data source and the delegate for their picker.

> **NOTE:** Here's a pop quiz: Is the picker data source part of the model, view, or controller portion of the application? It's a trick question. A data source sounds like it must be part of the model, but in fact, it's actually part of the controller. The data source isn't usually an object designed to hold data. In simple applications, the data source might hold data, but its true job is to retrieve data from the model and pass it along to the picker.

Let's fire up Xcode and get to it.

Setting Up the Tab Bar Framework

Although Xcode does provide a template for tab bar applications, we're going to build ours from scratch. It's not much extra work, and it's good practice.

Create a new project, select the *Empty Application* template again, and choose *Next* to go to the next screen. In the *Product Name* field, type *Pickers*. Make sure the checkbox that says *Use Core Data* is unchecked, and set the *Device Family* popup to *iPhone*. Then choose *Next* again, and Xcode will let you select the folder where you want to save your project.

We're going to walk you through the process of building the whole application, but at any step of the way, if you feel like challenging yourself by moving ahead of us, by all means go ahead. If you get stumped, you can always come back. If you don't feel like skipping ahead, that's just fine. We love the company.

Creating the Files

In the previous chapter, we created a root view controller (root controller for short) to manage the process of swapping our application's other views. We'll be doing that again this time, but we won't need to create our own root view controller class. Apple provides a very good class for managing tab bar views, so we're just going to use an instance of UITabBarController as our root controller.

First, we need to create five new classes in Xcode: the five view controllers that the root controller will swap in and out.

Expand the *Pickers* folder in the project navigator. There, you'll see the source code files that Xcode created to start off the project. Single-click the *Pickers* folder, and press ⌘N or select **File ➤ New ➤ New File…**

Select *Cocoa Touch* in the left pane of the new file assistant, and then select the icon for *UIViewController subclass* and click *Next* to continue. The next screen lets you give your new class a name. Enter *BIDDatePickerViewController* in the *Class* field. As always, when naming a new class file, carefully check your spelling. A typo here will cause your new class to be named incorrectly. You'll also see a control that lets you select or enter a superclass for your new class, which you should leave as *UIViewController*. Below that, you should see a checkbox labeled *With XIB for user* interface (see Figure 7–7). Make sure that's checked (and only that one; the *Targeted for iPad* option should be unchecked) before clicking *Next*.

Finally, you'll be shown a folder selection window, which lets you choose where the class should be saved. Choose the *Pickers* directory, which already contains the *BIDAppDelegate* class and a few other files. Make sure also that the *Group* popup has the *Pickers* folder selected, and that the target checkbox for *Pickers* is checked.

After you click the *Create* button, three new files will appear in your *Pickers* folder: *BIDDatePickerViewController.h*, *BIDDatePickerViewController.m*, and *BIDDatePickerViewController.xib*.

Repeat those steps four more times, using the names *BIDSingleComponentPickerViewController*, *BIDDoubleComponentPickerViewController*, *BIDDependentComponentPickerViewController*, and *BIDCustomPickerViewController*. Be sure to select the *Pickers* folder in the project navigator each time you create your new file, so the newly created files are bunched nicely together.

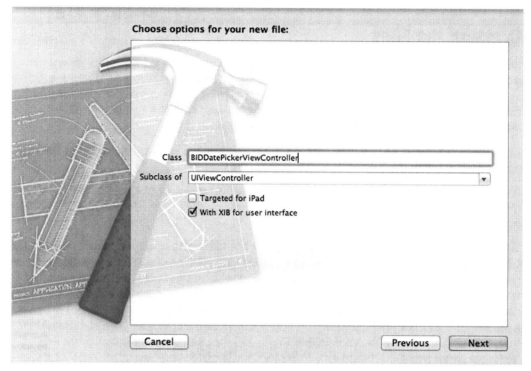

Figure 7–7. *When creating a subclass of UIViewController, Xcode will create the accompanying .xib file for you if you select the With XIB for user interface checkbox.*

Adding the Root View Controller

We're going to create our root view controller, which will be an instance of UITabBarController, in Interface Builder. Before we can do that, however, we should declare an outlet for it. Single-click *BIDAppDelegate.h*, and add the following code to it:

```
#import <UIKit/UIKit.h>

@interface BIDAppDelegate : UIResponder <UIApplicationDelegate>

@property (strong, nonatomic) IBOutlet UIWindow *window;
@property (strong, nonatomic) IBOutlet UITabBarController *rootController;

@end
```

Before we move to Interface Builder to create our root view controller, let's add the following code to *BIDAppDelegate.m*:

```
#import "BIDAppDelegate.h"

@implementation BIDAppDelegate

@synthesize window = _window;
@synthesize rootController;
```

```
- (BOOL)application:(UIApplication *)application
    didFinishLaunchingWithOptions:(NSDictionary *)launchOptions
{
    self.window = [[UIWindow alloc] initWithFrame:[[UIScreen mainScreen] bounds]];
    // Override point for customization after app launch
    [[NSBundle mainBundle] loadNibNamed:@"TabBarController" owner:self options:nil];
    [self.window addSubview:rootController.view];
    self.window.backgroundColor = [UIColor whiteColor];
    [self.window makeKeyAndVisible];
    return YES;
}

    .
    .
    .
```

There shouldn't be anything in this code that's a surprise to you. We're doing pretty much the same thing we did in the previous chapter, except that we're using a controller class provided by Apple instead of one we wrote ourselves. This time, we're also going to perform a new trick by loading a nib file containing the view controller, instead of creating the view controller directly in code. In the next section, we'll create that .xib file and configure it so that when it loads, our app delegate's rootController variable is connected to a UITabBarController, which is then ready to be inserted into our app's window.

Tab bars use icons to represent each of the tabs, so we should also add the icons we're going to use before editing the nib file for this class. You can find some suitable icons in the *07 Pickers/Tab Bar Icons/* folder of the project archive that accompanies this book. Add all five of the icons in that folder to the project. You can just drag the folder from the Finder and drop it on the *Pickers* folder in the project navigator. When asked, select *Create groups for any added folders*, and Xcode will add a *Tab Bar Icons* subfolder to the *Pickers* folder.

The icons you use should be 24 × 24 pixels and saved in *.png* format. The icon file should have a transparent background. Generally, medium-gray icons look the best on a tab bar. Don't worry about trying to match the appearance of the tab bar. Just as it does with the application icon, iOS will take your image and make it look just right.

Creating TabBarController.xib

Now, let's create the .xib file that will contain our tab bar controller. Select the *Pickers* folder in the project navigator, and press ⌘N to create a new file. When the standard file-creation assistant appears, select *User Interface* from the iOS section on the left, and then select the *Empty* template on the right and click *Next*. Leave *Device Family* set to *iPhone* on the next screen, hit *Next* again, and you'll come to the final screen, where the assistant asks you to name the file. Call it *TabBarController.xib*, making sure to use the exact same spelling we used in the code we entered earlier—otherwise, the app won't be able to locate and load the nib file. Make sure the *Pickers* directory, the *Pickers* group, and the *Pickers* target are selected. Those should all be selected by default, but it's always worth double-checking.

When you're all set, click *Create*. Xcode will create the file *TabBarController.xib*, and you'll see it appear in the project navigator. Select it, and the familiar Interface Builder editing view will appear.

At this point, this nib file is very much a blank slate. Let's remedy that by dragging a *Tab Bar Controller* from the object library (see Figure 7–8) over to the nib's main window.

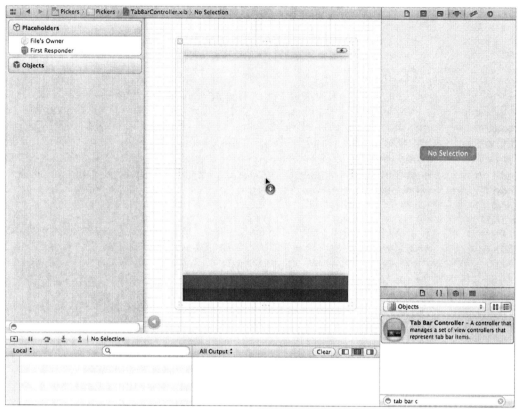

Figure 7–8. *Dragging a Tab Bar Controller from the library into the nib editor*

Once you drop the tab bar controller onto your nib's main window, a new window representing the UITabBarController will appear (see Figure 7–9), and the tab bar controller icon will appear in the Interface Builder dock. If you view the dock in list mode (click the triangle in a circle icon just to the right of the bottom of the dock), you can expand the tab bar controller icon to reveal the tab bar and the two view controllers and associated items that appear by default.

Figure 7–9. *The tab bar controller's window. Notice the tab bar at the bottom of the window, with two individual tabs. Also note the text in the view area, marking the view controller inside the tab bar controller.*

This tab bar controller will be our root controller. As a reminder, the root controller controls the very first view that the user will see when your program runs. It is responsible for switching the other views in and out. Since we'll connect each of our views to one of the tab bar tabs, the tab bar controller makes a logical choice as a root controller.

In the previous section, we added some code to the BIDAppDelegate class to load the nib we are currently creating and to use the rootController outlet to add the root controller's view to the application window. The problem right now is that this nib file doesn't yet know anything about the BIDAppDelegate class. It has no idea who its File's Owner should be, so we have no way to connect things together.

Open the identity inspector, and then select File's Owner in the dock. The identity inspector will show NSObject in the Custom Class's *Class* field. We need to change *NSObject* to *BIDAppDelegate*, marking the app delegate as the File's Owner, which will allow us to connect the rootController outlet to the new controller. Go ahead and type *BIDAppDelegate* or select it from the popup.

Press the enter key to make sure the new value is set, and then switch to the connections inspector, where you'll see that File's Owner now has an outlet named *rootController*, ready for connecting! There's no time like the present, so go ahead and drag from the *rootController* outlet's little connecting ring over to the *Tab Bar Controller* in the dock.

Our next step is to customize our tab bar so it reflects the five tabs shown in Figure 7–2. Each of those five tabs represents one of our five pickers.

In the nib editor, if the dock is not in list view, switch it over by clicking the triangle-in-a-circle icon just to the right of the bottom of the dock. Open the disclosure triangle to the left of *Tab Bar Controller* to reveal a *Tab Bar* and two *View Controller* entries. Next, open the disclosure triangles to the left of each *View Controller* to show the *Tab Bar Item* associated with each controller (see Figure 7–10). By opening up everything, you'll have a better understanding of what's happening as we customize this tab bar.

Figure 7–10. *The tab bar controller, opened all the way to show the items nested within*

Let's add three more *Tab Bar Items* to the tab bar. As you'll see, the associated *View Controllers* will be added automatically each time we drag over a new *Tab Bar Item*.

Bring up the object library (**View ➤ Utilities ➤ Show Object Library**). Locate a *Tab Bar Item* and drag it onto the tab bar (see Figure 7–11). Notice the insertion point. This tells you where on the tab bar your new item will end up. Since we will be customizing all of our tab bar items, it doesn't matter where this one lands.

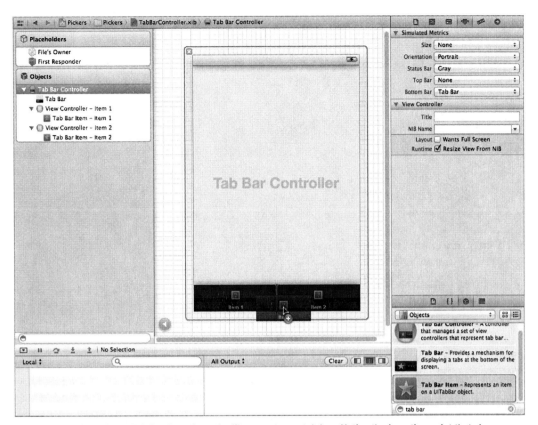

Figure 7–11. *Dragging a Tab Bar Item from the library onto our tab bar. Notice the insertion point that shows you where your new item will end up.*

Now drag out two more *Tab Bar Items*, so you have a total of five. If you take a look at your dock, you'll see that your tab bar now consists of five *View Controllers*, each with its own *Tab Bar Item*. Open the disclosure triangle to the left of each *View Controller* so you can see all of them (see Figure 7–12).

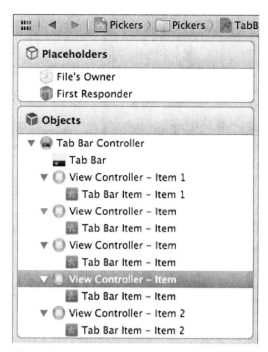

Figure 7–12. *The tab bar controller, opened all the way to show the five view controllers and their associated tab bar items*

Our next step is to customize each of the five view controllers. In the dock, select the first *View Controller*, and then bring up the attributes inspector (**View ➤ Utilities ➤ Show Attributes Inspector**). This is where we associate each tab's view controller with the appropriate nib.

In the attributes inspector, leave the *Title* field blank (see Figure 7–13). Tab bar view controllers don't use this title for anything. Specify a *NIB Name* of *BIDDatePickerViewController*. Do not include the *.xib* extension.

Just below the *NIB Name* field, the checkbox labeled *Wants Full Screen* indicates that the view that comes up when you select this tab will overlap and hide the tab bar. If you check this checkbox, you must provide an alternative mechanism for navigating off that tab. We will leave this value unchecked for all of our tabs. In addition, leave the *Resize View From NIB* checkbox checked. Since we'll design our views to be the size we want and to not need resizing, this last checkbox really won't matter.

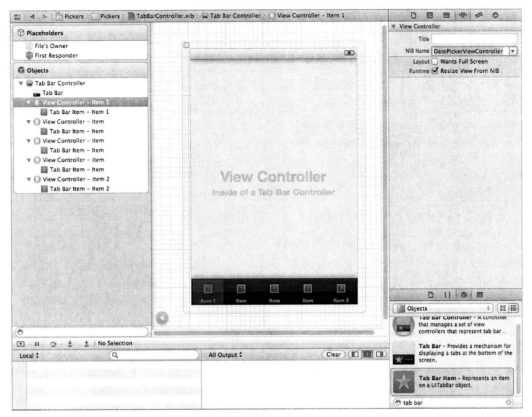

Figure 7–13. *We've selected the first of our five view controllers and associated the nib named BIDDatePickerViewController.xib with the controller. Note that we left off the extension .xib. This is automatically added to the nib name.*

While you're here, bring up the identity inspector for the view controller associated with the leftmost tab. In the *Custom Class* section of the inspector, change the class to *BIDDatePickerViewController*, and press return or tab to set it. You'll see that the name of the selected control in the dock changes to *Date Picker View Controller – Item 1,* mirroring the change you made.

Now repeat this same process for the next four view controllers. In the attributes inspector for each, make sure the checkboxes are correctly configured, and enter the nib names *BIDSingleComponentPickerViewController*, *BIDDoubleComponentPickerViewController*, *BIDDependentComponentPickerViewController*, and *BIDCustomPickerViewController*, respectively. For each view controller, you need to make sure to make changes in two places: Use the identity inspector to set the class name, and the attributes inspector to make the same change in the *NIB Name* field.

You've just made a lot of changes. Check your work and save it. Let's customize the five *Tab Bar Items* now, so they have the correct icon and label.

In the dock, select the *Tab Bar Item* that is a subitem of the *Date Picker View Controller*. Press ⌥⌘4 to return to the attributes inspector (see Figure 7–14).

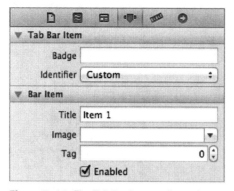

Figure 7–14. *The Tab Bar Item attributes inspector*

The first field in the *Tab Bar Item* section is labeled *Badge.* This can be used to put a red icon onto a tab bar item, similar to the red number placed on the *Mail* icon that tells you how many unread e-mail messages you have. We're not going to use the *Badge* field in this chapter, so you can leave it blank.

Under that, there's a popup button called *Identifier*. This field allows you to select from a set of commonly used tab bar item names and icons, such as *Favorites* and *Search*. If you select one of these, the tab bar will provide the name and icon for the item based on your selection. We're not using standard items, so leave this set to *Custom*.

The next two fields down are where we can specify a title and custom tab icon for a tab bar item. Change the *Title* from *Item 1* to *Date*. Next, click the *Image* combo box, and select the *clockicon.png* image. If you are using your own set of icons, select one of your *.png* files instead. For the rest of this chapter, we'll assume you used our resources. Adjust your thinking as necessary.

If you look over at the *Tab Bar Controller* window, you'll see that the leftmost tab bar item now reads *Date* and has a picture of a clock on it (see Figure 7–15).

Figure 7–15. *Our first tab bar item has changed to have a title of Date and an icon of a clock. Cool!*

Repeat this process for the other four tab bar items:

- Change the second Tab Bar Item to a Title of Single, and specify an Image of *singleicon.png*.

- Change the third Tab Bar Item to a Title of Double, and specify an Image of *doubleicon.png*.

- Change the fourth Tab Bar Item to a Title of Dependent, and specify an Image of *dependenticon.png*.

- Change the fifth Tab Bar Item to a Title of Custom, and specify an Image of *toolicon.png*.

Figure 7–16 shows our finished tab bar.

Figure 7–16. *Our finished tab bar, with all five titles and icons in place*

> **NOTE:** Don't worry about the view controller Title fields. We don't use them for this application. It doesn't matter whether they are blank or contain text. However, we *do* use the *tab bar item* Title fields. Don't confuse the two.

Before we move on to our next bit of nib editing, save your nib file.

We've been describing things here in terms of navigating the dock's list view in order to select an item, but it's also possible to select items in the graphical layout area, so direct your attention there for a moment. Double-click the *Tab Bar Controller* in the nib's window. This will open the *Tab Bar Controller* in the Interface Builder editing pane (if it's not open already). In that *Tab Bar Controller*, click one of the tab bar items while keeping your eye on the attributes inspector. The first time you click the tab bar item, you'll have selected that item's view controller. Click a second time, and you'll select the tab bar item itself.

We find selecting the item we want in the dock's list view to be much less confusing, but it is definitely worth knowing about the click once, click twice technique. This same clicking approach works with other nested nib elements. Experiment, and use the inspectors to make sure you've selected what you think you've selected.

The Initial Test Run

At this point, the tab bar and the content views should all be hooked up and working. Return to Xcode, compile and run, and your application should launch with a tab bar that functions. Click each of the tabs in turn. Each tab should be selectable.

There's nothing in the content views now, so the changes won't be very dramatic. But if everything went OK, the basic framework for your multiview application is now set up and working, and we can start designing the individual content views.

> **TIP:** If your simulator bursts into flames when you click one of the tabs, don't panic! Most likely, you've either missed a step or made a typo. Go back and check all the nib file names, make sure the connections are right, and make sure the class names are all set correctly.

If you want to make double sure everything is working, you can add a different label or some other object to each of the content views, and then relaunch the application. Then you should see the content of the different views change as you select different tabs.

Implementing the Date Picker

To implement the date picker, we'll need a single outlet and a single action. The outlet will be used to grab the value from the date picker. The action will be triggered by a button and will put up an alert to show the date value pulled from the picker. Single-click *BIDDatePickerViewController.h*, and add the following code:

```
#import <UIKit/UIKit.h>

@interface BIDDatePickerViewController : UIViewController

@property (strong, nonatomic) IBOutlet UIDatePicker *datePicker;
- (IBAction)buttonPressed;

@end
```

Save this file, and then click *BIDDatePickerViewController.xib* to edit the content view for our first tab.

The first thing we need to do is size the view so it accounts for the tab bar. Single-click the *View* icon and press ⌥⌘4 to bring up the attributes inspector. In the *Simulated Metrics* section, set the *Bottom Bar* popup to *Tab Bar*. This will cause Interface Builder to automatically reduce the view's height to 411 pixels and show a simulated tab bar.

Next, find a *Date Picker* in the library, and drag one over to the *View* window. Place the *Date Picker* at the top of the view, right up against the bottom of the status bar. It should take up the entire width of your content view and a good portion of the height. Don't use the blue guidelines for the picker; it's designed to fit snugly against the edges of the view (see Figure 7–17).

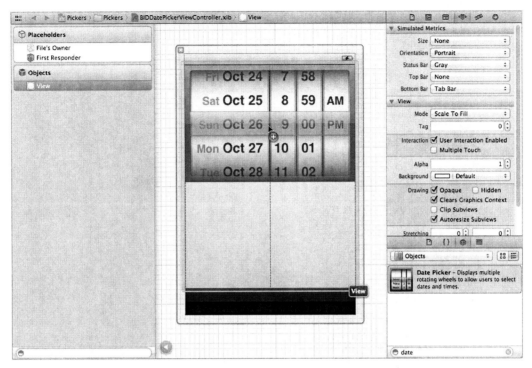

Figure 7–17. *We dragged a Date Picker from the library. Notice that it takes up the entire width of the view, and that we placed it at the top of the view, just below the status bar.*

Single-click the date picker if it's not already selected, and go back to the attributes inspector. As you can see in Figure 7–18, a number of attributes can be configured for a date picker. We're going to leave most of the values at their defaults (but feel free to play with the options when we're finished to see what they do). The one thing we will do is limit the range of the picker to reasonable dates. Look for the heading that says *Constraints*, and check the box that reads *Minimum Date.* Leave the value at the default of *1/1/1970.* Also check the box that reads *Maximum Date*, and set that value *to 12/31/2200.*

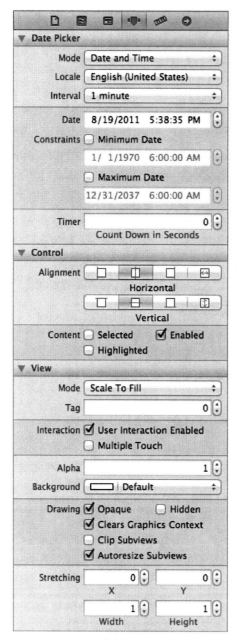

Figure 7–18. *The attributes inspector for a date picker. Set the minimum and maximum dates, but leave the rest of the settings at their default values.*

Next, grab a *Round Rect Button* from the library, and place it below the date picker. Double-click the button and give it a title of *Select*.

With the button still selected, press ⌥⌘6 to switch to the connections inspector. Drag from the circle next to the *Touch Up Inside* event over to the *File's Owner* icon, and connect to the **buttonPressed** action. Then control-drag from the *File's Owner* icon back to the date picker, and select the **datePicker** outlet. Finally, save your changes to the nib file, since we're finished with this part of the GUI.

Now we just need to implement BIDDatePickerViewController. Click *BIDDatePickerViewController.m* and start by adding the following code at the top of the file:

```
#import "BIDDatePickerViewController.h"

@implementation BIDDatePickerViewController
@synthesize datePicker;

- (IBAction)buttonPressed {
    NSDate *selected = [datePicker date];
    NSString *message = [[NSString alloc] initWithFormat:
        @"The date and time you selected is: %@", selected];
    UIAlertView *alert = [[UIAlertView alloc]
            initWithTitle:@"Date and Time Selected"
                  message:message
                 delegate:nil
        cancelButtonTitle:@"Yes, I did."
        otherButtonTitles:nil];
    [alert show];
}
.
.
.
```

Then add a bit of setup code to the viewDidLoad: method:

```
- (void)viewDidLoad {
    [super viewDidLoad];
    // Do any additional setup after loading the view from its nib.
    NSDate *now = [NSDate date];
    [datePicker setDate:now animated:NO];
}
.
.
.
```

Next, add one line to the existing viewDidUnload: method:

```
- (void)viewDidUnload {
    [super viewDidUnload];
    // Release any retained subviews of the main view.
    // e.g. self.myOutlet = nil;
    self.datePicker = nil;
}
```

Here, we first synthesize the accessor and mutator for our datePicker outlet, and then we add the implementation of buttonPressed and override viewDidLoad. In buttonPressed, we use our datePicker outlet to get the current date value from the date picker, and then we construct a string based on that date and use it to show an alert sheet.

In `viewDidLoad`, we create a new `NSDate` object. An `NSDate` object created this way will hold the current date and time. We then set `datePicker` to that date, which ensures that every time this view is loaded from the nib, the picker will reset to the current date and time.

Go ahead and build and run to make sure your date picker checks out. If everything went OK, your application should look like Figure 7–2 when it runs. If you choose the *Select* button, an alert sheet will pop up, telling you the date and time currently selected in the date picker.

> **NOTE:** The date picker does not allow you to specify seconds or a time zone. The alert displays the time with seconds and in Greenwich Mean Time (GMT). We could have added some code to simplify the string displayed in the alert, but isn't this chapter long enough already? If you're interested in customizing the formatting of the date, take a look at the `NSDateFormatter` class.

Implementing the Single-Component Picker

Our next picker lets the user select from a list of values. In this example, we're going to create an `NSArray` to hold the values we want to display in the picker.

Pickers don't hold any data themselves. Instead, they call methods on their data source and delegate to get the data they need to display. The picker doesn't really care where the underlying data lives. It asks for the data when it needs it, and the data source and delegate (which are often, in practice, the same object) work together to supply that data. As a result, the data could be coming from a static list, as we'll do in this section. It also could be loaded from a file or a URL, or even made up or calculated on the fly.

Declaring Outlets and Actions

As always, we need to make sure our outlets and actions are in place in our controller's header file before we start working on the GUI. In the project navigator, single-click *BIDSingleComponentPickerViewController.h*. This controller class will act as both the data source and the delegate for its picker, so we need to make sure it conforms to the protocols for those two roles. In addition, we need to declare an outlet and an action. Add the following code:

```
#import <UIKit/UIKit.h>

@interface BIDSingleComponentPickerViewController : UIViewController
    <UIPickerViewDelegate, UIPickerViewDataSource>

@property (strong, nonatomic) IBOutlet UIPickerView *singlePicker;
@property (strong, nonatomic) NSArray *pickerData;
- (IBAction)buttonPressed;

@end
```

We start by conforming our controller class to two protocols, UIPickerViewDelegate and UIPickerViewDataSource. After that, we declare an outlet for the picker and a pointer to an NSArray, which will be used to hold the list of items that will be displayed in the picker. Finally, we declare the action method for the button, just as we did for the date picker.

Building the View

Now select *BIDSingleComponentPickerViewController.xib* to edit the content view for the second tab in our tab bar. Click the *View* icon and press ⌥⌘4 to bring up the attributes inspector. Set the *Bottom Bar* to *Tab Bar* in the *Simulated Metrics* section. Next, bring over a *Picker View* from the library (see Figure 7–19), and add it to your nib's *View* window, placing it snugly into the top of the view, as you did with the date picker view.

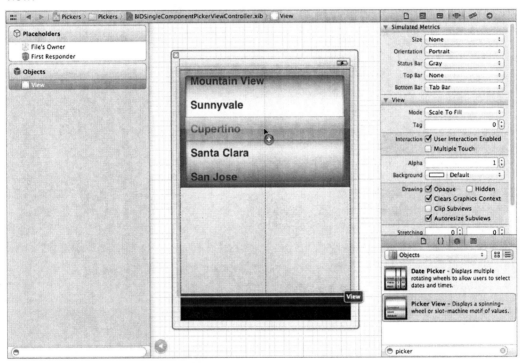

Figure 7–19. *Dragging a Picker View from the library onto our second view*

After placing the picker, control-drag from *File's Owner* to the picker, and select the **singlePicker** outlet.

Next, with the picker selected, press ⌥⌘6 to bring up the connections inspector. If you look at the connections available for the picker view, you'll see that the first two items are *dataSource* and *delegate*. If you don't see those outlets, make sure you have the picker selected, rather than the UIView that contains it! Drag from the circle next to *dataSource* to the *File's Owner* icon. Then drag from the circle next to *delegate* to the *File's Owner* icon.

Now this picker knows that the instance of the BIDSingleComponentPickerViewController class in the nib is its data source and delegate, and will ask it to supply the data to be displayed. In other words, when the picker needs information about the data it is going to display, it asks the BIDSingleComponentPickerViewController instance that controls this view for that information.

Drag a *Round Rect Button* to the view, double-click it, and give it a title of *Select*. Press return to commit the change. In the connections inspector, drag from the circle next to *Touch Up Inside* to the *File's Owner* icon, selecting the **buttonPressed** action. Now you've finished building the GUI for the second tab. Save the nib file, and let's get back to some coding.

Implementing the Controller As a Data Source and Delegate

To make our controller work properly as the picker's data source and delegate, we'll start with some code you should feel comfortable with, and then add a few methods that you've never seen before.

Single-click *BIDSingleComponentPickerViewController.m*, and add the following code at the beginning of the file:

```
#import "BIDSingleComponentPickerViewController.h"

@implementation BIDSingleComponentPickerViewController
@synthesize singlePicker;
@synthesize pickerData;

- (IBAction)buttonPressed {
    NSInteger row = [singlePicker selectedRowInComponent:0];
    NSString *selected = [pickerData objectAtIndex:row];
    NSString *title = [[NSString alloc] initWithFormat:
                        @"You selected %@!", selected];
    UIAlertView *alert = [[UIAlertView alloc] initWithTitle:title
                                    message:@"Thank you for choosing."
                                    delegate:nil
                            cancelButtonTitle:@"You're Welcome"
                            otherButtonTitles:nil];
    [alert show];
}
.
.
.

- (void)viewDidLoad {
    [super viewDidLoad];
    // Do any additional setup after loading the view from its nib.
    NSArray *array = [[NSArray alloc] initWithObjects:@"Luke", @"Leia",
            @"Han", @"Chewbacca", @"Artoo", @"Threepio", @"Lando", nil];
    self.pickerData = array;
}
.
.
.
```

These two methods should be familiar to you by now. The `buttonPressed` method is nearly identical to the one we used with the date picker.

Unlike the date picker, a regular picker can't tell us what data it holds, because it doesn't maintain the data. It hands off that job to the delegate and data source. Instead, we need to ask the picker which row is selected, and then grab the corresponding data from our `pickerData` array. Here is how we ask it for the selected row:

```
NSInteger row = [singlePicker selectedRowInComponent:0];
```

Notice that we needed to specify which component we want to know about. We have only one component in this picker, so we simply pass in 0, which is the index of the first component.

> **NOTE:** Did you notice that there is no asterisk between `NSInteger` and `row` in our request for the selected row? Throughout most of the iOS SDK, the prefix `NS` often indicates an Objective-C class from the Foundation framework, but this is one of the exceptions to that general rule. `NSInteger` is always defined as an integer datatype, either an `int` or a `long`. We use `NSInteger` rather than `int` or `long`, because with `NSInteger`, the compiler automatically chooses whichever size is best for the platform for which we are compiling. It will create a 32-bit `int` when compiling for a 32-bit processor and a longer 64-bit `long` when compiling for a 64-bit architecture. Currently, there is no 64-bit iOS device, but who knows? Someday in the future, there likely will be. You might also write classes for your iOS applications that you'll later want to recycle and use in Cocoa applications for Mac OS X, which does run on both 32- and 64-bit machines.

In `viewDidLoad`, we create an array with several objects so that we have data to feed the picker. Usually, your data will come from other sources, like a property list in your project's *Resources* folder. By embedding a list of items in our code the way we've done here, we are making it much harder on ourselves if we need to update this list or if we want to have our application translated into other languages. But this approach is the quickest and easiest way to get data into an array for demonstration purposes. Even though you won't usually create your arrays like this, you will almost always configure some form of access to your application's model objects here in the `viewDidLoad` method, so that you're not constantly going to disk or to the network every time the picker asks you for data.

> **TIP:** If you're not supposed to create arrays from lists of objects in your code as we just did in `viewDidLoad`, how should you do it? Embed the lists in property list files, and add those files to the *Resources* folder of your project. Property list files can be changed without recompiling your source code, which means there is little risk of introducing new bugs when you do so. You can also provide different versions of the list for different languages, as you'll see in Chapter 20. Property lists can be created using the Property List Editor application (*/Developer/Applications/Utilities/Property List Editor.app*) or directly in Xcode, which offers a template for creating a property list in the *Resource* section of the new file assistant, and supports the editing of property lists in the editor pane. Both `NSArray` and `NSDictionary` offer a method called `initWithContentsOfFile:` to allow you to initialize instances from a property list file, as we'll do later in this chapter when we implement the Dependent tab.

Next, insert the following new lines of code into the existing `viewDidUnload` method:

```
- (void)viewDidUnload {
    [super viewDidUnload];
    // Release any retained subviews of the main view.
    // e.g. self.myOutlet = nil;
    self.singlePicker = nil;
    self.pickerData = nil;
}
```

Notice that we set both `singlePicker` and `pickerData` to nil. In most cases, you'll set only outlets to nil and not other properties. However, setting `pickerData` to nil is appropriate here because the `pickerData` array will be re-created each time the view is reloaded, and we want to free up that memory when the view is unloaded. Anything that is created in the `viewDidLoad` method can be flushed in `viewDidUnload` because `viewDidLoad` will fire again when the view is reloaded.

Finally, insert the following new code at the end of the file:

```
.
.
.
#pragma mark -
#pragma mark Picker Data Source Methods

- (NSInteger)numberOfComponentsInPickerView:(UIPickerView *)pickerView {
    return 1;
}

- (NSInteger)pickerView:(UIPickerView *)pickerView
numberOfRowsInComponent:(NSInteger)component {
    return [pickerData count];
}

#pragma mark Picker Delegate Methods
- (NSString *)pickerView:(UIPickerView *)pickerView
            titleForRow:(NSInteger)row
```

```
        forComponent:(NSInteger)component {
    return [pickerData objectAtIndex:row];
}
```

@end

At the bottom of the file, we get into the new methods required to implement the picker. The first two methods are from the UIPickerViewDataSource protocol, and they are both required for all pickers (except date pickers). Here's the first one:

```
- (NSInteger)numberOfComponentsInPickerView:(UIPickerView *)pickerView {
    return 1;
}
```

Pickers can have more than one spinning wheel, or component, and this is how the picker asks how many components it should display. We want to display only one list this time, so we return a value of 1. Notice that a UIPickerView is passed in as a parameter. This parameter points to the picker view that is asking us the question, which makes it possible to have multiple pickers being controlled by the same data source. In our case, we know that we have only one picker, so we can safely ignore this argument because we already know which picker is calling us.

The second data source method is used by the picker to ask how many rows of data there are for a given component:

```
- (NSInteger)pickerView:(UIPickerView *)pickerView
numberOfRowsInComponent:(NSInteger)component {
    return [pickerData count];
}
```

#PRAGMA WHAT?

Did you notice the following lines of code from *BIDSingleComponentPickerViewController.m*?

```
#pragma mark -
#pragma mark Picker Data Source Methods
```

Any line of code that begins with #pragma is technically a compiler directive. More specifically, a #pragma marks a **pragmatic**, or compiler-specific, directive that won't necessarily work with other compilers or in other environments. If the compiler doesn't recognize the directive, it ignores it, though it may generate a warning. In this case, the #pragma directives are actually directives to the IDE, not the compiler, and they tell Xcode's editor to put a break in the popup menu of methods and functions at the top of the editor pane. The first one puts the break in the menu. The second creates a text entry containing whatever the rest of the line holds, which you can use as a sort of descriptive header for groups of methods in your source code.

Some of your classes, especially some of your controller classes, are likely to get rather long, and the methods and functions popup menu makes navigating around your code much easier. Putting in #pragma directives and logically organizing your code will make that popup more efficient to use.

Once again, we are told which picker view is asking and which component that picker is asking about. Since we know that we have only one picker and one component, we don't bother with either of the arguments, and simply return the count of objects from our sole data array.

After the two data source methods, we implement one delegate method. Unlike the data source methods, all of the delegate methods are optional. The term *optional* is a bit deceiving, because you do need to implement at least one delegate method. You will usually implement the method that we are implementing here. However, if you want to display something other than text in the picker, you must implement a different method instead, as you'll see when we get to the custom picker later in this chapter.

```
- (NSString *)pickerView:(UIPickerView *)pickerView
          titleForRow:(NSInteger)row
          forComponent:(NSInteger)component {
    return [pickerData objectAtIndex:row];
}
```

In this method, the picker is asking us to provide the data for a specific row in a specific component. We are provided with a pointer to the picker that is asking, along with the component and row that it is asking about. Since our view has one picker with one component, we simply ignore everything except the row argument and use that to return the appropriate item from our data array.

Go ahead and compile and run again. When the simulator comes up, switch to the second tab—the one labeled *Single*—and check out your new custom picker, which should look like Figure 7–3.

When you're done reliving all those *Star Wars* memories, come on back to Xcode and we'll show you how to implement a picker with two components. If you feel up to a challenge, this next content view is actually a good one for you to attempt on your own. You've already seen all the methods you'll need for this picker, so go ahead and take a crack at it. We'll wait here. You might want to start off with a good look at Figure 7–4, just to refresh your memory. When you're finished, read on, and you'll see how we tackled this problem.

Implementing a Multicomponent Picker

The next content pane will have a picker with two components, or wheels, each independent of the other. The left wheel will have a list of sandwich fillings, and the right wheel will have a selection of bread types. We'll write the same data source and delegate methods that we did for the single-component picker. We'll just need to write a little additional code in some of those methods to make sure we're returning the correct value and row count for each component.

Declaring Outlets and Actions

Single-click *BIDDoubleComponentPickerViewController.h*, and add the following code:

```
#import <UIKit/UIKit.h>

#define kFillingComponent 0
#define kBreadComponent    1

@interface BIDDoubleComponentPickerViewController : UIViewController
    <UIPickerViewDelegate, UIPickerViewDataSource>

@property (strong, nonatomic) IBOutlet UIPickerView *doublePicker;
@property (strong, nonatomic) NSArray *fillingTypes;
@property (strong, nonatomic) NSArray *breadTypes;

-(IBAction)buttonPressed;
@end
```

As you can see, we start out by defining two constants that will represent the two components, which is just to make our code easier to read. Components are assigned numbers, with the leftmost component being assigned zero and increasing by one each move to the right.

Next, we conform our controller class to both the delegate and data source protocols, and we declare an outlet for the picker, as well as for two arrays to hold the data for our two picker components. After declaring properties for each of our instance variables, we declare a single action method for the button, just as we did in the previous two content panes. Save this, and click *BIDDoubleComponentPickerViewController.xib* to open the nib file for editing.

Building the View

Select the *View* icon, and use the object attributes inspector to set the *Bottom Bar* to *Tab Bar* in the *Simulated Metrics* section.

Add a picker view and a button to the view, change the button label to *Select*, and then make the necessary connections. We're not going to walk you through it this time, but you can refer to the previous section if you need a step-by-step guide, since the two applications are identical in terms of the nib file. Here's a summary of what you need to do:

1. Connect the doublePicker outlet on *File's Owner* to the picker.

2. Connect the *DataSource* and *Delegate* connections on the picker view to *File's Owner* (use the connections inspector).

3. Connect the *Touch Up Inside* event of the button to the buttonPressed action on *File's Owner* (use the connections inspector).

Make sure you save your nib and close it before you dive back into the code. Oh, and dog-ear this page (or use a bookmark, if you prefer). You'll be referring to it in a bit.

Implementing the Controller

Select *BIDDoubleComponentPickerViewController.m,* and add the following code at the top of the file:

```
#import "BIDDoubleComponentPickerViewController.h"

@implementation BIDDoubleComponentPickerViewController
@synthesize doublePicker;
@synthesize fillingTypes;
@synthesize breadTypes;

-(IBAction)buttonPressed
{
    NSInteger fillingRow = [doublePicker selectedRowInComponent:
                            kFillingComponent];
    NSInteger breadRow = [doublePicker selectedRowInComponent:
                          kBreadComponent];


    NSString *bread = [breadTypes objectAtIndex:breadRow];
    NSString *filling = [fillingTypes objectAtIndex:fillingRow];

    NSString *message = [[NSString alloc] initWithFormat:
            @"Your %@ on %@ bread will be right up.", filling, bread];

    UIAlertView *alert = [[UIAlertView alloc] initWithTitle:
                                    @"Thank you for your order"
                                            message:message
                                            delegate:nil
                                cancelButtonTitle:@"Great!"
                                otherButtonTitles:nil];
    [alert show];
}
.
.
.
```

Next, add the following lines of code to the viewDidload method:

```
- (void)viewDidLoad {
    [super viewDidLoad];
    // Do any additional setup after loading the view from its nib.
    NSArray *fillingArray = [[NSArray alloc] initWithObjects:@"Ham",
                @"Turkey", @"Peanut Butter", @"Tuna Salad",
                @"Chicken Salad", @"Roast Beef", @"Vegemite", nil];
    self.fillingTypes = fillingArray;

    NSArray *breadArray = [[NSArray alloc] initWithObjects:@"White",
        @"Whole Wheat", @"Rye", @"Sourdough", @"Seven Grain", nil];
    self.breadTypes = breadArray;
}
```

Also, add the following lines of code to the existing `viewDidUnload` method:

```
- (void)viewDidUnload {
    [super viewDidUnload];
    // Release any retained subviews of the main view.
    // e.g. self.myOutlet = nil;
    self.doublePicker = nil;
    self.breadTypes = nil;
    self.fillingTypes = nil;
}
```

And add the delegate and data source methods at the bottom:

```
.
.
.
#pragma mark -
#pragma mark Picker Data Source Methods
- (NSInteger)numberOfComponentsInPickerView:(UIPickerView *)pickerView {
    return 2;
}

- (NSInteger)pickerView:(UIPickerView *)pickerView
numberOfRowsInComponent:(NSInteger)component {
    if (component == kBreadComponent)
        return [self.breadTypes count];

    return [self.fillingTypes count];
}

#pragma mark Picker Delegate Methods
- (NSString *)pickerView:(UIPickerView *)pickerView
          titleForRow:(NSInteger)row
        forComponent:(NSInteger)component {
    if (component == kBreadComponent)
        return [self.breadTypes objectAtIndex:row];
    return [self.fillingTypes objectAtIndex:row];
}

@end
```

The `buttonPressed` method is a bit more involved this time, but there's very little there that's new to you. We just need to specify which component we are talking about when we request the selected row using those constants we defined earlier, `kBreadComponent` and `kFillingComponent`.

```
NSInteger breadRow = [doublePicker selectedRowInComponent:
        kBreadComponent];
NSInteger fillingRow = [doublePicker selectedRowInComponent:
        kFillingComponent];
```

You can see here that using the two constants instead of 0 and 1 makes our code considerably more readable. From this point on, the `buttonPressed` method is fundamentally the same as the last one we wrote.

`viewDidLoad:` is also very similar to the version we wrote for the previous picker. The only difference is that we are loading two arrays with data rather than just one array. Again, we're just creating arrays from a hard-coded list of strings—something you generally won't do in your own applications.

When we get down to the data source methods, that's where things start to change a bit. In the first method, we specify that our picker should have two components rather than just one:

```
- (NSInteger)numberOfComponentsInPickerView:(UIPickerView *)pickerView {
    return 2;
}
```

This time, when we are asked for the number of rows, we need to check which component the picker is asking about and return the correct row count for the corresponding array.

```
- (NSInteger)pickerView:(UIPickerView *)pickerView
numberOfRowsInComponent:(NSInteger)component {
    if (component == kBreadComponent)
        return [self.breadTypes count];

    return [self.fillingTypes count];
}
```

Then, in our delegate method, we do the same thing. We check the component and use the correct array for the requested component to fetch and return the correct value.

```
- (NSString *)pickerView:(UIPickerView *)pickerView
            titleForRow:(NSInteger)row
          forComponent:(NSInteger)component {
    if (component == kBreadComponent)
        return [self.breadTypes objectAtIndex:row];
    return [self.fillingTypes objectAtIndex:row];
}
```

That wasn't so hard, was it? Compile and run your application, and make sure the *Double* content pane looks like Figure 7–4.

Notice that each wheel is completely independent of the other wheel. Turning one has no effect on the other. That's appropriate in this case. But there will be times when one component is dependent on another. A good example of this is in the date picker. When you change the month, the dial that shows the number of days in the month may need to change, because not all months have the same number of days. Implementing this isn't really hard once you know how, but it's not the easiest thing to figure out on your own, so let's do that next.

Implementing Dependent Components

We're picking up steam now. For this next section, we're not going to hold your hand quite as much when it comes to material we've already covered. Instead, we'll focus on the new stuff. Our new picker will display a list of US states in the left component and a list of ZIP codes in the right component that correspond to the state currently selected in the left.

We'll need a separate list of ZIP code values for each item in the left-hand component. We'll declare two arrays, one for each component, as we did last time. We'll also need an NSDictionary. In the dictionary, we're going to store an NSArray for each state (see Figure 7–20). Later, we'll implement a delegate method that will notify us when the picker's selection changes. If the value on the left changes, we will grab the correct array out of the dictionary and assign it to the array being used for the right-hand component. Don't worry if you didn't catch all that; we'll talk about it more as we get into the code.

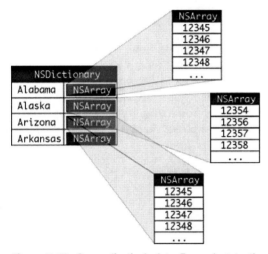

Figure 7–20. *Our application's data. For each state, there will be one entry in a dictionary with the name of the state as the key. Stored under that key will be an NSArray instance containing all the ZIP codes from that state.*

Add the following code to your *BIDDependentComponentPickerViewController.h* file:

```
#import <UIKit/UIKit.h>
#define kStateComponent    0
#define kZipComponent      1

@interface BIDDependentComponentPickerViewController : UIViewController
    <UIPickerViewDelegate, UIPickerViewDataSource>

@property (strong, nonatomic) IBOutlet UIPickerView *picker;
@property (strong, nonatomic) NSDictionary *stateZips;
@property (strong, nonatomic) NSArray *states;
@property (strong, nonatomic) NSArray *zips;

- (IBAction) buttonPressed;
@end
```

Now it's time to build the content view. That process will be almost identical to the previous two component views we built. If you get lost, flip back to the "Building the View" section for the single-component picker, and follow those step-by-step

instructions. Here's a hint: start off by opening
BIDDependentComponentPickerViewController.xib, and then repeat the same basic
steps you've done for all the other content views in this chapter. When you're finished,
save the nib.

OK, take a deep breath. Let's implement this controller class. This implementation may
seem a little gnarly at first. By making one component dependent on the other, we have
added a whole new level of complexity to our controller class. Although the picker
displays only two lists at a time, our controller class must know about and manage 51
lists. The technique we're going to use here actually simplifies that process. The data
source methods look almost identical to the one we implemented for the *DoublePicker*
view. All of the additional complexity is handled elsewhere, between `viewDidLoad` and a
new delegate method called `pickerView:didSelectRow:inComponent:`.

Before we write the code, we need some data to display. Up to now, we've created
arrays in code by specifying a list of strings. Because we didn't want you to need to type
in several thousand values, and because we figured we should show you the correct
way to do this, we're going to load the data from a property list. As we've mentioned,
both `NSArray` and `NSDictionary` objects can be created from property lists. We've
included a property list called *statedictionary.plist* in the project archive, under the *07
Pickers* folder.

Copy that file into the *Pickers* folder in your Xcode project. If you single-click the plist file in
the project window, you can see and even edit the data that it contains (see Figure 7–21).

Key	Type	Value
▶ Alabama	Array	(657 items)
▶ Alaska	Array	(251 items)
▶ Arizona	Array	(376 items)
▶ Arkansas	Array	(618 items)
▶ California	Array	(1757 items)
▶ Colorado	Array	(501 items)
▶ Connecticut	Array	(276 items)
▶ Delaware	Array	(68 items)
▶ Florida	Array	(972 items)
▶ Georgia	Array	(736 items)
▼ Hawaii	Array	(92 items)
Item 0	String	96701
Item 1	String	96703
Item 2	String	96704
Item 3	String	96705
Item 4	String	96706
Item 5	String	96707
Item 6	String	96708
Item 7	String	96710
Item 8	String	96712
Item 9	String	96713

Figure 7–21. *The statedictionary.plist file, showing our list of states. Within Hawaii, you can see the start of a list
of ZIP codes.*

Now, let's write some code. Add the following to
BIDDependentComponentPickerViewController.m, and then we'll break it down into
more digestible chunks:

```
#import "BIDDependentComponentPickerViewController.h"

@implementation BIDDependentComponentPickerViewController
@synthesize picker;
@synthesize stateZips;
@synthesize states;
@synthesize zips;

- (IBAction) buttonPressed {
    NSInteger stateRow = [picker selectedRowInComponent:kStateComponent];
    NSInteger zipRow = [picker selectedRowInComponent:kZipComponent];

    NSString *state = [self.states objectAtIndex:stateRow];
    NSString *zip = [self.zips objectAtIndex:zipRow];

    NSString *title = [[NSString alloc] initWithFormat:
                        @"You selected zip code %@.", zip];
    NSString *message = [[NSString alloc] initWithFormat:
                        @"%@ is in %@", zip, state];

    UIAlertView *alert = [[UIAlertView alloc] initWithTitle:title
                                                    message:message
                                                   delegate:nil
                                          cancelButtonTitle:@"OK"
                                          otherButtonTitles:nil];

    [alert show];
}
.
.
.
```

Then, add the following code to the existing viewDidLoad method:

```
- (void)viewDidLoad {
    [super viewDidLoad];
    // Do any additional setup after loading the view from its nib.

    NSBundle *bundle = [NSBundle mainBundle];
    NSURL *plistURL = [bundle URLForResource:@"statedictionary"
                                withExtension:@"plist"];


    NSDictionary *dictionary = [NSDictionary
                                dictionaryWithContentsOfURL:plistURL];
    self.stateZips = dictionary;

    NSArray *components = [self.stateZips allKeys];
    NSArray *sorted = [components sortedArrayUsingSelector:
                        @selector(compare:)];
    self.states = sorted;
```

```
    NSString *selectedState = [self.states objectAtIndex:0];
    NSArray *array = [stateZips objectForKey:selectedState];
    self.zips = array;
}
```

Next, add the following lines of code to the existing `viewDidUnload` method:

```
- (void)viewDidUnload {
    [super viewDidUnload];
    // Release any retained subviews of the main view.
    // e.g. self.myOutlet = nil;
    self.picker = nil;
    self.stateZips = nil;
    self.states = nil;
    self.zips = nil;
}
```

And, finally, add the delegate and data source methods at the bottom of the file:

```
    .
    .
    .
#pragma mark -
#pragma mark Picker Data Source Methods
- (NSInteger)numberOfComponentsInPickerView:(UIPickerView *)pickerView {
    return 2;
}

- (NSInteger)pickerView:(UIPickerView *)pickerView
numberOfRowsInComponent:(NSInteger)component {
    if (component == kStateComponent)
        return [self.states count];
    return [self.zips count];
}

#pragma mark Picker Delegate Methods
- (NSString *)pickerView:(UIPickerView *)pickerView
            titleForRow:(NSInteger)row
          forComponent:(NSInteger)component {
    if (component == kStateComponent)
        return [self.states objectAtIndex:row];
    return [self.zips objectAtIndex:row];
}

- (void)pickerView:(UIPickerView *)pickerView
      didSelectRow:(NSInteger)row
      inComponent:(NSInteger)component {
    if (component == kStateComponent) {
        NSString *selectedState = [self.states objectAtIndex:row];
        NSArray *array = [stateZips objectForKey:selectedState];
        self.zips = array;
        [picker selectRow:0 inComponent:kZipComponent animated:YES];
        [picker reloadComponent:kZipComponent];
    }
}

@end
```

There's no need to talk about the buttonPressed method, since it's fundamentally the same as the previous one. We should talk about the viewDidLoad method, though. There's some stuff going on there that you need to understand, so pull up a chair, and let's chat.

The first thing we do in this new viewDidLoad method is grab a reference to our application's main bundle.

```
NSBundle *bundle = [NSBundle mainBundle];
```

What is a bundle, you ask? Well, a **bundle** is just a special type of folder whose contents follow a specific structure. Applications and frameworks are both bundles, and this call returns a bundle object that represents our application.

One of the primary uses of NSBundle is to get to resources that you added to the *Resources* folder of your project. Those files will be copied into your application's bundle when you build your application. We've added resources like images to our projects, but up to now, we've used those only in Interface Builder. If we want to get to those resources in our code, we usually need to use NSBundle. We use the main bundle to retrieve the URL of the resource in which we're interested.

```
NSURL *plistURL = [bundle URLForResource:@"statedictionary"
                          withExtension:@"plist"];
```

This will return a URL containing the location of the *statedictionary.plist* file. We can then use that URL to create an NSDictionary object. Once we do that, the entire contents of that property list will be loaded into the newly created NSDictionary object, which we then assign to stateZips.

```
NSDictionary *dictionary = [NSDictionary
                          dictionaryWithContentsOfURL:plistURL];
self.stateZips = dictionary;
```

The dictionary we just loaded uses the names of the states as the keys and contains an NSArray with all the ZIP codes for that state as the values. To populate the array for the left-hand component, we get the list of all keys from our dictionary and assign those to the states array. Before we assign it, though, we sort it alphabetically.

```
NSArray *components = [self.stateZips allKeys];
NSArray *sorted = [components sortedArrayUsingSelector:
    @selector(compare:)];
self.states = sorted;
```

Unless we specifically set the selection to another value, pickers start with the first row (row 0) selected. In order to get the zips array that corresponds to the first row in the states array, we grab the object from the states array that's at index 0. That will return the name of the state that will be selected at launch time. We then use that state name to grab the array of ZIP codes for that state, which we assign to the zips array that will be used to feed data to the right-hand component.

```
NSString *selectedState = [self.states objectAtIndex:0];
NSArray *array = [stateZips objectForKey:selectedState];
self.zips = array;
```

The two data source methods are practically identical to the previous version. We return the number of rows in the appropriate array. The same is true for the first delegate method we implemented. The second delegate method is the new one, and it's where the magic happens.

```
- (void)pickerView:(UIPickerView *)pickerView
      didSelectRow:(NSInteger)row
       inComponent:(NSInteger)component {
    if (component == kStateComponent) {
        NSString *selectedState = [self.states objectAtIndex:row];
        NSArray *array = [stateZips objectForKey:selectedState];
        self.zips = array;
        [picker selectRow:0 inComponent:kZipComponent animated:YES];
        [picker reloadComponent:kZipComponent];
    }
}
```

In this method, which is called any time the picker's selection changes, we look at the component and see whether the left-hand component changed. If it did, we grab the array that corresponds to the new selection and assign it to the zips array. Then we set the right-hand component back to the first row and tell it to reload itself. By swapping the zips array whenever the state changes, the rest of the code remains pretty much the same as it was in the *DoublePicker* example.

We're not quite finished yet. Compile and run your application, and check out the *Dependent* tab (see Figure 7–22). Do you see anything there you don't like?

Figure 7–22. *Do we really want the two components to be of equal size? Notice the clipping of a long state name.*

The two components are equal in size. Even though the ZIP code will never be more than five characters long, it has been given equal billing with the state. Since state names like Mississippi and Massachusetts won't fit in half of the picker, this seems less than ideal. Fortunately, there's another delegate method we can implement to indicate how wide each component should be. We have about 295 pixels available to the picker components in portrait orientation, but for every additional component we add, we lose a little space to drawing the edges of the new component. You might need to experiment a bit with the values to get it to look right. Add the following method to the delegate section of *BIDDependentComponentPickerViewController.m*:

```
- (CGFloat)pickerView:(UIPickerView *)pickerView
    widthForComponent:(NSInteger)component {
    if (component == kZipComponent)
        return 90;
    return 200;
}
```

In this method, we return a number that represents how many pixels wide each component should be, and the picker will do its best to accommodate this. Save, compile, and run, and the picker on the *Dependent* tab will look more like the one shown in Figure 7–5.

By this point, you should be pretty darn comfortable with both pickers and tab bar applications. We have one more thing to show you about pickers, and we plan to have a little fun while doing it. Let's create a simple slot machine game.

Creating a Simple Game with a Custom Picker

Next up, we're going to create an actual working slot machine. Well, OK, it won't dispense silver dollars, but it does look pretty cool. Take a look back at Figure 7–6 before proceeding so you know what we're building.

Writing the Controller Header File

Add the following code to *BIDCustomPickerViewController.h* for starters:

```
#import <UIKit/UIKit.h>

@interface BIDCustomPickerViewController : UIViewController
      <UIPickerViewDataSource, UIPickerViewDelegate>

@property (strong, nonatomic) IBOutlet UIPickerView *picker;
@property (strong, nonatomic) IBOutlet UILabel *winLabel;
@property (strong, nonatomic) NSArray *column1;
@property (strong, nonatomic) NSArray *column2;
@property (strong, nonatomic) NSArray *column3;
@property (strong, nonatomic) NSArray *column4;
@property (strong, nonatomic) NSArray *column5;

- (IBAction)spin;
@end
```

We're declaring two outlets, one for a picker view and one for a label. The label will be used to tell users when they've won, which happens when they get three of the same symbol in a row.

We also create five pointers to NSArray objects. We'll use these to hold the image views containing the images we want the picker to draw. Even though we're using the same images in all five columns, we need separate arrays for each one with its own set of image views, because each view can be drawn in only one place in the picker at a time. We also declare an action method, this time called spin.

Building the View

Even though the picker in Figure 7–6 looks quite a bit fancier than the other ones we've built, there's actually very little difference in the way we'll design our nib. All the extra work is done in the delegate methods of our controller.

Make sure you've saved your new source code, and then select *BIDCustomPickerViewController.xib* in the project navigator to edit the GUI. Set the *Simulated Metrics* to simulate a tab bar at the bottom of the view, and then add a picker view, a label below that, and a button below that. Use the blue guideline toward the bottom of your view to place the bottom of your button, and center the label and button. Give the button a title of *Spin*.

Now, move your label so it lines up with the view's left guideline and touches the guideline below the bottom of the picker view. Next, resize the label so it goes all the way to the right guideline and down to the guideline above the top of the button.

With the label selected, bring up the attributes inspector. Set the *Alignment* to centered. Then click *Text Color* and set the color to something festive, like a bright fuchsia (we don't actually know what color that is, but it does sound festive).

Next, let's make the text a little bigger. Look for the *Font* setting in the inspector, and click the icon inside of it (it looks like the letter "T" inside a little box) to pop up the font selector. This control lets you switch from the device's standard system font to another if you like, or simply change the size. For now, just change the size to 48. After getting the text the way you want it, delete the word *Label*, since we don't want any text displayed until the first time the user wins.

After that, make all the connections to outlets and actions. You need to connect the *File's Owner*'s picker outlet to the picker view, the *File's Owner*'s winLabel outlet to the label, and the button's *Touch Up Inside* event to the spin action. After that, just make sure to specify the delegate and data source for the picker.

Oh, and there's one additional thing that you need to do. Select the picker, and bring up the attributes inspector. You need to uncheck the checkbox labeled *User Interaction Enabled* within the *View* settings so that the user can't manually change the dial and cheat. Once you've done all that, save the changes you've made to the nib file.

```
┌──────────────────────────────────────────────────────────────────────────┐
│                      Fonts Supported by iOS Devices                        │
└──────────────────────────────────────────────────────────────────────────┘
```

Be careful when using the fonts palette in Interface Builder for designing iOS interfaces. The attribute inspector's font selector will let you assign from a wide range of fonts, but not all iOS devices have the same set of fonts available. At the time of this writing, for instance, there are several fonts that are available on the iPad, but not on the iPhone or iPod touch. You should limit your font selections to one of the font families found on the iOS device you are targeting. This post on Jeff LaMarche's excellent iOS blog shows you how to grab this list programmatically: `http://iphonedevelopment.blogspot.com/2010/08/fonts-and-font-families.html`.

In a nutshell, create a view-based application and add this code to the method `application:didFinishLaunchingWithOptions:` in the application delegate:

```
for (NSString *family in [UIFont familyNames]) {
    NSLog(@"%@", family);
    for (NSString *font in [UIFont fontNamesForFamilyName:family]) {
        NSLog(@"\t%@", font);
    }
}
```

Run the project in the appropriate simulator, and your fonts will be displayed in the project's console log.

Adding Image Resources

Now we need to add the images that we'll be using in our game. We've included a set of six image files (*seven.png, bar.png, crown.png, cherry.png, lemon.png, and apple.png*) for you in the project archive under the *07 Pickers/Custom Picker Images* folder. Add all of those files to your project by dragging the entire folder into the *Pickers* folder in Xcode, just as you did for the tab bar icons. It's probably a good idea to copy them into the project folder when prompted to do so.

Implementing the Controller

We have a bunch of new stuff to cover in the implementation of this controller. Add the following code at the beginning of *BIDCustomPickerViewController.m* file:

```
#import "BIDCustomPickerViewController.h"

@implementation BIDCustomPickerViewController
@synthesize picker;
@synthesize winLabel;
@synthesize column1;
@synthesize column2;
@synthesize column3;
@synthesize column4;
@synthesize column5;

- (IBAction)spin {
    BOOL win = NO;
```

```
            int numInRow = 1;
            int lastVal = -1;
            for (int i = 0; i < 5; i++) {
                int newValue = random() % [self.column1 count];

                if (newValue == lastVal)
                    numInRow++;
                else
                    numInRow = 1;

                lastVal = newValue;
                [picker selectRow:newValue inComponent:i animated:YES];
                [picker reloadComponent:i];
                if (numInRow >= 3)
                    win = YES;
            }
            if (win)
                winLabel.text = @"WIN!";
            else
             winLabel.text = @"";
    }
    .
    .
    .
```

Then, insert the following code into the viewDidLoad method:

```
- (void)viewDidLoad {
    [super viewDidLoad];
    // Do any additional setup after loading the view from its nib.
    UIImage *seven = [UIImage imageNamed:@"seven.png"];
    UIImage *bar = [UIImage imageNamed:@"bar.png"];
    UIImage *crown = [UIImage imageNamed:@"crown.png"];
    UIImage *cherry = [UIImage imageNamed:@"cherry.png"];
    UIImage *lemon = [UIImage imageNamed:@"lemon.png"];
    UIImage *apple = [UIImage imageNamed:@"apple.png"];

    for (int i = 1; i <= 5; i++) {
        UIImageView *sevenView = [[UIImageView alloc] initWithImage:seven];
        UIImageView *barView = [[UIImageView alloc] initWithImage:bar];
        UIImageView *crownView = [[UIImageView alloc] initWithImage:crown];
        UIImageView *cherryView = [[UIImageView alloc]
                              initWithImage:cherry];
        UIImageView *lemonView = [[UIImageView alloc] initWithImage:lemon];
        UIImageView *appleView = [[UIImageView alloc] initWithImage:apple];
        NSArray *imageViewArray = [[NSArray alloc] initWithObjects:
                              sevenView, barView, crownView, cherryView,
                              lemonView, appleView, nil];

        NSString *fieldName =
            [[NSString alloc] initWithFormat:@"column%d", i];
        [self setValue:imageViewArray forKey:fieldName];
    }

    srandom(time(NULL));
```

```
}
```

Next, insert the following new lines into the `viewDidUnload` method:

```objc
- (void)viewDidUnload {
    [super viewDidUnload];
    // Release any retained subviews of the main view.
    // e.g. self.myOutlet = nil;
    self.picker = nil;
    self.winLabel = nil;
    self.column1 = nil;
    self.column2 = nil;
    self.column3 = nil;
    self.column4 = nil;
    self.column5 = nil;
}
```

Finally, add the following code to the end of the file:

```objc
    .
    .
    .
#pragma mark -
#pragma mark Picker Data Source Methods
- (NSInteger)numberOfComponentsInPickerView:(UIPickerView *)pickerView {
    return 5;
}

- (NSInteger)pickerView:(UIPickerView *)pickerView
    numberOfRowsInComponent:(NSInteger)component {
    return [self.column1 count];
}

#pragma mark Picker Delegate Methods
- (UIView *)pickerView:(UIPickerView *)pickerView
        viewForRow:(NSInteger)row
      forComponent:(NSInteger)component reusingView:(UIView *)view {
    NSString *arrayName = [[NSString alloc] initWithFormat:@"column%d",
        component+1];
    NSArray *array = [self valueForKey:arrayName];
    return [array objectAtIndex:row];
}

@end
```

There's a lot going on here, huh? Let's take the new stuff method by method.

The spin Method

The `spin` method fires when the user touches the *Spin* button. In it, we first declare a few variables that will help us keep track of whether the user has won. We'll use win to keep track of whether we've found three in a row by setting it to YES if we have. We'll use numInRow to keep track of how many of the same value we have in a row so far, and we will keep track of the previous component's value in lastVal so that we have a way to compare the current value to the previous value. We initialize `lastVal` to -1 because we know that value won't match any of the real values.

```
BOOL win = NO;
int numInRow = 1;
int lastVal = -1;
```

Next, we loop through all five components and set each one to a new, randomly generated row selection. We get the count from the `column1` array to do that, which is a shortcut we can use because we know that all five columns have the same number of values.

```
for (int i = 0; i < 5; i++) {
    int newValue = random() % [self.column1 count];
```

We compare the new value to the previous value and increment `numInRow` if it matches. If the value didn't match, we reset `numInRow` back to 1. We then assign the new value to `lastVal` so we'll have it to compare the next time through the loop.

```
if (newValue == lastVal)
    numInRow++;
else
    numInRow = 1;
lastVal = newValue;
```

After that, we set the corresponding component to the new value, telling it to animate the change, and we tell the picker to reload that component.

```
[picker selectRow:newValue inComponent:i animated:YES];
[picker reloadComponent:i];
```

The last thing we do each time through the loop is check whether we have three in a row, and set `win` to YES if we do.

```
    if (numInRow >= 3)
        win = YES;
}
```

Once we're finished with the loop, we set the label to say whether the spin was a win.

```
if (win)
    winLabel.text = @"Win!";
else
    winLabel.text = @"";
```

The viewDidLoad Method

The new version of `viewDidLoad` is somewhat scary looking, isn't it? Don't worry—once we break it down, it won't seem quite so much like the monster in your closet.

The first thing we do is load six different images. We do this using a convenience method on the `UIImage` class called `imageNamed:`.

```
UIImage *seven = [UIImage imageNamed:@"seven.png"];
UIImage *bar = [UIImage imageNamed:@"bar.png"];
UIImage *crown = [UIImage imageNamed:@"crown.png"];
UIImage *cherry = [UIImage imageNamed:@"cherry.png"];
```

```
UIImage *lemon = [UIImage imageNamed:@"lemon.png"];
UIImage *apple = [UIImage imageNamed:@"apple.png"];
```

Once we have the six images loaded, we then need to create instances of `UIImageView`, one for each image, for each of the five picker components. We do that in a loop.

```
for (int i = 1; i <= 5; i++) {
    UIImageView *sevenView = [[UIImageView alloc] initWithImage:seven];
    UIImageView *barView = [[UIImageView alloc] initWithImage:bar];
    UIImageView *crownView = [[UIImageView alloc] initWithImage:crown];
    UIImageView *cherryView = [[UIImageView alloc]
        initWithImage:cherry];
    UIImageView *lemonView = [[UIImageView alloc] initWithImage:lemon];
    UIImageView *appleView = [[UIImageView alloc] initWithImage:apple];
```

After we have the image views, we put them into an array. This array is the one that will be used to provide data to the picker for one of its five components.

```
NSArray *imageViewArray = [[NSArray alloc] initWithObjects:
                            sevenView, barView, crownView, cherryView,
                            lemonView, appleView, nil];
```

Now, we just need to assign this array to one of our five arrays. To do that, we create a string that matches the name of one of the arrays. The first time through the loop, this string will be `column1`, which is the name of the array we'll use to feed the first component in the picker. The second time through, it will equal `column2`, and so on.

```
NSString *fieldName = [[NSString alloc]
            initWithFormat:@"column%d", i];
```

Once we have the name of one of the five arrays, we can assign this array to that property using a very handy method called `setValue:forKey:`. This method lets you set a property based on its name. So, if we call this with a value of `"column1"`, it is exactly the same as calling the mutator method `setColumn1:`.

```
[self setValue:imageViewArray forKey:fieldName];
```

The last thing we do in this method is to seed the random number generator. If we don't do that, the game will play the same way every time, which gets kind of boring.

```
    srandom(time(NULL));
}
```

That wasn't so bad, was it? But, um, what do we do with those five arrays now that we've filled them with image views? If you scroll down through the code you just typed, you'll see that two data source methods look pretty much the same as before, but if you look down further into the delegate methods, you'll see that we're using a completely different delegate method to provide data to the picker. The one that we've used up to now returned an `NSString *`, but this one returns a `UIView *`.

Using this method instead, we can supply the picker with anything that can be drawn into a `UIView`. Of course, there are limitations on what will work here and look good at the same time, given the small size of the picker. But this method gives us a lot more freedom in what we display, although it is a bit more work.

```
- (UIView *)pickerView:(UIPickerView *)pickerView
    viewForRow:(NSInteger)row
```

```
        forComponent:(NSInteger)component
        reusingView:(UIView *)view {
```

This method returns one of the image views from one of the five arrays. To do that, we once again create an NSString with the name of one of the arrays. Because component is zero-indexed, we add one to it, which gives us a value between column1 and column5 and which will correspond to the component for which the picker is requesting data.

```
    NSString *arrayName = [[NSString alloc] initWithFormat:@"column%d",
        component+1];
```

Once we have the name of the array to use, we retrieve that array using a method called valueForKey:, which is the counterpart to the setValue:forKey: method that we used in viewDidLoad. Using it is the same as calling the accessor method for the property you specify. So, calling valueForKey: and specifying "column1" is the same as using the column1 accessor method. Once we have the correct array for the component, we just return the image view from the array that corresponds to the selected row.

```
    NSArray *array = [self valueForKey:arrayName];
    return [array objectAtIndex:row];
}
```

Wow, take a deep breath. You got through all of it in one piece, and now you get to take it for a spin.

Final Details

Our game is rather fun, especially when you think about how little effort it took to build it. Now let's improve it with a couple more tweaks. There are two things about this game right now that really bug us:

- It's so darn quiet. Slot machines aren't quiet!

- It tells us that we've won before the dials have finished spinning, which is a minor thing, but it does tend to eliminate the anticipation. To see this in action, run your application again. It is subtle, but the label really does appear before the wheels finish spinning.

The *07 Pickers/Custom Picker Sounds* folder in the project archive that accompanies the book contains two sound files: *crunch.wav* and *win.wav*. Add this folder to your project's *Pickers* folder. These are the sounds we'll play when the users tap the *Spin* button and when they win, respectively.

To work with sounds, we'll need access to the iOS Audio Toolbox classes. Insert this line at the top of *BIDCustomPickerViewController.m*:

```
#import <AudioToolbox/AudioToolbox.h>
```

Next, we need to add an outlet that will point to the button. While the wheels are spinning, we're going to hide the button. We don't want users tapping the button again until the current spin is all done. Add the following code to *BIDCustomPickerViewController.h*:

```
#import <UIKit/UIKit.h>

@interface BIDCustomPickerViewController : UIViewController
        <UIPickerViewDataSource, UIPickerViewDelegate>

@property (strong, nonatomic) IBOutlet UIPickerView *picker;
@property (strong, nonatomic) IBOutlet UILabel *winLabel;
@property (strong, nonatomic) NSArray *column1;
@property (strong, nonatomic) NSArray *column2;
@property (strong, nonatomic) NSArray *column3;
@property (strong, nonatomic) NSArray *column4;
@property (strong, nonatomic) NSArray *column5;
@property (strong, nonatomic) IBOutlet UIButton *button;

- (IBAction)spin;

@end
```

After you type that and save the file, click *BIDCustomPickerViewController.xib* to edit the nib. Once it's open, control-drag from *File's Owner* to the *Spin* button, and connect it to the new button outlet we just created. Save the nib.

Now, we need to do a few things in the implementation of our controller class. First, we need to synthesize the accessor and mutator for our new outlet. Open *BIDCustomPickerViewController.m* and add the following line:

```
@implementation BIDCustomPickerViewController
@synthesize picker;
@synthesize winLabel;
@synthesize column1;
@synthesize column2;
@synthesize column3;
@synthesize column4;
@synthesize column5;
@synthesize button;
.
.
.
```

We also need a couple of methods added to our controller class. Add the following two methods to *BIDCustomPickerViewController.m* as the first two methods in the class:

```
-(void)showButton {
    self.button.hidden = NO;
}

-(void)playWinSound {
    NSURL *soundURL = [[NSBundle mainBundle] URLForResource:@"win"
                                          withExtension:@"wav"];
    SystemSoundID soundID;
    AudioServicesCreateSystemSoundID((__bridge CFURLRef)soundURL, &soundID);
    AudioServicesPlaySystemSound(soundID);
    winLabel.text = @"WINNING!";
    [self performSelector:@selector(showButton) withObject:nil
        afterDelay:1.5];
}
```

The first method is used to show the button. As noted previously, we're going to hide the button when the user taps it, because if the wheels are already spinning, there's no point in letting them spin again until they've stopped.

The second method will be called when the user wins. The first line of this method asks the main bundle for the path to the sound called *win.wav*, just as we did when we loaded the property list for the Dependent picker view. Once we have the path to that resource, the next three lines of code load the sound file in and play it. Then we set the label to *WINNING!* and call the showButton method, but we call the showButton method in a special way using a method called performSelector:withObject:afterDelay:. This is a very handy method available to all objects. It lets you call the method some time in the future—in this case, one and a half seconds in the future—which will give the dials time to spin to their final locations before telling the user the result.

> **NOTE:** You may have noticed something a bit odd about the way we called the AudioServicesCreateSystemSoundID function. That function takes a URL as its first parameter, but it doesn't want an instance of NSURL. Instead, it wants a CFURLRef structure. Apple provides C interfaces to many common components—such as URLs, arrays, strings, and much more—via the Core Foundation framework. This allows even applications written entirely in C some access to the functionality that we normally use from Objective-C. The interesting thing is that these C components are "bridged" to their Objective-C counterparts, so that a CFURLRef is functionally equivalent to an NSURL pointer, for example. That means that certain kinds of objects created in Objective-C can be pushed over the bridge to use C APIs, and vice versa. This is accomplished by using a C language cast, putting the type you want your variable to be interpreted as inside parentheses before the variable name. Starting in iOS 5, with the use of ARC, the type name itself must be preceded by the keyword __bridge, which gives ARC a hint about how it should handle this Objective-C object as it passes into a C API call.

We also need to make some changes to the spin: method. We will write code to play a sound and to call the playerWon method if the player won. Make the following changes to the spin: method now:

```
- (IBAction)spin {
    BOOL win = NO;
    int numInRow = 1;
    int lastVal = -1;
    for (int i = 0; i < 5; i++) {
        int newValue = random() % [self.column1 count];

        if (newValue == lastVal)
            numInRow++;
        else
            numInRow = 1;

        lastVal = newValue;
        [picker selectRow:newValue inComponent:i animated:YES];
        [picker reloadComponent:i];
```

```
        if (numInRow >= 3)
            win = YES;
    }

    self.button.hidden = YES;
    NSString *path = [[NSBundle mainBundle] pathForResource:@"crunch"
        ofType:@"wav"];
    SystemSoundID soundID;
    AudioServicesCreateSystemSoundID(
        (__bridge CFURLRef)[NSURL fileURLWithPath:path], &soundID);
    AudioServicesPlaySystemSound (soundID);

    if (win)
        [self performSelector:@selector(playWinSound)
            withObject:nil
            afterDelay:.5];
    else
        [self performSelector:@selector(showButton)
            withObject:nil
            afterDelay:.5];

    winLabel.text = @"";

    if (win)
        winLabel.text = @"WIN!";
    else
        winLabel.text = @"";
}
```

The first line of code we added hides the *Spin* button. The next four lines play a sound to let the player know they've spun the wheels. Then, instead of setting the label to *WIN!* as soon as we know the user has won, we do something tricky. We call one of the two methods we just created, but we do it after a delay using performSelector:afterDelay:. If the user won, we call our playerWon method half a second into the future, which will give time for the dials to spin into place; otherwise, we just wait a half a second and reenable the *Spin* button.

The only thing left to do is to release our button outlet, so make the following changes to your viewDidUnload method:

```
- (void)viewDidUnload {
    [super viewDidUnload];
    // Release any retained subviews of the main view.
    // e.g. self.myOutlet = nil;
    self.picker = nil;
    self.winLabel = nil;
    self.column1 = nil;
    self.column2 = nil;
    self.column3 = nil;
    self.column4 = nil;
    self.column5 = nil;
    self.button = nil;
}
```

Linking in the Audio Toolbox Framework

If you try to compile now, you'll get a linking error. It turns out that the problem is with those functions we called to load and play sounds. Yeah, they're not in any of the frameworks that are linked in by default. A quick command-double-click on the AudioServicesCreateSystemSoundID function takes us to the header file where it's declared. If we scroll up to the top of that header file, we see this:

```
/*==========================================================================
      File: AudioToolbox/AudioServices.h

      Contains: API for general high level audio services.

      Copyright: (c) 2006 - 2008 by Apple Inc., all rights reserved.
...
```

This tells us that the function we're trying to call is part of the Audio Toolbox, so we need to manually link our project to that framework.

This is pretty easy to do. In the project navigator, click the *Pickers* target, which should be the top icon in our list. In the edit pane that appears, find the *TARGETS* area and click *Pickers*. In the pane that appears, click the *Build Phases* tab. In the *Build Phases* tab, open the *Link Binary With Libraries* disclosure triangle. Notice the *Add items* icon, which is a plus sign (see Figure 7–23).

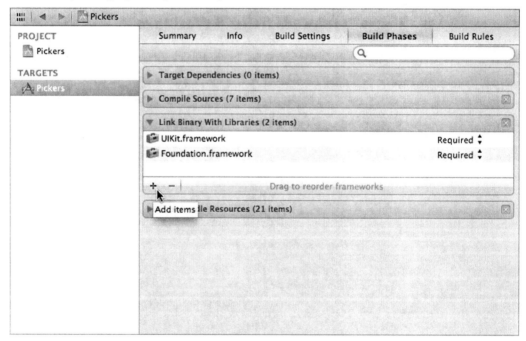

Figure 7–23. *To add a framework to your project, select the target in the project navigator, and then select the appropriate target in the pane that appears. Finally, select the Build Phases tab and open the Link Binary With Libraries disclosure triangle. Note that the cursor is hovering over the Add items plus sign icon.*

Click the *Add items* icon. A sheet will drop down, listing the available frameworks. Select *AudioToolbox.framework* and click the *Add* button (see Figure 7–24).

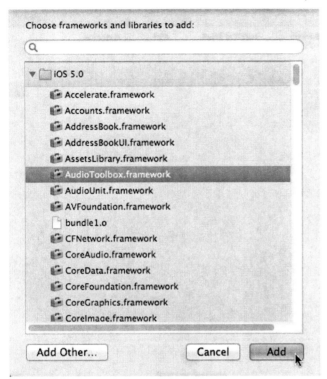

Figure 7–24. *Adding a framework to your project*

Your application should now link properly and, when it runs, the *Spin* button should play one sound, and a win should produce a winning sound. Hooray!

Final Spin

By now, you should be comfortable with tab bar applications and pickers. In this chapter, we built a full-fledged tab bar application containing five different content views from scratch. You learned how to use pickers in a number of different configurations, how to create pickers with multiple components, and even how to make the values in one component dependent on the value selected in another component. You also saw how to make the picker display images rather than just text.

Along the way, you learned about picker delegates and data sources, and saw how to load images, play sounds, create dictionaries from property lists, and link your project to additional frameworks. It was a long chapter, so congratulations on making it through! When you're ready to tackle table views, turn the page, and we'll keep going.

Chapter 8

Introduction to Table Views

In the next chapter, we're going to build a hierarchical navigation-based application similar to the Mail application that ships on iOS devices. Our application will allow the user to drill down into nested lists of data and edit that data. But before we can build that application, you need to master the concept of table views. And that's the goal of this chapter.

Table views are the most common mechanism used to display lists of data to the user. They are highly configurable objects that can be made to look practically any way you want them to look. Mail uses table views to show lists of accounts, folders, and messages, but table views are not limited to just the display of textual data. Table views are also used in the YouTube, Settings, and iPod applications, even though these applications have very different appearances (see Figure 8-1).

Figure 8–1. *Though they all look different, the Settings, iPod, and YouTube applications use table views to display their data.*

Table View Basics

Tables display lists of data. Each item in a table's list is a row. iOS tables can have an unlimited number of rows, constrained only by the amount of available memory. iOS tables can be only one column wide.

Table Views and Table View Cells

A table view is the view object that displays a table's data and is an instance of the class UITableView. Each visible row of the table is implemented by the class UITableViewCell. So, a table view is the object that displays the visible part of a table, and a table view cell is responsible for displaying a single row of the table (see Figure 8–2).

Figure 8–2. *Each table view is an instance of UITableView, and each visible row is an instance of UITableViewCell.*

Table views are not responsible for storing your table's data. They store only enough data to draw the rows that are currently visible. Table views get their configuration data from an object that conforms to the UITableViewDelegate protocol and their row data from an object that conforms to the UITableViewDataSource protocol. You'll see how all this works when we get into our sample programs later in the chapter.

As mentioned, all tables are implemented as a single column. But the YouTube application, shown on the right side of Figure 8–1, does give the appearance of having at least two columns, perhaps even three if you count the icons. But no, each row in the table is represented by a single UITableViewCell. Each UITableViewCell object can be configured with an image, some text, and an optional accessory icon, which is a small icon on the right side (we'll cover accessory icons in detail in the next chapter).

You can put even more data in a cell if you need to by adding subviews to UITableViewCell, using one of two basic techniques: adding subviews programmatically when creating the cell, or loading them from a nib file. You can lay out the table view cell out in any way you like and include any subviews you desire. So, the single-column limitation is far less limiting than it probably sounds at first. If this is confusing, don't worry—we'll show you how to use both of these techniques in this chapter.

Grouped and Plain Tables

Table views come in two basic styles:

■ **Grouped**: Each group in a grouped table is a set of rows embedded in a rounded rectangle, as shown in the leftmost picture in Figure 8–3. Note that a grouped table can consist of a single group.

■ **Plain**: Plain is the default style. Any table that doesn't feature rounded rectangles is a plain table view. When an index is used, this style is also referred to as **indexed**.

If your data source provides the necessary information, the table view will let the user navigate your list using an index that is displayed down the right side. Figure 8–3 shows a grouped table, a plain table without an index, and a plain table with an index (an indexed table).

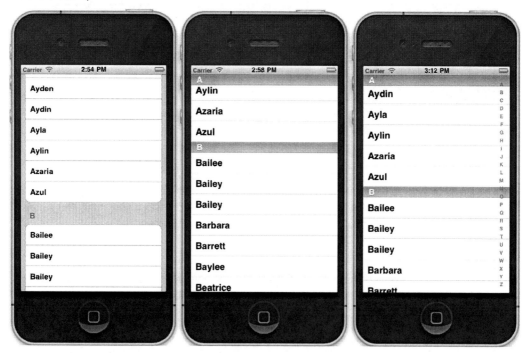

Figure 8–3. *The same table view displayed as a grouped table (left), a plain table without an index (middle), and a plain table with an index, also called an indexed table (right)*

Each division of your table is known to your data source as a **section**. In a grouped table, each group is a section. In an indexed table, each indexed grouping of data is a section. For example, in the indexed table shown in Figure 8–3, all the names beginning with *A* would be one section, those beginning with *B* another, and so on.

Sections have two primary purposes. In a grouped table, each section represents one group. In an indexed table, each section corresponds to one index entry. For example, if you wanted to display a list indexed alphabetically with an index entry for every letter, you would have 26 sections, each containing all the values that begin with a particular letter.

> **CAUTION:** Even though it is technically possible to create a grouped table with an index, you should not do so. The *iPhone Human Interface Guidelines* specifically state that grouped tables should not provide indexes.

Implementing a Simple Table

Let's look at the simplest possible example of a table view to get a feel for how it works. In this example, we're just going to display a list of text values.

Create a new project in Xcode. For this chapter, we're going back to the *Single View Application* template, so select that one. Call your project *Simple Table*, enter *BID* as the *Class Prefix*, and set the *Device Family* to *iPhone*. Be sure the *Use Storyboard* and *Include Unit Tests* checkboxes are unchecked.

Designing the View

In the project navigator, expand the *Simple Table* project and the *Simple Table* folder. This is such a simple application that we're not going to need any outlets or actions. Go ahead and select *BIDViewController.xib* to edit the GUI. If the *View* window isn't visible in the layout area, single-click its icon in the dock to open it. Then look in the object library for a *Table View* (see Figure 8–4), and drag that over to the *View* window.

The table view should automatically size itself to the height and width of the view. This is exactly what we want. Table views are designed to fill the entire width of the screen and as much of the height as isn't taken up by your application's navigation bars, toolbars, and tab bars.

After dropping the table view onto the *View* window and fitting it just below the status bar, it should still be selected. If it's not, single-click the table view to select it. Then press ⌥⌘6 to bring up the connections inspector. You'll notice that the first two available connections for the table view are the same as the first two for the picker view: *dataSource* and *delegate*. Drag from the circle next to each of those connections over to the *File's Owner* icon. By doing this, we are making our controller class both the data source and delegate for this table.

After setting the connections, save your nib file and get ready to dig into some UITableView code.

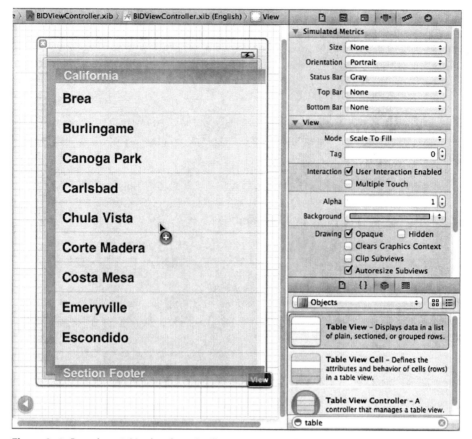

Figure 8–4. *Dragging a table view from the library onto our main view. Notice that the table view automatically resizes to the full size of the view.*

Writing the Controller

The next stop is our controller class's header file. Single-click *BIDViewController.h*, and add the following code:

```
#import <UIKit/UIKit.h>

@interface BIDViewController : UIViewController
      <UITableViewDelegate, UITableViewDataSource>

@property (strong, nonatomic) NSArray *listData;
@end
```

All we're doing here is conforming our class to the two protocols that are needed for it to act as the delegate and data source for the table view, and then declaring an array that will hold the data to be displayed.

Save your changes. Next, switch over to *BIDViewController.m*, and add the following code at the beginning of the file:

```objc
#import "BIDViewController.h"

@implementation BIDViewController
@synthesize listData;
.
.
.
- (void)viewDidLoad {
    [super viewDidLoad];
    // Do any additional setup after loading the view, typically from a nib.
    NSArray *array = [[NSArray alloc] initWithObjects:@"Sleepy", @"Sneezy",
        @"Bashful", @"Happy", @"Doc", @"Grumpy", @"Dopey", @"Thorin",
        @"Dorin", @"Nori", @"Ori", @"Balin", @"Dwalin", @"Fili", @"Kili",
        @"Oin", @"Gloin", @"Bifur", @"Bofur", @"Bombur", nil];
    self.listData = array;
}
.
.
.
```

Now, add the following line of code to the existing viewDidUnload method:

```objc
- (void)viewDidUnload {
    [super viewDidUnload];
    // Release any retained subviews of the main view.
    // e.g. self.myOutlet = nil;
    self.listData = nil;
}
```

Finally, add the following code at the end of the file:

```objc
.
.
.
#pragma mark -
#pragma mark Table View Data Source Methods
- (NSInteger)tableView:(UITableView *)tableView
    numberOfRowsInSection:(NSInteger)section {
    return [self.listData count];
}

- (UITableViewCell *)tableView:(UITableView *)tableView
        cellForRowAtIndexPath:(NSIndexPath *)indexPath {

    static NSString *SimpleTableIdentifier = @"SimpleTableIdentifier";

    UITableViewCell *cell = [tableView dequeueReusableCellWithIdentifier:
        SimpleTableIdentifier];
    if (cell == nil) {
        cell = [[UITableViewCell alloc]
            initWithStyle:UITableViewCellStyleDefault
            reuseIdentifier:SimpleTableIdentifier];
    }

    NSUInteger row = [indexPath row];
    cell.textLabel.text = [listData objectAtIndex:row];
```

```
        return cell;
}
```

`@end`

We added three methods to the controller. You should be comfortable with the first one, `viewDidLoad`, since we've done similar things in the past. We're simply creating an array of data to pass to the table. In a real application, this array would likely come from another source, such as a text file, property list, or URL.

If you scroll down to the end, you can see we added two data source methods. The first one, `tableView:numberOfRowsInSection:`, is used by the table to ask how many rows are in a particular section. As you might expect, the default number of sections is one, and this method will be called to get the number of rows in the one section that makes up the list. We just return the number of items in our array.

The next method probably requires a little explanation, so let's look at it more closely.

```
- (UITableViewCell *)tableView:(UITableView *)tableView
        cellForRowAtIndexPath:(NSIndexPath *)indexPath {
```

This method is called by the table view when it needs to draw one of its rows. Notice that the second argument to this method is an `NSIndexPath` instance. This is the mechanism that table views use to wrap the section and row into a single object. To get the row or the section out of an `NSIndexPath`, you just call either its `row` method or its `section` method, both of which return an `int`.

The first parameter, `tableView`, is a reference to the table doing the asking. This allows us to create classes that act as a data source for multiple tables.

Next, we declare a static string instance.

```
static NSString *SimpleTableIdentifier = @"SimpleTableIdentifier";
```

This string will be used as a key to represent the type of our table cell. Our table will use only a single type of cell.

A table view can display only a few rows at a time on the iPhone's small screen, but the table itself can conceivably hold considerably more. Remember that each row in the table is represented by an instance of `UITableViewCell`, a subclass of `UIView`, which means each row can contain subviews. With a large table, this could represent a huge amount of overhead if the table were to try to keep one table view cell instance for every row in the table, regardless of whether that row was currently being displayed. Fortunately, tables don't work that way.

Instead, as table view cells scroll off the screen, they are placed into a queue of cells available to be reused. If the system runs low on memory, the table view will get rid of the cells in the queue. But as long as the system has some memory available for those cells, it will hold on to them in case you want to use them again.

Every time a table view cell rolls off the screen, there's a pretty good chance that another one just rolled onto the screen on the other side. If that new row can just reuse one of the cells that has already rolled off the screen, the system can avoid the overhead associated with constantly creating and releasing those views. To take advantage of this mechanism, we'll ask the table view to give us a previously used cell of the specified type. Note that we're making use of the NSString identifier we declared earlier. In effect, we're asking for a reusable cell of type SimpleTableIdentifier.

```
UITableViewCell *cell = [tableView dequeueReusableCellWithIdentifier:
    SimpleTableIdentifier];
```

Now, it's completely possible that the table view won't have any spare cells (when it's being initially populated, for example), so we check cell after the call to see whether it's nil. If it is, we manually create a new table view cell using that identifier string. At some point, we'll inevitably reuse one of the cells we create here, so we need to make sure that we create it using SimpleTableIdentifier.

```
if (cell == nil) {
    cell = [[UITableViewCell alloc]
        initWithStyle:UITableViewCellStyleDefault
        reuseIdentifier:SimpleTableIdentifier];
}
```

Curious about UITableViewCellStyleDefault? Hold that thought. We'll get to it when we look at the table view cell styles.

We now have a table view cell that we can return for the table view to use. So, all we need to do is place whatever information we want displayed in this cell. Displaying text in a row of a table is a very common task, so the table view cell provides a UILabel property called textLabel that we can set in order to display strings. That just requires getting the correct string from our listData array and using it to set the cell's textLabel.

To get the correct value, however, we need to know which row the table view is asking for. We get that information from the indexPath variable, like so:

```
NSUInteger row = [indexPath row];
```

We use the row number of the table to get the corresponding string from the array, assign it to the cell's textLabel.text property, and then return the cell.

```
    cell.textLabel.text = [listData objectAtIndex:row];
    return cell;
}
```

That wasn't so bad, was it? Compile and run your application, and you should see the array values displayed in a table view (see Figure 8–5).

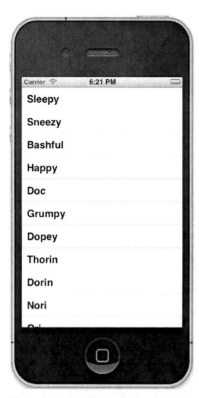

Figure 8–5. *The Simple Table application, in all its dwarven glory*

Adding an Image

It would be nice if we could add an image to each row. Guess we would need to create a subclass of UITableViewCell or add subviews in order to do that, huh? Actually, no, not if you can live with the image being on the left side of each row. The default table view cell can handle that situation just fine. Let's check it out.

In the project archive, in the *08 - Simple Table* folder, grab the file called *star.png*, and add it to your project's *Simple Table* folder. *star.png* is a small icon we prepared just for this project.

Next, let's get to the code. In the file *BIDViewController.m*, add the following code to the tableView:cellForRowAtIndexPath: method:

```
- (UITableViewCell *)tableView:(UITableView *)tableView
        cellForRowAtIndexPath:(NSIndexPath *)indexPath {

    static NSString *SimpleTableIdentifier = @" SimpleTableIdentifier ";

    UITableViewCell *cell = [tableView dequeueReusableCellWithIdentifier:
                                SimpleTableIdentifier];
    if (cell == nil) {
        cell = [[UITableViewCell alloc]
```

```
                    initWithStyle:UITableViewCellStyleDefault
                    reuseIdentifier:SimpleTableIdentifier];
    }

    UIImage *image = [UIImage imageNamed:@"star.png"];
    cell.imageView.image = image;

    NSUInteger row = [indexPath row];
    cell.textLabel.text = [listData objectAtIndex:row];

    return cell;
}
@end
```

Yep, that's it. Each cell has an imageView property. Each imageView has an image property, as well as a highlightedImage property. The image appears to the left of the cell's text and is replaced by the highlightedImage, if one is provided, when the cell is selected. You just set the cell's imageView.image property to whatever image you want to display.

If you compile and run your application now, you should get a list with a bunch of nice little star icons to the left of each row (see Figure 8–6). Of course, we could have included a different image for each row in the table. Or, with very little effort, we could have used one icon for all of Mr. Disney's dwarves and a different one for Mr. Tolkien's.

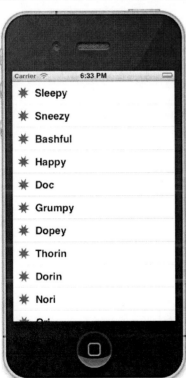

Figure 8–6. We used the cell's image property to add an image to each of the table view's cells.

If you like, make a copy of *star.png*, use your favorite graphics application to colorize it a bit, add it to the project, load it with `imageNamed:`, and use it to set `imageView.highlightedImage`. Now if you click a cell, your new image will be drawn. If you don't feel like coloring, use the *star2.png* icon we provided in the project archive.

> **NOTE:** `UIImage` uses a caching mechanism based on the file name, so it won't load a new image property each time `imageNamed:` is called. Instead, it will use the already cached version.

Using Table View Cell Styles

The work you've done with the table view so far has made use of the default cell style shown in Figure 8–6, represented by the constant `UITableViewCellStyleDefault`. But the `UITableViewCell` class includes several other predefined cell styles that let you easily add a bit more variety to your table views. These cell styles make use of three different cell elements:

- **Image**: If an image is part of the specified style, the image is displayed to the left of the cell's text.

- **Text label**: This is the cell's primary text. In the style we used earlier, `UITableViewCellStyleDefault`, the text label is the only text shown in the cell.

- **Detail text label**: This is the cell's secondary text, usually used as an explanatory note or label.

To see what these new style additions look like, add the following code to `tableView:cellForRowAtIndexPath:` in *BIDViewController.m*:

```objc
- (UITableViewCell *)tableView:(UITableView *)tableView
        cellForRowAtIndexPath:(NSIndexPath *)indexPath {

    static NSString *SimpleTableIdentifier = @"SimpleTableIdentifier";

    UITableViewCell *cell = [tableView dequeueReusableCellWithIdentifier:
                            SimpleTableIdentifier];
    if (cell == nil) {
        cell = [[UITableViewCell alloc]
            initWithStyle:UITableViewCellStyleDefault
            reuseIdentifier: SimpleTableIdentifier];
    }

    UIImage *image = [UIImage imageNamed:@"star.png"];
    cell.imageView.image = image;

    NSUInteger row = [indexPath row];
    cell.textLabel.text = [listData objectAtIndex:row];

    if (row < 7)
```

```
        cell.detailTextLabel.text = @"Mr. Disney";
    else
        cell.detailTextLabel.text = @"Mr. Tolkien";

    return cell;
}
```

All we've done here is set the cell's detail text. We use the string `@"Mr. Disney"` for the first seven rows and `@"Mr. Tolkien"` for the rest. When you run this code, each cell will look just the same as it did before (see Figure 8–7). That's because we are using the style `UITableViewCellStyleDefault`, which does not make use of the detail text.

Figure 8–7. *The default cell style shows the image and text label in a straight row.*

Now, change `UITableViewCellStyleDefault` to `UITableViewCellStyleSubtitle` and run the app again. With the subtitle style, both text elements are shown, one below the other (see Figure 8–8).

Figure 8–8. *The subtitle style shows the detail text in smaller, gray letters below the text label.*

Change `UITableViewCellStyleSubtitle` to `UITableViewCellStyleValue1`, and then build and run. This style places the text label and detail text label on the same line on opposite sides of the cell (see Figure 8–9).

Figure 8–9. *The style value 1 will place the text label on the left side in black letters and the detail text right-justified on the right side in blue letters.*

Finally, change `UITableViewCellStyleValue1` to `UITableViewCellStyleValue2`. This format is often used to display information along with a descriptive label. It doesn't show the cell's icon, but places the detail text label to the left of the text label (see Figure 8–10). In this layout, the detail text label acts as a label describing the type of data held in the text label.

Sneezy **Mr. Disney**

Figure 8–10. *The style value 2 does not display the image and places the detail text label in blue letters to the left of the text label.*

Now that you've seen the cell styles that are available, go ahead and change back to using `UITableViewCellStyleDefault` before continuing. Later in this chapter, you'll see how to customize the appearance of your table. But before you decide to do that, make sure you consider the available styles to see whether one of them will suit your needs.

You may have noticed that we made our controller both the data source and delegate for this table view, but up to now, we haven't actually implemented any of the methods from `UITableViewDelegate`. Unlike picker views, simpler table views don't require the use of a delegate in order to do their thing. The data source provides all the data needed to draw the table. The purpose of the delegate is to configure the appearance of the table view and to handle certain user interactions. Let's take a look at a few of the configuration options now. We'll discuss a few more in the next chapter.

Setting the Indent Level

The delegate can be used to specify that some rows should be indented. In the file *BIDViewController.m*, add the following method to your code, just above the @end declaration:

```
#pragma mark -
#pragma mark Table Delegate Methods

- (NSInteger)tableView:(UITableView *)tableView
    indentationLevelForRowAtIndexPath:(NSIndexPath *)indexPath {
        NSUInteger row = [indexPath row];
    return row;
}
```

This method sets the **indent level** for each row to its row number, so row 0 will have an indent level of 0, row 1 will have an indent level of 1, and so on. An indent level is simply an integer that tells the table view to move that row a little to the right. The higher the number, the further to the right the row will be indented. You might use this technique, for example, to indicate that one row is subordinate to another row, as Mail does when representing subfolders.

When you run the application again, you can see that each row is now drawn a little farther to the right than the last one (see Figure 8–11).

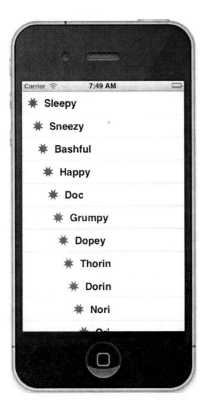

Figure 8–11. *Each row of the table is drawn with an indent level higher than the row before it.*

Handling Row Selection

The table's delegate can use two methods to determine if the user has selected a particular row. One method is called before the row is selected and can be used to prevent the row from being selected, or even change which row gets selected. Let's implement that method and specify that the first row is not selectable. Add the following method to the end of *BIDViewController.m*, just before the @end declaration:

```
- (NSIndexPath *)tableView:(UITableView *)tableView
      willSelectRowAtIndexPath:(NSIndexPath *)indexPath {
   NSUInteger row = [indexPath row];

   if (row == 0)
      return nil;

   return indexPath;
}
```

This method is passed indexPath, which represents the item that's about to be selected. Our code looks at which row is about to be selected. If the row is the first row, which is always index zero, then it returns nil, which indicates that no row should actually be selected. Otherwise, it returns indexPath, which is how we indicate that it's OK for the selection to proceed.

Before you compile and run, let's also implement the delegate method that is called after a row has been selected, which is typically where you'll actually handle the selection. This is where you take whatever action is appropriate when the user selects a row. In the next chapter, we'll use this method to handle the drill-downs, but in this chapter, we'll just put up an alert to show that the row was selected. Add the following method to the bottom of *BIDViewController.m*, just before the @end declaration again:

```
- (void)tableView:(UITableView *)tableView
        didSelectRowAtIndexPath:(NSIndexPath *)indexPath {
    NSUInteger row = [indexPath row];
    NSString *rowValue = [listData objectAtIndex:row];

    NSString *message = [[NSString alloc] initWithFormat:
        @"You selected %@", rowValue];
    UIAlertView *alert = [[UIAlertView alloc]
        initWithTitle:@"Row Selected!"
            message:message
            delegate:nil
    cancelButtonTitle:@"Yes I Did"
    otherButtonTitles:nil];
    [alert show];

    [tableView deselectRowAtIndexPath:indexPath animated:YES];
}
```

Once you've added this method, compile and run the app, and take it for a spin. See whether you can select the first row (you shouldn't be able to), and then select one of the other rows. The selected row should be highlighted, and then your alert should pop up, telling you which row you selected, while the selected row fades in the background (see Figure 8–12).

Figure 8–12. *In this example, the first row is not selectable, and an alert is displayed when any other row is selected. This was done using the delegate methods.*

Note that you can also modify the index path before you pass it back, which would cause a different row and/or section to be selected. You won't do that very often, as you should have a very good reason for changing the user's selection. In the vast majority of cases, when you use this method, you will either return indexPath unmodified to allow the selection or return nil to or disallow it.

Changing the Font Size and Row Height

Let's say that we want to change the size of the font being used in the table view. In most situations, you shouldn't override the default font; it's what users expect to see. But sometimes there are valid reasons to change the font. Add the following line of code to your tableView:cellForRowAtIndexPath: method:

```
- (UITableViewCell *)tableView:(UITableView *)tableView
        cellForRowAtIndexPath:(NSIndexPath *)indexPath
{
    static NSString *SimpleTableIdentifier = @"SimpleTableIdentifier";

    UITableViewCell *cell = [tableView dequeueReusableCellWithIdentifier:
                            SimpleTableIdentifier];
    if (cell == nil) {
```

```
        cell = [[UITableViewCell alloc]
            initWithStyle:UITableViewCellStyleDefault
            reuseIdentifier: SimpleTableIdentifier];
    }

    UIImage *image = [UIImage imageNamed:@"star.png"];
    cell.image = image;

    NSUInteger row = [indexPath row];
    cell.textLabel.text = [listData objectAtIndex:row];
    cell.textLabel.font = [UIFont boldSystemFontOfSize:50];

    if (row < 7)
        cell.detailTextLabel.text = @"Mr. Disney";
    else
        cell.detailTextLabel.text = @"Mr. Tolkein";
    return cell;
}
```

When you run the application now, the values in your list are drawn in a really large font size, but they don't exactly fit in the row (see Figure 8–13).

Figure 8–13. *Look how nice and big! But, um, it would be even nicer if we could see everything.*

Well, here comes the table view delegate to the rescue! The table view delegate can specify the height of the table view's rows. In fact, it can specify unique values for each row if you find that necessary. Go ahead and add this method to your controller class, just before @end:

```
- (CGFloat)tableView:(UITableView *)tableView
    heightForRowAtIndexPath:(NSIndexPath *)indexPath {
    return 70;
}
```

We've just told the table view to set the row height for all rows to 70 pixels tall. Compile and run, and your table's rows should be much taller now (see Figure 8–14).

Figure 8–14. *Changing the row size using the delegate*

There are more tasks that the delegate handles, but most of the remaining ones come into play when you start working with hierarchical data, which we'll do in the next chapter. To learn more, use the documentation browser to explore the UITableViewDelegate protocol and see what other methods are available.

Customizing Table View Cells

You can do a lot with table views right out of the box, but often, you will want to format the data for each row in ways that simply aren't supported by UITableViewCell directly. In those cases, there are two basic approaches: one that involves adding subviews to

UITableViewCell programmatically when creating the cell, and a second that involves loading a set of subviews from a nib file. Let's look at both techniques.

Adding Subviews to the Table View Cell

To show how to use custom cells, we're going to create a new application with another table view. In each row, we'll display two lines of information along with two labels (see Figure 8–15). Our application will display the name and color of a series of potentially familiar computer models, and we'll show both of those pieces of information in the same table cell by adding subviews to the table view cell.

Figure 8–15. *Adding subviews to the table view cell can give you multiline rows.*

Create a new Xcode project using the *Single View Application* template. Name the project *Cells*, and use the same settings as your last project. Click *BIDViewController.xib* to edit the nib file in Interface Builder.

Add a *Table View* to the main view, use the connections inspector to set its delegate and data source to *File's Owner* as we did for the Simple Table application, and then save the nib.

Creating a UITableViewCell Subclass

Up until this point, the standard table view cells we've been using have taken care of all the details of cell layout for us. Our controller code has been kept clear of the messy details about where to place labels and images, and has been able to just pass off the display values to the cell. This keeps presentation logic out of the controller, and that's a really good design to stick to. For this project, we're going to make a new cell subclass of our own that takes care of the details for the new layout, which will keep our controller as simple as possible.

Adding New Cells

Select the *Cells* folder in the project navigator, and press ⌘N to create a new file. In the assistant that pops up, choose *Objective-C class* from the *Cocoa Touch* section, and click *Next*. On the following screen, enter *BIDNameAndColorCell* as the name of the new class, select *UITableViewCell* in the *Subclass of* popup list, and click *Next* again. On the final screen, select the *Cells* folder that already contains your other source code, make sure *Cells* is chosen both in the *Group* and *Target* controls at the bottom, and click *Create*.

Select *BIDNameAndColorCell.h*, and add the following code:

```
#import <UIKit/UIKit.h>
@interface BIDNameAndColorCell : UITableViewCell

@property (copy, nonatomic) NSString *name;
@property (copy, nonatomic) NSString *color;

@end
```

Here, we've defined two properties that our controller will use to pass values to each cell. Note that instead of declaring the NSString properties with strong semantics, we're using copy. Doing so with NSString values is always a good idea, because there's a risk that the string value passed in to a property setter may actually be an NSMutableString, which the sender can modify later on, leading to problems. Copying each string that's passed in to a property gives us a stable, unchangeable snapshot of what the string contains at the moment the setter is called.

That's all we need to expose in the header, so let's move on to *BIDNameAndColorCell.m*. Add the following code at the top of the file:

```
#import "BIDNameAndColorCell.h"

#define kNameValueTag    1
#define kColorValueTag   2

@implementation BIDNameAndColorCell

@synthesize name;
@synthesize color;
    .
    .
    .
```

Notice that we've defined two constants. We're going to use these to assign tags to some of the subviews that we'll be adding to the table view cell. We'll add four subviews to the cell, and two of those need to be changed for every row. In order to do that, we need some mechanism that will allow us to retrieve the two fields from the cell when we go to update that cell with a particular row's data. If we set unique tag values for each label that we'll use again, we'll be able to retrieve them from the table view cell and set their values. We also declared the name and color properties, which the controller will use to set the values that should appear in the cell.

Now, edit the existing initWithStyle:reuseIdentifier: method to create the views that we'll need to display.

```
- (id)initWithStyle:(UITableViewCellStyle)style reuseIdentifier:(NSString
*)reuseIdentifier
{
    self = [super initWithStyle:style reuseIdentifier:reuseIdentifier];
    if (self) {
        // Initialization code
        CGRect nameLabelRect = CGRectMake(0, 5, 70, 15);
        UILabel *nameLabel = [[UILabel alloc] initWithFrame:nameLabelRect];
        nameLabel.textAlignment = UITextAlignmentRight;
        nameLabel.text = @"Name:";
        nameLabel.font = [UIFont boldSystemFontOfSize:12];
        [self.contentView addSubview: nameLabel];

        CGRect colorLabelRect = CGRectMake(0, 26, 70, 15);
        UILabel *colorLabel = [[UILabel alloc] initWithFrame:colorLabelRect];
        colorLabel.textAlignment = UITextAlignmentRight;
        colorLabel.text = @"Color:";
        colorLabel.font = [UIFont boldSystemFontOfSize:12];
        [self.contentView addSubview: colorLabel];

        CGRect nameValueRect = CGRectMake(80, 5, 200, 15);
        UILabel *nameValue = [[UILabel alloc] initWithFrame:
                                 nameValueRect];
        nameValue.tag = kNameValueTag;
        [self.contentView addSubview:nameValue];

        CGRect colorValueRect = CGRectMake(80, 25, 200, 15);
        UILabel *colorValue = [[UILabel alloc] initWithFrame:
                                 colorValueRect];
        colorValue.tag = kColorValueTag;
        [self.contentView addSubview:colorValue];
    }
    return self;
}
```

That should be pretty straightforward. We create four UILabels and add them to the table view cell. The table view cell already has a UIView subview called contentView, which it uses to group all of its subviews, much the way we grouped those two switches inside a UIView back in Chapter 4. As a result, we don't add the labels as subviews directly to the table view cell, but rather to its contentView.

```
[self.contentView addSubview:colorValue];
```

Two of these labels contain static text. The label nameLabel contains the text *Name:*, and the label colorLabel contains the text *Color:*. Those are just labels that we won't change. But we'll use the other two labels to display our row-specific data. Remember that we need some way of retrieving these fields later on, so we assign values to both of them. For example, we assign the constant kNameValueTag to nameValue's tag field.

```
nameValue.tag = kNameValueTag;
```

Now, to put the finishing touches on the BIDNameAndColorCell class, add these two setter methods just before the @end:

```
- (void)setName:(NSString *)n {
    if (![n isEqualToString:name]) {
        name = [n copy];
        UILabel *nameLabel = (UILabel *)[self.contentView viewWithTag:
                                         kNameValueTag];
        nameLabel.text = name;
    }
}

- (void)setColor:(NSString *)c {
    if (![c isEqualToString:color]) {
        color = [c copy];
        UILabel *colorLabel = (UILabel *)[self.contentView viewWithTag:
                                          kColorValueTag];
        colorLabel.text = color;
    }
}
```

You already know that using @synthesize, as we did at the top of the file, creates getter and setter methods for each property. Yet, here we're defining our own setters for both name and color! As it turns out, this is just fine. Any time a class defines its own getters or setters, those will be used instead of the default methods that @synthesize provides. In this class, we're using the default, synthesized getters, but defining our own setters, so that whenever we are passed new values for the name or color properties, we update the labels we created earlier.

Implementing the Controller's Code

Now, let's set up the simple controller to display values in our nice new cells. Start off by selecting *BIDViewController.h*, where you need to add the following code:

```
#import <UIKit/UIKit.h>

@interface BIDViewController : UIViewController
    <UITableViewDataSource, UITableViewDelegate>

@property (strong, nonatomic) NSArray *computers;
@end
```

In our controller, we need to set up some data to use, and then implement the table data source methods to feed that data to the table. Switch to *BIDViewController.m*, and add the following code at the beginning of the file:

```
#import "BIDViewController.h"
#import "BIDNameAndColorCell.h"

@implementation ViewController
@synthesize computers;
    .
    .
    .
- (void)viewDidLoad {
    [super viewDidLoad];
    // Do any additional setup after loading the view, typically from a nib.

    NSDictionary *row1 = [[NSDictionary alloc] initWithObjectsAndKeys:
                    @"MacBook", @"Name", @"White", @"Color", nil];
    NSDictionary *row2 = [[NSDictionary alloc] initWithObjectsAndKeys:
                    @"MacBook Pro", @"Name", @"Silver", @"Color", nil];
    NSDictionary *row3 = [[NSDictionary alloc] initWithObjectsAndKeys:
                    @"iMac", @"Name", @"Silver", @"Color", nil];
    NSDictionary *row4 = [[NSDictionary alloc] initWithObjectsAndKeys:
                    @"Mac Mini", @"Name", @"Silver", @"Color", nil];
    NSDictionary *row5 = [[NSDictionary alloc] initWithObjectsAndKeys:
                    @"Mac Pro", @"Name", @"Silver", @"Color", nil];

    self.computers = [[NSArray alloc] initWithObjects:row1, row2,
                    row3, row4, row5, nil];
}
    .
    .
    .
```

Of course, we need to be good memory citizens, so make the following changes to the existing viewDidUnload method:

```
- (void)viewDidUnload {
    [super viewDidUnload];
    // Release any retained subviews of the main view.
    // e.g. self.myOutlet = nil;
    self.computers = nil;
}
```

Then add this code at the end of the file, above the @end declaration:

```
    .
    .
    .
#pragma mark -
#pragma mark Table Data Source Methods
- (NSInteger)tableView:(UITableView *)tableView
    numberOfRowsInSection:(NSInteger)section {
    return [self.computers count];
}

- (UITableViewCell *)tableView:(UITableView *)tableView
    cellForRowAtIndexPath:(NSIndexPath *)indexPath {
    static NSString *CellTableIdentifier = @"CellTableIdentifier";

    BIDNameAndColorCell *cell = [tableView dequeueReusableCellWithIdentifier:
```

```
                            CellTableIdentifier];
    if (cell == nil) {
        cell = [[BIDNameAndColorCell alloc]
                    initWithStyle:UITableViewCellStyleDefault
                    reuseIdentifier:CellTableIdentifier];
    }

    NSUInteger row = [indexPath row];
    NSDictionary *rowData = [self.computers objectAtIndex:row];

    cell.name = [rowData objectForKey:@"Name"];
    cell.color = [rowData objectForKey:@"Color"];

    return cell;
}
```

@end

This version of `viewDidLoad` creates a series of dictionaries. Each dictionary contains the name and color information for one row in the table. The name for that row is held in the dictionary under the key Name, and the color is held under the key Color. We stick all the dictionaries into a single array, which is our data for this table.

> **NOTE:** Remember when Macs came in different colors, like beige, platinum, black, and white? And that's not to mention the original iMac and MacBook series, with their beautiful rainbow hues assortment. Now there's just one color: silver. Harrumph.

Let's focus on `tableView:cellForRowWithIndexPath:`, since that's where we're really getting into some new stuff. The first two lines of code are nearly identical to our earlier versions. We create an identifier and ask the table to dequeue a table view cell if it has one. The only difference here is that we declare the cell variable to be an instance of our `BIDNameAndColorCell` class instead of the standard `UITableViewCell`. That's so we can access the properties that we added specifically to our table view cell subclass.

If the table doesn't have any cells available for reuse, we need to create a new cell. This is essentially the same technique as before, except that here we're also using our custom class instead of `UITableViewCell`. We specify the default style, although the style actually doesn't matter, because we'll be adding our own subviews to display our data rather than using the provided ones.

```
cell = [[BIDNameAndColorCell alloc]
    initWithStyle:UITableViewCellStyleDefault
    reuseIdentifier:CellTableIdentifier];
```

Once we're finished creating our new cell, we use the `indexPath` argument that was passed in to determine which row the table is requesting a cell for, and then use that row value to grab the correct dictionary for the requested row. Remember that the dictionary has two key/value pairs: one with name and another with color.

```
NSUInteger row = [indexPath row];
NSDictionary *rowData = [self.computers objectAtIndex:row];
```

Now, all that's left to do is populate the cell with data from the chosen row, using the properties we defined in our subclass.

```
cell.name = [rowData objectForKey:@"Name"];
cell.color = [rowData objectForKey:@"Color"];
```

Compile and run your application. You should see a table of rows, each with two lines of data, as shown earlier in Figure 8–15.

Being able to add views to a table view cell provides a lot more flexibility than using the standard table view cell alone, but it can get a little tedious creating, positioning, and adding all the subviews programmatically. Gosh, it sure would be nice if we could design the table view cell graphically, using Xcode's nib editor. Well, we're in luck. As we mentioned earlier, you can use Interface Builder to design your table view cells, and then simply load the views from the nib file when you create a new cell.

Loading a UITableViewCell from a Nib

We're going to re-create that same two-line interface we just built in code using the visual layout capabilities that Xcode provides in Interface Builder. To do this, we'll create a new nib file that will contain the table view cell and lay out its views using Interface Builder. Then, when we need a table view cell to represent a row, instead of creating a standard table view cell, we'll just load the nib file and use the properties we already defined in our cell class to set the name and color. Besides making use of Interface Builder's visual layout, we'll also simplify our code in a few other places.

First, we'll make a few changes to the BIDNameAndColorCell class. Since we're going to wire things up in the nib editor, we'll add outlets to point out the labels that need to be accessed. Add these lines to the @interface declaration in *BIDNameAndColorCell.h*:

```
@interface BIDNameAndColorCell : UITableViewCell

@property (copy, nonatomic) NSString *name;
@property (copy, nonatomic) NSString *color;

@property (strong, nonatomic) IBOutlet UILabel *nameLabel;
@property (strong, nonatomic) IBOutlet UILabel *colorLabel;

@end
```

Now that we have these outlets, we don't need the tags anymore! Switch over to *BIDNameAndColorCell.m*, delete the tag definitions, and add method synthesis for our two new outlets:

```
#import "BIDNameAndColorCell.h"

#define kNameValueTag    1
#define kColorValueTag   2

@implementation BIDNameAndColorCell

@synthesize name;
@synthesize color;
```

```
@synthesize nameLabel;
@synthesize colorLabel;
```

Having those outlets available also means that both of our setters can be simplified by removing a couple of lines:

```
- (void)setName:(NSString *)n {
    if (![n isEqualToString:name]) {
        name = [n copy];
        UILabel *nameLabel = (UILabel *)[self.contentView viewWithTag:
                                                   kNameValueTag];
        nameLabel.text = name;
    }
}

- (void)setColor:(NSString *)c {
    if (![c isEqualToString:color]) {
        color = [c copy];
        UILabel *colorLabel = (UILabel *)[self.contentView viewWithTag:
                                                   kColorValueTag];
        colorLabel.text = color;
    }
}
```

And, last but not least, remember that setup we did in `initWithStyle:reuseIdentifier:`, where we created our labels? All that can go. In fact, you should just delete the entire method, since all that setup will now be done in Interface Builder.

After all that, you're left with a cell class that's even smaller and cleaner than before. Its only real function now is to shuffle data to the labels. Now we need to re-create the labels in Interface Builder.

Right-click the *Cells* folder in Xcode and select **New File. . .** from the contextual menu. In the left pane of the new file assistant, click *User Interface* (making sure to pick it in the *iOS* section, rather than the *Mac OS X* section). From the upper-right pane, select *Empty*, and then click *Next*. On the following screen, leave the *Device Family* popup set to *iPhone* and click *Next* once again. When prompted for a name, type *BIDNameAndColorCell.xib*. Make sure that the main project directory is selected in the file browser and that the *Cells* group is selected in the *Group* popup.

Designing the Table View Cell in Interface Builder

Next, select *BIDNameAndColorCell.xib* in the project navigator to open the file for editing. There are only two icons in this nib's dock: *File's Owner* and *First Responder*. Look in the library for a *Table View Cell* (see Figure 8–16), and drag one of those over to the GUI layout area.

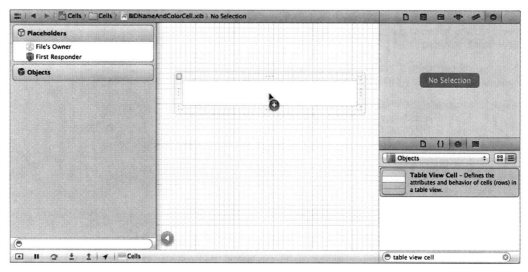

Figure 8–16. *We dragged a table view cell from the library into the nib editor's GUI layout area.*

Make sure the table view cell is selected, press ⌥⌘5 to bring up the size inspector, and in the *View* section, change the cell's height from *44* to *65*. That will give us a little more room to play with.

Next, press ⌥⌘4 to go to the attributes inspector (see Figure 8–17). One of the first fields you'll see there is *Identifier*. That's the reuse identifier that we've been using in our code. If this does not ring a bell, scan back through the chapter and look for CellTableIdentifier. Set the *Identifier* value to *CellTableIdentifier*.

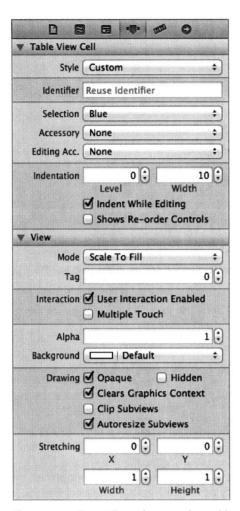

Figure 8–17. *The attributes inspector for a table view cell*

The idea here is that when we retrieve a cell for reuse, perhaps because of scrolling a new cell into view, we want to make sure we get the correct cell type. When this particular cell is instantiated from the nib file, its reuse identifier instance variable will be prepopulated with the NSString you entered in the *Identifier* field of the attributes inspector—*CellTableIdentifier* in this case.

Imagine a scenario where you created a table with a header and then a series of "middle" cells. If you scroll a middle cell into view, it's important that you retrieve a middle cell to reuse and not a header cell. The *Identifier* field lets you tag the cells appropriately.

Our next step is to edit our table cell's content view. Go to the library, drag out four *Label* controls, and place them in the content view, using Figure 8–18 as a guide. The labels will be too close to the top and bottom for those guidelines to be of much help, but the left guideline and the alignment guidelines should serve their purpose. Note that you can drag out one label, and then option-drag to create copies, if that approach is easier for you.

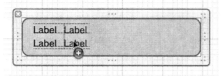

Figure 8–18. *The table view cell's content view, with four labels dragged in*

Next, double-click the upper-left label and change it to *Name:*. Then change the lower-left label to *Color:*.

Now, select both the *Name:* and *Color:* labels, and press the small *T* button in the attribute inspector's *Font* field. This will open a small panel containing a *Font* popup button. Click that, and choose *System Bold* as the typeface. If needed, select the two unchanged label fields on the right and drag them a little more to the right to give the design a bit of breathing room.

Finally, resize the two right-side labels so they stretch all the way to the right guideline. Figure 8–19 should give you a sense of our final cell content view.

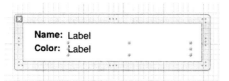

Figure 8–19. *The table view cell's content view with the left label names changed and set to bold, and with the right labels slightly moved and resized*

Now, we need to let Interface Builder know that this table view cell isn't just a normal cell, but rather our special subclass. Otherwise, we wouldn't be able to connect our outlets to the relevant labels. Select the table view cell, bring up the identity inspector by pressing ⌥⌘3, and choose *BIDNameAndColorCell* from the *Class* control.

Next, switch to the connections inspector (⌥⌘6), where you'll see the *colorLabel* and *nameLabel* outlets. Drag each of them to their corresponding label in the GUI.

Using the New Table View Cell

To use the cell we designed, we just need to make a few pretty simple changes to the `tableView:cellForRowAtIndexPath:` method in *BIDViewController.m*. We're going to add a bit and take a bit away.

```
-(UITableViewCell *)tableView:(UITableView *)tableView
        cellForRowAtIndexPath:(NSIndexPath *)indexPath {
    static NSString *CellTableIdentifier = @"CellTableIdentifier";
    static BOOL nibsRegistered = NO;
    if (!nibsRegistered) {
        UINib *nib = [UINib nibWithNibName:@"BIDNameAndColorCell" bundle:nil];
```

```
            [tableView registerNib:nib forCellReuseIdentifier:CellTableIdentifier];
            nibsRegistered = YES;
        }

        BIDNameAndColorCell *cell = [tableView dequeueReusableCellWithIdentifier:
                                    CellTableIdentifier];
        if (cell == nil) {
            cell = [[BIDNameAndColorCell alloc]
                        initWithStyle:UITableViewCellStyleDefault
                        reuseIdentifier:CellTableIdentifier];
        }

        NSUInteger row = [indexPath row];
        NSDictionary *rowData = [self.computers objectAtIndex:row];

        cell.name = [rowData objectForKey:@"Name"];
        cell.color = [rowData objectForKey:@"Color"];

        return cell;
    }
```

The first change you see here is the addition of a new static BOOL variable. This variable maintains its state across invocations of this method, and it is initialized to NO only the first time this method is called. This lets us insert a few lines that will be called only once (the first time this method is called), in order to register our nib with the table view. What does this mean?

Starting in iOS 5, a table view can keep track of which nib files are meant to be associated with particular reuse identifiers. UITableView's dequeueReusableCellWithIdentifier: method is now so smart that, even if there are no available cells, it can use this nib registry to load a new cell from a nib file. That means that as long as we've registered all the reuse identifiers we're going to use for a table view, its dequeueReusableCellWithIdentifier: method will always return a cell, and it never returns nil. Therefore, we can remove the lines that check for a nil cell value, since that will never happen.

There's one other addition we need to make. We already changed the height of our table view cell from the default value in *CustomCell.xib*, but that's not quite enough. We also need to inform the table view of that fact; otherwise, it won't leave enough space for the cell to display properly. The simplest way to do this is by adding a table view delegate method that lets us specify the value. Add the following new method to the bottom of the class definition in *BIDViewController.m*:

```
- (CGFloat)tableView:(UITableView *)tableView
        heightForRowAtIndexPath:(NSIndexPath *)indexPath {
    return 65.0; // Same number we used in Interface Builder
}
```

That's it. Build and run. Now your two-line table cells are based on your mad Interface Builder design skillz.

So, now that you've seen a couple of approaches, what do you think? Many people who delve into iOS development are somewhat confused at first by the focus on Interface

Builder, but as you've seen, it has a lot going for it. Besides having the obvious appeal of letting you visually design your GUI, this approach promotes the proper use of nib files, which helps you stick to the MVC architecture pattern. Also, you can make your application code simpler, more modular, and just plain easier to write. As our good buddy Mark Dalrymple says, "No code is the best code!"

Grouped and Indexed Sections

Our next project will explore another fundamental aspect of tables. We're still going to use a single table view—no hierarchies yet—but we'll divide data into sections. Create a new Xcode project using the *Single View Application* template again, this time calling it *Sections*.

Building the View

Open the *Sections* folders, and click *BIDViewController.xib* to edit the file. Drop a table view onto the *View* window, as we did before. Then press ⌥⌘6, and connect the *dataSource* and *delegate* connections to the *File's Owner* icon.

Next, make sure the table view is selected, and press ⌥⌘4 to bring up the attributes inspector. Change the table view's *Style* from *Plain* to *Grouped* (see Figure 8–20). You should see the change reflected in the sample table shown in the table view. Save your nib, and move along. (We discussed the difference between indexed and grouped styles at the beginning of the chapter.)

Figure 8–20. *The attributes inspector for the table view, showing the Style popup with Grouped selected*

Importing the Data

This project needs a fair amount of data to do its thing. To save you a few hours worth of typing, we've provided another property list for your tabling pleasure. Grab the file named *sortednames.plist* from the *08 Sections/Sections* subfolder in this book's project archive and add it to your project's *Sections* folder.

Once *sortednames.plist* is added to your project, single-click it just to get a sense of what it looks like (see Figure 8–21). It's a property list that contains a dictionary, with one entry for each letter of the alphabet. Underneath each letter is a list of names that start with that letter.

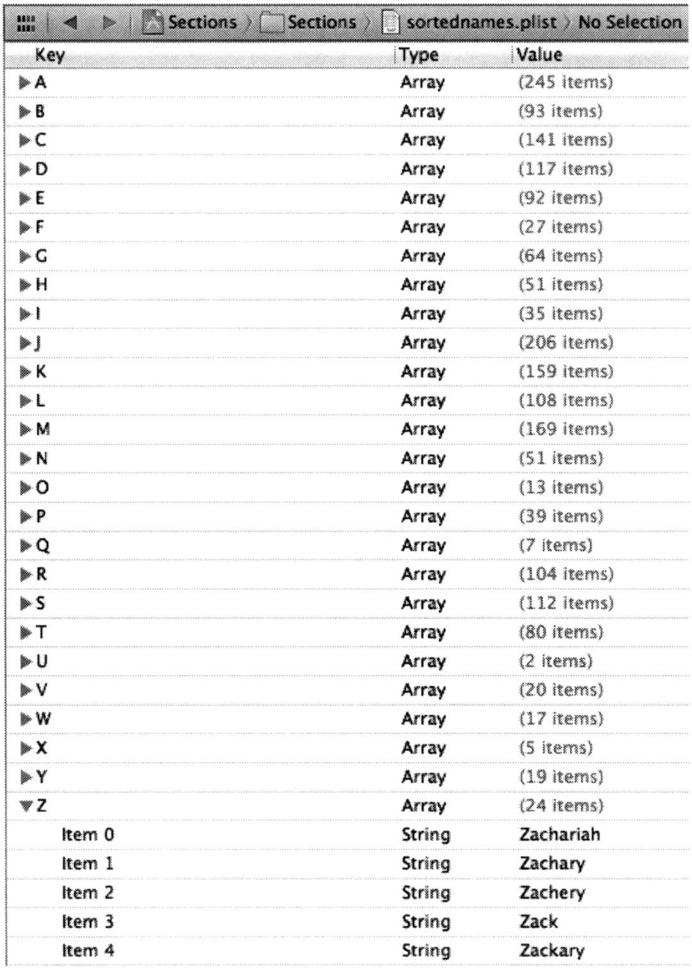

Key	Type	Value
▶ A	Array	(245 items)
▶ B	Array	(93 items)
▶ C	Array	(141 items)
▶ D	Array	(117 items)
▶ E	Array	(92 items)
▶ F	Array	(27 items)
▶ G	Array	(64 items)
▶ H	Array	(51 items)
▶ I	Array	(35 items)
▶ J	Array	(206 items)
▶ K	Array	(159 items)
▶ L	Array	(108 items)
▶ M	Array	(169 items)
▶ N	Array	(51 items)
▶ O	Array	(13 items)
▶ P	Array	(39 items)
▶ Q	Array	(7 items)
▶ R	Array	(104 items)
▶ S	Array	(112 items)
▶ T	Array	(80 items)
▶ U	Array	(2 items)
▶ V	Array	(20 items)
▶ W	Array	(17 items)
▶ X	Array	(5 items)
▶ Y	Array	(19 items)
▼ Z	Array	(24 items)
Item 0	String	Zachariah
Item 1	String	Zachary
Item 2	String	Zachery
Item 3	String	Zack
Item 4	String	Zackary

Figure 8–21. *The sortednames.plist property list file. We opened the letter Z to give you a sense of one of the dictionaries.*

We'll use the data from this property list to feed the table view, creating a section for each letter.

Implementing the Controller

Single-click the *BIDViewController.h* file, and add both an NSDictionary and an NSArray instance variable and corresponding property declarations. The dictionary will hold all of our data. The array will hold the sections sorted in alphabetical order. We also need to make the class conform to the UITableViewDataSource and UITableViewDelegate protocols.

```
#import <UIKit/UIKit.h>

@interface BIDViewController : UIViewController
    <UITableViewDataSource, UITableViewDelegate>

@property (strong, nonatomic) NSDictionary *names;
@property (strong, nonatomic) NSArray *keys;
@end
```

Now, switch over to *BIDViewController.m*, and add the following code to the beginning of that file:

```
#import "BIDViewController.h"

@implementation BIDViewController
@synthesize names;
@synthesize keys;
.
.
.
- (void)viewDidLoad {
    [super viewDidLoad];
    // Do any additional setup after loading the view, typically from a nib.
    NSString *path = [[NSBundle mainBundle] pathForResource:@"sortednames"
                                                      ofType:@"plist"];
    NSDictionary *dict = [[NSDictionary alloc]
                              initWithContentsOfFile:path];
    self.names = dict;

    NSArray *array = [[names allKeys] sortedArrayUsingSelector:
                          @selector(compare:)];
    self.keys = array;
}
.
.
.
```

Insert the following lines of code in the existing viewDidUnload method:

```
- (void)viewDidUnload {
    [super viewDidUnload];
    // Release any retained subviews of the main view.
    // e.g. self.myOutlet = nil;
    self.names = nil;
    self.keys = nil;
}
```

Now, add the following code at the end of the file, just above the @end declaration:

```
.
.
.
#pragma mark -
#pragma mark Table View Data Source Methods
- (NSInteger)numberOfSectionsInTableView:(UITableView *)tableView {
    return [keys count];
}
```

```
- (NSInteger)tableView:(UITableView *)tableView
        numberOfRowsInSection:(NSInteger)section {
    NSString *key = [keys objectAtIndex:section];
    NSArray *nameSection = [names objectForKey:key];
    return [nameSection count];
}

- (UITableViewCell *)tableView:(UITableView *)tableView
    cellForRowAtIndexPath:(NSIndexPath *)indexPath {
    NSUInteger section = [indexPath section];
    NSUInteger row = [indexPath row];

    NSString *key = [keys objectAtIndex:section];
    NSArray *nameSection = [names objectForKey:key];

    static NSString *SectionsTableIdentifier = @"SectionsTableIdentifier";
    UITableViewCell *cell = [tableView dequeueReusableCellWithIdentifier:
        SectionsTableIdentifier];
    if (cell == nil) {
        cell = [[UITableViewCell alloc]
            initWithStyle:UITableViewCellStyleDefault
            reuseIdentifier:SectionsTableIdentifier];
    }

    cell.textLabel.text = [nameSection objectAtIndex:row];
    return cell;
}

- (NSString *)tableView:(UITableView *)tableView
    titleForHeaderInSection:(NSInteger)section {
    NSString *key = [keys objectAtIndex:section];
    return key;
}

@end
```

Most of this isn't too different from what you've seen before. In the viewDidLoad method, we created an NSDictionary instance from the property list we added to our project and assigned it to names. After that, we grabbed all the keys from that dictionary and sorted them to give us an ordered NSArray with all the key values in the dictionary in alphabetical order. Remember that the NSDictionary uses the letters of the alphabet as its keys, so this array will have 26 letters, in order from *A* to *Z*, and we'll use that array to help us keep track of the sections.

Scroll down to the data source methods. The first one we added to our class specifies the number of sections. We didn't implement this method in the earlier example, because we were happy with the default setting of 1. This time, we're telling the table view that we have one section for each key in our dictionary.

```
- (NSInteger)numberOfSectionsInTableView:(UITableView *)tableView {
    return [keys count];
}
```

The next method calculates the number of rows in a specific section. In the previous example, we had only one section, so we just returned the number of rows in our array. This time, we need to break it down by section. We can do this by retrieving the array that corresponds to the section in question and returning the count from that array.

```
- (NSInteger)tableView:(UITableView *)tableView
        numberOfRowsInSection:(NSInteger)section {
    NSString *key = [keys objectAtIndex:section];
    NSArray *nameSection = [names objectForKey:key];
    return [nameSection count];
}
```

In our `tableView:cellForRowAtIndexPath:` method, we need to extract both the section and row from the index path, and use those to determine which value to use. The section will tell us which array to pull out of the `names` dictionary, and then we can use the row to figure out which value from that array to use. Everything else in that method is basically the same as the version in the Simple Table application we built earlier in the chapter.

The method `tableView:titleForHeaderInSection` allows you to specify an optional header value for each section, and we simply return the letter for this group.

```
- (NSString *)tableView:(UITableView *)tableView
    titleForHeaderInSection:(NSInteger)section {
    NSString *key = [keys objectAtIndex:section];
    return key;
}
```

Compile and run the project, and revel in its grooviness. Remember that we changed the table's *Style* to *Grouped*, so we ended up with a grouped table with 26 sections, which should look like Figure 8–22.

As a contrast, let's change our table view back to the plain style and see what a plain table view with multiple sections looks like. Select *BIDViewController.xib* to edit the file in Interface Builder again. Select the table view, and use the attributes inspector to switch the view to *Plain*. Save the project, and then build and run it—same data, different grooviness (see Figure 8–23).

Figure 8–22. *A grouped table with multiple sections*

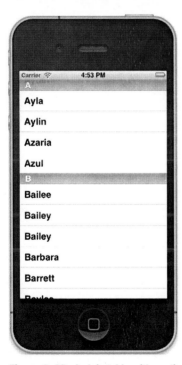

Figure 8–23. *A plain table with sections and no index*

Adding an Index

One problem with our current table is the sheer number of rows. There are 2,000 names in this list. Your finger will get awfully tired looking for Zachariah or Zayne, not to mention Zoie.

One solution to this problem is to add an index down the right side of the table view. Now that we've set our table view style back to *Plain*, that's relatively easy to do. Add the following method to the bottom of *BIDViewController.m*, just above the @end:

```
- (NSArray *)sectionIndexTitlesForTableView:(UITableView *)tableView {
    return keys;
}
```

Yep, that's it. In this method, the delegate is asking for an array of the values to display in the index. You must have more than one section in your table view to use the index, and the entries in this array must correspond to those sections. The returned array must have the same number of entries as you have sections, and the values must correspond to the appropriate section. In other words, the first item in this array will take the user to the first section, which is section 0.

Compile and run the app again, and you'll have yourself a nice index (see Figure 8–24).

Figure 8–24. *The table view with an index*

Implementing a Search Bar

The index is helpful, but even so, we still have a whole lot of names here. If we want to see whether the name Arabella is in the list, for example, we'll need to scroll for a while even after using the index. It would be nice if we could let the user pare down the list by specifying a search term, wouldn't it? That would be darn user-friendly. Well, it's a bit of extra work, but it's not too bad. We're going to implement a standard iOS search bar, like the one shown in Figure 8–25.

Figure 8–25. *The application with a search bar added to the table*

Rethinking the Design

Before we set about adding a search bar, we need to put some thought into our approach. Currently, we have a dictionary that holds a series of arrays, one for each letter of the alphabet. The dictionary is immutable, which means we can't add or delete values from it, and so are the arrays that it holds. We also need to retain the ability to get back to the original dataset when the user hits cancel or erases the search term.

The solution is to create two dictionaries: an immutable dictionary to hold the full dataset and a mutable copy from which we can remove rows. The delegate and data sources will read from the mutable dictionary, and when the search criteria change or the search is canceled, we can refresh the mutable dictionary from the immutable one. Sounds like a plan. Let's do it.

> **CAUTION:** This next project is a bit advanced and may cause a distinct burning sensation if taken too quickly. If some of these concepts give you a headache, retrieve your copy of *Learn Objective-C on the Mac* by Mark Dalrymple and Scott Knaster (Apress, 2009) and review the bits about categories and mutability.

A Deep Mutable Copy

To use our new approach, there's one problem we'll need to solve. NSDictionary conforms to the NSMutableCopying protocol, which returns an NSMutableDictionary, but that method creates what's called a **shallow** copy. This means that when you call the mutableCopy method, it will create a new NSMutableDictionary object that has all the objects that the original dictionary had. They won't be copies; they will be the same actual objects. This would be fine if, say, we were dealing with a dictionary storing strings, because removing a value from the copy wouldn't do anything to the original. Since we have a dictionary full of arrays, however, if we were to remove objects from the arrays in the copy, we would also be removing them from the arrays in the original, because both the copies and the original point to the same objects. In this particular case, the original arrays are immutable, so you couldn't actually remove objects from them anyway, but our intention is to illustrate the point.

In order to deal with this properly, we need to be able to make a deep mutable copy of a dictionary full of arrays. That's not too hard to do, but where should we put this functionality?

If you said, "in a category," then great, now you're thinking with portals! If you didn't, don't worry, it takes a while to get used to this language. Categories, in case you've forgotten, allow you to add more methods to existing objects without subclassing them. Categories are frequently overlooked by folks new to Objective-C, because they're a feature most other languages don't have.

With categories, we can add a method to NSDictionary to do a deep copy, returning an NSMutableDictionary with the same data but not containing the same actual objects.

> **NOTE:** Before you move on to this next series of steps, consider making a backup copy of your project. This way, you'll make sure you have a working version to go back to if things go south with this next set of changes.

In your project window, select the *Sections* folder, and press ⌘N to create a new file. When the new file assistant comes up, select *Cocoa Touch* from the very top of the *iOS* section. In the right-hand panel, select *Objective-C category*, since that's just what we want to create, and click *Next*. On the following screen, name your protocol

MutableDeepCopy, and enter *NSDictionary* in the *Category on* field. Then click *Next* once again. On the final screen, make sure *Sections* is selected in the file browser, *Group* popup, and *Target* control.

Put the following code in *NSDictionary+MutableDeepCopy.h*:

```
#import <Foundation/Foundation.h>

@interface NSDictionary (MutableDeepCopy)
- (NSMutableDictionary *)mutableDeepCopy;
@end
```

Flip over to *NSDictionary+MutableDeepCopy.m*, and add the implementation:

```
#import "NSDictionary+MutableDeepCopy.h"

@implementation NSDictionary (MutableDeepCopy)

- (NSMutableDictionary *)mutableDeepCopy {
    NSMutableDictionary *returnDict = [[NSMutableDictionary alloc]
        initWithCapacity:[self count]];
    NSArray *keys = [self allKeys];
    for (id key in keys) {
        id oneValue = [self valueForKey:key];
        id oneCopy = nil;

        if ([oneValue respondsToSelector:@selector(mutableDeepCopy)])
            oneCopy = [oneValue mutableDeepCopy];
        else if ([oneValue respondsToSelector:@selector(mutableCopy)])
            oneCopy = [oneValue mutableCopy];
        if (oneCopy == nil)
            oneCopy = [oneValue copy];
        [returnDict setValue:oneCopy forKey:key];
    }
    return returnDict;
}
@end
```

This method creates a new mutable dictionary and then loops through all the keys of the original dictionary, making mutable copies of each array it encounters. Since this method will behave just as if it were part of NSDictionary, any reference to self is a reference to the dictionary on which this method is being called. The method first attempts to make a deep mutable copy, and if the object doesn't respond to the mutableDeepCopy message, it tries to make a mutable copy. If the object doesn't respond to the mutableCopy message, it falls back on making a regular copy to ensure that all the objects contained in the dictionary are copied. By doing it this way, if we were to have a dictionary that contained dictionaries (or other objects that supported deep mutable copies), the contained ones would also get deep-copied.

For a few of you, this might be the first time you've seen this syntax in Objective-C:

```
for (id key in keys)
```

Objective-C 2.0 introduced a feature called fast enumeration. **Fast enumeration** is a language-level replacement for NSEnumerator. It allows you to quickly iterate through a collection, such as an NSArray, without the hassle of creating additional objects or loop variables.

All of the delivered Cocoa collection classes—including NSDictionary, NSArray, and NSSet—support fast enumeration, and you should use this syntax any time you need to iterate over a collection. It will ensure that you get the most efficient loop possible.

If we include the *NSDictionary+MutableDeepCopy.h* header file in one of our other classes, we'll be able to call mutableDeepCopy on any NSDictionary object we like. Let's take advantage of that now.

Updating the Controller Header File

Next, we need to add some outlets to our controller class header file. We'll add an outlet for the table view. Up until now, we haven't needed a pointer to the table view outside the data source methods. But for our search bar implementation, we need one to tell the table to reload itself based on the result of the search. We're also going to add an outlet to a search bar, which is a control used for, well, searching.

In addition to those two outlets, we'll add another dictionary. The existing dictionary and array are both immutable objects, and we need to change both of them to the corresponding mutable version, so the NSArray becomes an NSMutableArray and the NSDictionary becomes an NSMutableDictionary.

We won't need any new action methods in our controller, but we will use a couple of new methods. For now, just declare them, and we'll talk about them in detail after you enter the code.

We also must conform our class to the UISearchBarDelegate protocol. We'll need to become the search bar's delegate in addition to being the table view's delegate.

Make the following changes to *BIDViewController.h*:

```
#import <UIKit/UIKit.h>

@interface ViewController : UIViewController
<UITableViewDataSource, UITableViewDelegate, UISearchBarDelegate>

@property (strong, nonatomic) NSDictionary *names;
@property (strong, nonatomic) NSArray *keys;
@property (strong, nonatomic) IBOutlet UITableView *table;
@property (strong, nonatomic) IBOutlet UISearchBar *search;
@property (strong, nonatomic) NSDictionary *allNames;
@property (strong, nonatomic) NSMutableDictionary *names;
@property (strong, nonatomic) NSMutableArray *keys;
- (void)resetSearch;
- (void)handleSearchForTerm:(NSString *)searchTerm;
@end
```

Here's what we just did:

- The outlet `table` will point to our table view.

- The outlet `search` will point to the search bar.

- The dictionary `allNames` will hold the full dataset.

- The dictionary `names` will hold the dataset that matches the current search criteria.

- `keys` will hold the index values and section names.

If you're clear on everything, let's forge ahead and modify our view.

Modifying the View

Every table view allows for the possibility of placing a view at the top end of the table view, above any content. That header view will scroll with the rest of the content. A perfect example of this is the `UISearchBar`. If you drag a *Search Bar* from the library and place it just above the table view, Interface Builder will size the search bar so it fits at the top of the table view and scrolls with the table view.

Unfortunately, the `UISearchBar` does not play well with the right-hand indexing feature. If you take a look at Figure 8–26, you'll see why. Notice that the index overlaps the search bar, covering a bit of the *Cancel* button.

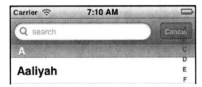

Figure 8–26. *In the current version of our application, the search bar's Cancel button is overlapped by the index.*

Fortunately, there is a solution, and we can implement it without changing a line of our existing code, entirely by configuring the view hierarchy in Interface Builder. The idea is to put the search bar inside a plain old `UIView`. That way, the table view will make sure the `UIView` fills up the space, but the `UIView`'s contents will still be presented just the way we set them up.

Select *BIDViewController.xib* to edit the file in Xcode's Interface Builder, and select the table view in the editing area. Use the object library to find a *View*, and drag it to the top of the table view.

You're trying to drop the view into the table view's header section, a special part of the table view that lies before the first section. The way to do this is to place the view at the top of the table view. Before you release the mouse button, you should see a rounded, blue rectangle at the top of the table view (see Figure 8–27). That's your indication that if you drop the view now, it will go into the table header. Release the mouse button to drop the view once you see that blue rectangle.

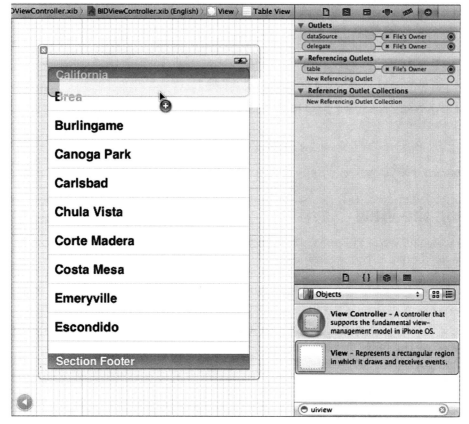

Figure 8–27. *Dropping a view onto the table view. Notice the rounded rectangle that appears at the top of the table view, indicating that the view will be added to the table's header.*

Now, grab a *Search Bar* from the library, and drop it straight onto the view you just added. You'll see the familiar blue rectangle, and find that the search bar fits perfectly within the view (see Figure 8–28).

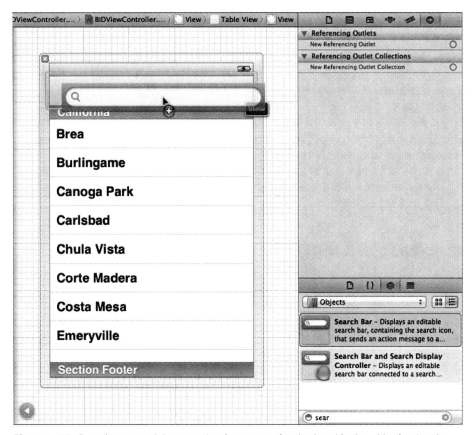

Figure 8–28. *Dropping a search bar onto the view we previously placed in the table view header*

Next, let's resize the search bar to make room for the index. First, select the search bar, and then grab the resizing handle on the right edge of the search bar and drag it to the left by 25 pixels. That should give us enough clearance for the table view's index column. You'll see a little, floating size panel while you're dragging. The search bar starts out at 320 pixels wide, so resize it down to 295 pixels wide (see Figure 8–29).

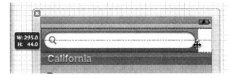

Figure 8–29. *We grabbed the right edge of the search bar and dragged it to the left 25 pixels. Notice the tool tip on the left that helps you tell when you've resized the search bar to 295 pixels wide.*

So far so good. Our next step is to do something about that unsightly white gap to the right of the search bar. Fortunately, there's another class whose background looks just like `UISearchBar` that we can use to fill in that space.

The navigation bar (UINavigationBar) is normally used to contain navigation elements (and you'll learn more about that in Chapter 9), but at its heart it is, after all, a subclass of UIView. This means it can be placed on screen and resized just like anything else.

Locate a *Navigation Bar* in the library, and drag it into the view at the top of the table view. You'll see that it also fills the entire UIView, obscuring the search bar. Double-click the *Title* text and delete it, leaving just the gradient background. Now go back to the dock and select the navigation bar (deleting the text may have left the Navigation Item selected, which is not what you want). With the navigation bar selected, grab the resize handle from its left edge and drag it to the right until the navigation bar is just 25 pixels wide (the same size we carved out of the search bar). You'll see that the gap is now covered, and the same smooth gradient appears all the way across the screen (see Figure 8–30). The illusion is complete!

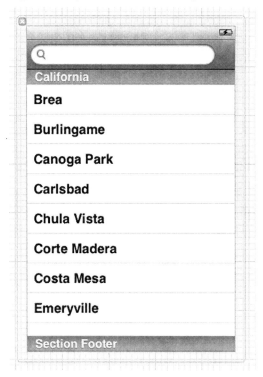

Figure 8–30. *We inserted a navigation bar in the table view header's view. We deleted its title and resized it to be 25 pixels wide. This looks good, and our search bar will not collide with our index.*

Next, control-drag from the *File's Owner* icon to the table view, and select the *table* outlet. Repeat this procedure with the search bar, and select the *search* outlet. Single-click the search bar, and go to the attributes inspector by pressing ⌥⌘**4**. It should look like Figure 8–31.

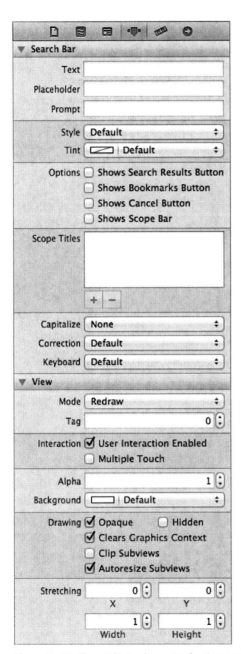

Figure 8–31. *The attributes inspector for the search bar*

Type *search* in the *Placeholder* field. The word *search* will appear, very lightly, in the search field.

A bit farther down, in the *Options* section, you'll find a series of checkboxes for adding a search results button or a bookmarks button at the far-right end of the search bar.

These buttons don't do anything on their own (except toggle when the user taps them), but you could use them to let the delegate set up some different display content depending on the status of the toggle buttons.

Leave those first two unchecked, but do check the box that says *Shows Cancel Button*. A *Cancel* button will appear to the right of the search field. The user can tap this button to cancel the search. The final checkbox is for enabling the scope bar, which is a series of connected buttons designed to let the user pick from various categories of searchable things (as specified by the series of *Scope Titles* beneath it). We're not going to use the scope functionality, so just leave those parts alone.

Below the checkboxes and *Scope Titles*, set the *Correction* popup menu to *No* to indicate that the search bar should not try to correct the user's spelling.

Switch to the connections inspector by pressing ⌥⌘6, and drag from the *delegate* connection to the *File's Owner* icon to tell this search bar that our view controller is also the search bar's delegate.

That should be everything we need here, so make sure to save your work before moving on. Now, let's dig into some code.

Modifying the Controller Implementation

The changes to accommodate the search bar are fairly drastic. Make the following modifications to *BIDViewController.m*:

```
#import "BIDViewController.h"
#import "NSDictionary+MutableDeepCopy.h"

@implementation ViewController
@synthesize names;
@synthesize keys;
@synthesize table;
@synthesize search;
@synthesize allNames;

#pragma mark -
#pragma mark Custom Methods
- (void)resetSearch {
    self.names = [self.allNames mutableDeepCopy];
    NSMutableArray *keyArray = [[NSMutableArray alloc] init];
    [keyArray addObjectsFromArray:[[self.allNames allKeys]
        sortedArrayUsingSelector:@selector(compare:)]];
    self.keys = keyArray;
}

- (void)handleSearchForTerm:(NSString *)searchTerm {
    NSMutableArray *sectionsToRemove = [[NSMutableArray alloc] init];
    [self resetSearch];

    for (NSString *key in self.keys) {
        NSMutableArray *array = [names valueForKey:key];
        NSMutableArray *toRemove = [[NSMutableArray alloc] init];
```

```
        for (NSString *name in array) {
            if ([name rangeOfString:searchTerm
                            options:NSCaseInsensitiveSearch].location == NSNotFound)
                [toRemove addObject:name];
        }
        if ([array count] == [toRemove count])
            [sectionsToRemove addObject:key];

        [array removeObjectsInArray:toRemove];
    }
    [self.keys removeObjectsInArray:sectionsToRemove];
    [table reloadData];
}
.
.
.
- (void)viewDidLoad {
    [super viewDidLoad];
    // Do any additional setup after loading the view, typically from a nib.
    NSString *path = [[NSBundle mainBundle] pathForResource:@"sortednames"
        ofType:@"plist"];
    NSDictionary *dict = [[NSDictionary alloc]
        initWithContentsOfFile:path];
    self.names = dict;
    self.allNames = dict;

    NSArray *array = [[names allKeys] sortedArrayUsingSelector:
        @selector(compare:)];
    self.keys = array;

    [self resetSearch];
    [table reloadData];
    [table setContentOffset:CGPointMake(0.0, 44.0) animated:NO];
}

- (void)viewDidUnload {
    [super viewDidUnload];
    // Release any retained subviews of the main view.
    // e.g. self.myOutlet = nil;
    self.table = nil;
    self.search = nil;
    self.allNames = nil;
    self.names = nil;
    self.keys = nil;
}
.
.
.
#pragma mark -
#pragma mark Table View Data Source Methods
- (NSInteger)numberOfSectionsInTableView:(UITableView *)tableView {
    return [keys count];
    return ([keys count] > 0) ? [keys count] : 1;
}

- (NSInteger)tableView:(UITableView *)aTableView
        numberOfRowsInSection:(NSInteger)section {
```

```
        if ([keys count] == 0)
            return 0;
        NSString *key = [keys objectAtIndex:section];
        NSArray *nameSection = [names objectForKey:key];
        return [nameSection count];
}

- (UITableViewCell *)tableView:(UITableView *)aTableView
        cellForRowAtIndexPath:(NSIndexPath *)indexPath {
    NSUInteger section = [indexPath section];
    NSUInteger row = [indexPath row];

    NSString *key = [keys objectAtIndex:section];
    NSArray *nameSection = [names objectForKey:key];

    static NSString *sectionsTableIdentifier = @"SectionsTableIdentifier";

    UITableViewCell *cell = [aTableView dequeueReusableCellWithIdentifier:
        sectionsTableIdentifier];
    if (cell == nil) {
        cell = [[[UITableViewCell alloc] initWithFrame:CGRectZero
            reuseIdentifier: sectionsTableIdentifier] autorelease];
    }

    cell.textLabel.text = [nameSection objectAtIndex:row];
    return cell;
}

- (NSString *)tableView:(UITableView *)tableView
    titleForHeaderInSection:(NSInteger)section {
    if ([keys count] == 0)
        return nil;

    NSString *key = [keys objectAtIndex:section];
    return key;
}

- (NSArray *)sectionIndexTitlesForTableView:(UITableView *)tableView {
    return keys;
}

#pragma mark -
#pragma mark Table View Delegate Methods
- (NSIndexPath *)tableView:(UITableView *)tableView
    willSelectRowAtIndexPath:(NSIndexPath *)indexPath {
    [search resignFirstResponder];
    return indexPath;
}

#pragma mark -
#pragma mark Search Bar Delegate Methods
- (void)searchBarSearchButtonClicked:(UISearchBar *)searchBar {
    NSString *searchTerm = [searchBar text];
    [self handleSearchForTerm:searchTerm];
}
```

```
- (void)searchBar:(UISearchBar *)searchBar
    textDidChange:(NSString *)searchTerm {
    if ([searchTerm length] == 0) {
        [self resetSearch];
        [table reloadData];
        return;
    }
    [self handleSearchForTerm:searchTerm];
}

- (void)searchBarCancelButtonClicked:(UISearchBar *)searchBar {
    search.text = @"";
    [self resetSearch];
    [table reloadData];
    [searchBar resignFirstResponder];
}
@end
```

Wow, are you still with us after all that typing? Let's break it down and see what we just did. We'll start with the first new method we added.

Copying Data from allNames

Our new resetSearch method will be called any time the search is canceled or the search term changes.

```
- (void)resetSearch {
    self.names = [self.allNames mutableDeepCopy];
    NSMutableArray *keyArray = [[NSMutableArray alloc] init];
    [keyArray addObjectsFromArray:[[self.allNames allKeys]
            sortedArrayUsingSelector:@selector(compare:)]];
    self.keys = keyArray;
}
```

All this method does is create a mutable copy of allNames, assign it to names, and then refresh the keys array so it includes all the letters of the alphabet.

We need to refresh the keys array because if a search eliminates all values from a section, we need to get rid of that section, too. Otherwise, the screen would be filled up with headers and empty sections, and it wouldn't look good. We also don't want to provide an index to something that doesn't exist, so as we cull the names based on the search terms, we also cull the empty sections.

Implementing the Search

The other new method we added is the actual search.

```
- (void)handleSearchForTerm:(NSString *)searchTerm {
    NSMutableArray *sectionsToRemove = [[NSMutableArray alloc] init];
    [self resetSearch];

    for (NSString *key in self.keys) {
        NSMutableArray *array = [names valueForKey:key];
```

```
        NSMutableArray *toRemove = [[NSMutableArray alloc] init];
        for (NSString *name in array) {
            if ([name rangeOfString:searchTerm
                options:NSCaseInsensitiveSearch].location == NSNotFound)
                    [toRemove addObject:name];
        }

        if ([array count] == [toRemove count])
            [sectionsToRemove addObject:key];

        [array removeObjectsInArray:toRemove];
    }
    [self.keys removeObjectsInArray:sectionsToRemove];
    [table reloadData];
}
```

Although we'll kick off the search in the search bar delegate methods, we pulled handleSearchForTerm: into its own method, since we're going to need to use the exact same functionality in two different delegate methods. By embedding the search in the handleSearchForTerm: method, we consolidate the functionality into a single place so it's easier to maintain, and then we just call this new method as required. Since this is the real meat (or tofu, if you prefer) of this section, let's break this method down into smaller chunks.

First, we create an array that's going to hold the empty sections as we find them. We use this array to remove those empty sections later, because it is not safe to remove objects from a collection while iterating through that collection. Since we are using fast enumeration, attempting to do that will raise an exception. So, since we won't be able to remove keys while we're iterating through them, we store the sections to be removed in an array, and after we're finished enumerating, we remove all the objects at once. After allocating the array, we reset the search.

```
NSMutableArray *sectionsToRemove = [[NSMutableArray alloc] init];
[self resetSearch];
```

Next, we enumerate through all the keys in the newly restored keys array.

```
for (NSString *key in self.keys) {
```

Each time through the loop, we grab the array of names that corresponds to the current key and create another array to hold the values we need to remove from the names array. Remember that we're removing names and sections, so we must keep track of which keys are empty as well as which names don't match the search criteria.

```
NSMutableArray *array = [names valueForKey:key];
NSMutableArray *toRemove = [[NSMutableArray alloc] init];
```

Next, we iterate through all the names in the current array. So, if we're currently working through the key of *A*, this loop will enumerate through all the names that begin with *A*.

```
for (NSString *name in array) {
```

Inside this loop, we use one of NSString's methods that returns the location of a substring within a string. We specify an option of NSCaseInsensitiveSearch to tell it we don't care about the search term's case—in other words, *A* is the same as *a*. The value returned by this method is an NSRange struct with two members: location and length. If the search term was not found, the location will be set to NSNotFound, so we just check for that. If the NSRange that is returned contains NSNotFound, we add the name to the array of objects to be removed later.

```
        if ([name rangeOfString:searchTerm
            options:NSCaseInsensitiveSearch].location == NSNotFound)
                [toRemove addObject:name];
    }
```

After we've looped through all the names for a given letter, we check to see whether the array of names to be removed is the same length as the array of names. If it is, we know this section is now empty, and we add it to the array of keys to be removed later.

```
    if ([array count] == [toRemove count])
        [sectionsToRemove addObject:key];
```

Next, we actually remove the nonmatching names from this section's arrays.

```
        [array removeObjectsInArray:toRemove];
    }
```

Finally, we remove the empty sections, release the array used to keep track of the empty sections, and tell the table to reload its data.

```
    [self.keys removeObjectsInArray:sectionsToRemove];
    [sectionsToRemove release];
    [table reloadData];
}
```

Changes to viewDidLoad

Down in viewDidLoad, we made a few changes. First, we now load the property list into the allNames dictionary instead of the names dictionary and delete the code that loads the keys array, because that is now done in the resetSearch method. We then call the resetSearch method, which populates the names mutable dictionary and the keys array for us. After that, we call reloadData on our tableView. In the normal flow of the program, reloadData will be called before the user ever sees the table, so most of the time it's not necessary to call it in viewDidLoad:. However, in order for the line after it, setContentOffset:animated:, to work, we need to make sure that the table is all set up, which we do by calling reloadData on the table.

```
    [table reloadData];
    [table setContentOffset:CGPointMake(0.0, 44.0) animated:NO];
```

So, what does setContentOffset:animated: do? Well, it does exactly what it sounds like. It offsets the contents of the table—in our case, by 44 pixels, the height of the search bar. This causes the search bar to be scrolled off the top when the table first comes up. In effect, we are "hiding" the search bar up there at the top, to be discovered by users the first time they scroll all the way up. This is similar to the way that Mail, Contacts, and other standard iOS applications support searching. Users don't see the search bar at first, but can bring it into view with a simple downward swipe.

Hiding the search bar bears a certain risk in that the user might not discover the search functionality at first, or perhaps not at all! However, this is a risk shared among a wide variety of iOS apps, and this usage of the search bar is now so common that there's no real reason to show anything more explicit. We'll talk more about this in the "Adding a Magnifying Glass to the Index" section, coming up soon.

Changes to Data Source Methods

If you skip down to the data source methods, you'll see we made a few minor changes there. Because the names dictionary and keys array are still being used to feed the data source, these methods are basically the same as they were before.

We did need to account for the fact that table views always have a minimum of one section, and yet the search could potentially exclude all names from all sections. So, we added a little code to check for the situation where all sections were removed. In those cases, we feed the table view a single section with no rows and a blank name. This avoids any problems and doesn't give any incorrect feedback to the user.

Adding a Table View Delegate Method

Below the data source methods, we've added a single delegate method. If the user clicks a row while using the search bar, we want the keyboard to go away. We accomplish this by implementing tableView:willSelectRowAtIndexPath: and telling the search bar to resign first responder status, which will cause the keyboard to retract. Next, we return indexPath unchanged.

```
- (NSIndexPath *)tableView:(UITableView *)tableView
    willSelectRowAtIndexPath:(NSIndexPath *)indexPath {
    [search resignFirstResponder];
    return indexPath;
}
```

We could also have done this in tableView:didSelectRowAtIndexPath:, but because we're doing it here, the keyboard retracts a bit sooner.

Adding Search Bar Delegate Methods

The search bar has a number of methods that it calls on its delegate. When the user taps return or the search key on the keyboard, searchBarSearchButtonClicked: will be called. Our version of this method grabs the search term from the search bar and calls our search method, which will remove the nonmatching names from names and the empty sections from keys.

```
- (void)searchBarSearchButtonClicked:(UISearchBar *)searchBar {
    NSString *searchTerm = [searchBar text];
    [self handleSearchForTerm:searchTerm];
}
```

The searchBarSearchButtonClicked: method should be implemented any time you use a search bar.

We also implement another search bar delegate method, which requires a bit of caution. This next method implements a live search. Every time the search term changes, regardless of whether the user has selected the search button or tapped return, we redo the search. This behavior is very user-friendly, as the users can see the results change while typing. If users pare the list down far enough on the third character, they can stop typing and select the row they want.

You can easily hamstring the performance of your application by implementing live search, especially if you're displaying images or have a complex data model. In this case, with 2,000 strings and no images or accessory icons, things actually work pretty well, even on a first-generation iPhone or iPod touch.

> **CAUTION:** Do not assume that snappy performance in the simulator translates to snappy performance on your device. If you're going to implement a live search like this, you need to test extensively on actual hardware to make sure your application stays responsive. When in doubt, don't use the live search feature. Your users will likely be perfectly happy tapping the search button.

To handle a live search, implement the search bar delegate method `searchBar:textDidChange:` like so:

```
- (void)searchBar:(UISearchBar *)searchBar
    textDidChange:(NSString *)searchTerm {
    if ([searchTerm length] == 0) {
        [self resetSearch];
        [table reloadData];
        return;
    }
    [self handleSearchForTerm:searchTerm];
}
```

Notice that we check for an empty string. If the string is empty, we know all names are going to match it, so we simply reset the search and reload the data, without bothering to enumerate over all the names.

Last, we implement a method that allows us to be notified when the user clicks the *Cancel* button on the search bar.

```
- (void)searchBarCancelButtonClicked:(UISearchBar *)searchBar {
    search.text = @"";
    [self resetSearch];
    [table reloadData];
    [searchBar resignFirstResponder];
}
```

When the user clicks *Cancel*, we set the search term to an empty string, reset the search, and reload the data so that all names are showing. We also tell the search bar to yield first responder status, so that the keyboard drops away and the user can resume working with the table view.

If you haven't done so already, fire up our app and try out the search functionality. Remember that the search bar is scrolled just off the top of the screen, so drag down to bring it into view. Click in the search field and start typing. The name list should trim to match the text you type (Figure 8–32). It works, right?

Figure 8–32. *Our Sections app in all its glory. As promised, the index no longer steps on the Cancel button. Nice!*

For our next bit of tinkering, how about making that index disappear when you tap on the search field? This is not mandatory—it's strictly a design decision—but worth knowing how to do.

First, let's add a property variable to keep track of whether the user is currently using the search bar. Add the following to *BIDViewController.h*:

```
@interface ViewController : UIViewController
<UITableViewDataSource, UITableViewDelegate, UISearchBarDelegate>

@property (strong, nonatomic) IBOutlet UITableView *table;
@property (strong, nonatomic) IBOutlet UISearchBar *search;
@property (strong, nonatomic) NSDictionary *allNames;
@property (strong, nonatomic) NSMutableDictionary *names;
@property (strong, nonatomic) NSMutableArray *keys;
@property (assign, nonatomic) BOOL isSearching;
- (void)resetSearch;
- (void)handleSearchForTerm:(NSString *)searchTerm;
@end
```

Save the file, and let's shift our attention to *BIDViewController.m*. First, add a method synthesizer for the new property:

```
@implementation ViewController
@synthesize names;
@synthesize keys;
@synthesize table;
@synthesize search;
@synthesize allNames;
@synthesize isSearching;
```

Next, we need to modify the `sectionIndexTitlesForTableView:` method to return `nil` if the user is searching:

```
- (NSArray *)sectionIndexTitlesForTableView:(UITableView *)tableView {
    if (isSearching)
        return nil;
    return keys;
}
```

We need to implement a new delegate method to set `isSearching` to `YES` when searching begins. Add the following method to the search bar delegate methods section of *BIDViewController.m*:

```
- (void)searchBarTextDidBeginEditing:(UISearchBar *)searchBar {
    isSearching = YES;
    [table reloadData];
}
```

This method is called when the search bar is tapped. In it, we set `isSearching` to `YES`, and then tell the table to reload itself, which causes the index to disappear. We also need to remember to set `isSearching` to `NO` when the user is finished searching. There are two ways a user can finish searching: by pressing the *Cancel* button or by tapping a row in the table. Therefore, we need to add code to the `searchBarCancelButtonClicked:` method:

```
- (void)searchBarCancelButtonClicked:(UISearchBar *)searchBar {
    isSearching = NO;
    search.text = @"";
    [self resetSearch];
    [table reloadData];
    [searchBar resignFirstResponder];
}
```

We also need to make that change to the `tableView:willSelectRowAtIndexPath:` method:

```
- (NSIndexPath *)tableView:(UITableView *)tableView
  willSelectRowAtIndexPath:(NSIndexPath *)indexPath {
    [search resignFirstResponder];
    isSearching = NO;
    search.text = @"";
    [tableView reloadData];
    return indexPath;
}
```

Now, try it again. You'll see that when you tap the search bar, the index disappears until you're finished searching.

Adding a Magnifying Glass to the Index

Because we offset the table view's content, the search bar is not visible when the application first launches, but a quick flick down brings the search bar into view so it can be used. It is also acceptable to put a search bar above the table view, rather than in it, so that the bar is always visible, but this eats up valuable screen real estate. Having the search bar scroll with the table uses the iPhone's small screen more efficiently, and the user can always get to the search bar quickly by tapping in the status bar at the top of the screen.

The problem is that not everyone knows that tapping in the status bar takes you to the top of the current table. The ideal solution would be to put a magnifying glass at the top of the index the way that the Contacts application does (see Figure 8–33). And guess what? We can actually do just that. iOS includes the ability to place a magnifying glass in a table index. Let's do that now for our application.

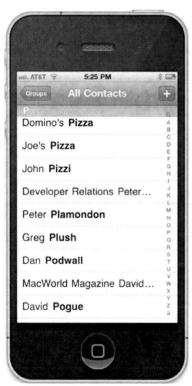

Figure 8–33. *The Contacts application has a magnifying glass icon in the index that takes you to the search bar. Prior to iOS 3, this was not available to other applications, but now it is.*

Only three steps are involved in adding the magnifying glass:

■ Add a special value to our keys array to indicate that we want the magnifying glass.

■ Prevent iOS from printing a section header in the table for that special value.

■ Tell the table to scroll to the top when that item is selected.

Let's tackle these tasks in order.

Adding the Special Value to the Keys Array

To add the special value to our keys array, all we need to do is add one line of code to the resetSearch method:

```
- (void)resetSearch {
    self.names = [self.allNames mutableDeepCopy];
    NSMutableArray *keyArray = [[NSMutableArray alloc] init];
    [keyArray addObject:UITableViewIndexSearch];
    [keyArray addObjectsFromArray:[[self.allNames allKeys]
            sortedArrayUsingSelector:@selector(compare:)]];
    self.keys = keyArray;
}
```

Suppressing the Section Header

Now, we need to suppress that value from coming up as a section title. We do that by adding a check in the existing tableView:titleForHeaderInSection: method, and return nil when it asks for the title for the special search section:

```
- (NSString *)tableView:(UITableView *)tableView
    titleForHeaderInSection:(NSInteger)section {
    if ([keys count] == 0)
        return nil;

    NSString *key = [keys objectAtIndex:section];
    if (key == UITableViewIndexSearch)
        return nil;
    return key;
}
```

Telling the Table View What to Do

Finally, we need to tell the table view what to do when the user taps the magnifying glass in the index. When the user taps the magnifying glass, the delegate method tableView:sectionForSectionIndexTitle:atIndex: is called, if it is implemented.

Add this method to the bottom of *BIDViewController.m*, just above the @end:

```
- (NSInteger)tableView:(UITableView *)tableView
        sectionForSectionIndexTitle:(NSString *)title
        atIndex:(NSInteger)index {
```

```
        NSString *key = [keys objectAtIndex:index];
        if (key == UITableViewIndexSearch) {
            [tableView setContentOffset:CGPointZero animated:NO];
            return NSNotFound;
        } else return index;
}
```

To tell it to go to the search box, we must do two things. First, we need to get rid of the content offset we added earlier, and then we must return NSNotFound. When the table view gets this response, it knows to scroll up to the top. So, now that we've removed the offset, it will scroll to the search bar rather than to the top section.

> **NOTE:** In the tableView:sectionForSectionIndexTitle:atIndex: method, we used a special constant called CGPointZero, which represents the point (0, 0) in the coordinate system. It's a handy, readable thing, but that constant requires the use of the Core Graphics framework. When you build your project, if you get a link error complaining about a reference to _CGPointZero, you'll know that Xcode did not include that framework by default, and you'll need to add it yourself. To add this framework, go to the project navigator and select the top-level *Sections* item. Next, click the *Build Phases* tab at the top of the main pane. Then expand the *Link Binary With Libraries* section, click the plus button, select *CoreGraphics.framework* from the list that appears, and click the *Add* button.

Now you can build and run your app. And there you have it—live searching in an iPhone table, with a magnifying glass in the index!

> **TIP:** iOS includes even more cool search stuff. Interested? Go to the documentation browser and do a search for UISearchDisplay to read up on UISearchDisplayController and UISearchDisplayDelegate. You'll likely find this material much easier to understand once you've made your way through Chapter 9.

Putting It All on the Table

Well, how are you doing? This was a pretty hefty chapter, and you've learned a ton! You should have a very solid understanding of the way that flat tables work. You should know how to customize tables and table view cells, as well as how to configure table views. You also saw how to implement a search bar, which is a vital tool in any iOS application that presents large volumes of data. Make sure you understand everything we did in this chapter, because we're going to build on it.

We're going to continue working with table views in the next chapter. You'll learn how to use them to present hierarchical data. You'll see how to create content views that allow the user to edit data selected in a table view, as well as how to present checklists in tables, embed controls in table rows, and delete rows.

Chapter 9

Navigation Controllers and Table Views

In the previous chapter, you mastered the basics of working with table views. In this chapter, you'll get a whole lot more practice, because we're going to explore **navigation controllers.**

Table views and navigation controllers work hand in hand. Strictly speaking, a navigation controller doesn't need a table view in order to do its thing. As a practical matter, however, when you implement a navigation controller, you almost always implement at least one table, and usually several, because the strength of the navigation controller lies in the ease with which it handles complex hierarchical data. On the iPhone's small screen, hierarchical data is best presented using a succession of table views.

In this chapter, we're going to build an application progressively, just as we did with the Pickers application back in Chapter 7. We'll get the navigation controller and the first view controller working, and then we'll start adding more controllers and layers to the hierarchy. Each view controller we create will reinforce some aspect of table use or configuration:

- How to drill down from table views into child tables

- How to drill down from table views into content views, where detailed data can be viewed and even edited

- How to use a table list to allow the user to select from multiple values

- How to use edit mode to allow rows to be deleted from a table view

That's a lot, isn't it? Well, let's get started with an introduction to navigation controllers.

Navigation Controller Basics

The main tool you'll use to build hierarchical applications is UINavigationController. UINavigationController is similar to UITabBarController in that it manages, and swaps in and out, multiple content views. The main difference between the two is that `UINavigationController` is implemented as a stack, which makes it well suited to working with hierarchies.

Do you already know everything there is to know about stacks? Scan through the following subsection, and we'll meet you at the beginning of the next subsection, "A Stack of Controllers." If you're new to stacks, continue reading. Fortunately, stacks are a pretty easy concept to grasp.

Stacky Goodness

A **stack** is a commonly used data structure that works on the principle of last in, first out. Believe it or not, a Pez dispenser is a great example of a stack. Ever try to load one? According to the little instruction sheet that comes with each and every Pez dispenser, there are a few easy steps. First, unwrap the pack of Pez candy. Second, open the dispenser by tipping its head straight back. Third, grab the stack (notice the clever way we inserted the word "stack" in there!) of candy, holding it firmly between your pointer finger and thumb, and insert the column into the open dispenser. Fourth, pick up all the little pieces of candy that flew all over the place because these instructions just never work.

OK, so far this example has not been particularly useful. But what happens next is. As you pick up the pieces and jam them, one at a time, into the dispenser, you are working with a stack. Remember that we said a stack was last in, first out? That also means first in, last out. The first piece of Pez you push into the dispenser will be the last piece that pops out. The last piece of Pez you push in will be the first piece you pop out. A computer stack follows the same rules:

- When you add an object to a stack, it's called a **push**. You push an object onto the stack.

- The first object you push onto the stack is called the **base** of the stack.

- The last object you pushed onto the stack is called the **top** of the stack (at least until it is replaced by the next object you push onto the stack).

- When you remove an object from the stack, it's called a **pop.** When you pop an object off the stack, it's always the last one you pushed onto the stack. Conversely, the first object you push onto the stack will always be the last one you pop off the stack.

A Stack of Controllers

A navigation controller maintains a stack of view controllers. Any kind of view controller is fair game for the stack. When you design your navigation controller, you'll need to specify the very first view the user sees. As we've discussed in previous chapters, that view is called the **root view controller**, or just **root controller**, and is the base of the navigation controller's stack of view controllers. As the user selects the next view to display, a new view controller is pushed onto the stack, and the view it controls appears. We refer to these new view controllers as **subcontrollers.** As you'll see, this chapter's application, Nav, is made up of a navigation controller and six subcontrollers.

Take a look at Figure 9–1. Notice the **navigation button** in the upper-left corner of the current view. The navigation button is similar to a web browser's back button. When the user taps that button, the current view controller is popped off the stack, and the previous view becomes the current view.

Navigation Button  Title

Figure 9–1. *The Settings application uses a navigation controller. In the upper left is the navigation button used to pop the current view controller off the stock, returning you to the previous level of the hierarchy. The title of the current content view controller is also displayed.*

We love this design pattern. It allows us to build complex hierarchical applications iteratively. We don't need to know the entire hierarchy to get things up and running. Each controller only needs to know about its child controllers so it can push the appropriate new controller object onto the stack when the user makes a selection. You can build up a large application from many small pieces this way, which is exactly what we're going to do in this chapter.

The navigation controller is really the heart and soul of many iPhone apps, but when it comes to iPad apps, the navigation controller plays a more marginal role. A typical example of this is the Mail app, which features a hierarchical navigation controller to let the user navigate among all their mail servers, folders, and messages. In the iPad version of Mail, the navigation controller never fills the screen, but appears either as a sidebar or a temporary popover window. We'll dig into that usage a little later, when we cover iPad-specific GUI functionality in Chapter 11.

Nav, a Hierarchical Application in Six Parts

The application we're about to build will show you how to do most of the common tasks associated with displaying a hierarchy of data. When the application launches, you'll be presented with a list of options (see Figure 9–2).

Figure 9–2. *This chapter application's top-level view. Note the accessory icons on the right side of the view. This particular type of accessory icon is called a disclosure indicator. It tells the user that touching that row drills down to another table view.*

Each of the rows in this top-level view represents a different view controller that will be pushed onto the navigation controller's stack when that row is selected. The icons on the right side of each row are called **accessory icons.** This particular accessory icon (the gray arrow) is called a **disclosure indicator**, because it lets the user know that touching that row drills down to another table view.

Meet the Subcontrollers

Before we start building the Nav application, let's take a quick look at each of the views displayed by our six subcontrollers.

The Disclosure Button View

Touching the first row of the table shown in Figure 9–2 will bring up the child view shown in Figure 9–3.

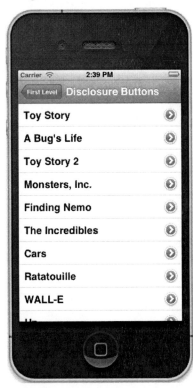

Figure 9–3. *The first of the Nav application's six subcontrollers implements a table in which each row contains a detail disclosure button.*

The accessory icon to the right of each row in Figure 9–3 is a bit different. Each of these icons is known as a **detail disclosure button**. Tapping the detail disclosure button should allow the user to view, and perhaps edit, more detailed information about the current row.

Unlike the disclosure indicator, the detail disclosure button is not just an icon—it's a control that the user can tap. This means that you can have two different options available for a given row: one action triggered when the user selects the row and another action triggered when the user taps the disclosure button.

A good example of the proper use of the detail disclosure button is found in the iPhone's Phone application. Selecting a person's row from the *Favorites* tab places a call to the person whose row you touched, but selecting the disclosure button next to a name takes you to detailed contact information. The YouTube application offers another great example. Selecting a row plays a video, but tapping the detail disclosure button takes you to more detailed information about the video. In the Contacts application, the list of contacts does not feature detail disclosure buttons, even though selecting a row does take you to a detail view. Since there is only one option available for each row in the Contacts application, no accessory icon is displayed.

Here's a recap of when to use disclosure indicators and detail disclosure buttons:

- If you want to offer a single choice for a row tap, don't use an accessory icon if a row tap will *only* lead to a more detailed view of that row.

- Mark the row with a disclosure indicator (gray arrow) if a row tap will lead to a new view (*not* a detail view).

- If you want to offer two choices for a row, mark the row with a detail disclosure button. This allows the user to tap on the row for a new view or the disclosure button for more details.

The Checklist View

The second of our application's six subcontrollers is shown in Figure 9–4. This is the view that appears when you select *Check One* in Figure 9–2.

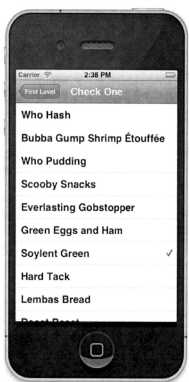

Figure 9–4. *The second of the Nav application's six subcontrollers allows you to select one row from many.*

This view comes in handy when you want to present a list from which only one item can be selected. This approach is to iOS what radio buttons are to Mac OS X. These lists use a check mark to mark the currently selected row.

The Rows Control View

The third of our application's six subcontrollers is shown in Figure 9–5. This view features a tappable button in each row's **accessory view.** The accessory view is the far-right part of the table view cell that usually holds the accessory icon, but it can be used for other things. When we get to this part of our application, you'll see how to create controls in the accessory view.

Figure 9–5. *The third of the Nav application's six subcontrollers adds a button to the accessory view of each table view cell.*

The Movable Rows View

The fourth of our application's six subcontrollers is shown in Figure 9–6. In this view, we'll let the user rearrange the order of the rows in a list by having the table enter edit mode (more on this when we get to it in code later in this chapter).

Figure 9–6. *The fourth of the Nav application's six subcontrollers lets the user rearrange rows in a list by touching and dragging the move icon. Recognize the rhyme?*

The Deletable Rows View

The fifth of our application's six subcontrollers is shown in Figure 9–7. In this view, we're going to demonstrate another use of edit mode by allowing the user to delete rows from our table.

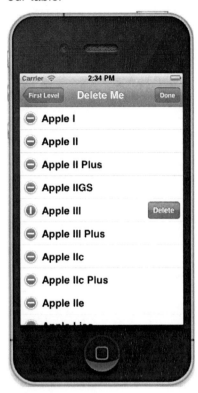

Figure 9–7. *The fifth of the Nav application's six subcontrollers implements edit mode to allow the user to delete items from the table.*

The Editable Detail View

The sixth and last of our application's subcontrollers is shown in Figure 9–8. It shows an editable detail view using a grouped table. This technique for detail views is used widely by the applications that ship on the iPhone.

Figure 9–8. *The sixth and last of the Nav application's subcontrollers implements an editable detail view using a grouped table.*

We have so very much to do. Let's get started!

The Nav Application's Skeleton

Xcode offers a perfectly good template for creating navigation-based applications, and you will likely use it much of the time when you need to create hierarchical applications. However, we're not going to use that template today. Instead, we'll construct our navigation-based application from the ground up so you get a feel for how everything fits together. It's not really much different from the way we built the tab bar controller in Chapter 7, so you shouldn't have any problems keeping up.

In Xcode, press ⌘⇧N to create a new project, select *Empty Application* from the iOS *Application* template list, and then click *Next* to continue. Set *Nav* as the *Product Name*, *com.apress* as the *Company Identifier*, and *BID* as the *Class Prefix*. Make sure that *Use*

Core Data and *Include Unit Tests* are not checked, that *Use Automatic Reference Counting* is checked, and that *Device Family* is set to *iPhone*.

As you'll see if you select the project navigator and open the *Nav* folder, this template gives you an application delegate and not much else. At this point, there are no view controllers or navigation controllers.

To make this app run, we'll need to add a navigation controller, which includes a navigation bar. We'll also need to add a series of views and view controllers for the navigation bar to show. The first of these views is the top-level view shown in Figure 9–2.

Each row in that top-level view is tied to a child view controller, as shown in Figures 9–3 through 9–8. Don't worry about the specifics. You'll see how those connections work as you make your way through the chapter.

Creating the Top-Level View Controller

In this chapter, we're going to subclass `UITableViewController` instead of `UIViewController` for our table views. When we subclass `UITableViewController`, we inherit some nice functionality from that class that will create a table view with no need for a nib file. We can provide a table view in a nib, as we did in the previous chapter, but if we don't, `UITableViewController` will create a table view automatically. This table view will take up the entire space available and will connect the appropriate outlets in our controller class, making our controller class the delegate and data source for that table. When all you need for a specific controller is a table, subclassing `UITableViewController` is the way to go.

We'll create one class called `BIDFirstLevelController` that represents the first level in the navigation hierarchy. That's the table that contains one row for each of the second-level table views. Those second-level table views will each be represented by the `BIDSecondLevelViewController` class. You'll see how all this works as you make your way through the chapter.

In your project window, select the *Nav* folder in the project navigator, and then press ⌘N or select **File ➤ New ➤ New File...**. When the new file assistant comes up, select *Cocoa Touch*, select *Objective-C class,* and then click *Next*. On the next screen, enter *BIDFirstLevelController* in the *Class* field, and type *UITableViewController* in the *Subclass of* field. As always, be sure to check your spelling carefully before you click *Next*. Then make sure the *Nav* folder or group is selected in the file browser, *Group*, and *Target* controls before clicking *Create*.

You may have noticed an entry named *UIViewController* in the file template selector. That option provides you with a number of empty "stub" methods as a starting point to build a view controller, and even lets you pick a subclass of `UIViewController`, such as `UITableViewController`, with even more empty methods just waiting for you to plug in additional functionality. When creating your own applications, feel free to use those templates. We didn't use any of the view controller templates here, so we wouldn't need to spend time sorting through all the unneeded template methods, working out where to insert or delete code. By creating a plain Objective-C object, and simply setting its superclass to UITableViewController, we get a smaller, more manageable file.

Once the files have been created, single-click *BIDFirstLevelController.h,* and take a look at it.

```
#import <UIKit/UIKit.h>

@interface BIDFirstLevelController : UITableViewController

@end
```

Since the class we chose to subclass from is a UIKit class, Xcode handily imported UIKit instead of just Foundation. The two files we just created contain the controller class for the top-level view, as shown in Figure 9–2. Our next step is to set up our navigation controller.

Setting Up the Navigation Controller

Our goal here is to edit the application delegate to add our navigation controller's view to the application window.

Let's start by editing *BIDAppDelegate.h* to add a property, navController, to point to our navigation controller:

```
#import <UIKit/UIKit.h>

@interface BIDAppDelegate : UIResponder <UIApplicationDelegate>

@property (strong, nonatomic) UIWindow *window;
@property (strong, nonatomic) UINavigationController *navController;
@end
```

Next, we need to hop over to the implementation file, where we'll import a header for the view controller class we just created and add the @synthesize statement for navController. In the application:didFinishLaunchingWithOptions: method, we'll create navController, set it up with the initial view controller that it's going to display, and add its view as a subview of our application's window so that it is shown to the user. We'll explain each of those steps in a moment. For now, select *BIDAppDelegate.m,* and make the following changes:

```
#import "BIDAppDelegate.h"
#import "BIDFirstLevelController.h"

@implementation BIDAppDelegate

@synthesize window = _window;
@synthesize navController;

#pragma mark -
#pragma mark Application lifecycle

- (BOOL)application:(UIApplication *)application
        didFinishLaunchingWithOptions:(NSDictionary *)launchOptions {
```

```
    self.window = [[UIWindow alloc] initWithFrame:[[UIScreen mainScreen] bounds]];
    // Override point for customization after application launch

    BIDFirstLevelController *first = [[BIDFirstLevelController alloc]
        initWithStyle:UITableViewStylePlain];
    self.navController = [[UINavigationController alloc]
        initWithRootViewController:first];
    [self.window addSubview:navController.view];

    self.window.backgroundColor = [UIColor whiteColor];
    [self.window makeKeyAndVisible];

    return YES;
}

    .
    .
    .

@end
```

The lines we added to the application:didFinishLaunchingWithOptions: method deserve some attention. The first thing we did was to create an instance of BIDFirstLevelController.

```
    BIDFirstLevelController *first = [[BIDFirstLevelController alloc]
        initWithStyle:UITableViewStylePlain];
```

Since BIDFirstLevelController is a subclass of UITableViewController, it can use the methods defined there, including the handy initWithStyle: method, which lets you create a controller whose table view will be either plain or grouped (whichever you choose), without the need for a nib file. Many iOS applications are built around table views whose appearance is dictated entirely by the cells they contain, and don't need any nib customization for the table views themselves. Therefore, using the initWithStyle: method is a common shortcut for instantiating table view controllers without much ado.

Next, we created an instance of the navigation controller.

```
    self.navController = [[UINavigationController alloc]
        initWithRootViewController:first];
```

Here, we see that UINavigationController, like UITableViewController, has its own special initializer. Here, the initWithRootViewController: method lets us pass in the top-level controller that the navigation control should use to display its initial content—in this case, the BIDFirstLevelController referenced by the first variable.

Finally, we added navController's view to our window in order to display it.

```
    [self.window addSubview:navController.view];
```

It's worth taking a moment to think about this. What exactly is the view we're passing with the addSubview: method? It's a composite view provided by the navigation controller, which contains a combination of two things: the navigation bar at the top of the screen (which usually contains some sort of title and often a back button of some kind on the left), and the content of whatever the navigation controller's current view controller wants to display. In our case, the lower part of the display will be filled with the table view associated with the BIDFirstLevelController instance we created a few lines ago.

You'll learn more about how to control what the navigation controller shows in the navigation bar as we go forward. You'll also gain an understanding of how the navigation controller shifts focus from one subordinate view controller to another. For now, we've laid enough groundwork here that we can start defining what our own custom view controllers are going to do.

Now, we need a list of rows for our BIDFirstLevelController to display. In the previous chapter, we used simple arrays of strings to populate our table rows. In this application, the first-level view controller will manage a list of its subcontrollers, which we will be building throughout the chapter.

When we were designing this application, we decided that we wanted our first-level view controller to display an icon to the left of each of its subcontroller names. Instead of adding a UIImage property to every subcontroller, we'll create a subclass of UITableViewController that has a UIImage property to hold the row icon. We will then subclass this new class instead of subclassing UITableViewController directly. As a result, all of our subclasses will get that UIImage property for free, which will make our code much cleaner.

> **NOTE:** We will never actually create an instance of our new UITableViewController subclass. It exists solely to let us add a common item to the rest of the controllers we're going to write. In many languages, we would declare this as an **abstract class,** but Objective-C doesn't include any syntax to support abstract classes. We can make classes that aren't intended to be instantiated, but the Objective-C compiler won't actually prevent us from writing code that creates instances of such a class, the way that the compilers for many other languages might. Objective-C is much more permissive than most other popular languages, and this can be a little hard to get used to.

Single-click the *Nav* folder in Xcode, and then press ⌘**N** to bring up the new file assistant. Select *Cocoa Touch* from the left pane, select *Objective-C class,* and click *Next*. On the next screen, name the new class *BIDSecondLevelViewController*, and enter *UITableViewController* for *Subclass of*. Then click *Next* again, and go on and save the class files as usual. Once the new files are created, select *BIDSecondLevelViewController.h*, and make the following changes:

```
#import <UIKit/UIKit.h>

@interface BIDSecondLevelViewController : UITableViewController

@property (strong, nonatomic) UIImage *rowImage;
@end
```

Over in *BIDSecondLevelViewController.m*, add the following line of code:

```
#import "BIDSecondLevelViewController.h"

@implementation BIDSecondLevelViewController
@synthesize rowImage;
@end
```

Any controller class that we want to implement as a second-level controller—in other words, any controller that the user can navigate to directly from the first table shown in our application—should subclass BIDSecondLevelViewController instead of UITableViewController. Because we're subclassing BIDSecondLevelViewController, all of those classes will have a property they can use to store a row image, and we can write our code in BIDFirstLevelController before we've actually written any concrete second-level controller classes by using BIDSecondLevelViewController as a placeholder.

Let's implement our BIDFirstLevelController class now. Be sure to save the changes you made to BIDSecondLevelViewController. Then make these changes to *BIDFirstLevelController.h*:

```
#import <UIKit/UIKit.h>

@interface BIDFirstLevelController : UITableViewController
@property (strong, nonatomic) NSArray *controllers;
@end
```

The array we just added will hold the instances of the second-level view controllers. We'll use it to feed data to our table.

Add the following code to *BIDFirstLevelController.m,* and then come on back and gossip with us, 'K?

```
#import "BIDFirstLevelController.h"
#import "BIDSecondLevelViewController.h"

@implementation BIDFirstLevelController
@synthesize controllers;

- (void)viewDidLoad {
    [super viewDidLoad];
    self.title = @"First Level";
    NSMutableArray *array = [[NSMutableArray alloc] init];
    self.controllers = array;
}

- (void)viewDidUnload {
    [super viewDidUnload];
    self.controllers = nil;
}

#pragma mark -
#pragma mark Table Data Source Methods
- (NSInteger)tableView:(UITableView *)tableView
 numberOfRowsInSection:(NSInteger)section {
    return [self.controllers count];
}
```

```objc
- (UITableViewCell *)tableView:(UITableView *)tableView
        cellForRowAtIndexPath:(NSIndexPath *)indexPath {

    static NSString *FirstLevelCell = @"FirstLevelCell";
    UITableViewCell *cell = [tableView dequeueReusableCellWithIdentifier:
                                FirstLevelCell];
    if (cell == nil) {
        cell = [[UITableViewCell alloc]
                    initWithStyle:UITableViewCellStyleDefault
                    reuseIdentifier: FirstLevelCell];
    }
    // Configure the cell
    NSUInteger row = [indexPath row];
    BIDSecondLevelViewController *controller =
        [controllers objectAtIndex:row];
    cell.textLabel.text = controller.title;
    cell.imageView.image = controller.rowImage;
    cell.accessoryType = UITableViewCellAccessoryDisclosureIndicator;
    return cell;
}

#pragma mark -
#pragma mark Table View Delegate Methods
- (void)tableView:(UITableView *)tableView
        didSelectRowAtIndexPath:(NSIndexPath *)indexPath {
    NSUInteger row = [indexPath row];
    BIDSecondLevelViewController *nextController = [self.controllers
                                            objectAtIndex:row];
    [self.navigationController pushViewController:nextController
                                    animated:YES];
}

@end
```

First, notice that we've imported that new *BIDSecondLevelViewController.h* header file.
Doing that lets us use the `BIDSecondLevelViewController` class in our code so that the
compiler will know about the `rowImage` property.

Next comes the `viewDidLoad` method. The first thing we do is set `self.title`. A
navigation controller knows what to display in the title of its navigation bar by asking the
currently active controller for its title. Therefore, it's important to set the title for all
controller instances in a navigation-based application, so the users know where they are
at all times.

We then create a mutable array and assign it to the `controllers` property we declared
earlier. Later, when we're ready to add rows to our table, we will add view controllers to
this array, and they will show up in the table automatically. Selecting any row will
automatically cause the corresponding controller's view to be presented to the user.

> **TIP:** Did you notice that our `controllers` property is declared as an `NSArray`, but that we're creating an `NSMutableArray`? It's perfectly acceptable to assign a subclass to a property like this. In this case, we use the mutable array in `viewDidLoad` to make it easier to add new controllers in an iterative fashion, but we leave the property declared as an immutable array as a message to other code that it shouldn't be modifying this array.

The final piece of the `viewDidLoad` method is the call to `[super viewDidLoad]`. We do this because we are subclassing `UITableViewController`. You should always call `[super viewDidLoad]` when you override the `viewDidLoad` method, because there's no way to know if your parent class does something important in its own `viewDidLoad` method.

The `tableView:numberOfRowsInSection:` method here is identical to ones you've seen before. It simply returns the count from our array of controllers. The `tableView:cellForRowAtIndexPath:` method is also very similar to ones we've written in the past. It gets a dequeued cell, or creates a new one if none exists, and then grabs the controller object from the array corresponding to the row being asked about. It then sets the cell's `textLabel` and `image` properties using the `title` and `rowImage` from that controller. Note that in this case, since we are using one of `UITableViewCell`'s built-in styles instead of laying out a subclass of our own in a nib file, we have no nib file to register with the table view, and therefore can't rely on the `dequeue...` method returning anything. So, we need to include the check for `nil` and the resulting cell-creation code, as you've seen before.

Notice that we are assuming the object retrieved from the array is an instance of `BIDSecondLevelViewController` and are assigning the controller's `rowImage` property to a `UIImage`. This step will make more sense when we declare and add the first concrete second-level controller to the array.

The last method we added is the most important one here, and it's the only functionality that's truly new. You've seen the `tableView:didSelectRowAtIndexPath:` method before—it's the one that is called after a user taps a row. If tapping a row needs to trigger a drill-down, this is how we do it. First, we get the row from `indexPath`.

```
NSUInteger row = [indexPath row];
```

Next, we grab the correct controller from our array that corresponds to that row.

```
BIDSecondLevelViewController *nextController =
    [self.controllers objectAtIndex:row];
```

Then we use our `navigationController` property, which points to our application's navigation controller, to push the next controller—the one we pulled from our array— onto the navigation controller's stack.

```
[self.navigationController pushViewController:nextController
                                    animated:YES];
```

That's really all there is to it. Each controller in the hierarchy needs to know only about its children. When a row is selected, the active controller is responsible for getting or creating a new subcontroller, setting its properties if necessary (it's not necessary here), and then pushing that new subcontroller onto the navigation controller's stack. Once you've done that, everything else is handled automatically by the navigation controller.

At this point, the application skeleton is complete. Save all your files, and build and run the app. If all is well, the application should launch, and a navigation bar with the title *First Level* should appear. Since our array is currently empty, no rows will display at this point (see Figure 9–9).

Figure 9–9. *The application skeleton in action*

Adding the Images to the Project

Now, we're ready to start developing the second-level views. Before we do that, go grab the folder of image icons from the *09 Nav* source code archive directory. You'll find a folder called *Images* with eight *.png* images: six that will act as row images and an additional two that we'll use to make a button look nice later in the chapter.

In the project navigator, make sure you can see the *Nav* folder. Then drag the *Images* folder from the Finder to that *Nav* folder (*not* the *Nav* target just above the *Nav* folder) to add the images to the project.

First Subcontroller: The Disclosure Button View

Let's implement the first of our second-level view controllers. To do that, we'll need to create a subclass of BIDSecondLevelViewController.

In the project navigator, select the *Nav* folder and press ⌘**N** to bring up the new file assistant. Select *Cocoa Touch* in the left pane, and then select *Objective-C class* and click *Next*. On the following screen, name the class *BIDDisclosureButtonController* and enter *BIDSecondLevelViewController* for *Subclass of*. Remember to check your spelling! This class will manage the table of movie names that will be displayed when the user clicks the *Disclosure Buttons* item from the top-level view (see Figure 9–3).

Creating the Detail View

When the user clicks any movie title, the application will drill down into another view that will report which row was selected. So, we also need to create a detail view for the user to drill down to. Repeat the steps we just took to create another *Objective-C class* called *BIDDisclosureDetailController*, this time using *UIViewController* as the superclass. Again, be sure to check your spelling.

> **NOTE:** Just a reminder: BIDDisclosureButtonController keeps track of the table of movie names, while BIDDisclosureDetailController manages the next level down, which is the detail view that is pushed on the navigation stack when a specific movie is selected.

The detail view will hold just a single label that we can set. It won't be editable; we'll just use it to show how to pass values into a child controller. Because this controller will not be responsible for a table view, we also need a nib file to go along with the controller class. Before we create the nib, let's quickly add the outlet for the label. Make the following changes to *BIDDisclosureDetailController.h*:

```
#import <UIKit/UIKit.h>

@interface BIDDisclosureDetailController : UIViewController

@property (strong, nonatomic) IBOutlet UILabel *label;
@property (copy, nonatomic) NSString *message;
@end
```

Why, pray tell, are we adding both a label and a string? Remember the concept of lazy loading? Well, view controllers use lazy loading behind the scenes as well. When we create our controller, it won't load its nib file until it is actually displayed. When the controller is pushed onto the navigation controller's stack, we can't count on there being a label to set. If the nib file has not been loaded, label will just be a pointer set to nil. But it's OK. Instead, we'll set message to the value we want, and in the viewWillAppear: method, we'll set label based on the value in message.

Why are we using viewWillAppear: to do our updating instead of using viewDidLoad, as we've done in the past? The problem is that viewDidLoad is called only the first time the controller's view is loaded. But in our case, we are reusing the BIDDisclosureDetailController's view. No matter which fine Pixar flick you pick, when you tap the disclosure button, the detail message appears in the same BIDDisclosureDetailController view. If we used viewDidLoad to manage our updates, that view would be updated only the first time the BIDDisclosureDetailController view appeared. When we picked our second fine Pixar flick, we would still see the detail message from the first fine Pixar flick (try saying that ten times fast)—not good. Since viewWillAppear: is called every time a view is about to be drawn, we'll be fine using it for our updating.

Going back to the property declarations, you may notice that the message property is declared using the copy keyword instead of strong. What's up with that? Why should we be copying strings willy-nilly? The reason is the potential existence of mutable strings.

Imagine we had declared the property using strong, and an outside piece of code passed in an instance of NSMutableString to set the value of the message property. This is something that often happens when you're dealing with strings entered by the user in a user interface object. If that original caller later decides to change the content of that string, the BIDDisclosureDetailController instance will end up in an inconsistent state, where the value of message and the value displayed in the text field aren't the same! Using copy eliminates that risk, since calling copy on any NSString (including subclasses that are mutable) always gives us an immutable copy. Also, we don't need to worry about the performance impact too much. As it turns out, sending copy to any immutable string instance doesn't actually copy the string. Instead, it returns the same string object, after increasing its reference count. In effect, calling copy on an immutable string is the same as calling retain, which is what ARC might do behind the scenes anytime you set a strong property. So, it works out just fine for everyone, since the object can never change.

Add the following code to *BIDDisclosureDetailController.m*:

```
#import "BIDDisclosureDetailController.h"

@implementation BIDDisclosureDetailController
@synthesize label;
@synthesize message;

- (void)viewWillAppear:(BOOL)animated {
    label.text = message;
    [super viewWillAppear:animated];
}

- (void)viewDidUnload {
    self.label = nil;
    self.message = nil;
    [super viewDidUnload];
}

@end
```

That's all pretty straightforward, right? Now, let's create the nib to go along with this source code. Be sure you've saved your source changes.

Select the *Nav* folder in the project navigator, and press ⌘N to create another new file. This time, select *User Interface* from the *iOS* section on the left pane and *View* from the upper right. Then click *Next*. On the next screen, set *Device Family* to *iPhone*. Move to the next screen, and name this file *BIDDisclosureDetail.xib*. This file will implement the view seen when the user taps one of the movie buttons.

Select *BIDDisclosureDetail.xib* in the project navigator to open the file for editing. Once it's open, single-click *File's Owner*, and press ⌥⌘3 to bring up the identity inspector. Change the underlying class to *BIDDisclosureDetailController*. Now control-drag from the *File's Owner* icon to the *View* icon, and select the *view* outlet to establish a link from the controller to its view.

Drag a *Label* from the library, and place it on the *View* window, centering the label both vertically and horizontally. It doesn't need to be perfectly centered. Resize the label so it stretches from the left blue guideline to the right blue guideline, and then use the attributes inspector (⌥⌘4) to change the text alignment to centered. Control-drag from *File's Owner* to the label, and select the *label* outlet. Save your changes.

Modifying the Disclosure Button Controller

For this example, our table of movies will base its data on rows from an array, so we will declare an NSArray named list to serve that purpose. We also need to declare a property to hold one instance of our child controller, which will point to an instance of the BIDDisclosureDetailController class we just built. We could allocate a new instance of that controller class every time the user taps a detail disclosure button, but it's more efficient to create one and then keep reusing it. Make the following changes to *BIDDisclosureButtonController.h*:

```
#import "BIDSecondLevelViewController.h"

@interface BIDDisclosureButtonController : BIDSecondLevelViewController
@property (strong, nonatomic) NSArray *list;
@end
```

Now we get to the juicy part. Add the following code to *BIDDisclosureButtonController.m.* We'll talk about what's going on afterward.

```
#import "BIDDisclosureButtonController.h"
#import "BIDAppDelegate.h"
#import "BIDDisclosureDetailController.h"

@interface BIDDisclosureButtonController ()
@property (strong, nonatomic) BIDDisclosureDetailController *childController;
@end

@implementation BIDDisclosureButtonController

@synthesize list;
@synthesize childController;
```

```objc
- (void)viewDidLoad {
    [super viewDidLoad];
    NSArray *array = [[NSArray alloc] initWithObjects:@"Toy Story",
                      @"A Bug's Life", @"Toy Story 2", @"Monsters, Inc.",
                      @"Finding Nemo", @"The Incredibles", @"Cars",
                      @"Ratatouille", @"WALL-E", @"Up", @"Toy Story 3",
                      @"Cars 2", @"Brave", nil];
    self.list = array;
}

- (void)viewDidUnload {
    [super viewDidUnload];
    self.list = nil;
    self.childController = nil;
}

#pragma mark -
#pragma mark Table Data Source Methods
- (NSInteger)tableView:(UITableView *)tableView
 numberOfRowsInSection:(NSInteger)section {
    return [list count];
}

- (UITableViewCell *)tableView:(UITableView *)tableView
        cellForRowAtIndexPath:(NSIndexPath *)indexPath {

    static NSString * DisclosureButtonCellIdentifier =
    @"DisclosureButtonCellIdentifier";

    UITableViewCell *cell = [tableView dequeueReusableCellWithIdentifier:
                             DisclosureButtonCellIdentifier];
    if (cell == nil) {
        cell = [[UITableViewCell alloc]
                initWithStyle:UITableViewCellStyleDefault
                reuseIdentifier: DisclosureButtonCellIdentifier];
    }
    NSUInteger row = [indexPath row];
    NSString *rowString = [list objectAtIndex:row];
    cell.textLabel.text = rowString;
    cell.accessoryType = UITableViewCellAccessoryDetailDisclosureButton;
    return cell;
}

#pragma mark -
#pragma mark Table Delegate Methods
- (void)tableView:(UITableView *)tableView
didSelectRowAtIndexPath:(NSIndexPath *)indexPath {
    UIAlertView *alert = [[UIAlertView alloc] initWithTitle:
          @"Hey, do you see the disclosure button?"
          message:@"If you're trying to drill down, touch that instead"
          delegate:nil
          cancelButtonTitle:@"Won't happen again"
          otherButtonTitles:nil];
```

```
    [alert show];
}

- (void)tableView:(UITableView *)tableView
accessoryButtonTappedForRowWithIndexPath:(NSIndexPath *)indexPath {
    if (childController == nil) {
        childController = [[BIDDisclosureDetailController alloc]
                            initWithNibName:@"BIDDisclosureDetail" bundle:nil];
    }
    childController.title = @"Disclosure Button Pressed";
    NSUInteger row = [indexPath row];
    NSString *selectedMovie = [list objectAtIndex:row];
    NSString *detailMessage = [[NSString alloc]
            initWithFormat:@"You pressed the disclosure button for %@.",
            selectedMovie];
    childController.message = detailMessage;
    childController.title = selectedMovie;
    [self.navigationController pushViewController:childController
                                    animated:YES];
}

@end
```

Right near the top of that big chunk, you may have noticed the following @interface declaration, just where you may have expected an @implementation section to start instead:

```
@interface BIDDisclosureButtonController ()
@property (strong, nonatomic) BIDDisclosureDetailController *childController;
@end
```

This kind of category declaration, where the parentheses are empty rather than containing the name of the category you're declaring, is called a **class extension**. This is a handy place to declare properties and methods that will be in the main @implementation section containing your class, but that you don't want to show up in the public header file.

A class extension is a good place to put a property for the childController. We are using this property internally in our class and don't want to expose it to others, so we don't advertise its existence by declaring it in the header.

By now, you should be fairly comfortable with pretty much everything up to and including the three data source methods we just wrote. Let's look at the two delegate methods we added, which you haven't seen before.

The first method, tableView:didSelectRowAtIndexPath:, is called when the row is selected. It puts up a polite little alert telling the user to tap the disclosure button instead of selecting the row. If the user actually taps the detail disclosure button, the other one of our new delegate methods, tableView:accessoryButtonTappedForRowWithIndexPath:, is called.

The first thing we do in tableView:accessoryButtonTappedForRowWithIndexPath: is check the childController instance variable to see if it's nil. If it is, we have not yet allocated and initialized a new instance of `BIDDetailDisclosureController`, so we do that next.

```
if (childController == nil)
    childController = [[BIDDisclosureDetailController alloc]
                        initWithNibName:@"BIDDisclosureDetail" bundle:nil];
```

This gives us a new controller that we can push onto the navigation stack, just as we did earlier in `BIDFirstLevelController`. Before we push it onto the stack, though, we need to give it some text to display.

```
childController.title = @"Disclosure Button Pressed";
```

In this case, we set `message` to reflect the row whose disclosure button was tapped. We also set the new view's title based on the selected row.

```
NSUInteger row = [indexPath row];
NSString *selectedMovie = [list objectAtIndex:row];
NSString *detailMessage = [[NSString alloc]
        initWithFormat:@"You pressed the disclosure button for %@.",
        selectedMovie];
childController.message = detailMessage;
childController.title = selectedMovie;
```

Finally, we push the detail view controller onto the navigation stack.

```
[self.navigationController pushViewController:childController
                                    animated:YES];
```

And, with that, our first second-level controller is complete, as is our detail controller. The only remaining task is to create an instance of our second-level controller and add it to `BIDFirstLevelController`'s controllers.

Adding a Disclosure Button Controller Instance

Select *BIDFirstLevelController.m*. Up at the top of the file, we'll need to add one line of code to import the header file for our new class. Insert this line directly above the `@implementation` declaration:

`#import "BIDDisclosureButtonController.h"`

Then insert the following code in the `viewDidLoad` method:

```
- (void)viewDidLoad {
    [super viewDidLoad];
    self.title = @"First Level";
    NSMutableArray *array = [[NSMutableArray alloc] init];

    // Disclosure Button
    BIDDisclosureButtonController *disclosureButtonController =
        [[BIDDisclosureButtonController alloc]
          initWithStyle:UITableViewStylePlain];
    disclosureButtonController.title = @"Disclosure Buttons";
    disclosureButtonController.rowImage = [UIImage
        imageNamed:@"disclosureButtonControllerIcon.png"];
    [array addObject:disclosureButtonController];
```

```
        self.controllers = array;
}
```

All that we're doing is creating a new instance of BIDDisclosureButtonController. We specify UITableViewStylePlain to indicate that we want a normal table, not a grouped table. Next, we set the title and the image to one of the *.png* files we added to our project, add the controller to the array, and release the controller.

Save your changes, and try building. If everything went as planned, your project should compile and then launch in the simulator. When it comes up, there should be just a single row (see Figure 9–10).

Figure 9–10. *Our application after adding the first of six second-level controllers*

If you touch the one row, it will take you down to the BIDDisclosureButtonController table view we just implemented (see Figure 9–11).

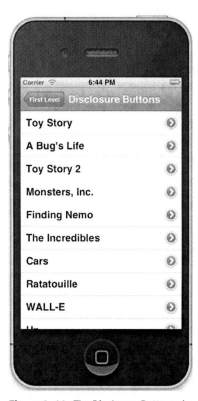

Figure 9–11. *The Disclosure Buttons view*

Notice that the title that we set for our controller is now displayed in the navigation bar, and the title of the view controller we were previously using (*First Level*) is contained in a navigation button. Tapping that button will take you back up to the first level. Select any row in this table, and you will get a gentle reminder that the detail disclosure button is there for drilling down (see Figure 9–12).

Figure 9–12. *Selecting the row does not drill down when there is a detail disclosure button visible.*

If you touch the detail disclosure button itself, you drill down into the BIDDisclosureDetailController view (see Figure 9–13). This view shows information that we passed into it. Even though this is a simple example, the same basic technique is used anytime you show a detail view.

Figure 9–13. *The detail view*

Notice that when you drill down to the detail view, the title again changes, as does the back button, which now takes you to the previous view instead of the root view.

That finishes up the first view controller. Do you see now how the design Apple used here with the navigation controller makes it possible to build your application in small chunks? That's pretty cool, isn't it?

Second Subcontroller: The Checklist

The next second-level view we're going to implement is another table view. But this time, we'll use the accessory icon to let the user select one and only one item from the list. We'll use the accessory icon to place a check mark next to the currently selected row, and we'll change the selection when the user touches another row.

Since this view is a table view and it has no detail view, we don't need a new nib, but we do need to create another subclass of BIDSecondLevelViewController. Select the *Nav* folder in the project navigator in Xcode, and then select File ➤ New ➤ New File. . . or press ⌘N. Select *Cocoa Touch* on the left, *Objective-C class* on the right, and click *Next*. Then name your new class *BIDCheckListController*, enter *BIDSecondLevelViewController* for *Subclass of*, and click the *Next* button. On the final screen, make sure the *Nav* folder, *Group*, and *Target* are selected (just as you've done for the other classes in this project).

Creating the Checklist View

To present a checklist, we need a way to keep track of which row is currently selected. We'll declare an NSIndexPath property to track the last row selected. Single-click *BIDCheckListController.h*, and make the following changes:

```
#import "BIDSecondLevelViewController.h"

@interface BIDCheckListController : BIDSecondLevelViewController

@property (strong, nonatomic) NSArray *list;
@property (strong, nonatomic) NSIndexPath *lastIndexPath;
@end
```

Then switch over to *BIDCheckListController.m*, and add the following code:

```
#import "BIDCheckListController.h"

@implementation BIDCheckListController
@synthesize list;
@synthesize lastIndexPath;

- (void)viewDidLoad {
    [super viewDidLoad];
    NSArray *array = [[NSArray alloc] initWithObjects:@"Who Hash",
        @"Bubba Gump Shrimp Étouffée", @"Who Pudding", @"Scooby Snacks",
        @"Everlasting Gobstopper", @"Green Eggs and Ham", @"Soylent Green",
        @"Hard Tack", @"Lembas Bread", @"Roast Beast", @"Blancmange", nil];
    self.list = array;
}

- (void)viewDidUnload {
    [super viewDidUnload];
    self.list = nil;
    self.lastIndexPath = nil;
}

#pragma mark -
#pragma mark Table Data Source Methods
- (NSInteger)tableView:(UITableView *)tableView
 numberOfRowsInSection:(NSInteger)section {
    return [list count];
}

- (UITableViewCell *)tableView:(UITableView *)tableView
        cellForRowAtIndexPath:(NSIndexPath *)indexPath {
    static NSString *CheckMarkCellIdentifier = @"CheckMarkCellIdentifier";

    UITableViewCell *cell = [tableView dequeueReusableCellWithIdentifier:
                             CheckMarkCellIdentifier];
    if (cell == nil) {
        cell = [[UITableViewCell alloc]
            initWithStyle:UITableViewCellStyleDefault
            reuseIdentifier:CheckMarkCellIdentifier];
```

```
    }
    NSUInteger row = [indexPath row];
    NSUInteger oldRow = [lastIndexPath row];
    cell.textLabel.text = [list objectAtIndex:row];
    cell.accessoryType = (row == oldRow && lastIndexPath != nil) ?
    UITableViewCellAccessoryCheckmark : UITableViewCellAccessoryNone;

    return cell;
}

#pragma mark -
#pragma mark Table Delegate Methods
- (void)tableView:(UITableView *)tableView
didSelectRowAtIndexPath:(NSIndexPath *)indexPath {
    int newRow = [indexPath row];
    int oldRow = (lastIndexPath != nil) ? [lastIndexPath row] : -1;

    if (newRow != oldRow) {
        UITableViewCell *newCell = [tableView cellForRowAtIndexPath:
                                        indexPath];
        newCell.accessoryType = UITableViewCellAccessoryCheckmark;

        UITableViewCell *oldCell = [tableView cellForRowAtIndexPath:
                                        lastIndexPath];
        oldCell.accessoryType = UITableViewCellAccessoryNone;
        lastIndexPath = indexPath;
    }
    [tableView deselectRowAtIndexPath:indexPath animated:YES];
}

@end
```

Let's start with the tableView:cellForRowAtIndexPath: method, which has a few new things worth noticing. The first several lines should be familiar to you.

```
static NSString *CheckMarkCellIdentifier = @"CheckMarkCellIdentifier";

UITableViewCell *cell = [tableView dequeueReusableCellWithIdentifier:
    CheckMarkCellIdentifier];
if (cell == nil) {
    cell = [[UITableViewCell alloc]
            initWithStyle:UITableViewCellStyleDefault
            reuseIdentifier:CheckMarkCellIdentifier];
}
```

Next is where things get interesting. First, we extract the row from this cell and from the current selection.

```
NSUInteger row = [indexPath row];
NSUInteger oldRow = [lastIndexPath row];
```

We grab the value for this row from our array and assign it to the cell's title.

```
cell.textLabel.text = [list objectAtIndex:row];
```

Then we set the accessory to show either a check mark or nothing, depending on whether the two rows are the same. In other words, if the table is requesting a cell for a row that is the currently selected row, we set the accessory icon to be a check mark; otherwise, we set it to be nothing. Notice that we also check lastIndexPath to make sure it's not nil. We do this because a nil lastIndexPath indicates no selection. However, calling the row method on a nil object will return a 0, which is a valid row, but we don't want to put a check mark on row 0 when, in reality, there is no selection.

```
cell.accessoryType = (row == oldRow && lastIndexPath != nil) ?
    UITableViewCellAccessoryCheckmark : UITableViewCellAccessoryNone;
```

Now skip down to the last method. You've seen the tableView:didSelectRowAtIndexPath: method before, but we're doing something new here. We grab not only the row that was just selected, but also the row that was previously selected.

```
int newRow = [indexPath row];
int oldRow = [lastIndexPath row];
```

We do this so if the new row and the old row are the same, we don't bother making any changes.

```
if (newRow != oldRow) {
```

Next, we grab the cell that was just selected and assign a check mark as its accessory icon.

```
UITableViewCell *newCell = [tableView
    cellForRowAtIndexPath:indexPath];
newCell.accessoryType = UITableViewCellAccessoryCheckmark;
```

We then grab the previously selected cell, and we set its accessory icon to none.

```
UITableViewCell *oldCell = [tableView cellForRowAtIndexPath:
    lastIndexPath];
oldCell.accessoryType = UITableViewCellAccessoryNone;
```

After that, we store the index path that was just selected in lastIndexPath, so we'll have it the next time a row is selected.

```
    lastIndexPath = indexPath;
}
```

When we're finished, we tell the table view to deselect the row that was just selected, because we don't want the row to stay highlighted. We've already marked the row with a check mark, so leaving it blue would just be a distraction.

```
    [tableView deselectRowAtIndexPath:indexPath animated:YES];
}
```

Adding a Checklist Controller Instance

Our next task is to add an instance of this controller to BIDFirstLevelController's controllers array. Start off by importing the new header file, adding this line just after all the other #import statements at the top of the file:

```
#import "BIDCheckListController.h"
```

Then create an instance of BIDCheckListController by adding the following code to the viewDidLoad method in *BIDFirstLevelController.m*:

```
- (void)viewDidLoad {
    [super viewDidLoad];
    self.title = @"First Level";
    NSMutableArray *array = [[NSMutableArray alloc] init];

    // Disclosure Button
    BIDDisclosureButtonController *BIDDisclosureButtonController =
        [[BIDDisclosureButtonController alloc]
        initWithStyle:UITableViewStylePlain];
    BIDDisclosureButtonController.title = @"Disclosure Buttons";
    BIDDisclosureButtonController.rowImage = [UIImage imageNamed:
        @"BIDDisclosureButtonControllerIcon.png"];
    [array addObject:BIDDisclosureButtonController];

    // Checklist
    BIDCheckListController *checkListController = [[BIDCheckListController alloc]
        initWithStyle:UITableViewStylePlain];
    checkListController.title = @"Check One";
    checkListController.rowImage = [UIImage imageNamed:
        @"checkmarkControllerIcon.png"];
    [array addObject:checkListController];

    self.controllers = array;
}
```

Well, what are you waiting for? Save your changes, compile, and run. If everything went smoothly, the application launched again in the simulator, and there was much rejoicing. This time there will be two rows (see Figure 9–14).

Figure 9–14. *Two second-level controllers and two rows. What a coincidence!*

If you touch the *Check One* row, it will take you down to the view controller we just implemented (see Figure 9–15). When it first comes up, no rows will be selected and no check marks will be visible. If you tap a row, a check mark will appear. If you then tap a different row, the check mark will switch to the new row. Huzzah!

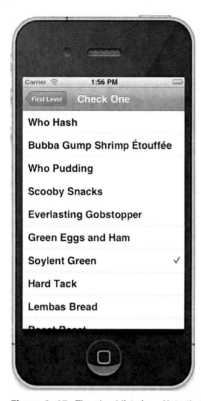

Figure 9–15. *The checklist view. Note that only a single item can be checked at a time. Soylent Green, anyone?*

Third Subcontroller: Controls on Table Rows

In the previous chapter, we showed you how to add subviews to a table view cell to customize its appearance. However, we didn't put any active controls into the content view; it had only labels. Now let's see how to add controls to a table view cell.

In our example, we'll add a button to each row, but the same technique will work with most controls. We'll add the control to the accessory view, which is the area on the right side of each row where you found the accessory icons covered earlier in the chapter.

To add another row to our BIDFirstLevelController's table, we need another second-level controller. You know the drill: select the *Nav* folder in the project navigator, and then press ⌘N or select **File** ➤ **New** ➤ **New File. . ..** Select *Cocoa Touch*, select *Objective-C class*, and click *Next*. Name the class *BIDRowControlsController*, and enter *BIDSecondLevelViewController* for *Subclass of*. Save the file in the *Nav* folder, with *Nav* selected for both *Target* and *Group*, as usual. Just as with the previous subcontroller, this controller can be completely implemented with a single table view; no nib file is necessary.

Creating the Row Controls View

Single-click *BIDRowControlsController.h*, and make the following changes:

```
#import "BIDSecondLevelViewController.h"

@interface BIDRowControlsController : BIDSecondLevelViewController

@property (strong, nonatomic) NSArray *list;
- (IBAction)buttonTapped:(id)sender;
@end
```

Not much there, huh? We change the parent class and create an array to hold our table data. Then we define a property for that array and declare an action method that will be called when the row buttons are pressed.

> **NOTE:** Strictly speaking, we don't need to declare the buttonTapped: method an action method by specifying IBAction, since we won't be triggering it from controls in a nib file. Since it is an action method and will be called by a control, however, it's still a good idea to use the IBAction keyword, since it signals our intent to future readers of this code.

Switch over to *BIDRowControlsController.m,* and make the following changes:

```
#import "BIDRowControlsController.h"

@implementation BIDRowControlsController
@synthesize list;

- (IBAction)buttonTapped:(id)sender {
    UIButton *senderButton = (UIButton *)sender;
    UITableViewCell *buttonCell =
        (UITableViewCell *)[senderButton superview];
    NSUInteger buttonRow = [[self.tableView
        indexPathForCell:buttonCell] row];
    NSString *buttonTitle = [list objectAtIndex:buttonRow];
    UIAlertView *alert = [[UIAlertView alloc]
                    initWithTitle:@"You tapped the button"
                    message:[NSString stringWithFormat:
                        @"You tapped the button for %@", buttonTitle]
                    delegate:nil
                    cancelButtonTitle:@"OK"
                    otherButtonTitles:nil];
    [alert show];
}

- (void)viewDidLoad {
    [super viewDidLoad];
    NSArray *array = [[NSArray alloc] initWithObjects:@"R2-D2",
            @"C3PO", @"Tik-Tok", @"Robby", @"Rosie", @"Uniblab",
            @"Bender", @"Marvin", @"Lt. Commander Data",
            @"Evil Brother Lore", @"Optimus Prime", @"Tobor", @"HAL",
```

```
                @"Orgasmatron", nil];
      self.list = array;
}

- (void)viewDidUnload {
    [super viewDidUnload];
    self.list = nil;
}

#pragma mark -
#pragma mark Table Data Source Methods
- (NSInteger)tableView:(UITableView *)tableView
        numberOfRowsInSection:(NSInteger)section {
    return [list count];
}

- (UITableViewCell *)tableView:(UITableView *)tableView
        cellForRowAtIndexPath:(NSIndexPath *)indexPath {
    static NSString *ControlRowIdentifier = @"ControlRowIdentifier";

    UITableViewCell *cell = [tableView
        dequeueReusableCellWithIdentifier:ControlRowIdentifier];
    if (cell == nil) {
        cell = [[UITableViewCell alloc]
                    initWithStyle:UITableViewCellStyleDefault
                    reuseIdentifier:ControlRowIdentifier];
        UIImage *buttonUpImage = [UIImage imageNamed:@"button_up.png"];
        UIImage *buttonDownImage = [UIImage imageNamed:@"button_down.png"];
        UIButton *button = [UIButton buttonWithType:UIButtonTypeCustom];
        button.frame = CGRectMake(0.0, 0.0, buttonUpImage.size.width,
            buttonUpImage.size.height);
        [button setBackgroundImage:buttonUpImage
            forState:UIControlStateNormal];
        [button setBackgroundImage:buttonDownImage
            forState:UIControlStateHighlighted];
        [button setTitle:@"Tap" forState:UIControlStateNormal];
        [button addTarget:self action:@selector(buttonTapped:)
            forControlEvents:UIControlEventTouchUpInside];
        cell.accessoryView = button;
    }
    NSUInteger row = [indexPath row];
    NSString *rowTitle = [list objectAtIndex:row];
    cell.textLabel.text = rowTitle;

    return cell;
}

#pragma mark -
#pragma mark Table Delegate Methods
- (void)tableView:(UITableView *)tableView
        didSelectRowAtIndexPath:(NSIndexPath *)indexPath {
    NSUInteger row = [indexPath row];
    NSString *rowTitle = [list objectAtIndex:row];
    UIAlertView *alert = [[UIAlertView alloc]
```

```
                    initWithTitle:@"You tapped the row."
                    message:[NSString
                    stringWithFormat:@"You tapped %@.", rowTitle]
                    delegate:nil
                    cancelButtonTitle:@"OK"
                    otherButtonTitles:nil];
        [alert show];
        [tableView deselectRowAtIndexPath:indexPath animated:YES];
}

@end
```

Let's begin with our new action method. The first thing we do is declare a new UIButton variable and set it to sender. This is just so we don't need to cast sender multiple times throughout our method.

```
UIButton *senderButton = (UIButton *)sender;
```

Next, we get the button's superview, which is the table view cell for the row it's in, and we use that to determine the row that was pressed and to retrieve the title for that row.

```
UITableViewCell *buttonCell =
    (UITableViewCell *)[senderButton superview];
NSUInteger buttonRow = [[self.tableView
    indexPathForCell:buttonCell] row];
NSString *buttonTitle = [list objectAtIndex:buttonRow];
```

Then we show an alert, saying that the user pressed the button.

```
UIAlertView *alert = [[UIAlertView alloc]
                initWithTitle:@"You tapped the button"
                message:[NSString stringWithFormat:
                    @"You tapped the button for %@", buttonTitle]
                delegate:nil
                cancelButtonTitle:@"OK"
                otherButtonTitles:nil];
[alert show];
```

Everything from there to tableView:cellForRowAtIndexPath: should be familiar to you, so skip down to that method, which is where we set up the table view cell with the button. The method starts as usual. We declare an identifier and then use it to request a reusable cell.

```
static NSString *ControlRowIdentifier = @"ControlRowIdentifier";
UITableViewCell *cell = [tableView
    dequeueReusableCellWithIdentifier:ControlRowIdentifier];
```

If there are no reusable cells, we create one.

```
if (cell == nil) {
    cell = [[UITableViewCell alloc]
                initWithStyle:UITableViewCellStyleDefault
                reuseIdentifier:ControlRowIdentifier];
```

To create the button, we load in two of the images that were in the *Images* folder you imported earlier. One will represent the button in the normal state; the other will represent the button in its highlighted state—in other words, when the button is being tapped.

```
UIImage *buttonUpImage = [UIImage imageNamed:@"button_up.png"];
UIImage *buttonDownImage = [UIImage imageNamed:@"button_down.png"];
```

Next, we create a button. Because the buttonType property of UIButton is declared read-only, we need to create the button using the factory method buttonWithType:. If we created it using alloc and init, we wouldn't be able to change the button's type to UIButtonTypeCustom, which we need to do in order to use the custom button images.

```
UIButton *button = [UIButton buttonWithType:UIButtonTypeCustom];
```

Next, we set the button's size to match the images, assign the images for the two states, and give the button a title.

```
button.frame = CGRectMake(0.0, 0.0, buttonUpImage.size.width,
    buttonUpImage.size.height);
[button setBackgroundImage:buttonUpImage
    forState:UIControlStateNormal];
[button setBackgroundImage:buttonDownImage
    forState:UIControlStateHighlighted];
[button setTitle:@"Tap" forState:UIControlStateNormal];
```

Finally, we tell the button to call our action method on the touch up inside event and assign it to the cell's accessory view.

```
[button addTarget:self action:@selector(buttonTapped:)
    forControlEvents:UIControlEventTouchUpInside];
cell.accessoryView = button;
```

Everything else in the tableView:cellForRowAtIndexPath: method is just as we've done it in the past.

The last method we implemented is tableView:didSelectRowAtIndexPath:, which is the delegate method that is called after the user selects a row. All we do here is find out which row was selected and grab the appropriate title from our array.

```
NSUInteger row = [indexPath row];
NSString *rowTitle = [list objectAtIndex:row];
```

Then we create another alert to inform the user that a row was tapped, but not the button.

```
UIAlertView *alert = [[UIAlertView alloc]
                    initWithTitle:@"You tapped the row."
                    message:[NSString
                    stringWithFormat:@"You tapped %@.", rowTitle]
                    delegate:nil
                    cancelButtonTitle:@"OK"
                    otherButtonTitles:nil];
[alert show];
[tableView deselectRowAtIndexPath:indexPath animated:YES];
```

Adding a Rows Control Controller Instance

Now, all we need to do is add this controller to the array in BIDFirstLevelController.
Single-click *BIDFirstLevelController.m*, and import the header file for the
BIDRowControlsController class by adding the following line of code just before the
@implementation line:

```
#import "BIDRowControlsController.h"
```

Then move on and add the following code to viewDidLoad:

```
- (void)viewDidLoad {
    [super viewDidLoad];
    self.title = @"Root Level";
    NSMutableArray *array = [[NSMutableArray alloc] init];

    // Disclosure Button
    BIDDisclosureButtonController *BIDDisclosureButtonController =
        [[BIDDisclosureButtonController alloc]
        initWithStyle:UITableViewStylePlain];
    BIDDisclosureButtonController.title = @"Disclosure Buttons";
    BIDDisclosureButtonController.rowImage = [UIImage
        imageNamed:@"BIDDisclosureButtonControllerIcon.png"];
    [array addObject:BIDDisclosureButtonController];
    [BIDDisclosureButtonController release];

    // Checklist
    BIDCheckListController *checkListController = [[BIDCheckListController alloc]
            initWithStyle:UITableViewStylePlain];
    checkListController.title = @"Check One";
    checkListController.rowImage = [UIImage
        imageNamed:@"checkmarkControllerIcon.png"];
    [array addObject:checkListController];
    [checkListController release];

    // Table Row Controls
    BIDRowControlsController *rowControlsController =
        [[BIDRowControlsController alloc]
        initWithStyle:UITableViewStylePlain];
    rowControlsController.title = @"Row Controls";
    rowControlsController.rowImage = [UIImage imageNamed:
        @"rowControlsIcon.png"];
    [array addObject:rowControlsController];

    self.controllers = array;
}
```

Save everything, and compile it. This time, you should see yet another row when your
application launches (see Figure 9–16).

Figure 9–16. *The Row Controls controller added to the root level controller*

If you tap this new row, it will take you down to a new list where every row has a button control on the right side of the row. Tapping either the button or the row will show an alert telling you which one you tapped (see Figure 9–17).

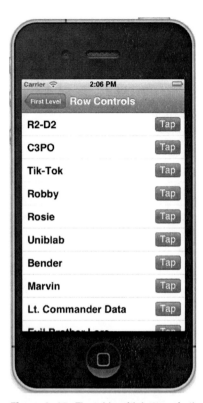

Figure 9–17. *The table with buttons in the accessory view*

Tapping a row anywhere but on its switch will display an alert telling you whether the switch for that row is turned on or off.

At this point, you should be getting pretty comfortable with how this all works, so let's try a slightly more difficult case, shall we? Next, we'll take a look at how to allow the user to reorder the rows in a table.

> **NOTE:** How are you doing? Hanging in there? We know this chapter is a bit of a marathon, with a lot of stuff to absorb. At this point, you've already accomplished a lot. Why not take a break, and grab a Fresca and a pastel de Belém? We'll do the same. Come back when you're refreshed and ready to move on.

Fourth Subcontroller: Movable Rows

Moving and deleting rows, as well as inserting rows at a specific spot in the table, are tasks that can be implemented fairly easily. All three are implemented by turning on something called **editing mode**, which is done using the setEditing:animated: method on the table view.

The setEditing:animated: method takes two Boolean values. The first indicates whether you are turning on or off editing mode, and the second indicates whether the table should animate the transition. If you set editing to the mode it's already in (in other words, turning it on when it's already on or off when it's already off), the transition will not be animated, regardless of what you specify in the second parameter.

Once editing mode is turned on, a number of new delegate methods come into play. The table view uses them to ask if a certain row can be moved or edited, and again to notify you if the user actually does move or edit a specific row. It sounds more complex than it is. Let's see it in action in our movable row controller.

Because we don't need to display a detail view, this view controller can be implemented without a nib and with just a single controller class. Select the *Nav* folder in the project navigator in Xcode, and then press ⌘N or select **File ➤ New ➤ New File....** Select *Cocoa Touch*, select *Objective-C class*, and click *Next*. Then enter *BIDMoveMeController* as the class name, and enter *BIDSecondLevelViewController* in the *Subclass of* control. Click *Next* again, and save the class files as usual.

Creating the Movable Row View

In our header file, we need two things. First, we need a mutable array to hold our data and keep track of the order of the rows. It must be mutable because we need to be able to move items around as we are notified of moves. We also need an action method to toggle edit mode on and off. The action method will be called by a navigation bar button that we will create.

Single-click *BIDMoveMeController.h,* and make the following changes:

```
#import "BIDSecondLevelViewController.h"

@interface BIDMoveMeController : BIDSecondLevelViewController

@property (strong, nonatomic) NSMutableArray *list;
- (IBAction)toggleMove;
@end
```

Now, switch over to *BIDMoveMeController.m*, and add the following code:

```
#import "BIDMoveMeController.h"

@implementation BIDMoveMeController
@synthesize list;

- (IBAction)toggleMove {
    [self.tableView setEditing:!self.tableView.editing animated:YES];

    if (self.tableView.editing)
        [self.navigationItem.rightBarButtonItem setTitle:@"Done"];
    else
        [self.navigationItem.rightBarButtonItem setTitle:@"Move"];
}
```

```objc
- (void)viewDidLoad {
    [super viewDidLoad];
    if (list == nil) {
        NSMutableArray *array = [[NSMutableArray alloc] initWithObjects:
                    @"Eeny", @"Meeny", @"Miney", @"Moe", @"Catch", @"A",
                    @"Tiger", @"By", @"The", @"Toe", nil];
        self.list = array;
    }

    UIBarButtonItem *moveButton = [[UIBarButtonItem alloc]
                            initWithTitle:@"Move"
                            style:UIBarButtonItemStyleBordered
                            target:self
                            action:@selector(toggleMove)];
    self.navigationItem.rightBarButtonItem = moveButton;
}

#pragma mark -
#pragma mark Table Data Source Methods
- (NSInteger)tableView:(UITableView *)tableView
        numberOfRowsInSection:(NSInteger)section {
    return [list count];
}

- (UITableViewCell *)tableView:(UITableView *)tableView
        cellForRowAtIndexPath:(NSIndexPath *)indexPath {

    static NSString *MoveMeCellIdentifier = @"MoveMeCellIdentifier";
    UITableViewCell *cell = [tableView
        dequeueReusableCellWithIdentifier:MoveMeCellIdentifier];
    if (cell == nil) {
        cell = [[UITableViewCell alloc]
                    initWithStyle:UITableViewCellStyleDefault
                    reuseIdentifier:MoveMeCellIdentifier];
        cell.showsReorderControl = YES;
    }
    NSUInteger row = [indexPath row];
    cell.textLabel.text = [list objectAtIndex:row];

    return cell;
}

- (UITableViewCellEditingStyle)tableView:(UITableView *)tableView
            editingStyleForRowAtIndexPath:(NSIndexPath *)indexPath {
    return UITableViewCellEditingStyleNone;
}

- (BOOL)tableView:(UITableView *)tableView
        canMoveRowAtIndexPath:(NSIndexPath *)indexPath {
    return YES;
}

- (void)tableView:(UITableView *)tableView
```

```
        moveRowAtIndexPath:(NSIndexPath *)fromIndexPath
        toIndexPath:(NSIndexPath *)toIndexPath {
        NSUInteger fromRow = [fromIndexPath row];
        NSUInteger toRow = [toIndexPath row];

        id object = [list objectAtIndex:fromRow];
        [list removeObjectAtIndex:fromRow];
        [list insertObject:object atIndex:toRow];
    }

    @end
```

Let's take this one step at a time. The first code we added is the implementation of our action method.

```
- (IBAction)toggleMove {
    [self.tableView setEditing:!self.tableView.editing animated:YES];

    if (self.tableView.editing)
        [self.navigationItem.rightBarButtonItem setTitle:@"Done"];
    else
        [self.navigationItem.rightBarButtonItem setTitle:@"Move"];
}
```

All that we're doing here is toggling edit mode and then setting the button's title to an appropriate value. Easy enough, right?

The next method we touched is viewDidLoad. The first part of that method doesn't do anything you haven't seen before. It checks to see if list is nil, and if it is (meaning this is the first time this method has been called), it creates a mutable array filled with values, so our table has some data to show. After that, though, there is something new.

```
    UIBarButtonItem *moveButton = [[UIBarButtonItem alloc]
            initWithTitle:@"Move"
            style:UIBarButtonItemStyleBordered
            target:self
            action:@selector(toggleMove)];
    self.navigationItem.rightBarButtonItem = moveButton;
```

Here, we're creating a button bar item, which is a button that will sit on the navigation bar. We give it a title of *Move* and specify a constant, UIBarButtonItemStyleBordered, to indicate that we want a standard bordered bar button. The last two arguments, target and action, tell the button what to do when it is tapped. By passing self as the target and giving it a selector to the toggleMove method as the action, we are telling the button to call our toggleMove method whenever the button is tapped. As a result, any time the user taps this button, editing mode will be toggled. After we create the button, we add it to the right side of the navigation bar, and then release it.

Unlike most view controllers we create, this one does not have a viewDidUnload method. That's intentional. We have no outlets, and if we were to flush our list array, we would lose any reordering that the user had done when the view is flushed, which we don't want to happen. Therefore, since we have nothing to do in the viewDidUnload method, we don't bother to override it.

Now, skip down to the `tableView:cellForRowAtIndexPath:` method we just added. Did you notice the following new line of code?

```
cell.showsReorderControl = YES;
```

Standard accessory icons can be specified by setting the `accessoryType` property of the cell. But the reorder control is not a standard accessory icon. It's a special case that's shown only when the table is in edit mode. To enable the reorder control, we need to set a property on the cell itself. Note, though, that setting this property to `YES` doesn't actually display the reorder control until the table is put into edit mode. Everything else in this method is stuff we've done before.

The next new method is short but important. In our table view, we want to be able to reorder the rows, but we don't want the user to be able to delete or insert rows. As a result, we implement the method `tableView:editingStyleForRowAtIndexPath:`. This method allows the table view to ask if a specific row can be deleted or if a new row can be inserted at a specific spot. By returning `UITableViewCellEditingStyleNone` for each row, we are indicating that we don't support inserts or deletes for any row.

Next comes the method `tableView:canMoveRowAtIndexPath:`. This method is called for each row, and it gives you the chance to disallow the movement of specific rows. If you return `NO` from this method for any row, the reorder control will not be shown for that row, and the user will be unable to move it from its current position. We want to allow full reordering, so we just return `YES` for every row.

The last method, `tableView:moveRowAtIndexPath:fromIndexPath:`, is the one that will actually be called when the user moves a row. The two parameters besides `tableView` are both `NSIndexPath` instances that identify the row that was moved and the row's new position. The table view has already moved the rows in the table, so the user is seeing the correct display, but we need to update our data model to keep the two in sync and avoid causing display problems.

First, we retrieve the row that needs to be moved. Then we retrieve the row's new position.

```
NSUInteger fromRow = [fromIndexPath row];
NSUInteger toRow = [toIndexPath row];
```

We now need to remove the specified object from the array and reinsert it at its new location.

```
id object = [list objectAtIndex:fromRow];
[list removeObjectAtIndex:fromRow];
```

After we've removed it, we need to reinsert it into the specified new location.

```
[list insertObject:object atIndex:toRow];
```

Well, there you have it. We've implemented a table that allows reordering of rows.

Adding a Move Me Controller Instance

Now, we just need to add an instance of this new class to `BIDFirstLevelController`'s array of controllers. You're probably comfortable doing this by now, but we'll walk you through it just to keep you company.

In *BIDFirstLevelController.m,* import the new view's header file by adding the following line of code just before the @implementation declaration:

```
#import "BIDMoveMeController.h"
```

Now, add the following code to the viewDidLoad method in the same file:

```
- (void)viewDidLoad {
    [super viewDidLoad];
    self.title = @"First Level";
    NSMutableArray *array = [[NSMutableArray alloc] init];

    // Disclosure Button
    BIDDisclosureButtonController *BIDDisclosureButtonController =
        [[BIDDisclosureButtonController alloc]
        initWithStyle:UITableViewStylePlain];
    BIDDisclosureButtonController.title = @"Disclosure Buttons";
    BIDDisclosureButtonController.rowImage = [UIImage
        imageNamed:@"BIDDisclosureButtonControllerIcon.png"];
    [array addObject:BIDDisclosureButtonController];

    // Checklist
    BIDCheckListController *checkListController = [[BIDCheckListController alloc]
        initWithStyle:UITableViewStylePlain];
    checkListController.title = @"Check One";
    checkListController.rowImage = [UIImage
        imageNamed:@"checkmarkControllerIcon.png"];
    [array addObject:checkListController];

    // Table Row Controls
    BIDRowControlsController *rowControlsController =
        [[BIDRowControlsController alloc]
        initWithStyle:UITableViewStylePlain];
    rowControlsController.title = @"Row Controls";
    rowControlsController.rowImage = [UIImage imageNamed:
        @"rowControlsIcon.png"];
    [array addObject:rowControlsController];

    // Move Me
    BIDMoveMeController *moveMeController = [[BIDMoveMeController alloc]
        initWithStyle:UITableViewStylePlain];
    moveMeController.title = @"Move Me";
    moveMeController.rowImage = [UIImage imageNamed:@"moveMeIcon.png"];
    [array addObject:moveMeController];

    self.controllers = array;
}
```

OK, let's go ahead and compile this bad boy and see what shakes out. If everything went smoothly, our application will launch in the simulator with (count 'em) four rows in the root-level table (see Figure 9–18). If you click the new one, called *Move Me*, you'll go to a table whose rows make up a familiar childhood choosing rhyme (see Figure 9–6).

Figure 9–18. *The Move Me row has been added to our table.*

To reorder the rows, click the *Move* button in the upper-right corner, and the reorder controls should appear. If you tap a row's reorder control and then drag, the row should move as you drag, as in Figure 9–6. Once you are happy with the row's new position, release the drag. The row should settle into its new position nicely. You can even navigate back up to the top level and come back down, and your rows will be right where you left them. If you quit and return, they will be restored to their original order, but don't worry; in a few chapters, we'll show you how to save and restore data on a more permanent basis.

NOTE: If you find you have a bit of trouble making contact with the move control, don't panic. This gesture actually requires a little patience. Try holding the mouse button clicked (if you're in the simulator) or your finger pressed on the control (if you're on a device) a bit longer before moving it, in order to make the drag-to-reorder gesture work.

Now let's move on to the fifth subcontroller, which demonstrates another use of edit mode. This time, we'll allow the user to delete our precious rows. Gasp!

Fifth Subcontroller: Deletable Rows

Letting users delete rows isn't really significantly harder than letting them move rows. Let's take a look at that process.

Instead of creating an array from a hard-coded list of objects, we're going to load a property list file this time, just to save some typing. You can grab the file called *computers.plist* out of the *09 Nav* folder in the projects archive that accompanies this book and add it to the *Nav* folder of your Xcode project.

Select the *Nav* folder in the project navigator, and then press ⌘N or select File ➤ New ➤ New File.... Select *Cocoa Touch*, select *Objective-C class,* and click *Next*. Name your new class *BIDDeleteMeController*, and enter *BIDSecondLevelViewController* for *Subclass of.*

Creating the Deletable Rows View

The changes we're going to make to *BIDDeleteMeController.h* should look familiar, as they're nearly identical to the ones we made in the movable rows view controller we just built. Go ahead and make these changes now:

```
#import "BIDSecondLevelViewController.h"

@interface BIDDeleteMeController : BIDSecondLevelViewController

@property (strong, nonatomic) NSMutableArray *list;
- (IBAction)toggleEdit:(id)sender;
@end
```

No surprises here, right? We declare a mutable array to hold our data and an action method to toggle edit mode.

In the previous controller, we used edit mode to let the users reorder rows. In this version, edit mode will be used to let them delete rows. You can actually combine both in the same table if you like. We separated them so the concepts would be a bit easier to follow, but the delete and reorder operations do play nicely together.

A row that can be reordered will display the reorder icon anytime that the table is in edit mode. When you tap the red, circular icon on the left side of the row (see Figure 9–7), the *Delete* button will pop up, obscuring the reorder icon, but only temporarily.

Switch over to *BIDDeleteMeController.m,* and add the following code:

```
#import "BIDDeleteMeController.h"

@implementation BIDDeleteMeController
@synthesize list;

- (IBAction)toggleEdit:(id)sender {
```

```objc
    [self.tableView setEditing:!self.tableView.editing animated:YES];

    if (self.tableView.editing)
        [self.navigationItem.rightBarButtonItem setTitle:@"Done"];
    else
        [self.navigationItem.rightBarButtonItem setTitle:@"Delete"];
}

- (void)viewDidLoad {
    [super viewDidLoad];
    if (list == nil) {
        NSString *path = [[NSBundle mainBundle]
            pathForResource:@"computers" ofType:@"plist"];
        NSMutableArray *array = [[NSMutableArray alloc]
                                 initWithContentsOfFile:path];
        self.list = array;
    }
    UIBarButtonItem *editButton = [[UIBarButtonItem alloc]
                                   initWithTitle:@"Delete"
                                   style:UIBarButtonItemStyleBordered
                                   target:self
                                   action:@selector(toggleEdit:)];
    self.navigationItem.rightBarButtonItem = editButton;
}

#pragma mark -
#pragma mark Table Data Source Methods
- (NSInteger)tableView:(UITableView *)tableView
        numberOfRowsInSection:(NSInteger)section {
    return [list count];
}

- (UITableViewCell *)tableView:(UITableView *)tableView
         cellForRowAtIndexPath:(NSIndexPath *)indexPath {
    static NSString *DeleteMeCellIdentifier = @"DeleteMeCellIdentifier";

    UITableViewCell *cell = [tableView dequeueReusableCellWithIdentifier:
                             DeleteMeCellIdentifier];

    if (cell == nil) {
        cell = [[UITableViewCell alloc]
            initWithStyle:UITableViewCellStyleDefault
            reuseIdentifier:DeleteMeCellIdentifier];
    }
    NSInteger row = [indexPath row];
    cell.textLabel.text = [self.list objectAtIndex:row];
    return cell;
}

#pragma mark -
#pragma mark Table View Data Source Methods
- (void)tableView:(UITableView *)tableView
    commitEditingStyle:(UITableViewCellEditingStyle)editingStyle
    forRowAtIndexPath:(NSIndexPath *)indexPath {
```

```
        NSUInteger row = [indexPath row];
        [self.list removeObjectAtIndex:row];
        [tableView deleteRowsAtIndexPaths:[NSArray arrayWithObject:indexPath]
                    withRowAnimation:UITableViewRowAnimationAutomatic];
}

@end
```

Here, the new action method, `toggleEdit:`, is pretty much the same as our previous version. It sets edit mode to on if it's currently off and vice versa, and then sets the button's title as appropriate. The `viewDidLoad` method is also similar to the one from the previous view controller and, again, we do not have a `viewDidUnload` method because we have no outlets and we want to preserve changes made to our mutable array in edit mode. The only difference is that we're loading our array from a property list rather than feeding it a hard-coded list of strings. The property list we're using is a flat array of strings containing a variety of computer model names that might be a bit familiar. We also assign the name *Delete* to the edit button, to make the button's effect obvious to the user.

The two data source methods contain nothing new, but the last method in the class is something you've never seen before, so let's take a closer look at it.

```
- (void)tableView:(UITableView *)tableView
    commitEditingStyle:(UITableViewCellEditingStyle)editingStyle
    forRowAtIndexPath:(NSIndexPath *)indexPath {
```

This method is called by the table view when the user has made an edit, which means a deletion or an insertion. The first argument is the table view on which a row was edited. The second parameter, `editingStyle`, is a constant that tells us what kind of edit just happened. Currently, three editing styles are defined:

- **UITableViewCellEditingStyleNone**: We used this style in the previous controller to indicate that a row can't be edited. The option `UITableViewCellEditingStyleNone` will never be passed into this method, because it is used to indicate that editing is not allowed for this row.

- **UITableViewCellEditingStyleDelete**: This is the default option. We ignore this parameter, because the default editing style for rows is the delete style, so we know that every time this method is called, it will be requesting a delete. You can use this parameter to allow both inserts and deletes within a single table.

- **UITableViewCellEditingStyleInsert**: This is generally used when you need to let the user insert rows at a specific spot in a list. In a list whose order is maintained by the system, such as an alphabetical list of names, the user will usually tap a toolbar or navigation bar button to ask the system to create a new object in a detail view. Once the user is finished specifying the new object, the system will place in the appropriate row.

The last parameter, indexPath, tells us which row is being edited. For a delete, this index path represents the row to be deleted. For an insert, it represents the index where the new row should be inserted.

> **NOTE:** We won't be covering the use of inserts, but the insert functionality works in fundamentally the same way as the delete functionality we are about to implement. The only difference is that instead of deleting the specified row from your data model, you need to create a new object and insert it at the specified spot.

In our method, we first retrieve the row that is being edited from indexPath.

```
NSUInteger row = [indexPath row];
```

Then we remove the object from the mutable array we created earlier.

```
[self.list removeObjectAtIndex:row];
```

Finally, we tell the table to delete the row, specifying the constant UITableViewRowAnimationAutomatic, which sets the animation so that the row disappears as either the rows below or the rows above appear to slide over it. The table view will decide the direction of the sliding animation on its own, depending on which row is being deleted.

```
[tableView deleteRowsAtIndexPaths:[NSArray arrayWithObject:indexPath]
    withRowAnimation:UITableViewRowAnimationAutomatic];
}
```

> **NOTE:** Several types of animation are available for table views. You can look up UITableViewRowAnimation in Xcode's document browser to see what other animations are available.

And that's all she wrote, folks. That's the whole enchilada for this class.

Adding a Delete Me Controller Instance

Now, let's add an instance of the new controller to our root view controller and try it out. In *BIDFirstLevelController.m,* we first need to import our new controller class's header file, so add the following line of code directly before the @implementation declaration:

```
#import "BIDDeleteMeController.h"
```

Next, add the following code to the viewDidLoad method:

```
- (void)viewDidLoad {
    [super viewDidLoad];
    self.title = @"First Level";
    NSMutableArray *array = [[NSMutableArray alloc] init];

    // Disclosure Button
    BIDDisclosureButtonController *disclosureButtonController =
```

```
        [[BIDDisclosureButtonController alloc]
            initWithStyle:UITableViewStylePlain];
    disclosureButtonController.title = @"Disclosure Buttons";
    disclosureButtonController.rowImage = [UIImage imageNamed:
        @"disclosureButtonControllerIcon.png"];
    [array addObject:disclosureButtonController];

    // Checklist
    BIDCheckListController *checkListController = [[BIDCheckListController alloc]
        initWithStyle:UITableViewStylePlain];
    checkListController.title = @"Check One";
    checkListController.rowImage = [UIImage imageNamed:
        @"checkmarkControllerIcon.png"];
    [array addObject:checkListController];

    // Table Row Controls
    RowControlsController *rowControlsController =
        [[RowControlsController alloc]
        initWithStyle:UITableViewStylePlain];
    rowControlsController.title = @"Row Controls";
    rowControlsController.rowImage = [UIImage imageNamed:
        @"rowControlsIcon.png"];
    [array addObject:rowControlsController];

    // Move Me
    BIDMoveMeController *moveMeController = [[BIDMoveMeController alloc]
        initWithStyle:UITableViewStylePlain];
    moveMeController.title = @"Move Me";
    moveMeController.rowImage = [UIImage imageNamed:@"moveMeIcon.png"];
    [array addObject:moveMeController];

    // Delete Me
    BIDDeleteMeController *deleteMeController = [[BIDDeleteMeController alloc]
        initWithStyle:UITableViewStylePlain];
    deleteMeController.title = @"Delete Me";
    deleteMeController.rowImage = [UIImage imageNamed:@"deleteMeIcon.png"];
    [array addObject:deleteMeController];

    self.controllers = array;
}
```

Save everything, compile, and let her rip. When the simulator comes up, the root level will now have—can you guess?—five rows. If you select the new *Delete Me* row, you'll be presented with a list of computer models (see Figure 9–19). How many of these have you owned?

Figure 9–19. *The Delete Me view when it first launches. Recognize any of these computers?*

Notice that we again have a button on the right side of the navigation bar, this time labeled *Delete*. If you tap that, the table enters edit mode, which looks like Figure 9–20.

Figure 9–20. *The Delete Me view in edit mode*

Next to each editable row is a little icon that looks a little like a Do Not Enter street sign. If you tap the icon, it rotates sideways, and a button labeled *Delete* appears (see Figure 9–7). Tapping that button will cause its row to be deleted, both from the underlying model as well as from the table, using the animation style we specified.

When you implement edit mode to allow deletions, you get additional functionality for free. Swipe your finger horizontally across a row. Look at that! The *Delete* button comes up for just that row, just as in the Mail application.

We're coming around the bend, now, and the finish line is in sight, albeit still a little ways in the distance. If you're still with us, give yourself a pat on the back, or have someone do it for you. This is a long, tough chapter.

Sixth Subcontroller: An Editable Detail Pane

The next concept we're going to explore is how to implement a reusable editable detail view. You may notice as you look through the various applications that come on your iPhone that many of them, including the Contacts application, implement their detail views as a grouped table (see Figure 9–21).

Figure 9–21. *An example of a grouped table view being used to present an editable table view*

Let's look at how to do this now. Before we begin, we need some data to show, and we need more than just a list of strings. In the previous two chapters, when we needed more complex data, such as with the multiline table in Chapter 8 or the ZIP codes picker in Chapter 7, we used an NSArray to hold a bunch of NSDictionary instances filled with our data. That works fine and is very flexible, but it's a little hard to work with. For this table's data, let's create a custom Objective-C data object to hold the individual instances that will be displayed in the list.

Creating the Data Model Object

The property list we'll be using in this section of the application contains data about the US presidents: each president's name, his party, the year he took office, and the year he left office. Let's create the class to hold that data.

Once again, single-click the *Nav* folder in Xcode to select it, and then press ⌘N to bring up the new file assistant. Select *Cocoa Touch* from the left pane, *Objective-C class* from the right pane, and click *Next*. Then name the new class *BIDPresident*, and select *NSObject* for *Subclass of*.

Click *BIDPresident.h,* and make the following changes:

```
#import <Foundation/Foundation.h>
```

```
#define kPresidentNumberKey          @"President"
#define kPresidentNameKey            @"Name"
#define kPresidentFromKey            @"FromYear"
#define kPresidentToKey              @"ToYear"
#define kPresidentPartyKey           @"Party"

@interface BIDPresident : NSObject
@interface BIDPresident : NSObject <NSCoding>

@property int number;
@property (nonatomic, copy) NSString *name;
@property (nonatomic, copy) NSString *fromYear;
@property (nonatomic, copy) NSString *toYear;
@property (nonatomic, copy) NSString *party;
@end
```

The five constants will be used to identify the fields when they are read from the file
system. Conforming this class to the NSCoding protocol is what allows this object to be
written to and created from files. The rest of the new stuff we've added to this header file
is there to implement the properties needed to hold our data. Switch over to
BIDPresident.m, and make these changes:

```
#import "BIDPresident.h"

@implementation BIDPresident
@synthesize number;
@synthesize name;
@synthesize fromYear;
@synthesize toYear;
@synthesize party;

#pragma mark -
#pragma mark NSCoding
- (void)encodeWithCoder:(NSCoder *)coder {
    [coder encodeInt:self.number forKey:kPresidentNumberKey];
    [coder encodeObject:self.name forKey:kPresidentNameKey];
    [coder encodeObject:self.fromYear forKey:kPresidentFromKey];
    [coder encodeObject:self.toYear forKey:kPresidentToKey];
    [coder encodeObject:self.party forKey:kPresidentPartyKey];
}

- (id)initWithCoder:(NSCoder *)coder {
    if (self = [super init]) {
        number = [coder decodeIntForKey:kPresidentNumberKey];
        name = [coder decodeObjectForKey:kPresidentNameKey];
        fromYear = [coder decodeObjectForKey:kPresidentFromKey];
        toYear = [coder decodeObjectForKey:kPresidentToKey];
        party = [coder decodeObjectForKey:kPresidentPartyKey];
    }
    return self;
}

@end
```

Don't worry too much about the encodeWithCoder: and initWithCoder: methods. We'll be covering those in more detail in Chapter 13. All you need to know for now is that these two methods are part of the NSCoding protocol, which can be used to save objects to disk and load them back in. The encodeWithCoder: method encodes our object to be saved. initWithCoder: is used to create new objects from the saved file. These methods will allow us to create BIDPresident objects from a property list archive file. Everything else in this class should be fairly self-explanatory.

We've provided you with a property list file that contains data for all the US presidents and can be used to create new instances of the BIDPresident object we just wrote. We will be using this in the next section, so you don't need to type in a whole bunch of data. Grab the *Presidents.plist* file from the *09 Nav* folder in the project archive, and add it to the *Nav* folder of your project.

Now, we're ready to write our two controller classes.

Creating the Detail View List Controller

For this part of the application, we're going to need two new controllers: one that will show the list to be edited, and another to view and edit the details of the item selected in that list. Since both of these view controllers will be based on tables, we won't need to create any nib files, but we will need two separate controller classes. Let's create the files for both classes now and then implement them.

Select the *Nav* folder in the project navigator, and then press ⌘N or select File ➤ New ➤ New File…. Select *Cocoa Touch,* select *Objective-C class*, and click *Next*. Name the new class *BIDPresidentsViewController*, and enter *BIDSecondLevelViewController* for *Subclass of*. Be sure to check your spelling.

Repeat the same process a second time using the name *BIDPresidentDetailController*, but that time use *UITableViewController* in the *Subclass of* field.

> **NOTE:** In case you were wondering, BIDPresidentDetailController is singular (as opposed to BIDPresidentsDetailController) because it deals with the details of a single president. Yes, we actually had a fistfight about that little detail, but one intense paintball session later, we are friends again.

Let's create the view controller that shows the list of presidents first. Single-click *BIDPresidentsViewController.h,* and make the following changes:

```
#import "BIDSecondLevelViewController.h"

@interface BIDPresidentsViewController : BIDSecondLevelViewController

@property (strong, nonatomic) NSMutableArray *list;
@end
```

Then switch over to *BIDPresidentsViewController.m* and make the following changes:

```objc
#import "BIDPresidentsViewController.h"
#import "BIDPresidentDetailController.h"
#import "BIDPresident.h"

@implementation BIDPresidentsViewController
@synthesize list;

- (void)viewDidLoad {
    [super viewDidLoad];
    NSString *path = [[NSBundle mainBundle] pathForResource:@"Presidents"
                                                     ofType:@"plist"];

    NSData *data;
    NSKeyedUnarchiver *unarchiver;

    data = [[NSData alloc] initWithContentsOfFile:path];
    unarchiver = [[NSKeyedUnarchiver alloc] initForReadingWithData:data];
    NSMutableArray *array = [unarchiver decodeObjectForKey:@"Presidents"];
    self.list = array;
    [unarchiver finishDecoding];
}

- (void)viewWillAppear:(BOOL)animated {
    [super viewWillAppear:animated];
    [self.tableView reloadData];
}

#pragma mark -
#pragma mark Table Data Source Methods
- (NSInteger)tableView:(UITableView *)tableView
 numberOfRowsInSection:(NSInteger)section {
    return [list count];
}

- (UITableViewCell *)tableView:(UITableView *)tableView
         cellForRowAtIndexPath:(NSIndexPath *)indexPath {

    static NSString *PresidentListCellIdentifier =
        @"PresidentListCellIdentifier";

    UITableViewCell *cell = [tableView
        dequeueReusableCellWithIdentifier:PresidentListCellIdentifier];
    if (cell == nil) {
        cell = [[UITableViewCell alloc]
            initWithStyle:UITableViewCellStyleSubtitle
            reuseIdentifier:PresidentListCellIdentifier];
    }
    NSUInteger row = [indexPath row];
    BIDPresident *thePres = [self.list objectAtIndex:row];
    cell.textLabel.text = thePres.name;
    cell.detailTextLabel.text = [NSString stringWithFormat:@"%@ - %@",
        thePres.fromYear, thePres.toYear];
    return cell;
}
```

```
#pragma mark -
#pragma mark Table Delegate Methods
- (void)tableView:(UITableView *)tableView
didSelectRowAtIndexPath:(NSIndexPath *)indexPath {
    NSUInteger row = [indexPath row];
    BIDPresident *prez = [self.list objectAtIndex:row];

    BIDPresidentDetailController *childController =
    [[BIDPresidentDetailController alloc] initWithStyle:UITableViewStyleGrouped];

    childController.title = prez.name;
    childController.president = prez;

    [self.navigationController pushViewController:childController
        animated:YES];
}

@end
```

Most of the code you just entered is stuff you've seen before. One new thing is in the viewDidLoad method, where we used an NSKeyedUnarchiver method to create an array full of instances of the BIDPresident class from our property list file. It's not important that you understand exactly what's going on there, as long as you know that we're loading an array full of Presidents.

First, we get the path for the property file.

```
NSString *path = [[NSBundle mainBundle] pathForResource:@"Presidents"
    ofType:@"plist"];
```

Next, we declare a data object that will temporarily hold the encoded archive and an NSKeyedUnarchiver, which we'll use to actually restore the objects from the archive.

```
NSData *data;
NSKeyedUnarchiver *unarchiver;
```

We load the property list into data, and then use data to initialize unarchiver.

```
data = [[NSData alloc] initWithContentsOfFile:path];
unarchiver = [[NSKeyedUnarchiver alloc] initForReadingWithData:data];
```

Now, we decode an array from the archive. The key @"Presidents" is the same value that was used to create this archive.

```
NSMutableArray *array = [unarchiver decodeObjectForKey:@"Presidents"];
```

We then assign this decoded array to our list property, and finalize the decoding process.

```
self.list = array;
[unarchiver finishDecoding];
```

We also need to tell our tableView to reload its data in the viewWillAppear: method. If the user changes something in the detail view, we need to make sure that the parent view shows that new data. Rather than testing for a change, we force the parent view to reload its data and redraw each time it appears.

```
- (void)viewWillAppear:(BOOL)animated {
    [super viewWillAppear:animated];
    [self.tableView reloadData];
}
```

There's one other change from the last time we created a detail view. It's in the last method, `tableView:didSelectRowAtIndexPath:`. When we created the Disclosure Button view, we reused the same child controller every time and just changed its values. That's relatively easy to do when you have a nib with outlets. When you're using a table view to implement your detail view, the methods that fire the first time and the ones that fire subsequent times are different. Also, the table cells that are used to display and change the data are reused. The combination of these two details means your code can get very complex if you're trying to make it behave exactly the same way every time and to make sure that you are able to keep track of all the changes. Therefore, it's well worth the bit of additional overhead from allocating and releasing new controller objects to reduce the complexity of our controller class.

Let's look at the detail controller, because that's where the bulk of the new stuff is this time. This new controller is pushed onto the navigation stack when the user taps one of the rows in the `BIDPresidentsViewController` table to allow data entry for that president. Let's implement the detail view now.

Creating the Detail View Controller

Please fasten your seatbelts, ladies and gentlemen. We're expecting a little turbulence ahead. Air sickness bags are located in the seat pocket in front of you.

This next controller is just a little on the gnarly side, but we'll get through it safely. Please remain seated. Single-click *BIDPresidentDetailController.h,* and make the following changes:

```
#import <UIKit/UIKit.h>

@class BIDPresident;
#define kNumberOfEditableRows      4
#define kNameRowIndex              0
#define kFromYearRowIndex          1
#define kToYearRowIndex            2
#define kPartyIndex                3

#define kLabelTag                  4096

@interface BIDPresidentDetailController : UITableViewController
        <UITextFieldDelegate>

@property (strong, nonatomic) BIDPresident *president;
@property (strong, nonatomic) NSArray *fieldLabels;
@property (strong, nonatomic) NSMutableDictionary *tempValues;
@property (strong, nonatomic) UITextField *currentTextField;

- (IBAction)cancel:(id)sender;
- (IBAction)save:(id)sender;
```

```
- (IBAction)textFieldDone:(id)sender;
@end
```

What the heck is going on here? This is new. In all our previous table view examples, each table row corresponded to a single row in an array. The array provided all the data the table needed. For example, our table of Pixar movies was driven by an array of strings, each string containing the title of a single Pixar movie.

Our presidents example features two different tables. One is a list of president names, and it is driven by an array with one president per row. The second table implements a detail view of a selected president. Since this table has a fixed number of fields, instead of using an array to supply data to this table, we define a series of constants we will use in our table data source methods. These constants define the number of editable fields, along with the index value for the row that will hold each of those properties.

There's also a constant called `kLabelTag` that we'll use to retrieve the `UILabel` from the cell so that we can set the label correctly for the row. Shouldn't there be another tag for the `UITextField`? Normally, yes, but we will need to use the `tag` property of the text field for another purpose. We'll use another slightly less convenient mechanism to retrieve the text field when we need to set its value. Don't worry if that seems confusing; everything should become clear when we actually write the code.

You should notice that this class conforms to three protocols this time: the table data source and delegate protocols (which this class inherits because it's a subclass of `UITableViewController`) and a new one, `UITextFieldDelegate`. By conforming to `UITextFieldDelegate`, we'll be notified when a user makes a change to a text field so that we can save the field's value. This application doesn't have enough rows for the table to ever need to scroll, but in many applications, a text field could scroll off the screen, and perhaps be deallocated or reused. If the text field is lost, the value stored in it is lost, so saving the value when the user makes a change is the way to go.

Down a little further, we declare a pointer to a `BIDPresident` object. This is the object that we will actually be editing using this view, and it's set in the `tableView:didSelectRowAtIndexPath:` of our parent controller based on the row selected there. When the user taps the row for Thomas Jefferson, for example, the `BIDPresidentsViewController` will create an instance of the `BIDPresidentDetailController`. The `BIDPresidentsViewController` will then set the `president` property of that instance to the object that represents Thomas Jefferson, and push the newly created instance of `BIDPresidentDetailController` onto the navigation stack.

The second instance variable, `fieldLabels`, is an array that holds a list of labels that correspond to the constants kNameRowIndex, kFromYearRowIndex, kToYearRowIndex, and kPartyIndex. For example, kNameRowIndex is defined as 0. So, the label for the row that shows the president's name is stored at index 0 in the `fieldLabels` array. You'll see this in action when we get to it in code.

Next, we define a mutable dictionary, `tempValues`, that will hold values from fields the user changes. We don't want to make the changes directly to the `president` object because if the user selects the *Cancel* button, we need the original data so we can go back to it. Instead, we will store any value that is changed in our new mutable dictionary, tempValues. For example, if the user edited the *Name:* field and then tapped the *Party:* field to start editing that one, the BIDPresidentDetailController would be notified at that time that the *Name:* field had been edited, because it is the text field's delegate.

When the BIDPresidentDetailController is notified of the change, it stores the new value in the dictionary using the name of the property it represents as the key. In our example, we would store a change to the *Name:* field using the key @"name". That way, regardless of whether users save or cancel, we have the data we need to handle it. If the users cancel, we just discard this dictionary; if they save, we copy the changed values over to president.

Next up is a pointer to a UITextField, named currentTextField. The moment the users click in one of the BIDPresidentDetailController text fields, currentTextField is set to point to that text field. Why do we need this text field pointer? We have an interesting timing problem, and currentTextField is the solution.

Users can take one of two basic paths to finish editing a text field. First, they can touch another control or text field that becomes first responder. In this case, the text field that was being edited loses first responder status, and the delegate method textFieldDidEndEditing: is called. textFieldDidEndEditing: takes the new value of the text field and stores it in tempValues.

The second way that users can finish editing a text field is by tapping the *Save* or *Cancel* button. When they do this, the save: or cancel: action method is called. In both methods, the BIDPresidentDetailController view must be popped off the stack, since both the save and cancel actions end the editing session. This presents a problem. The save: and cancel: action methods do not have a simple way of finding the just-edited text field to save the data.

The delegate method textFieldDidEndEditing: does have access to the text field, since the text field is passed in as a parameter. That's where currentTextField comes in. The cancel: action method ignores currentTextField, since the user did not want to save changes, so the changes can be lost without causing any problems. But the save: method does care about those changes and needs a way to save them.

Since currentTextField is maintained as a pointer to the current text field being edited, save: uses that pointer to copy the value in the text field to tempValues. Now, save: can do its job and pop the BIDPresidentDetailController view off the stack, which will bring our list of presidents back to the top of the stack. When the view is popped off the stack, the text field and its value are lost. But since we've saved that sucker already, all is cool.

Single-click *BIDPresidentDetailController.m,* and make the following changes:

```
#import "BIDPresidentDetailController.h"
#import "BIDPresident.h"

@implementation BIDPresidentDetailController
@synthesize president;
@synthesize fieldLabels;
```

```objectivec
@synthesize tempValues;
@synthesize currentTextField;

- (IBAction)cancel:(id)sender {
    [self.navigationController popViewControllerAnimated:YES];
}

- (IBAction)save:(id)sender {
    if (currentTextField != nil) {
        NSNumber *tagAsNum = [NSNumber numberWithInt:currentTextField.tag];
        [tempValues setObject:currentTextField.text forKey:tagAsNum];
    }
    for (NSNumber *key in [tempValues allKeys]) {
        switch ([key intValue]) {
            case kNameRowIndex:
                president.name = [tempValues objectForKey:key];
                break;
            case kFromYearRowIndex:
                president.fromYear = [tempValues objectForKey:key];
                break;
            case kToYearRowIndex:
                president.toYear = [tempValues objectForKey:key];
                break;
            case kPartyIndex:
                president.party = [tempValues objectForKey:key];
            default:
                break;
        }
    }
    [self.navigationController popViewControllerAnimated:YES];

    NSArray *allControllers = self.navigationController.viewControllers;
    UITableViewController *parent = [allControllers lastObject];
    [parent.tableView reloadData];
}

- (IBAction)textFieldDone:(id)sender {
    [sender resignFirstResponder];
}

#pragma mark -
- (void)viewDidLoad {
    [super viewDidLoad];
    NSArray *array = [[NSArray alloc] initWithObjects:@"Name:", @"From:",
                        @"To:", @"Party:", nil];
    self.fieldLabels = array;

    UIBarButtonItem *cancelButton = [[UIBarButtonItem alloc]
                                    initWithTitle:@"Cancel"
                                    style:UIBarButtonItemStylePlain
                                    target:self
                                    action:@selector(cancel:)];
    self.navigationItem.leftBarButtonItem = cancelButton;
```

```
        UIBarButtonItem *saveButton = [[UIBarButtonItem alloc]
                                        initWithTitle:@"Save"
                                        style:UIBarButtonItemStyleDone
                                        target:self
                                        action:@selector(save:)];
        self.navigationItem.rightBarButtonItem = saveButton;

        NSMutableDictionary *dict = [[NSMutableDictionary alloc] init];
        self.tempValues = dict;
}

#pragma mark -
#pragma mark Table Data Source Methods
- (NSInteger)tableView:(UITableView *)tableView
 numberOfRowsInSection:(NSInteger)section {
    return kNumberOfEditableRows;
}

- (UITableViewCell *)tableView:(UITableView *)tableView
        cellForRowAtIndexPath:(NSIndexPath *)indexPath {
    static NSString *PresidentCellIdentifier = @"PresidentCellIdentifier";

    UITableViewCell *cell = [tableView dequeueReusableCellWithIdentifier:
                            PresidentCellIdentifier];
    if (cell == nil) {

        cell = [[UITableViewCell alloc]
            initWithStyle:UITableViewCellStyleDefault
            reuseIdentifier:PresidentCellIdentifier];
        UILabel *label = [[UILabel alloc] initWithFrame:
                    CGRectMake(10, 10, 75, 25)];
        label.textAlignment = UITextAlignmentRight;
        label.tag = kLabelTag;
        label.font = [UIFont boldSystemFontOfSize:14];
        [cell.contentView addSubview:label];

        UITextField *textField = [[UITextField alloc] initWithFrame:
                                    CGRectMake(90, 12, 200, 25)];
        textField.clearsOnBeginEditing = NO;
        [textField setDelegate:self];
        textField.returnKeyType = UIReturnKeyDone;
        [textField addTarget:self
                    action:@selector(textFieldDone:)
            forControlEvents:UIControlEventEditingDidEndOnExit];
        [cell.contentView addSubview:textField];
    }
    NSUInteger row = [indexPath row];

    UILabel *label = (UILabel *)[cell viewWithTag:kLabelTag];
    UITextField *textField = nil;
    for (UIView *oneView in cell.contentView.subviews) {
        if ([oneView isMemberOfClass:[UITextField class]])
            textField = (UITextField *)oneView;
    }
```

```
        label.text = [fieldLabels objectAtIndex:row];
        NSNumber *rowAsNum = [NSNumber numberWithInt:row];
        switch (row) {
            case kNameRowIndex:
                if ([[tempValues allKeys] containsObject:rowAsNum])
                    textField.text = [tempValues objectForKey:rowAsNum];
                else
                    textField.text = president.name;
                break;
            case kFromYearRowIndex:
                if ([[tempValues allKeys] containsObject:rowAsNum])
                    textField.text = [tempValues objectForKey:rowAsNum];
                else
                    textField.text = president.fromYear;
                break;
            case kToYearRowIndex:
                if ([[tempValues allKeys] containsObject:rowAsNum])
                    textField.text = [tempValues objectForKey:rowAsNum];
                else
                    textField.text = president.toYear;
                break;
            case kPartyIndex:
            if ([[tempValues allKeys] containsObject:rowAsNum])
                textField.text = [tempValues objectForKey:rowAsNum];
            else
                textField.text = president.party;
            default:
                break;
        }
        if (currentTextField == textField) {
            currentTextField = nil;
        }
        textField.tag = row;
        return cell;
}

#pragma mark -
#pragma mark Table Delegate Methods
- (NSIndexPath *)tableView:(UITableView *)tableView
  willSelectRowAtIndexPath:(NSIndexPath *)indexPath {
    return nil;
}

#pragma mark Text Field Delegate Methods
- (void)textFieldDidBeginEditing:(UITextField *)textField {
    self.currentTextField = textField;
}

- (void)textFieldDidEndEditing:(UITextField *)textField {
    NSNumber *tagAsNum = [NSNumber numberWithInt:textField.tag];
    [tempValues setObject:textField.text forKey:tagAsNum];
}

@end
```

The first new method is our `cancel:` action method. This is called, appropriately enough, when the user taps the *Cancel* button. When the *Cancel* button is tapped, the current view will be popped off the stack, and the previous view will rise to the top of the stack. Ordinarily, that job would be handled by the navigation controller, but a little later in the code, we're going to manually set the left bar button item. This means we're replacing the button that the navigation controller uses for that purpose. We can pop the current view off the stack by getting a reference to the navigation controller and telling it to do just that.

```
- (IBAction)cancel:(id)sender {
    [self.navigationController popViewControllerAnimated:YES];
}
```

The next method is `save:`, which is called when the user taps the *Save* button. When the *Save* button is tapped, the values that the user has entered have already been stored in the `tempValues` dictionary, unless the keyboard is still visible and the cursor is still in one of the text fields. In that case, there may be changes to that text field that have not yet been put into our `tempValues` dictionary. To account for this, the first thing the `save:` method does is check to see if there is a text field that is currently being edited. Whenever the user starts editing a text field, we store a pointer to that text field in `currentTextField`. If `currentTextField` is not nil, we grab its value and stick it in `tempValues`.

```
if (currentTextField != nil) {
    NSNumber *tfKey = [NSNumber numberWithInt:currentTextField.tag];
    [tempValues setObject:currentTextField.text forKey:tfKey];
}
```

We then use fast enumeration to step through all the key values in the dictionary, using the row numbers as keys. We can't store raw datatypes like `int` in an `NSDictionary`, so we create `NSNumber` objects based on the row number and use those instead. We use `intValue` to turn the number represented by key back into an `int`, and then use a `switch` on that value using the constants we defined earlier and assign the appropriate value from the `tempValues` array back to the designated field on our `president` object.

```
for (NSNumber *key in [tempValues allKeys]) {
    switch ([key intValue]) {
        case kNameRowIndex:
            president.name = [tempValues objectForKey:key];
            break;
        case kFromYearRowIndex:
            president.fromYear = [tempValues objectForKey:key];
            break;
        case kToYearRowIndex:
            president.toYear = [tempValues objectForKey:key];
            break;
        case kPartyIndex:
            president.party = [tempValues objectForKey:key];
        default:
            break;
    }
}
```

Now, our `president` object has been updated, and we need to move up a level in the view hierarchy. Tapping a *Save* or *Done* button on a detail view should generally bring the user back up to the previous level, so we grab our application delegate and use its navController outlet to pop ourselves off the navigation stack, sending the user back up to the list of presidents:

```
[self.navigationController popViewControllerAnimated:YES];
```

There's one other thing we need to do here: tell our parent view's table to reload its data. Because one of the fields that the user can edit is the name field, which is displayed in the BIDPresidentsViewController table, if we don't have that table reload its data, it will continue to show the old value.

```
UINavigationController *navController = [delegate navController];
NSArray *allControllers = navController.viewControllers;
UITableViewController *parent = [allControllers lastObject];
[parent.tableView reloadData];
```

The third action method will be called when the user taps the *Done* button on the keyboard. Without this method, the keyboard won't retract when the user taps *Done*. This approach isn't strictly necessary in our application, since the four rows that can be edited here fit in the area above the keyboard. That said, you'll need this method if you add a row or in a future application that requires more screen real estate. It's a good idea to keep the behavior consistent from application to application, even if doing so is not critical to your application's functionality.

```
- (IBAction)textFieldDone:(id)sender {
    [sender resignFirstResponder];
}
```

The viewDidLoad method doesn't contain anything too surprising. We create the array of field names and assign it the fieldLabels property.

```
NSArray *array = [[NSArray alloc] initWithObjects:@"Name:",
        @"From:", @"To:", @"Party:", nil];
self.fieldLabels = array;
```

Next, we create two buttons and add them to the navigation bar. We put the *Cancel* button in the left bar button item spot, which supplants the navigation button put there automatically. We put the *Save* button in the right spot and assign it the style UIBarButtonItemStyleDone. This style was specifically designed for this occasion—as a button users tap when they are happy with their changes and ready to leave the view. A button with this style will be blue instead of gray, and it usually will carry a label of *Save* or *Done*.

```
UIBarButtonItem *cancelButton = [[UIBarButtonItem alloc]
        initWithTitle:@"Cancel"
        style:UIBarButtonItemStylePlain
        target:self
        action:@selector(cancel:)];
self.navigationItem.leftBarButtonItem = cancelButton;

UIBarButtonItem *saveButton = [[UIBarButtonItem alloc]
        initWithTitle:@"Save"
        style:UIBarButtonItemStyleDone
```

```
                target:self
                action:@selector(save:)];
    self.navigationItem.rightBarButtonItem = saveButton;
```

Finally, we create a new mutable dictionary and assign it to `tempValues` so that we have a place to stick the changed values. If we made the changes directly to the `president` object, we would have no easy way to roll back to the original data if the user tapped *Cancel*.

```
    NSMutableDictionary *dict = [[NSMutableDictionary alloc] init];
    self.tempValues = dict;
```

We can skip over the first data source method, as there is nothing new under the sun there. We do need to stop and chat about `tableView:cellForRowAtIndexPath:`, however, because there are a few gotchas there. The first part of the method is exactly like every other `tableView:cellForRowAtIndexPath:` method we've written.

```
- (UITableViewCell *)tableView:(UITableView *)tableView
    cellForRowAtIndexPath:(NSIndexPath *)indexPath {
    static NSString *PresidentCellIdentifier = @"PresidentCellIdentifier";

    UITableViewCell *cell = [tableView dequeueReusableCellWithIdentifier:
        PresidentCellIdentifier];
    if (cell == nil) {
        cell = [[UITableViewCell alloc] initWithFrame:CGRectZero
                    reuseIdentifier:PresidentCellIdentifier];
```

When we create a new cell, we create a label, make it right-aligned and bold, and assign it a tag so that we can retrieve it again later. Next, we add it to the cell's `contentView` and release it.

```
        UILabel *label = [[UILabel alloc] initWithFrame:
            CGRectMake(10, 10, 75, 25)];
        label.textAlignment = UITextAlignmentRight;
        label.tag = kLabelTag;
        label.font = [UIFont boldSystemFontOfSize:14];
        [cell.contentView addSubview:label];
```

After that, we create a new text field. The user actually types in this field. We set it so it does not clear the current value when editing, so we don't lose the existing data, and we set `self` as the text field's delegate. By setting the text field's delegate to `self`, we can get notified by the text field when certain events occur by implementing appropriate methods from the `UITextFieldDelegate` protocol. As you'll see soon, we've implemented two text field delegate methods in this class. Those methods will be called by the text fields on all rows when the user begins and ends editing the text they contain. We also set the keyboard's **return key type**, which is how we specify the text for the key in the bottom right of the keyboard. The default value is *Return*, but since we have only single-line fields, we want the key to say *Done* instead, so we pass `UIReturnKeyDone`.

```
        UITextField *textField = [[UITextField alloc] initWithFrame:
            CGRectMake(90, 12, 200, 25)];
        textField.clearsOnBeginEditing = NO;
            [textField setDelegate:self];
        textField.returnKeyType = UIReturnKeyDone;
```

After that, we tell the text field to call our `textFieldDone:` method on the *Did End on Exit* event. This is exactly the same thing as dragging from the *Did End on Exit* event in the connections inspector in Interface Builder to *File's Owner* and selecting an action method. Since we don't have a nib file, we must do it programmatically, but the result is the same.

When we're finished configuring the text field, we add it to the cell's content view. Notice, however, that we did not set a tag before we added it to that view.

```
    [textField addTarget:self
                action:@selector(textFieldDone:)
                forControlEvents:UIControlEventEditingDidEndOnExit];
    [cell.contentView addSubview:textField];
}
```

At this point, we know that we have either a brand-new cell or a reused cell, but we don't know which. The first thing we do is figure out which row this darn cell is going to represent.

```
    NSUInteger row = [indexPath row];
```

Next, we need to get a reference to the label and the text field from inside this cell. The label is easy; we just use the tag we assigned to it to retrieve it from `cell`.

```
UILabel *label = (UILabel *)[cell viewWithTag:kLabelTag];
```

The text field, however, isn't going to be quite as easy, because we need the tag in order to tell our text field delegates which text field is calling them. So, we're going to rely on the fact that there's only one text field that is a subview of our cell's `contentView`. We'll use fast enumeration to work through all of its subviews, and when we find a text field, we assign it to the pointer we declared a moment earlier. When the loop is finished, the `textField` pointer should be pointing to the one and only text field contained in this cell.

```
    UITextField *textField = nil;

    for (UIView *oneView in cell.contentView.subviews) {
        if ([oneView isMemberOfClass:[UITextField class]])
            textField = (UITextField *)oneView;
    }
```

Now that we have pointers to both the label and the text field, we can assign them the correct values based on which field from the `president` object this row represents. Once again, the label gets its value from the `fieldLabels` array.

```
    label.text = [fieldLabels objectAtIndex:row];
```

To assign the value to the text field, we need to first check to see if there is a value in the `tempValues` dictionary corresponding to this row. If there is, we assign it to the text field. If there isn't any corresponding value in `tempValues`, we know there have been no changes entered for this field, so we assign this field the corresponding value from `president`.

```
    NSNumber *rowAsNum = [NSNumber numberWithInt:row];
    switch (row) {
            case kNameRowIndex:
                if ([[tempValues allKeys] containsObject:rowAsNum])
```

```
                    textField.text = [tempValues objectForKey:rowAsNum];
                else
                    textField.text = president.name;
                break;
            case kFromYearRowIndex:
                if ([[tempValues allKeys] containsObject:rowAsNum])
                    textField.text = [tempValues objectForKey:rowAsNum];
                else
                    textField.text = president.fromYear;
                break;
            case kToYearRowIndex:
                if ([[tempValues allKeys] containsObject:rowAsNum])
                    textField.text = [tempValues objectForKey:rowAsNum];
                else
                    textField.text = president.toYear;
                break;
            case kPartyIndex:
                if ([[tempValues allKeys] containsObject:rowAsNum])
                    textField.text = [tempValues objectForKey:rowAsNum];
                else
                    textField.text = president.party;
            default:
                break;
    }
```

If the field we're using is the one that is currently being edited, that's an indication that the value we're holding in currentTextField is no longer valid, so we set currentTextField to nil. If the text field did get released or reused, our text field delegate would have been called, and the correct value would already be in the tempValues dictionary.

```
    if (currentTextField == textField) {
        currentTextField = nil;
    }
```

Next, we set the text field's tag to the row it represents, which will allow us to know which field is calling our text field delegate methods. And finally, we return the cell.

```
    textField.tag = row;
    return cell;
}
```

We do implement one table delegate method this time, which is tableView:willSelectRowAtIndexPath:. Remember that this method is called before a row is selected and gives us a chance to disallow the row selection. In this view, we never want a row to appear selected. We need to know that the user selected a row so we can place a check mark next to it, but we don't want the row to actually be highlighted. Don't worry. A row doesn't need to be selected for a text field on that row to be editable, so this method just keeps the row from staying highlighted after it is touched.

```
- (NSIndexPath *)tableView:(UITableView *)tableView
    willSelectRowAtIndexPath:(NSIndexPath *)indexPath {
    return nil;
}
```

All that's left now are the two text field delegate methods. The first one we implement, `textFieldDidBeginEditing:`, is called whenever a text field for which we are the delegate becomes first responder. So, if the user taps a field and the keyboard pops up, we get notified. In this method, we store a pointer to the field currently being edited so that we have a way to get to the last changes made before the *Save* button was tapped.

```
- (void)textFieldDidBeginEditing:(UITextField *)textField {
    self.currentTextField = textField;
}
```

The last method we wrote is called when the user stops editing a text field by tapping a different text field or pressing the *Done* button, or when another field became the first responder, which will happen, for example, when the user navigates back up to the list of presidents. Here, we save the value from that field in the `tempValues` dictionary so that we will have the changes if the user taps the *Save* button to confirm the changes.

```
- (void)textFieldDidEndEditing:(UITextField *)textField {
    NSNumber *tagAsNum = [NSNumber numberWithInt:textField.tag];
    [tempValues setObject:textField.text forKey:tagAsNum];
}
```

And that's it. We're finished with these two view controllers.

Adding an Editable Detail View Controller Instance

Now, all we need to do is add an instance of this class to the top-level view controller. You know how to do this by now. Single-click *BIDFirstLevelController.m*.

First, import the header from the new second-level view by adding the following line of code directly before the @implementation declaration:

```
#import "BIDPresidentsViewController.h"
```

Then add the following code to the viewDidLoad method:

```
- (void)viewDidLoad {
    [super viewDidLoad];
    self.title = @"Top Level";
    NSMutableArray *array = [[NSMutableArray alloc] init];

    // Disclosure Button
    BIDDisclosureButtonController *BIDDisclosureButtonController =
        [[BIDDisclosureButtonController alloc]
            initWithStyle:UITableViewStylePlain];
    BIDDisclosureButtonController.title = @"Disclosure Buttons";
    BIDDisclosureButtonController.rowImage = [UIImage
        imageNamed:@"BIDDisclosureButtonControllerIcon.png"];
    [array addObject:BIDDisclosureButtonController];

    // Checklist
    BIDCheckListController *checkListController = [[BIDCheckListController alloc]
            initWithStyle:UITableViewStylePlain];
    checkListController.title = @"Check One";
    checkListController.rowImage = [UIImage
        imageNamed:@"checkmarkControllerIcon.png"];
```

```
    [array addObject:checkListController];

    // Table Row Controls
    RowControlsController *rowControlsController =
        [[RowControlsController alloc]
        initWithStyle:UITableViewStylePlain];
    rowControlsController.title = @"Row Controls";
    rowControlsController.rowImage =
        [UIImage imageNamed:@"rowControlsIcon.png"];
    [array addObject:rowControlsController];

    // Move Me
    BIDMoveMeController *moveMeController = [[BIDMoveMeController alloc]
        initWithStyle:UITableViewStylePlain];
    moveMeController.title = @"Move Me";
    moveMeController.rowImage = [UIImage imageNamed:@"moveMeIcon.png"];
    [array addObject:moveMeController];
    [moveMeController release];

    // Delete Me
    BIDDeleteMeController *deleteMeController = [[BIDDeleteMeController alloc]
            initWithStyle:UITableViewStylePlain];
    deleteMeController.title = @"Delete Me";
    deleteMeController.rowImage = [UIImage imageNamed:@"deleteMeIcon.png"];
    [array addObject:deleteMeController];

    // BIDPresident View/Edit
    BIDPresidentsViewController *presidentsViewController =
        [[BIDPresidentsViewController alloc]
        initWithStyle:UITableViewStylePlain];
    presidentsViewController.title = @"Detail Edit";
    presidentsViewController.rowImage = [UIImage imageNamed:
        @"detailEditIcon.png"];
    [array addObject:presidentsViewController];

    self.controllers = array;
}
```

Save everything, sigh deeply, hold your breath, and then build that sucker. If everything is in order, the simulator will launch, and a sixth and final row will appear, just like the one in Figure 9–2. If you click the new row, you'll be taken to a list of US presidents (see Figure 9–22).

Figure 9–22. *Our sixth and final subcontroller presents a list of US presidents. Tap one of the presidents, and you'll be taken to a detail view (or a secret service agent will wrestle you to the ground).*

Tapping any of the rows will take you down to the detail view that we just built (see Figure 9–8), and you'll be able to edit the values. If you select the *Done* button in the keyboard, the keyboard should retract. Tap one of the editable values, and the keyboard will reappear. Make some changes and tap *Cancel*, and the application will pop back to the list of presidents. If you revisit the president you just canceled out of, your changes will be gone. On the other hand, if you make some changes and tap *Save*, your changes will be reflected in the parent table, and when you come back into the detail view, the new values will still be there.

But There's One More Thing. . .

There's one more bit of polish we need to add to make our application behave the way it should. In the version we just built, the keyboard incorporates a *Done* button that, when tapped, makes the keyboard retract. That behavior is proper if there are other controls on the view that the user might need to access. Since every row on this table view is a text field, however, we need a slightly different solution. The keyboard should feature a *Return* button instead of a *Done* button. When tapped, that button should take the user to the next row's text field.

In order to accomplish this, our first step is to replace the *Done* button with a *Return* button. We can do this by deleting a single line of code from *BIDPresidentDetailController.m*. In the tableView:cellForRowAtIndexPath: method, delete the following line of code:

```
- (UITableViewCell *)tableView:(UITableView *)tableView
        cellForRowAtIndexPath:(NSIndexPath *)indexPath {
    static NSString *PresidentCellIdentifier = @"PresidentCellIdentifier";
    UITableViewCell *cell = [tableView dequeueReusableCellWithIdentifier:
        PresidentCellIdentifier];
    if (cell == nil) {

        cell = [[UITableViewCell alloc] initWithFrame:CGRectZero
                    reuseIdentifier:PresidentCellIdentifier];
        UILabel *label = [[UILabel alloc] initWithFrame:
            CGRectMake(10, 10, 75, 25)];
        label.textAlignment = UITextAlignmentRight;
        label.tag = kLabelTag;
        label.font = [UIFont boldSystemFontOfSize:14];
        [cell.contentView addSubview:label];

        UITextField *textField = [[UITextField alloc] initWithFrame:
            CGRectMake(90, 12, 200, 25)];
        textField.clearsOnBeginEditing = NO;
        [textField setDelegate:self];
        textField.returnKeyType = UIReturnKeyDone;
        [textField addTarget:self
                    action:@selector(textFieldDone:)
                    forControlEvents:UIControlEventEditingDidEndOnExit];
        [cell.contentView addSubview:textField];
    }
    NSUInteger row = [indexPath row];
...
```

The next step isn't quite as straightforward. In our textFieldDone: method, instead of simply telling sender to resign first responder status, we need to somehow figure out what the next field should be and tell that field to become the first responder. Replace your current version of textFieldDone: with the following new version, and then we'll chat about how it works.

```
- (IBAction)textFieldDone:(id)sender {
    UITableViewCell *cell =
        (UITableViewCell *)[[sender superview] superview];
    UITableView *table = (UITableView *)[cell superview];
    NSIndexPath *textFieldIndexPath = [table indexPathForCell:cell];
    NSUInteger row = [textFieldIndexPath row];
    row++;
    if (row >= kNumberOfEditableRows) {
        row = 0;
    }
    NSIndexPath *newPath = [NSIndexPath indexPathForRow:row inSection:0];
    UITableViewCell *nextCell = [self.tableView
        cellForRowAtIndexPath:newPath];
    UITextField *nextField = nil;
    for (UIView *oneView in nextCell.contentView.subviews) {
```

```
        if ([oneView isMemberOfClass:[UITextField class]])
            nextField = (UITextField *)oneView;
    }
    [nextField becomeFirstResponder];
}
```

Unfortunately, cells don't know which row they represent. The table view, however, does know which row a given cell is currently representing. So, we get a reference to the table view cell. We know that the text field that is triggering this action method is a subview of the table cell view's content view, so we just need to get sender's superview's superview (now say *that* ten times fast).

If that sounded confusing, think of it this way: In this case, sender is the text field being edited. Our sender's superview is the content view that groups the text field and its label. And sender's superview's superview is the cell that encompasses that content view.

```
UITableViewCell *cell = (UITableViewCell *)[[(UIView *)sender
    superview] superview];
```

We also need access to the cell's enclosing table view, which is easy enough, since it's the superview of the cell.

```
UITableView *table = (UITableView *)[cell superview];
```

We then ask the table which row the cell represents. The response is an NSIndexPath, and we get the row from that.

```
NSIndexPath *textFieldIndexPath = [table indexPathForCell:cell];
NSUInteger row = [textFieldIndexPath row];
```

Next, we increment row by one, which represents the next row in the table. If incrementing the row number puts us beyond the last one, we reset row to 0.

```
row++;
if (row >= kNumberOfEditableRows) {
    row = 0;
}
```

Then we build a new NSIndexPath to represent the next row, and use that index path to get a reference to the cell currently representing the next row.

```
NSIndexPath *newPath = [NSIndexPath indexPathForRow:row inSection:0];
UITableViewCell *nextCell = [self.tableView
    cellForRowAtIndexPath:newPath];
```

Note that instead of using alloc and init methods to create the NSIndexPath, we're using a special factory method that exists just for the purpose of creating an index path that points out a row in a UITableView. The normal way of creating an NSIndexPath otherwise involves first creating a C array, and then passing it to the initWithIndexes:length: method along with the length of the array. What we're doing here is much more straightforward.

For the text field, we're already using tag for another purpose, so we need to loop through the subviews of the cell's content view to find the text field rather than using tag to retrieve it.

```
    UITextField *nextField = nil;
    for (UIView *oneView in nextCell.contentView.subviews) {
        if ([oneView isMemberOfClass:[UITextField class]])
            nextField = (UITextField *)oneView;
}
```

Finally, we can tell that new text field to become the first responder.

```
[nextField becomeFirstResponder];
```

Now, compile and run. This time, when you drill down to the detail view, tapping the *Return* button will take you to the next field in the table, which will make entering data much easier for your users.

Breaking the Tape

This chapter was a marathon, and if you're still standing, you should feel pretty darn good about yourself. Dwelling on these mystical table view and navigation controller objects is important because they are the backbone of a great many iOS applications, and their complexity can definitely get you into trouble if you don't truly understand them.

As you start building your own tables, check back to this chapter and the previous one, and don't be afraid of Apple's documentation, either. Table views are extraordinarily complex, and we could never cover every conceivable permutation, but you should now have a very good set of table view building blocks you can use as you design and build your own applications. As always, feel free to reuse this code in your own applications. It's a gift from us to you. Enjoy!

In the next chapter, we're going to introduce you to Storyboards, one of the biggest new features iOS 5 brings to developers. That's right, the Storyboards concept isn't really an end-user feature, but rather a set of enhancements to Xcode and new APIs in UIKit, which allow developers to design the structure of a complex navigation-based application in a whole new way. Storyboards will make your job much easier and even more fun!

Chapter 10

Storyboards

Over the course of the past several chapters, you've become intimately familiar with nib files, the UITableView class, and navigating between views using UINavigationViewController. Together, these building blocks make up a robust, flexible tool kit for building mobile apps, as evidenced by the hundreds of thousands of iPhone and iPad applications that have been created in the past few years.

However, there's always room for improvement. As more and more people have been introduced to these tools, some of their shortcomings have started to feel like a burden. As you've been learning, you may have wondered about some of these issues yourself:

- The delegate/dataSource pattern for specifying UITableView content is great for creating dynamic tables, but if you know in advance precisely what your table will contain, isn't it all a bit wordy? What if you could specify table contents in a more declarative manner, skipping all the call and response that the current system entails?

- The use of nib files for storing freeze-dried object graphs is great, as far as it goes. But if your app has more than one view controller, as almost all apps do, switching between them always requires a certain amount of manual labor in the form of redundant, boilerplate code. What if this could be streamlined?

- In a complex application with many view controllers, it can be hard to see the big picture. Communications and transitions between controllers are codified within every controller class. Not only does this make it hard to read an app's source code, requiring you to study delegate and action methods in each controller to see how it connects to other controllers, but it also makes the app's code more fragile. What if you had a way to describe interactions between view controllers at a level outside the controllers themselves, letting you see the entire flow of data and interactions in one place?

If you've been concerned about any of these issues, you're in luck! As it turns out, Apple has been concerned about them, too. It has included with the iOS 5 SDK a new system called **Storyboards** that aims to solve each of these problems.

Storyboards build on the familiar nib concept, and are edited in the same way, using Xcode's Interface Builder. But unlike nibs, storyboards also allow you to work with multiple views, each attached to its own view controller, in a single visual workspace. You can configure how transitions between view controllers should occur, and you can also configure a table view with a fixed set of predefined cells. In this chapter, we'll explore some of these new capabilities so you can get a feel for storyboards, see how they differ from nib files, and figure out where to work storyboards into your own apps.

> **NOTE:** As you'll see, storyboards are awesome. It's important to know, though, that at the moment at least, storyboards will run only on devices sporting iOS 5 and above. That will be less of an issue as more and more people move to iOS 5 over time.

Creating a Simple Storyboard

Let's start off with a simple project to demonstrate some of the basic characteristics of storyboards. In Xcode, select File ➤ New ➤ New Project... to create a new project. Select *Single View Application* from the *iOS Application group*, and click *Next*. Name the project *Simple Storyboard*, click to enable the *Use Storyboard* checkbox, and click *Next* again. Finally, select the directory where you want to save your project and click *Create*.

Once the project is created, look at the project navigator. You'll see some familiar sights, such as the *BIDAppDelegate* and *BIDViewController* class files. You'll also notice that there's no *BIDViewController.xib*, but there is a *MainStoryboard.storyboard* file. The storyboard file isn't named after the view controller because, unlike nibs, a storyboard is designed to be able to easily represent content for several view controllers.

Select *MainStoryboard.storyboard*, and you'll see that Xcode switches into the familiar Interface Builder display, as shown in Figure 10-1. Some subtle differences exist between the nib editing interface we've used up to this point and the storyboard editing interface. For example, the storyboard Interface Builder editor doesn't contain an icon mode when working with a storyboard. Instead, clicking the disclosure triangle to the right of the bottom of the dock makes the entire dock collapse and disappear.

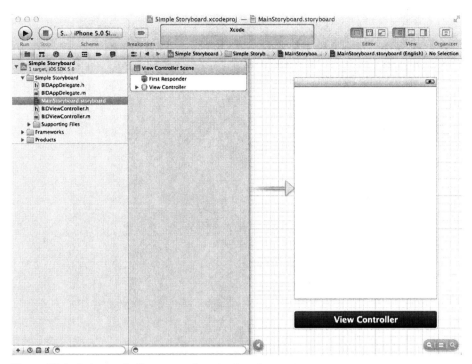

Figure 10-1. *A view controller, as seen in a storyboard*

Another example involves the first responder and view controller icons. If you select *View Controller* in the dock, the *First Responder* and *View Controller* icons will appear below the view controller, in addition to appearing in the dock (see Figure 10-2). In a storyboard, each view and its corresponding view controller always appear together in this way, and together they are referred to as a **scene**.

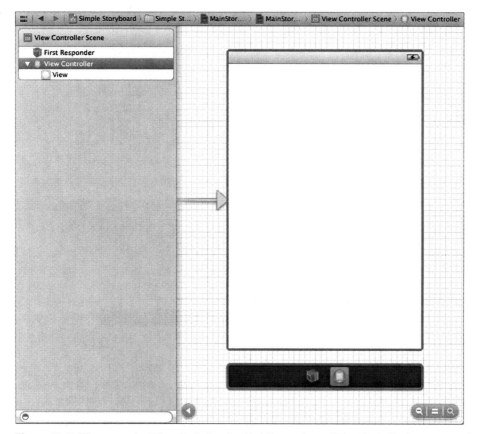

Figure 10-2. *When you select View Controller, the First Responder and View Controller icons appear below the view controller.*

You'll also see that the view shown in the editing area has a big arrow pointing to it. This will come in handy later in this chapter when we create storyboards containing multiple views. The arrow points out which view controller is the initial view controller that should be loaded and displayed when the app loads this storyboard. When you actually have multiple views in your storyboard, you simply drag the arrow around to point to the correct initial view controller. Right now, we have only one view. If you try dragging the arrow around, you'll see that it always pops back to where it started as soon as you release the mouse button.

The other big change to the editing area is that you can zoom in and out with a set of controls at the lower right of the editing pane. This is handy when dealing with a lot of view controllers in a storyboard, since it allows you to see several view controllers and the connections between them all at once. Note that when you're zoomed out, Interface Builder won't let you drag any objects from the object library into your views. You also can't select any of the objects inside your views when you're zoomed out. So, overall, this isn't really a useful mode for editing your views, but it is a good way to get a sense of the overall picture.

Now, let's add a label to our view. Make sure you're zoomed all the way in, and drag a *Label* from the object library to the center of the view. Double-click the label to select its text, and change the text to *Simple*. Run your application, and you'll see your app launch and display the label you just created.

You've seen some template-generated apps before, but with storyboards, things are a little different. Let's take a look at the rest of our project to see what's happening with the storyboard-based app behind the scenes.

In the project navigator, select *BIDViewController.m*, and take a look through the code. Except for the autorotation method, all the methods in this file send a message to the superclass and then return. Certainly, there's no mention of our storyboard here.

Move on to *BIDAppDelegate.m*. You'll see a series of empty methods. Direct your attention to `application:didFinishLaunchingWithArguments:`, which looks quite a bit different from how the same method has been implemented in our other apps. In the apps we've created up to this point, this method has contained code to create a `UIWindow`, perhaps open a nib file, and more. This one, however, is pretty empty! So, how does our app know that it's supposed to load our storyboard, and how does its initial view get put into a window? Bottom line: the key to this question lies in the target settings.

Go to the project navigator and select the topmost *Simple Storyboard* item, which represents the project itself. Make sure the *Simple Storyboard* target is also selected, along with the *Summary* tab at the top, and look for the *iPhone / iPod Deployment Info* section, where you'll see that *MainStoryboard* is configured as the *Main Storyboard* (see Figure 10-3).

As it turns out, that's all the configuration required to make your app automatically create a window, load the storyboard and its initial view, create the initial view controller specified in the storyboard, and hook it all up. That means, among other things, that the application delegate gets to be a little simpler, since the creation of the window and the initial view are all taken care of for you. All of the wiring is done behind the scenes, but if you enable storyboarding in a project, you get this simplification for free.

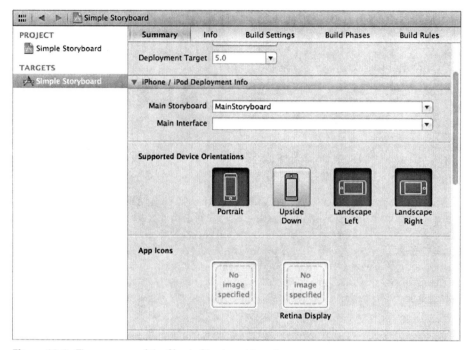

Figure 10-3. *The summary of the Simple Storyboard target*

Dynamic Prototype Cells

As you may recall from Chapter 8, iOS 5 lets you create a nib file containing a UITableViewCell and whatever objects you want the cell to contain, and use a unique identifier to register that cell with the table view. Then, at runtime, you can ask the table view to give you a cell based on an identifier, and if the identifier matches your earlier registration, that's what you'll get.

With storyboards, this concept is cranked up a notch. Now, instead of creating separate nib files for each type of cell, you can create them all in a single storyboard, directly inside the table view where they will be presented! Let's see how this is done.

Dynamic Table Content, Storyboard-Style

We're going to make a controller that displays a list of items. Depending on the content of each item, we'll display them with a fairly plain style or with a more eye-catching style to alert the users that they need to pay special attention to the item. Just to make things concrete, let's imagine these items are to-do list entries, and we want to alert the users to entries that are urgent. To keep things simple, we'll use a plain UITableViewCell for each displayable cell instead of defining any cell subclasses. For more complicated displays in a real app, you would probably want to create your own cell subclasses to use here. In either case, the setup and workflow are the same.

Since you've already made a new project and hardly changed a thing, let's stick with our *Simple Storyboard* project. Make a new class in the Xcode project window. Select File ➤ New ➤ New File..., then in the *Cocoa Touch* section choose *UIViewController subclass*, and then click *Next*. Name the class *BIDTaskListController*, and select *UITableViewController* from the *Subclass of* popup. Click to turn off the checkbox for creating a new XIB file, since we are going to work within the storyboard instead.

After the class files are created, switch back to *MainStoryboard.storyboard*, where we'll put an instance of our new controller (and, of course, a view to match). Grab a *Table View Controller* from the object library and pull it out into the editing area, placing it to the right of the controller you started with. It should now look something like Figure 10-4. You'll likely see a warning, complaining that "prototype table cells must have reused identifiers," but no worries; we'll address this soon.

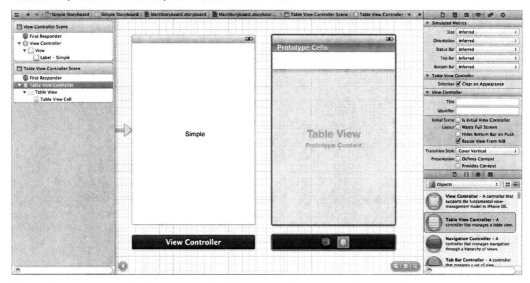

Figure 10-4. *The new table view controller, just to the right of the original view controller*

Now we need to configure the table view controller to be an instance of our controller class, not the default `UITableViewController`. Select the newly inserted table view controller, making sure you're selecting the controller itself, not the table view it contains. The easiest way to do this is to click the icon bar below the table view or to click the *Table View Controller* row in the dock. You'll know you've got it when both the table view and the icon bar are outlined in blue. Open the identity inspector, and change the controller's class to *BIDTaskListController* so that the table view knows where to get its data.

Editing Prototype Cells

You'll see that the table view has a one-item group at the top labeled *Prototype Cells*. This is where you can graphically lay out cells however you like, and give them unique identifiers for later retrieval in your code.

Start by selecting the one blank cell that is already there and opening the attributes inspector. Set this cell's *Identifier* to *plainCell*, and then drag a *Label* from the object library directly into the cell itself. Drag the label over near the left edge so that the blue guideline pops up, and then resize its width, dragging the right edge out toward the right edge of the cell until the blue guideline appears over there. Finally, with the label selected, use the object inspector to set its tag to *1* (see Figure 10-5). This will let us find the label from within our code.

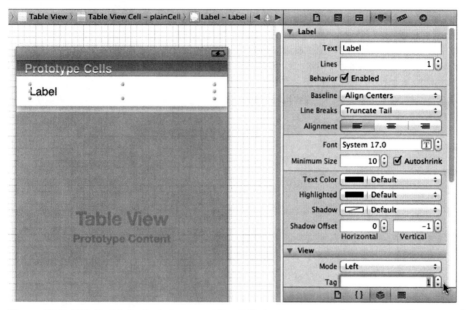

Figure 10-5. *With the label selected, we used the attributes inspector to change the label's tag to 1. Note that the Tag field is in the View section toward the bottom of the inspector. The cursor is pointing at the field.*

Now, select the table view cell itself (not the label it contains), and choose **Edit ➤ Duplicate**. This places a new copy of the cell directly below the original.

> **NOTE:** Selecting the table view cell can be tricky. In the version of Xcode we used when we wrote this chapter, we needed to actually click the table cell in order to be able to duplicate it. Selecting the cell in the dock was not enough. Likely this will change over time.

With the new cell selected, use the object inspector to set its identifier to *attentionCell*. Then select the new cell's label, and use the attribute inspector to change the label's *Text Color* field to red, and set its *Font* to *System Bold*.

Now we have two prototype cells that are ready to use in this table view. Before we implement the code to populate this table, we need to make one more change to the storyboard. Remember that big, floating arrow that points to our original view? Drag that over to point to our new view instead. Save your storyboard.

Good Old Table View Data Source

Head over to *BIDTaskListController.m*, where we're going to add some code to populate our table. This is mostly pretty standard table view stuff that you've seen plenty of times before, so we're going to whizz through it and skip explanations for all but the newest bits. Start by adding these bold lines near the top of the file:

```
#import "BIDTaskListController.h"

@interface BIDTaskListController ()
@property (strong, nonatomic) NSArray *tasks;
@end

@implementation BIDTaskListController

@synthesize tasks;
```

This just sets up a property to contain a list of items we want to show.

Now, insert the following code in viewDidLoad to populate the tasks property. Note that we've left out the comments in this code.

```
- (void)viewDidLoad
{
    [super viewDidLoad];

    self.tasks = [NSArray arrayWithObjects:
                    @"Walk the dog",
                    @"URGENT:Buy milk",
                    @"Clean hidden lair",
                    @"Invent miniature dolphins",
                    @"Find new henchmen",
                    @"Get revenge on do-gooder heroes",
                    @"URGENT: Fold laundry",
                    @"Hold entire world hostage",
                    @"Manicure",
                    nil];
}
```

And, of course, we need to be a good citizen and clear that out when our view is no longer going to be shown:

```
- (void)viewDidUnload
{
    [super viewDidUnload];
    // Release any retained subviews of the main view.
    // e.g. self.myOutlet = nil;

    self.tasks = nil;
}
```

Now, we get to the meat of the controller, and implement the methods that give the table view some content. Start with the simple methods that tell the table view how many sections and rows there are:

```
- (NSInteger)numberOfSectionsInTableView:(UITableView *)tableView
{
#warning Potentially incomplete method implementation.
    // Return the number of sections.
    return 0;
    return 1;
}

- (NSInteger)tableView:(UITableView *)tableView numberOfRowsInSection:(NSInteger)section
{
#warning Incomplete method implementation.
    // Return the number of rows in the section.
    return 0;
    return [tasks count];
}
```

Next, replace the content of the method that populates each cell:

```
- (UITableViewCell *)tableView:(UITableView *)tableView
cellForRowAtIndexPath:(NSIndexPath *)indexPath
{
    static NSString *CellIdentifier = @"Cell";

    UITableViewCell *cell = [tableView dequeueReusableCellWithIdentifier:
                                CellIdentifier];
    if (cell == nil) {
        cell = [[UITableViewCell alloc] initWithStyle:UITableViewCellStyleDefault
                                        reuseIdentifier:CellIdentifier];
    }

    NSString *identifier = nil;
    NSString *task = [self.tasks objectAtIndex:indexPath.row];
    NSRange urgentRange = [task rangeOfString:@"URGENT"];
    if (urgentRange.location == NSNotFound) {
        identifier = @"plainCell";
    } else {
        identifier = @"attentionCell";
    }
    UITableViewCell *cell = [tableView dequeueReusableCellWithIdentifier:identifier];

    // Configure the cell...

    UILabel *cellLabel = (UILabel *)[cell viewWithTag:1];
    cellLabel.text = task;

    return cell;
}
```

We start off here by grabbing a task from our array, and checking to see whether it contains the string "URGENT". This is not an especially advanced algorithm for finding urgent to-do list items, but it will do for now. We use the presence or absence of the test word to decide which cell we want to load, and pick our cell identifier accordingly.

Back in Chapter 8, we showed you how to tell a table view that it can find a cell for a given identifier inside a nib file with a particular name. Putting dynamic cell prototypes inside a table view in a storyboard works similarly, with the difference that you don't need to write any code to tell the table view about these cell prototypes. Instead, any table view loaded from a storyboard automatically has access to its associated cell prototypes. Just as with the registered-nib variety, the table view will create such cells on the fly when you ask it using the `dequeueReusableCellWithIdentifier:` method, so there's no need to check for a `nil` return value there, which makes the rest of this method pretty simple.

Click the *Run* button to see this in action (see Figure 10-6). The app launches and displays our entire task list, with the URGENT-flagged items shown in red. Note that the code doesn't know or care about the color or font shown in each cell. In fact, it knows almost nothing about the cells it uses! It assumes only that each cell contains a `UILabel` with a tag of 1. That way, the GUI is pretty well decoupled from the controller code, which means that we can easily change the look of the table view cells without messing with the source code.

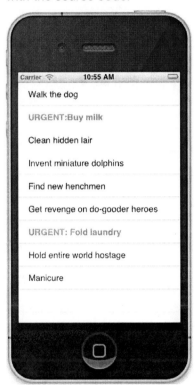

Figure 10-6. *Running our custom cell storyboard app. Notice the use of the two different cell types. Best get to cleaning that hidden lair, eh?*

Will It Load?

It's probably worth mentioning here that the creation of the `BIDTaskListController` and its corresponding view happened in a different way than what you may be used to. Up until now, every time we've wanted to display a view controller, we've created it explicitly in code somewhere, typically telling it to load its view from a nib file (except for table view controllers, which have mostly been created without a nib, which triggers them to just create their own view from scratch).

In the simplest apps, we've created a view controller and loaded its nib inside the app delegate. In other cases, we've created them in response to a user action, to then push them onto a navigation stack. In the current example, however, the view controller is created automatically when the app launches, just as the template-generated BIDViewController was at the beginning. This bit of magic happens during app launch, but there are ways of programmatically pulling any view controller out of a storyboard while the app is running. We'll get into that a little later in the chapter.

Static Cells

Next, let's take a look at the new table view configuration that storyboards allow you to use: static cells. Up until now, all the table view cells you've seen (in this chapter and previous chapters) have been created and populated dynamically within a controller's UITableViewDataSource methods. That's great for displaying lists whose size may vary at runtime, but sometimes you know exactly what you want to display. If the number of items and the kinds of cells are known entirely in advance, implementing a dataSource can feel like a chore.

Fortunately, storyboards provide an alternative to dynamic cells: static cells! You can now define a set of cells inside a table view, and they will be displayed in the running app exactly as they are in Interface Builder. Note that these cells are only "static" in the sense that the existence of the cells themselves will be consistent every time you run the app. Their content can change on the fly, however. In fact, you can connect outlets to them in order to access them and set their content from the controller. Let's get started!

Select the *Simple Storyboard* folder in the Xcode project browser, and choose File ➤ New ➤ New File.... In the file creation assistant's *iOS / Cocoa Touch* section, select *UIViewController subclass* and click *Next*. Name the new class *BIDStaticCellsController*, select *UITableViewController* in the *Subclass of* popup, make sure the checkbox for creating a matching XIB file is switched off, and create your new files. We'll make some changes to this controller's implementation later, after configuring the GUI.

Go back into *MainStoryboard.storyboard*, and pull another *UITableViewController* from the library, adding it alongside the two you already have. Now, go ahead and move the big arrow—the one that points out the initial view controller—so that it points to our new controller.

If you're concerned about the fact that we now have three view controllers here, but only one of them is ever displayed, don't worry! When building a real application, we wouldn't normally leave our storyboard full of discarded views and controllers, but for now, we're just doing some exploratory programming. Later in the chapter, we'll show you how to make use of multiple view controllers in a single storyboard.

So, let's get back to those static cells.

Going Static

Select the table view you just dragged in with the controller, and open the attributes inspector. Click the popup at the very top, labeled *Content*, and change it from *Dynamic Prototypes* to *Static Cells*. Doing this changes the basic functioning of this table view, even though it still looks pretty much the same. Now, any cells you add will be created as is, in the order you specify, when this table view and its controller are loaded from the storyboard.

To give this table a slightly different look than the task list, use the *Style* popup to select *Grouped*. The table view starts off with just one section, and you'll see that it now gets the rounded appearance of a typical grouped table view. Click to select the section (not one of the cells, but the section itself), and the object inspector will display a few items you can set there as well. Set the number of rows to 2, and set the header to *Silliest Clock Ever*, since that's what this controller is going to be.

Now, select the first cell, and use the attributes inspector to set its *Style* to *Left Detail*. This is one of the built-in table view cell styles that you've seen before. It presents a descriptive label on the left side of the cell and a larger label to contain the value you actually want to display. Double-click to select the text of the label on the left, and change it to *The Date*. Repeat the same steps for the second cell, changing its text to *The Time* (see Figure 10-7).

Figure 10-7. *Changes to our static cells in our new table view*

We're going to create a pair of outlets to connect our controller directly to the detail labels, so that we can set their values when the app runs. First, select the new view controller in the dock, bring up the identity inspector, and change the view controller's

Class to *BIDStaticCellsController*. Press return to commit the change. You should see the view controller's name change to *Static Cells Controller* in the dock.

Select the *Static Cells Controller* by clicking its icon in the dock. Bring up the assistant editor and make sure it is showing the *BIDStaticCellsController.h* file.

Select the right-hand label in the table view's first cell (next to *The Date* label), and then control-drag from it over to the header file, releasing the mouse button anywhere between the @interface line and the @end line. In the popup that appears, set its *Name* to *dateLabel*, and leave the default values for the rest. Then do the same for the second cell of the table view, but name that one *timeLabel*. With these simple steps, you've created new outlet properties and connected them properly all at once.

Next, let's switch over to *BIDStaticCellsController.m* to make it display date and time values.

So Long, Good Old Table View Data Source

Over in *BIDStaticCellsController.m*, we need to delete our familiar friends, the dataSource methods. All three of them must go! Otherwise, our table view may get confused about what it's really meant to display. Simply remove these methods completely, including the content between the curly braces (which we're not showing here).

```
- (NSInteger)numberOfSectionsInTableView:(UITableView *)tableView
{
    ...
}

- (NSInteger)tableView:(UITableView *)tableView numberOfRowsInSection:(NSInteger)section
{
    ...
}

- (UITableViewCell *)tableView:(UITableView *)tableView
cellForRowAtIndexPath:(NSIndexPath *)indexPath
{
    ...
}
```

With these methods out of the way, making our silly clock show the date and time is as simple as inserting the following few lines at the bottom of the viewDidLoad method:

```
- (void)viewDidLoad
{
    [super viewDidLoad];

    // Some comments you can safely ignore right now!

    NSDate *now = [NSDate date];
    dateLabel.text = [NSDateFormatter localizedStringFromDate:now
                                       dateStyle:NSDateFormatterLongStyle
                                       timeStyle:NSDateFormatterNoStyle];
```

```
timeLabel.text = [NSDateFormatter localizedStringFromDate:now
                                    dateStyle:NSDateFormatterNoStyle
                                    timeStyle:NSDateFormatterLongStyle];
}
```

Here, we're grabbing the current date, and then using a handy NSDateFormatter class method to pull out date and time values separately, passing each off to the appropriate label. And that's all we need to do! Click the *Run* button, and you'll see our dazzling new clock in fine working order (see Figure 10-8).

Figure 10-8. *Our silliest clock in all its glory*

As you've seen, for what we wanted to do here, using a table view with static cells can be a whole lot simpler than taking the dataSource-based approach, since our design lets us use a fixed set of cells. If you want to display a more open-ended dataset, you'll need to use the dataSource methods, but static cells provide a much simpler way to create certain types of displays, such as menus and details.

You Say Segue, I Say Segue

It's time to move on to another new piece of functionality that Apple built in to iOS 5: the **segue**. With segues, you can use Interface Builder to define how one scene will transition to another. This works only in storyboards, not nibs.

The idea is that you can use a single storyboard to represent the majority of your app's GUI, with all scenes and their interconnecting segues represented in the graphical layout view. This has a couple of nice side effects. One is that you can see and edit the entire flow of your application in one place, which makes working with the big picture a lot easier. Also, this ends up eliminating some code from your view controllers, as you'll see soon.

> **NOTE**: We've noticed some confusion about the word *segue* and its pronunciation. It's an English word (borrowed from Italian) that means about the same as *transition*. It's mostly used in certain journalistic and musical contexts, and is relatively uncommon. It's pronounced "seg-way," just like the Segway transportation devices (whose name, it should now be clear, is basically a trademarkable respelling of *segue*).

Segues are most useful in the context of a navigation-based application—the kind that you're quite familiar with after the past few chapters. Here, we're not going to build anything as large as the app we created in Chapter 9, but we'll demonstrate how you could start down that road using a combination of segues and static tables.

Creating Segue Navigator

Use Xcode to create a new iOS application project, choosing the *Empty Application* template. Name the project *Seg Nav*. This creates the same empty app that you've seen in earlier chapters, including the non-storyboard code for creating a window in the app delegate. Since we're going to create a storyboard and want it to autoload, as with our previous project, select *BIDAppDelegate.m*, find the `application:didFinishLaunchingWithOptions:` method, and delete all but the last line of the method.

```
- (BOOL)application:(UIApplication *)application
didFinishLaunchingWithOptions:(NSDictionary *)launchOptions
{
    self.window = [[UIWindow alloc] initWithFrame:[[UIScreen mainScreen] bounds]];
    // Override point for customization after application launch.
    self.window.backgroundColor = [UIColor whiteColor];
    [self.window makeKeyAndVisible];
    return YES;
}
```

Select the *Seg Nav* folder in the project navigator, and choose **File ➤ New ➤ New File....** to create a new file. From the *User Interface* section, choose *Storyboard* and click *Next*. Name the new file *MainStoryboard.storyboard*.

After the file is created, we need to configure our project so that the storyboard will be loaded when the app launches. Select the icon representing the *Seg Nav* project itself at the top of the project navigator. In the target's *Summary* tab, you'll find a popup button that lets you set the *Main Storyboard*. Set it to *MainStoryboard* by clicking the popup and picking *MainStoryboard* (which is, conveniently, the only item in the list).

Filling the Blank Slate

Select *MainStoryboard.storyboard* in the project navigator. You'll see the familiar layout area appear, but now it's completely empty. Remedy that by finding a *Navigation Controller* in the object library and dragging it into the layout area to create our app's initial scene.

As you may recall, the UINavigationController class doesn't display any content itself. It shows just the navigation bar. So, when you drop the *UINavigationController* into your storyboard, Interface Builder does you a favor, creating a *UIViewController* and its view at the same time. You'll see the two controllers side by side, with a special sort of arrow pointing from the navigation controller to the view controller. This arrow represents the navigation controller's rootViewController property, and it is connected to the view controller so that you have something to load.

Now, you could start populating that view with some content, but in this case, it would be instructive to create a storyboard-based app that mimics some of the behavior of the Nav app we built in Chapter 9. So, let's insert a table view controller into the mix as our start screen. To do that, select the *view controller* on the right (the one that's being pointed to, as opposed to the navigation controller that is pointed from), and press the delete key to delete it. Then drag a *Table View Controller* from the object library to the layout area, placing it to the right of the navigation controller, just as the original one was.

At this point, the navigation controller doesn't know where to find its rootViewController, so we need to reestablish that connection with the new view controller. Control-drag from the navigation controller to the table view controller, and you'll see a popup that looks like the same popup you'e seen before when connecting outlets and actions (see Figure 10-9). This time, however, instead of showing you outlets or actions, it displays a group of connections labeled *Storyboard Segues*. This contains an item called *Relationship - rootViewController*, followed by three items labeled *Push*, *Modal*, and *Custom*. Those final three choices are used for creating segues between scenes, and we'll get to them soon. For now, select the *rootViewController* entry to establish that relationship.

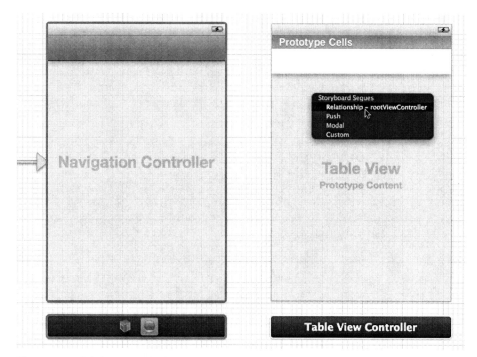

Figure 10-9. *This image shows the popup that appears as the result of control-dragging from the navigation controller to the new table view controller. We are selecting the item that establishes the table view controller as the root view controller.*

Now, let's make this table view show a menu of items. We want it to work about the same as the root table view in the Nav app from Chapter 9. However, with static table view cells, this is going to be a lot easier!

Use the dock to select the table view (as opposed to the table view controller), and use the attributes inspector to change its *Content* to *Static Cells*. You'll see that the table view immediately acquires three cells. We're going to use only two, so select one of the cells and delete it.

Select each of the two remaining table view cells, and use the inspector to change its *Style* to *Basic* so they acquire a title. Then change the titles to *Single view* and *Sub-menu*, respectively (see Figure 10-10).

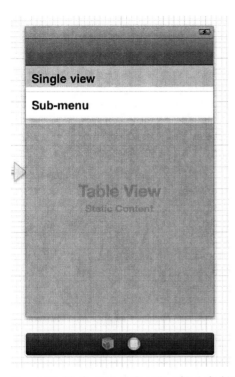

Figure 10-10. *Our two static cells, with their titles changed*

Now, we'll provide some titling so that the table view plays nicely with the navigation controller. Select the navigation item (it looks like an empty toolbar) that is shown at the top of the table view. Use the attributes inspector to set its *Title* to *Segue Navigator* and its *Back Button* to *Seg Nav*. Note the *Back Button* setting doesn't define a value that will be displayed while this view is shown; rather, it defines the value that a subsequent view controller will show in its back button, leading back to this root view.

Run the app to get a sense of how things are going so far. You should see the root table view that you just created, containing two cells with the titles you just specified.

That's actually all we need to do for this root view, and we did it all in Interface Builder, without writing a single line of code! You may recall that in the Nav app in Chapter 9, we had a custom UITableViewController subclass that contained code to allocate and initialize every other view controller. Every time we added a new second-level controller, we needed to import its header and add code to create an instance of it. Thanks to storyboards, we don't need any of that.

As we add each second-level controller to our storyboard, we'll connect a root table view cell to it using a segue in Interface Builder. Our root controller class doesn't need to know about any of the other controllers, since it's not directly involved in creating or displaying them. That's why we can get by with a plain UITableViewController here, instead of making a subclass of our own.

First Transition

It's time to create our first segue. Start by dragging a *View Controller* from the object library into the layout area, dropping it somewhere to the right of the other views. We're not going to display any special content here, because we really just want to jump into a segue. That being said, you need to have some way of visually confirming that the correct view is being shown, so drag a *UILabel* from the object library into the new view, change its text to *Single View*, and center the label vertically and horizontally in the view.

Now comes the segue-creation magic. Control-drag from the *Single view* cell in the middle table view controller to the new view you just added. You'll see the *Storyboard Segues* popup again, this time without the `rootViewController` relationship you saw previously. It contains only the three kinds of segues that Interface Builder supports: *Push*, *Modal*, and *Custom*. Choose *Push*, because here we want to implement the standard navigation controller model involving pushing controllers onto a stack. Once you do this, you'll see a new connection arrow appear, pointing to our new controller from the root view controller. You'll also see that the *Single view* row in the root table view has acquired a disclosure arrow on the right side, to let the user know that there's something more to be discovered by tapping that cell.

Run your app again, and you'll see that the top cell now contains the disclosure arrow you just saw, and tapping the row takes you to the new view that you just set up. You'll see the label you added, the title at the top of screen, and the back button, which is labeled *Seg Nav* (remember that we configured that a page or two ago). Press the back button to make sure it takes you back to the root view.

So, now we have a fair amount of view navigation going on, without having written any code at all. This elimination of boring glue code is one of the biggest wins involved when using segues. But, of course, you can't eliminate all your code. Let's see what it takes to make a GUI that displays a list of items in a normal, dynamically sized table view, and lets you select one of those items to drill down into more detail. Note that this will require a bit of code so the list controller can pass the selected item off to the detail controller. With segues, that sort of thing can be handled a bit more simply than before, as you'll soon see.

A Slightly More Useful Task List

We're going to make a slightly more capable version of the task list that we created in the Simple Storyboard example earlier. In this version, you'll see a list of tasks, and also be able to tap on one to edit it. Fortunately, we can build on what we created earlier to make this go pretty easily.

Open the *Simple Storyboard* project you created earlier in the chapter, and drag the *BIDTaskListController.h* and *BIDTaskListController.m* files from that project onto the *Seg Nav* folder in the *Seg Nav* project. When Xcode asks, make sure you turn on the checkbox telling Xcode to copy the files into the project. This will copy the files from the

Simple Storyboard project folder on your hard drive into the *Seq Nav* folder on your hard drive and add the two files to the *Seq Nav* project.

Next, select the *Seg Nav* folder in the project navigator, and choose **File ➤ New ➤ New File…** to make a new file. Select the *Cocoa Touch* section on the left, select *UIViewController subclass*, and click *Next*. Name this class *BIDTaskDetailController*, and choose *UIViewController* as its superclass. Do not have Xcode create the accompanying nib file. Create this class, but don't edit it in any way just yet. We'll get back to it soon enough, after setting up the GUI.

Switch over to *MainStoryboard.storyboard*. It's time to create the next scene in our storyboard, which will be the task list display. Drag a *UITableViewController* from the object library to the layout area, placing it to the right of the other controllers. Use the identity inspector to change the new controller's class to *BIDTaskListController*.

In the dock, select the table view associated with the new controller. Now, open the attributes inspector and take a look at the *Content* popup menu. Since the table view we just added will display a dynamic list of items, we're going to leave it set to *Dynamic Prototypes* instead of switching to *Static Cells*.

We'll use the same two prototype cells we described earlier in the "Dynamic Prototype Cells" section. Either flip back there to see how those were configured or open *Simple Storyboard* and copy both cells, pasting them into your new table. To recap, the critical configuration points are making sure that each cell has a label, each label has its tag set to *1*, and the cells have *plainCell* and *attentionCell* set as their identifiers.

If you want to copy the two cells from *Simple Storyboard* instead of re-creating them, switch over to the *Simple Storyboard* project and find the two table view cells in the *Task List Controller Scene*. They should be labeled *plainCell* and *attentionCell*. Close the disclosure triangles, and select both cells (that way, you get the labels, too). Copy, then switch over to the *Seg Nav* project, select the new table view, and paste. Your cells and labels should appear in your new controller. Delete the original cell, leaving you with two prototype cells instead of three.

Viewing Task Details

Now, we're going to add another scene to manage the detail display that's shown when the user selects a row in the task list. Drag a *View Controller* (*not* a *Table View Controller*) from the object library to the layout area, placing it to the right of the previous scene. Use the identity inspector to change the controller's class to *BIDTaskDetailController*.

This scene is all about editing the details of a chosen task, which in our simple case, means just editing a string. To accomplish this, we'll use a UITextView, the full-featured multiline text editor employed throughout iOS. Drag one of those from the object library into the new scene. You'll see that it snaps into place, filling the entire space. Since we want to let users edit this text, filling the screen isn't what we want. We'll be better off with a smaller text view that just fills the space that the on-screen keyboard doesn't cover. Grab the resize handle in the center of the bottom edge, and drag it upward until the text view is just 200 pixels tall. A coordinate display appears at the top of the view when you start dragging, to make this easier.

Next, we need to add an outlet to the controller so that it can find this text view to set and retrieve the string it contains. We can create and connect that outlet in one step using Interface Builder's drag-to-code feature, so let's do that. First bring up the assistant editor. With the *BIDTaskDetailController* or any part of its view selected in the layout editor, the assistant editor should show the controller's header file, *BIDTaskDetailController.h*. If not, then the assistant editor probably isn't in automatic mode. You can fix that using the jump bar above the assistant editor, clicking the icon just to the right of the arrows, and selecting *Automatic*.

Now, select the text view, and control-drag from it to the code, dropping it anywhere between the @interface and @end lines. In the popup that appears, make sure the *Connection* type is set to *Outlet*, and name it *textView*. Also make sure *Storage* is set to *Weak*, and then click *Connect*. This creates a new outlet in the *BIDTaskDetailController.h*, and synthesizes the getter and setter in *BIDTaskDetailController.m*.

Make More Segues, Please

With the basic GUIs for our new scenes set up, it's time to connect them using segues. Start by heading back to our root table view (the one with the *Simple View* and *Sub-menu* cells). Select the *Sub-menu* cell, and control-drag from it to the task list controller. It is probably easiest to do this in the dock. Note that this control-drag will be skipping over the *Single View* scene. In the popup that appears, select *Push* so that the segue will push the task list scene onto the stack when *Sub-menu* is tapped.

Now, select the first prototype cell in the task list controller scene (it's called *Table View Cell – plainCell* in the dock, and you might need to open a disclosure triangle to see it), and control-drag from there to the *Task Detail Controller*, again selecting *Push*. Do the same for the second prototype cell (it's called *Table View Cell – attentionCell* in the dock), so that both are connected to the *Task Detail Controller*. This will show up as two arrows pointing from the task list to the task detail. Each arrow represents a segue, and each comes from a specific prototype cell. When you run the app later, all the cells that are created for the task list are duplicated from one of these prototypes, so each of them will have a segue that enables them to create and activate a task detail controller.

At this point, our GUI configuration is complete. All we need to do is implement a few methods to pass the selected task from the list view to the detail view, then back the other way after the user has had a chance to view and edit the task.

Passing a Task from the List

Select *BIDTaskListController.m*, and add this method to the class at the end of the file, just above the @end:

```
- (void)prepareForSegue:(UIStoryboardSegue *)segue sender:(id)sender {
    UIViewController *destination = segue.destinationViewController;
```

```
    if ([destination respondsToSelector:@selector(setDelegate:)]) {
        [destination setValue:self forKey:@"delegate"];
    }
    if ([destination respondsToSelector:@selector(setSelection:)]) {
        // prepare selection info
        NSIndexPath *indexPath = [self.tableView indexPathForCell:sender];
        id object = [self.tasks objectAtIndex:indexPath.row];
        NSDictionary *selection = [NSDictionary dictionaryWithObjectsAndKeys:
                                    indexPath, @"indexPath",
                                    object, @"object",
                                    nil];
        [destination setValue:selection forKey:@"selection"];
    }
}
```

This new prepareForSegue:sender: method is called in our controller whenever our controller's view is about to be replaced by another controller's view as a result of a segue being activated. In our case, this means that when any of the cells in our table view is selected, the cell will activate its associated segue, and this method will be called in our controller. This gives us a chance to prepare data to pass along to the next controller. In the past, we did this sort of thing in a table view delegate method that was called when a row was selected, but this new way is a bit more flexible. We could, for instance, replace our table view cells with buttons, and as long as those buttons use segues to launch other view controllers, this method will be called in the same way as it is now.

We have some new things here, so let's walk through the prepareForSegue:sender: method together. Through the segue parameter that we're given, we can access the destinationViewController (the one that's going to be displayed in a moment) and the sourceViewController (the one that's about to be removed from the display). In this case, we'll use the destinationViewController property to configure the detail view controller, so we put it into a local variable for easy access:

```
    UIViewController *destination = segue.destinationViewController;
```

Next, we configure the destination's delegate, if it has one, to point back at us. That will let the destination send us back data when it's finished. Right now, our detail display doesn't have a delegate property, but it will later.

```
    if ([destination respondsToSelector:@selector(setDelegate:)]) {
        [destination setValue:self forKey:@"delegate"];
    }
```

KEY-VALUE CODING

Instead of calling a setDelegate: method, we're using something called key-value coding (KVC), which allows us to use getters and setters on any object indirectly, using strings instead of method names. KVC is a core feature of the Cocoa Touch frameworks, and its main methods—setValue:forKey: and valueForKey:—are built into NSObject, so they're available everywhere. We're not covering this topic in detail in this book, but are using it briefly here.

Why are we using KVC here instead of just setting the delegate directly? One of the benefits of KVC is that it frees us from knowing the specifics of other classes' interfaces, which leads to less tightly coupled code. If we wanted to call the method directly, we would need to declare an interface of some kind that included the `setDelegate:` method, and cast our destination variable to be a type that implemented that method. Using KVC, our code doesn't need to know anything about the `setDelegate:` method (apart from the fact that the receiver responds to it), so we don't need to have any declared interface to it. `BIDTaskListController` doesn't import `BIDTaskDetailController`'s header, and, in fact, doesn't even know that it exists, and that's a good thing! In general, the fewer dependencies you have between classes, the better. Properly applied, KVC can help you move your code in that direction.

Then we put the selected task and the index of the selected row into a dictionary to pass along to the detail view controller. We need to include the row index here so that later on, when the detail view is done, its controller can pass us back the same index so we know which task to change. Otherwise, we would just get back a string and have no idea where it belongs in our list.

```
if ([destination respondsToSelector:@selector(setSelection:)]) {
    // prepare selection info
    NSIndexPath *indexPath = [self.tableView indexPathForCell:sender];
    id object = [self.tasks objectAtIndex:indexPath.row];
    NSDictionary *selection = [NSDictionary dictionaryWithObjectsAndKeys:
                                    indexPath, @"indexPath",
                                    object, @"object",
                                    nil];
    [destination setValue:selection forKey:@"selection"];
}
```

Just as when we set the destination's `delegate`, we set its `selection` using KVC. Right now, our detail view controller doesn't have a `selection` property either, but we're about to change that.

Handling Task Details

Select *BIDTaskDetailViewController.h*. You'll see the `textView` property that was added earlier when you dragged an outlet from the storyboard. Now, add two more properties, as shown here:

```
#import <UIKit/UIKit.h>

@interface BIDTaskDetailController : UIViewController
@property (weak, nonatomic) IBOutlet UITextView *textView;
@property (copy, nonatomic) NSDictionary *selection;
@property (weak, nonatomic) id delegate;

@end
```

Note that the `selection` property specifies copy storage (which is normally what we use whenever we're dealing with value-based classes such as `NSString` and `NSDictionary`), but `delegate` specifies weak storage. We need to use weak storage for the `delegate` property so that we don't accidentally retain our delegate, who may already be retaining us! In our case, we happen to know that the delegate isn't retaining this object, but the standard pattern used throughout Cocoa Touch is to make sure that delegates aren't retained, and there's no reason for us to do anything different here.

Switch to *BIDTaskDetailViewController.m*, and add some code near the top to synthesize getters and setters for our new properties:

```
@implementation BIDTaskDetailController
@synthesize textView;
@synthesize selection;
@synthesize delegate;
.
.
.
```

Scroll down a bit, and you'll see that the viewDidLoad method is commented out by default. Remove the comment markers, and insert the code shown here:

```
- (void)viewDidLoad
{
    [super viewDidLoad];
    textView.text = [selection objectForKey:@"object"];
    [textView becomeFirstResponder];
}
```

By the time this code is called, the segue is already underway, and the list view controller has set up our selection property. We pull out the value it contains, and pass it along to the text view. Then we tell the text view to become the *First Responder*, effectively saying, "tag, you're it," which will make the on-screen keyboard appear immediately.

Run your app, navigate your way into the task list, and select a task. You should see its value appear in an editable text field. So far, so good.

Passing Back Details

Now all that's left is to bring the user's edits back into the list. Unfortunately, our detail view won't be sent the prepareForSegue:sender: method when the user hits the back button. That method is called only when a segue is pushing a new controller onto the stack, not when it's popping one off. Instead, we'll use a standard UIViewController method to implement something similar to what we did in BIDTaskListController. Add this method just after viewDidLoad:

```
- (void)viewWillDisappear:(BOOL)animated {
    [super viewWillDisappear:animated];

    if ([delegate respondsToSelector:@selector(setEditedSelection:)]) {
        // finish editing
        [textView endEditing:YES];
        // prepare selection info
        NSIndexPath *indexPath = [selection objectForKey:@"indexPath"];
        id object = textView.text;
        NSDictionary *editedSelection = [NSDictionary dictionaryWithObjectsAndKeys:
                                          indexPath, @"indexPath",
```

```
                                      object, @"object",
                                      nil];
            [delegate setValue:editedSelection forKey:@"editedSelection"];
        }
    }
}
```

Here, we're setting our delegate's `editedSelection` property, if it has one, again using KVC. As before, this frees us from needing to know anything in particular about the other controller class. The only piece of this that might seem new to you is the call to `[textView endEditing:YES]`, which simply forces the text view to finish any editing the user may have been doing so that its text value (which we grab a few lines later) is up to date.

Making the List Receive the Details

The final piece in this puzzle is ready to be put in place. We need to go back to the list view controller to make sure it can receive the `editedSelection` and do something reasonable with it. Return to *BIDTaskListController.m*, and make the following changes to the class extension at the top:

```
@interface BIDTaskListController ()
@property (strong, nonatomic) NSArray *tasks;
@property (strong, nonatomic) NSMutableArray *tasks;
@property (copy, nonatomic) NSDictionary *editedSelection;
@end
    .
    .
    .
```

The first change is there to turn the `tasks` property into a mutable array, so that we can change it when the user edits a task. The `editedSelection` property will contain the edited value passed back from the detail controller, as we just described.

Next, synthesize a getter and setter for `editedSelection`:

```
    .
    .
    .
@implementation BIDTaskListController

@synthesize tasks;
@synthesize editedSelection;
```

Then make a slight change to `viewDidLoad`, replacing our old `NSArray` usage with `NSMutableArray`:

```
- (void)viewDidLoad
{
    [super viewDidLoad];
    .
    .
    .
    self.tasks = [NSArray arrayWithObjects:
```

```
    self.tasks = [NSMutableArray arrayWithObjects:
.
.
.
```

Finally, we're going to implement a custom setter for the editedSelection property. This will replace the setter that was implicitly created by the earlier @synthesize declaration. You can place this method at the end of the file, just before @end:

```
- (void)setEditedSelection:(NSDictionary *)dict {
    if (![dict isEqual:editedSelection]) {
        editedSelection = dict;
        NSIndexPath *indexPath = [dict objectForKey:@"indexPath"];
        id newValue = [dict objectForKey:@"object"];
        [tasks replaceObjectAtIndex:indexPath.row withObject:newValue];
        [self.tableView reloadRowsAtIndexPaths:[NSArray arrayWithObject:indexPath]
                        withRowAnimation:UITableViewRowAnimationAutomatic];
    }
}
```

This method pulls out the index and value (representing the edited item) that were passed back to us. It then puts the new value into the correct place in our tasks array, and reloads the corresponding cell so that the correct value is displayed.

Run your app again, navigate to the task list, pick a task, and edit it. When you then tap the back button, you'll see that the edited value replaces the old value in the list. Moreover, since the affected cell is actually reloaded, the cell's type can change from *plainCell* to *attentionCell* and vice versa, depending on the edited value. Try adding the word *URGENT* to a task that doesn't already have it, or removing the word from one that does to see what happens.

If Only We Could End with a Smooth Transition

Now that you've been flipping through storyboards, have you flipped for them? We think they're great for all kinds of navigation-based applications, and we're going to use them a few more times in this book. As we mentioned at the beginning of the chapter, a downside to using storyboards at the time of this writing is that they work only with iOS 5 and later, which means that you're limited to people who have upgraded or have just purchased a new device. Over time, this situation will change as more and more people upgrade to iOS 5.

Moving on, it's time to consider even more navigation issues as you learn about iPad-specific view controllers in Chapter 11.

iPad Considerations

From a technical standpoint, programming for the iPad is pretty much the same as programming for any other iOS platform. Apart from the screen size, there's very little that differentiates a 3G iPad from an iPhone, or a Wi-Fi iPad from an iPod touch. In spite of the fundamental similarities between iPhone and iPad, from the user's point of view these devices are really quite different. Fortunately, Apple had the good sense to recognize this fact from the outset and equip the iPad with additional UIKit components that help create applications that better utilize the iPad's screen size and usage patterns. In this chapter, you'll learn how to use these components. Let's get started!

Split Views and Popovers

In Chapter 9, you spent a lot of time dealing with app navigation based on selections in table views, where each selection causes the top-level view, which fills the entire screen, to slide to the left and bring in the next view in the hierarchy, perhaps yet another table view. Plenty of iPhone and iPod touch apps work this way, both among Apple's own apps and third-party apps.

One typical example is Mail, which lets you drill down through servers and folders until you finally make your way to a message. Technically, this approach can work on the iPad as well, but it leads to a user interaction problem.

On a screen the size of the iPhone or iPod touch, having a screen-sized view slide away to reveal another screen-sized view works well. On a screen the size of the iPad, however, that same interaction feels a little wrong, a little exaggerated, and even a little overwhelming. In addition, consuming such a large display with a single table view is inefficient in most cases. As a result, you'll see that the built-in iPad apps do not actually behave that way. Instead, any drill-down navigation functionality, like that used in Mail, is relegated to a narrow column whose contents slide left or right as the user drills down or backs out. With the iPad in landscape mode, the navigation column is in a fixed position on the left, with the content of the selected item displayed on the right. This is what's called, in the iPad world, a **split view** (see Figure 11–1).

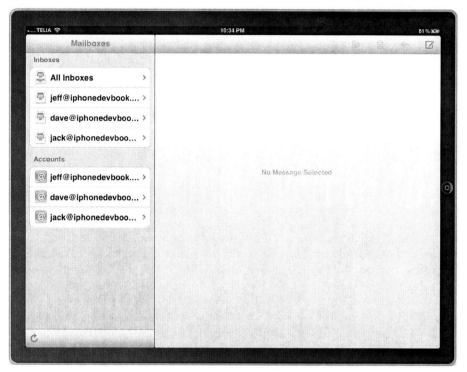

Figure 11–1. *This iPad, in landscape mode, is showing a split view. The navigation column is on the left. Tap an item in the navigation column—in this case, a specific mail account—and that item's content is displayed in the area on the right.*

The left side of the split view is always 320 points wide (the same width as an iPhone in its vertical position). The split view itself, with navigation and content side by side, appears only in landscape mode. If you turn the device to a portrait orientation, the split view is still in play, but it's no longer visible in the same way. The navigation view loses its permanent location and can be activated only by pressing a toolbar button, which causes the navigation view to pop up in a view that floats in front of everything else on the screen (see Figure 11–2). This is what's called a **popover**.

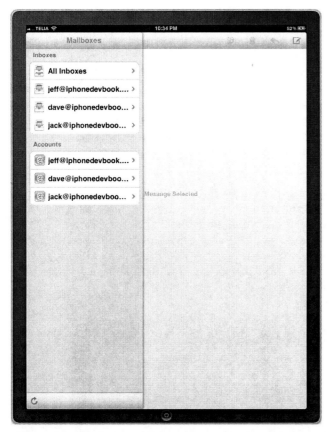

Figure 11–2. *This iPad, in portrait mode, does not show the same split view as seen in landscape mode. Instead, the information that made up the left side of the split view in landscape mode is embedded in a popover. Mmmm, popovers.*

In this chapter's example project, you'll see how to create an iPad application that uses both a split view and a popover.

Creating a SplitView Project

We're going to start off pretty easily, by taking advantage of one of Xcode's predefined templates to create a split view project. We'll build an app that presents a slightly different take on Chapter 9's presidential app, listing all the US presidents and showing the Wikipedia entry for whichever one you select.

Go to Xcode and select **File ➤ New ➤ New Project....** From the iOS *Application* group, select *Master-Detail Application* and click Next. On the next screen, name the new project *Presidents*, set the *Class Prefix* to *BID*, and switch *Device Family* to *iPad*. Make sure the *Use Storyboard* and *Use Automatic Reference Counting* checkboxes are checked, but that the *Use Core Data* and *Include Unit Tests* checkboxes are unchecked. Click *Next*, choose the location for your project, and then click *Create*. Xcode will do its usual thing, creating a handful of classes and storyboard files for you, and then showing the project. If it's not already open, expand the *Presidents* folder, and take a look at what it contains.

From the start, the project contains an app delegate (as usual), a class called `BIDMasterViewController`, and a class called `BIDDetailViewController`. Those two view controllers represent, respectively, the views that will appear on the left and right sides of the split view. `BIDMasterViewController` defines the top level of a navigation structure, and `BIDDetailViewController` defines what's displayed in the larger area when a navigation element is selected. When the app launches, both of these are contained inside a split view, which, as you may recall, does a bit of shape-shifting as the device is rotated.

To see what this particular application template gives you in terms of functionality, build the app and run it in the simulator. Switch between landscape mode (see Figure 11–3) and portrait mode (see Figure 11–4), and you'll see the split view in action. In landscape mode, the split view works by showing the navigation view on the left and the detail view on the right. In portrait mode, the detail view occupies most of the picture, with the navigation elements confined to the popover, which is brought into view with the press of the button in the top left of the view.

Figure 11–3. *The default Master-Detail Application template in landscape mode. Note the similarity between this figure and Figure 11–1.*

Figure 11–4. *The default Master-Detail Application template in portrait mode with the popover showing. Note the similarity between this figure and Figure 11–2.*

We're going to build on this to make the president-presenting app we want, but first let's dig into what's already there.

The Storyboard Defines the Structure

Right off the bat, you have a pretty complex set of view controllers in play:

- A split view controller that contains all the elements
- A navigation controller to handle what's happening on the left side of the split
- A master view controller (displaying a master list of items) inside the navigation controller
- A detail view controller on the right

In the default Master-Detail Application template that we used, these view controllers are set up and interconnected primarily in the main storyboard file, rather than in code.

Apart from doing GUI layout, Interface Builder really shines as a way of letting you connect different components without writing a bunch of code just to establish relationships. Let's dig into the project's storyboard to see how things are set up.

Select *MainStoryboard.storyboard* to open it in Interface Builder. This storyboard really has a lot of stuff going on. You'll definitely want to turn on list view (you can refer back to Chapter 10 for a review of storyboard basics) for the best results (see Figure 11–5).

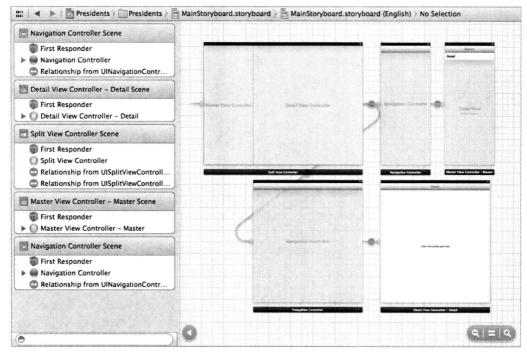

Figure 11–5. *MainStoryboard.storyboard open in Interface Builder. This complex object hierarchy is best viewed in list mode.*

To get a better sense of how these controllers relate to one another, open the connections inspector, and then spend some time clicking each of the view controllers in turn.

The split view controller and both of the navigation controller objects each has one or more connections to other controllers from the start, as shown in the *Storyboard Segues* section of the connections inspector. In Chapter 10, you gained some familiarity with these sorts of connections, including the rootViewController relationship that each UINavigationController has. Here, you'll find that the UISplitViewController actually has two relationships connected to other controllers: masterViewController and detailViewController. These are used to tell the UISplitViewController what it should use for the narrow strip it displays on the left or in a popup (the masterViewController), and what it should use for the larger display area (the detailViewController).

At this point, the content of *MainStoryboard.storyboard* is really a definition of how the app's various controllers are interconnected. As in most cases where you're using storyboards, this eliminates a lot of code, which is usually a good thing. If you're the kind of person who likes to see all such configuration done in code, you're free to do so, but for this example, we're going to stick with what Xcode has provided.

The Code Defines the Functionality

One of the main reasons for keeping the view controller interconnections in a storyboard is that they don't clutter up your source code with configuration information that doesn't need to be there. What's left is just the code that defines the actual functionality.

Let's look at what we have as a starting point. Xcode defined several classes for us when the project was created, and we're going to peek into each of them before we start making any changes.

The App Delegate

First up is *BIDAppDelegate.h*, which looks something like this:

```
#import <UIKit/UIKit.h>

@interface BIDAppDelegate : UIResponder <UIApplicationDelegate>

@property (strong, nonatomic) UIWindow *window;

@end
```

This is pretty similar to several other application delegates you've seen in this book so far.

Switch over to the implementation in *BIDAppDelegate.m*, which looks something like the following (we've deleted most comments and empty methods here for the sake of brevity):

```
#import "BIDAppDelegate.h"

@implementation BIDAppDelegate

@synthesize window = _window;

- (BOOL)application:(UIApplication *)application
    didFinishLaunchingWithOptions:(NSDictionary *)launchOptions {
    // Override point for customization after application launch.
    UISplitViewController *splitViewController =
        (UISplitViewController *)self.window.rootViewController;
    UINavigationController *navigationController =
        [splitViewController.viewControllers lastObject];
    splitViewController.delegate = (id)navigationController.topViewController;
    return YES;
}

@end
```

This bunch of code really does just one thing: sets the `UISplitViewController`'s `delegate` property, pointing it at the controller for the main part of the display (the view labeled *Detail* in Figure 11–5). Later in this chapter, when we dig into split views, we'll explore the logic behind that `UISplitViewController` delegate connection. But why make this connection here in code, instead of having it hooked up directly in the storyboard? After all, elimination of boring code—"connect this thing to that thing"—is really one of the main benefits of both nibs and storyboards. And you've seen us hook up delegates in nib files plenty of times, so why can't we do that here?

To understand why using a storyboard to make the connections can't really work here, you need to consider how a storyboard differs from a nib. A nib file is really a frozen object graph. When you load a nib into a running application, the objects it contains all "thaw out" and spring into existence, including all interconnections specified in the nib file. A storyboard, however, is something more than that.

Imagine each scene in a storyboard corresponding to a nib file. When you add in the metadata describing how the scenes are connected via segues, you end up with a storyboard. Unlike a single nib, a complex storyboard is not normally loaded all at once. Instead, any activity that causes a new scene to come into focus will end up loading that particular scene's frozen object graph from the storyboard. This means that the objects you see when looking at a storyboard won't necessarily all exist at the same time.

Since Interface Builder has no way of knowing which scenes will coexist, it actually forbids you from making any outlet or target/action connections from an object in one scene to an object in another scene. In fact, the only connections it allows you to make from one scene to another are segues.

But don't just take our word for it, try it out yourself! First, select the *Split View Controller* in the storyboard (you'll find it within the dock in the *Split View Controller Scene*). Now, bring up the connections inspector, and try to drag out a connection from the *delegate* outlet to another view controller or object. You can drag all over the layout view and the list view, and you won't find any spot that highlights (which would indicate it was ready to accept a drag). In fact, the only item that you'll see accepting your drag is the *First Responder* proxy contained within the same scene as the *Split View Controller* (and that's not what we want in this case).

So, we'll need to connect the delegate outlet from our `UISplitViewController` to its destination in code. Referring back to *BIDAppDelegate.m*, that sequence starts off like this:

```
UISplitViewController *splitViewController =
    (UISplitViewController *)self.window.rootViewController;
```

This lets us grab the window's `rootViewController`, which you may recall is pointed out in the storyboard by the free-floating arrow directed at our `UISplitViewController` instance. Then comes this:

```
UINavigationController *navigationController =
    [splitViewController.viewControllers lastObject];
```

On this line, we dig into the `UISplitViewController`'s viewControllers array. We happen to know that it always has exactly two view controllers: one for the left side and one for the right (more on that later). So, we grab the one for the right side, which will contain our detail view. Finally, we see this:

```
splitViewController.delegate = (id)navigationController.topViewController;
```

This last line simply assigns the detail view controller to the delegate.

All in all, this extra bit of code is a small price to pay, considering how much other code is eliminated by our use of storyboards.

The Master View Controller

Now, let's take a look at `BIDMasterViewController`, which controls the setup of the table view containing the app's navigation. *BIDMasterViewController.h* looks like this:

```
#import <UIKit/UIKit.h>

@class BIDDetailViewController;

@interface BIDMasterViewController : UITableViewController

@property (strong, nonatomic) BIDDetailViewController *detailViewController;

@end
```

And its corresponding *BIDMasterViewController.m* file looks like this (after removing the noncode bits and methods that do nothing but call their superclass's implementation):

```
#import "BIDMasterViewController.h"

#import "BIDDetailViewController.h"

@implementation BIDMasterViewController

@synthesize detailViewController = _detailViewController;

- (void)awakeFromNib
{
    self.clearsSelectionOnViewWillAppear = NO;
    self.contentSizeForViewInPopover = CGSizeMake(320.0, 600.0);
    [super awakeFromNib];
}
.
.
.
- (void)viewDidLoad
{
    [super viewDidLoad];
    // Do any additional setup after loading the view, typically from a nib.
    self.detailViewController = (BIDDetailViewController
*)[[self.splitViewController.viewControllers lastObject] topViewController];
    [self.tableView selectRowAtIndexPath:[NSIndexPath indexPathForRow:0 inSection:0]
animated:NO scrollPosition:UITableViewScrollPositionMiddle];
}
```

```
      .
      .
      .
- (BOOL)shouldAutorotateToInterfaceOrientation:
(UIInterfaceOrientation)interfaceOrientation
{
    // Return YES for supported orientations
    return YES;
}
      .
      .
      .

@end
```

A fair amount of configuration is happening here, but fortunately, Xcode provides it as part of the split view template. This code contains a few things that are relevant to the iPad that you may not have come across before.

First, the awakeFromNib method starts off with this:

```
    self.clearsSelectionOnViewWillAppear = NO;
```

The clearsSelectionOnViewWillAppear property is defined in the UITableViewController class (our superclass), and lets us tweak the controller's behavior a bit. By default, UITableViewController is set up to deselect all rows each time it's displayed. That may be OK in an iPhone app, where each table view is usually displayed on its own, but in an iPad app featuring a split view, you probably don't want that selection to disappear. To revisit an earlier example, consider the Mail app. The user selects a message on the left side, and expects that selection to remain there, even if the message list disappears (due to rotating the iPad or closing the popover containing the list). This line fixes that.

The awakeFromNib method also includes a line that sets the view's contentSizeForViewInPopover property. Chances are you can guess what this does: it sets the size of the view if this view controller should happen to be used to provide the display for a popover controller. This rectangle must be at least 320 pixels wide, but apart from that, you can set the size pretty much however you like. We'll talk more about popover issues later in this chapter.

The next point of interest here is the viewDidLoad method. In previous chapters, when you implemented a table view controller that responds to a user row selection, you typically responded to the user selecting a row by creating a new view controller and pushing it onto the navigation controller's stack. In this app, however, the view controller we want to show is already in place from the start, and it will be reused each time the user makes a selection on the left. It's the instance of BIDDetailViewController contained in the storyboard file. Here, we're grabbing that BIDDetailViewController instance, anticipating that we'll want to use it later, when we have some content to display.

The final thing worth mentioning is the shouldAutorotateToInterfaceOrientation: method. Typically, in an iPhone app, you would use this method to specify whether a particular orientation was suitable for your purposes. In an iPad app, however, the recommendation is generally to let your users choose for themselves which way is up. Unless you're making a game, where you want to force the display to a specific orientation, iPad apps nearly always want this method to return YES.

The Detail View Controller

The final class created for us by Xcode is BIDDetailViewController, which takes care of the actual display of the item the user chooses. Here's what *BIDDetailViewController.h* looks like:

```
#import <UIKit/UIKit.h>

@interface BIDDetailViewController : UIViewController <UISplitViewControllerDelegate>

@property (strong, nonatomic) id detailItem;

@property (strong, nonatomic) IBOutlet UILabel *detailDescriptionLabel;

@end
```

Apart from the detailItem property that we've seen referenced before (in the BIDMasterViewController class), BIDDetailViewController also has an outlet for connecting to a label in the storyboard (detailDescriptionLabel).

Switch over to *BIDDetailViewController.m*, where you'll find the following (once again, somewhat abridged):

```
#import "BIDDetailViewController.h"

@interface BIDDetailViewController ()
@property (strong, nonatomic) UIPopoverController *masterPopoverController;
- (void)configureView;
@end

@implementation BIDDetailViewController

@synthesize detailItem = _detailItem;
@synthesize detailDescriptionLabel = _detailDescriptionLabel;
@synthesize masterPopoverController = _masterPopoverController;

- (void)setDetailItem:(id)newDetailItem
{
    if (_detailItem != newDetailItem) {
        _detailItem = newDetailItem;

        // Update the view.
        [self configureView];
    }

    if (self.masterPopoverController != nil) {
        [self.masterPopoverController dismissPopoverAnimated:YES];
    }
}

- (void)configureView
```

```
{
    // Update the user interface for the detail item.

    if (self.detailItem) {
        self.detailDescriptionLabel.text = [self.detailItem description];
    }
}
.
.
.
- (void)viewDidLoad
{
    [super viewDidLoad];
    // Do any additional setup after loading the view, typically from a nib.
    [self configureView];
}
.
.
.
- (BOOL)shouldAutorotateToInterfaceOrientation:
(UIInterfaceOrientation)interfaceOrientation
{
    // Return YES for supported orientations
    return YES;
}

- (void)splitViewController:(UISplitViewController *)splitController
willHideViewController:(UIViewController *)viewController
withBarButtonItem:(UIBarButtonItem *)barButtonItem
forPopoverController:(UIPopoverController *)popoverController
{
    barButtonItem.title = NSLocalizedString(@"Master", @"Master");
    [self.navigationItem setLeftBarButtonItem:barButtonItem animated:YES];
    self.masterPopoverController = popoverController;
}

- (void)splitViewController:(UISplitViewController *)splitController
willShowViewController:(UIViewController *)viewController
invalidatingBarButtonItem:(UIBarButtonItem *)barButtonItem
{
    // Called when the view is shown again in the split view, invalidating the button
and popover controller.
    [self.navigationItem setLeftBarButtonItem:nil animated:YES];
    self.masterPopoverController = nil;
}

@end
```

Much of this should look familiar to you, but this class contains a few items worth going over. The first of these is something called a **class extension**, declared near the top of the file:

```
@interface BIDDetailViewController ()
@property (strong, nonatomic) UIPopoverController *masterPopoverController;
- (void)configureView;
@end
```

We've talked a bit about class extensions before, but their purpose is worth mentioning again. Creating a class extension lets you define some methods and properties that are going to be used within your class but that you don't want to expose to other classes in a header file. Here, we've declared a popoverController property, which will make use of the instance variable we declared earlier, and a utility method, which will be called whenever we need to update the display. We still haven't told you what the masterPopoverController property is meant to be used for, but we're getting there!

Just a bit farther down, you'll see this method:

```
- (void)setDetailItem:(id)newDetailItem
{
    if (_detailItem != newDetailItem) {
        _detailItem = newDetailItem;

        // Update the view.
        [self configureView];
    }

    if (self.masterPopoverController != nil) {
        [self.masterPopoverController dismissPopoverAnimated:YES];
    }
}
```

The setDetailItem: method may seem surprising to you. We did, after all, define detailItem as a property, and we synthesized it to create the getter and setter for us, so why create a setter in code? In this case, we need to be able to react whenever the user calls the setter (by selecting a row in the master list on the left) so that we can update the display, and this is a good way to do it. The first part of the method seems pretty straightforward, but at the end it diverges into a call to dismiss the current masterPopoverController, if there is one. Where in the world is that hypothetical masterPopupController coming from? Scroll down a bit, and you'll see that this method contains the answer:

```
- (void)splitViewController:(UISplitViewController *)splitController
willHideViewController:(UIViewController *)viewController
withBarButtonItem:(UIBarButtonItem *)barButtonItem
forPopoverController:(UIPopoverController *)popoverController
{
    barButtonItem.title = NSLocalizedString(@"Master", @"Master");
    [self.navigationItem setLeftBarButtonItem:barButtonItem animated:YES];
    self.masterPopoverController = popoverController;
}
```

This is a delegate method for UISplitViewController. It's called when the split view controller is no longer going to show the left side of the split view as a permanent fixture (that is, when the iPad is rotated to portrait orientation). The first thing this method does is configure the title displayed in barButtonItem's title, using the NSLocalizedString function, which gives you a chance to make use of text strings in other languages, if you've prepared any. We'll talk more about localization issues in Chapter 21, but for now, all you need to know is that one parameter is basically a key that the function uses to retrieve a localized string from a dictionary, and the other is a fallback value that will be used in case no other value is found.

The split view controller calls this method in the delegate when the left side of the split is about to disappear, and passes in a couple of interesting items: a UIBarButtonItem and a UIPopoverController. The UIPopoverController is already preconfigured to contain whatever was in the left side of the split view, and the UIBarButtonItem is set up to display that very same popover. This means that if our GUI contains a UIToolBar or a UINavigationItem (the standard toolbar presented by UINavigationController), we just need to add the button item to it in order let the user bring up the navigation view, wrapped inside a popover, with a single tap on the button item.

In this case, since this controller is itself wrapped inside a UINavigationController, we have immediate access to a UINavigationItem where we can place the button item. If our GUI didn't contain a UINavigationItem or a UIToolbar, we would still have the popover controller passed in, which we could assign to some other element of our GUI so it could pop open the popover for us. We're also handed the wrapped UIViewController itself (BIDMasterViewController, in this example) in case we would rather present its contents in some other way.

So, that's where the popover controller comes from. You may not be too surprised to learn that the next method effectively takes it away:

```
- (void)splitViewController:(UISplitViewController *)splitController
willShowViewController:(UIViewController *)viewController
invalidatingBarButtonItem:(UIBarButtonItem *)barButtonItem
{
    // Called when the view is shown again in the split view, invalidating the button
and popover controller.
    [self.navigationItem setLeftBarButtonItem:nil animated:YES];
    self.masterPopoverController = nil;
}
```

This method is called when the user switches back to landscape orientation. At that point, the split view controller wants to once again draw the left-side view in a permanent position, so it tells us to get rid of the UIBarButtonItem we were given previously.

That concludes our overview of what Xcode's Master-Detail Application template gives you. It might be a lot to absorb at a glance, but, ideally, by presenting it a piece at a time, we've helped you understand how all the pieces fit together.

Here Come the Presidents

Now that you've seen the basic layout of our project, it's time to fill in the blanks and turn this autogenerated app into something all our own. Start by looking in the book's source code archive, where the folder *11 – Presidents* contains a file called *PresidentList.plist*. Drag that file into your project's *Presidents* folder in Xcode to add it to the project, making sure that the checkbox telling Xcode to copy the file itself is in the on state. This plist file contains information about all the US presidents so far, consisting of just the name and Wikipedia entry URL for each of them.

Now, let's look at the BIDMasterViewController class and see how we need to modify it to handle the presidential data properly. It's going to be a simple matter of loading the list of presidents, presenting them in the table view, and passing a URL to the detail view for display. In *BIDMasterViewController.h*, add the bold line shown here:

```
#import <UIKit/UIKit.h>

@class BIDDetailViewController;

@interface BIDMasterViewController : UITableViewController

@property (strong, nonatomic) BIDDetailViewController *detailViewController;
@property (strong, nonatomic) NSArray *presidents;
@end
```

Then switch to *BIDMasterViewController.m*, where the changes are a little more involved (but still not too bad). Start off by synthesizing the presidents property near the top of the file:

```
@implementation BIDMasterViewController

@synthesize detailViewController = _detailViewController;
@synthesize presidents;
```

Next, update the viewDidLoad method, adding a few lines to load the list of presidents:

```
- (void)viewDidLoad {
    [super viewDidLoad];
    // Do any additional setup after loading the view, typically from a nib.

    NSString *path = [[NSBundle mainBundle] pathForResource:@"PresidentList"
        ofType:@"plist"];
    NSDictionary *presidentInfo = [NSDictionary dictionaryWithContentsOfFile:path];
    self.presidents = [presidentInfo objectForKey:@"presidents"];

    self.detailViewController = (BIDDetailViewController
*)[[self.splitViewController.viewControllers lastObject] topViewController];
    [self.tableView selectRowAtIndexPath:[NSIndexPath indexPathForRow:0 inSection:0]
animated:NO scrollPosition:UITableViewScrollPositionMiddle];
}
```

Complete the "bookkeeping" part of this class by making the following changes to the viewDidUnload method farther down:

```
- (void)viewDidUnload {
    [super viewDidUnload];
    // Release any retained subviews of the main view.
    // e.g. self.myOutlet = nil;

    self.presidents = nil;
}
```

Now, at some point while looking at this class, you may have been surprised to notice that BIDMasterViewController, despite being a table view controller for displaying a list of items, doesn't implement any table view delegate or data source methods! Since we're subclassing from UITableViewController, which implements the required methods in some basic way, we can get away with that for a while, but now it's time to start filling in the code so that we can present some data.

Let's start by letting the table view know how many sections it's going to have. Add this method at the bottom of the file *BIDMasterViewController.m*, just before the @end:

```
- (NSInteger)numberOfSectionsInTableView:(UITableView *)tableView {
    return 1;
}
```

Technically, this method isn't really necessary. If it's not present, the table view will assume 1 as the default number of sections. However, as a point of consistency, including this method makes sense. With the method in place, the number of sections in the table view is explicit and clear.

Next, add the method that tells the table view how many rows to display:

```
- (NSInteger)tableView:(UITableView *)tableView numberOfRowsInSection:(NSInteger)section
{
    return [self.presidents count];
}
```

After that, write the the tableView:cellForRowAtIndexPath: method to make each cell display a president's name:

```
- (UITableViewCell *)tableView:(UITableView *)tableView
        cellForRowAtIndexPath:(NSIndexPath *)indexPath {
    static NSString *Identifier = @"Master List Cell";
    UITableViewCell *cell = [tableView dequeueReusableCellWithIdentifier:Identifier];
    if (!cell) {
        cell = [[UITableViewCell alloc]
                initWithStyle:UITableViewCellStyleDefault
                reuseIdentifier:Identifier];
    }

    // Configure the cell.
    NSDictionary *president = [self.presidents objectAtIndex:indexPath.row];
    cell.textLabel.text = [president objectForKey:@"name"];

    return cell;
}
```

Finally, it's time to implement tableView:didSelectRowAtIndexPath: to pass the URL to the detail view controller, as follows:

```
- (void)tableView:(UITableView *)aTableView
        didSelectRowAtIndexPath:(NSIndexPath *)indexPath {
    NSDictionary *president = [self.presidents objectAtIndex:indexPath.row];
    NSString *urlString = [president objectForKey:@"url"];
    self.detailViewController.detailItem = urlString;
}
```

That's all we need to do for BIDMasterViewController. But before we can run this app, we need to make a change to the storyboard. As you will recall from Chapter 10, a UITableView can be configured to display a fixed set of cells instead of a dynamic collection pulled from the data source. As it turns out, the UITableView in the default storyboard created with this project is configured just that way, so we need to change that.

Select *MainStoryboard.storyboard* and open the attributes inspector, and then click your way to the UITableView. To do this, find the *Master View Controller – Master Scene* in the dock list. Open the disclosure triangle to the left of the *Master View Controller – Master* item, which will reveal the *Table View* item. Alternatively, you can type *table view* in the search field at the bottom of the dock. No matter how you got there, select that item.

Once the table view is selected, the attributes inspector will display *Table View* at the top of the first group of controls. The very first option in that group is *Content*, which you should now change from *Static Cells* to *Dynamic Prototypes*. While you're there, go ahead and reduce the number of *Prototype Cells* to *0*, as well. We won't be using the one that's there by default, and leaving unused, unidentified cell prototypes here will just generate an annoying compiler warning later.

At this point, you can build and run the app. Tap the *Master* button in the upper-left corner to bring up a popover with a list of presidents (see Figure 11–6). Tap a president's name to display that president's Wikipedia page URL in the detail view.

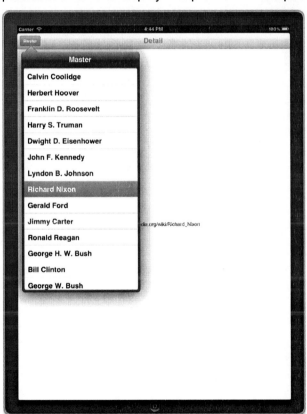

Figure 11–6. *Our first run of the Presidents app. Note that we tapped the Master button to bring up the popover. Tap a president's name, and the link to that president's Wikipedia entry will be displayed.*

Let's finish this section by making the detail view do something a little more useful with the URL. Start with *BIDDetailViewContoller.h*, where we'll add an outlet for a web view to display the Wikipedia page for the selected president. Add the bold line shown here:

```
#import <UIKit/UIKit.h>

@interface BIDDetailViewController : UIViewController < UISplitViewControllerDelegate>

@property (strong, nonatomic) id detailItem;

@property (strong, nonatomic) IBOutlet UILabel *detailDescriptionLabel;
@property (weak, nonatomic) IBOutlet UIWebView *webView;

@end
```

Then switch to *BIDDetailViewController.m*, where we have a bit more to do (though really, not too much). Start near the top of the class's @implementation block, adding synthesis for the webView property:

```
@synthesize detailItem = _detailItem;
@synthesize detailDescriptionLabel = _detailDescriptionLabel;
@synthesize masterPopoverController = _masterPopoverController;
@synthesize webView;
```

Then scroll down to the configureView method, and add the methods shown in bold here:

```
- (void)configureView
{
    // Update the user interface for the detail item.
    NSURL *url = [NSURL URLWithString:self.detailItem];
    NSURLRequest *request = [NSURLRequest requestWithURL:url];
    [self.webView loadRequest:request];

    if (self.detailItem) {
        self.detailDescriptionLabel.text = [self.detailItem description];
    }
}
```

These new lines are all we need to get our web view to load the requested page.

Next, move on down to the splitViewController:
willHideViewController:withBarButtonItem:forPopoverController: method, where we're simply going to give the UIBarButtonItem a more relevant title:

```
barButtonItem.title = NSLocalizedString(@"Master", @"Master");
barButtonItem.title = NSLocalizedString(@"Presidents", @"Presidents");
```

All that's left now is to clean up after ourselves in the viewDidUnload method:

```
- (void)viewDidUnload {
    // Release any retained subviews of the main view.
    // e.g. self.myOutlet = nil;
    self.webView = nil;
}
```

Believe it or not, these few edits are all the code we need to write at this point.

The final changes we need to make are in *MainStoryboard.storyboard*. Open it for editing, find the detail view at the lower right, and start by taking care of the label in the GUI (whose text reads "Detail view content goes here").

Start by selecting the label. You might find it easiest to select the label in the dock list, in the section labeled *Detail View Controller – Detail Scene*. You'll find it quickly by typing *label* in the dock's search field.

Once the label is selected, drag the label to the top of the window. Note that the label should run from the left to right blue guideline and fit snugly under the toolbar. This label is being repurposed to show the current URL. But when the application launches, before the user has chosen a president, we want this field to give the user a hint about what to do.

Double-click the label, and change it to *Select a President*. You should also use the size inspector to make sure that the label's position is anchored to both the left and right sides, as well as the top edge, and that it allows horizontal resizing so that it can adjust itself between the landscape and portrait orientations (see Figure 11–7).

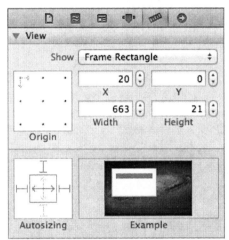

Figure 11–7. *The size inspector, showing the settings for the "Select a President" label*

Next, use the library to find a *UIWebView* and drag it into the space below the label you just moved. After dropping the web view there, use the resize handles to make it fill the rest of the view below the label. Make it go from the left edge to the right edge, and from the blue guideline just below the bottom of the label all the way to the very bottom of the window. Then use the size inspector to anchor the web view to all four edges, and allow it to resize both horizontally and vertically (see Figure 11–8).

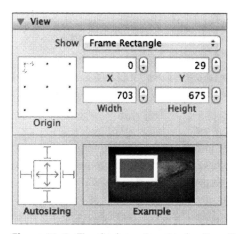

Figure 11–8. *The size inspector, showing the settings for the web view*

We have one last bit of trickery to perform. To hook up the outlet you created, control-drag from the *Detail View Controller* icon (in the *Detail View Controller – Detail* section in the dock, just below the *First Responder* icon) to our new web view (same section, just below the label), and connect the *webView* outlet. Save your changes, and you're finished!

Now, you can build and run the app, and it will let you see the Wikipedia entries for each of the presidents. Rotate the display between the two orientations, and you'll see how the split view controller takes care of everything for you, with a little help from the detail view controller for handling the toolbar item required for showing a popover (just as in the original app before we made our changes).

The final change to make in this section is strictly a cosmetic one. When you run this app in landscape orientation, the heading above the navigation view on the left is still *Master*. Switch to portrait orientation, tap the *Presidents* toolbar button, and you'll see the same heading.

To fix the heading, open *MainStoryboard.storyboard*, double-click the navigation bar above the table view at the upper right, double-click the text shown there, and change it to *Presidents* (see Figure 11–9). Save the storyboard, build and run the app, and you should see your change in place.

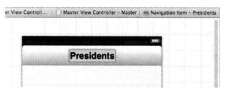

Figure 11–9. *The current state of MainStoryboard.storyboard. We've changed the title of the master detail view's table view to Presidents.*

Creating Your Own Popover

There's still one piece of iPad GUI technology that we haven't dealt with in quite enough detail yet: the creation and display of your own popover. So far, we've had a UIPopoverController handed to us from a UISplitView delegate method, which let us keep track of it in an instance variable so we could force it to go away, but popovers really come in handy when you want to present your own view controllers.

To see how this works, we're going to add a popover to be activated by a permanent toolbar item (unlike the one that the UISplitView delegate method gives us, which is meant to come and go). This popover will display a table view containing a list of languages. If the user picks a language from the list, the web view will load whatever Wikipedia entry was already showing in the new language. This will be simple enough to do, since switching from one language to another in Wikipedia is just a matter of changing a small piece of the URL that contains an embedded country code.

> **NOTE:** Both uses of popovers in this example are in the service of showing a UITableView, but don't let that mislead you—UIPopoverController can be used to handle the display of any view controller content you like! We're sticking with table views for this example because it's a common use case, it's easy to show in a relatively small amount of code, and it's something with which you should already be quite familiar.

Start off by right-clicking the *Presidents* folder in Xcode and selecting **New File...** from the contextual menu. When the assistant appears, select *Cocoa Touch*, then select *UIViewController subclass*, and then click *Next*. On the next screen, name the new class *BIDLanguageListController* and select *UITableViewController* from the *Subclass of* field. Turn on the checkbox next to *Targeted for iPad,* and turn off the checkbox next to *With XIB for user interface*. Click *Next*, double-check the location where you're saving the file, and click *Create*.

The BIDLanguageListController is going to be a pretty standard table view controller class. It will display a list of items and let the detail view controller know when a choice is made by using a pointer back to the detail view controller. Edit *BIDLanguageListController.h*, adding the bold lines shown here:

```
#import <UIKit/UIKit.h>

@class BIDDetailViewController;

@interface BIDLanguageListController : UITableViewController

@property (weak, nonatomic) BIDDetailViewController *detailViewController;
@property (strong, nonatomic) NSArray *languageNames;
@property (strong, nonatomic) NSArray *languageCodes;

@end
```

These additions define a pointer back to the detail view controller (which we'll set from code in the detail view controller itself when we're about to display the language list), as well as a pair of arrays for containing the values that will be displayed (English, French, and so on) and the underlying values that will be used to build an URL from the chosen language (en, fr, and so on).

If you copied and pasted this code from the book's source archive (or e-book) into your own project or typed it yourself a little sloppily, you may not have noticed an important difference in how the detailViewController property was declared earlier. Unlike most properties that reference an object pointer, we declared this one using weak instead of strong. This is something that we must do in order to avoid a retain cycle.

What's a **retain cycle**? It's a situation where a set of two or more objects have retained one another in a circular fashion. Each object has a retain counter of one or higher and will therefore never release the pointers it contains, so they will never be deallocated either. Most potential retain cycles can be avoided by carefully considering the creation of your objects, often by trying to figure out who "owns" whom. In this sense, an instance of BIDDetailViewController owns an instance of BIDLanguageListController, because it's the BIDDetailViewController that actually creates the BIDLanguageListController in order to get a piece of work done. Whenever you have a pair of objects that each needs to refer to one another, you'll usually want the owner object to retain the other object, while the other object should specifically not retain its owner. Since we're using the ARC feature that Apple introduced in Xcode 4.2, the compiler does most of the work for us. Instead of paying attention to the details about releasing and retaining objects, all we need to do is declare a property with the weak keyword instead of strong. ARC will do the rest!

Now, switch over to *BIDLanguageListController.m* to implement the following changes. At the top of the file, start by importing the header for BIDDetailViewController, and then synthesize getters and setters for the properties you declared:

```
#import "BIDLanguageListController.h"
#import "BIDDetailViewController.h"

@implementation BIDLanguageListController

@synthesize languageNames;
@synthesize languageCodes;
@synthesize detailViewController;
.
.
.
```

Then scroll down a bit to the viewDidLoad method, and add a bit of setup code:

```
- (void)viewDidLoad {
    [super viewDidLoad];

    self.languageNames = [NSArray arrayWithObjects:@"English", @"French",
        @"German", @"Spanish", nil];
    self.languageCodes = [NSArray arrayWithObjects:@"en", @"fr", @"de", @"es", nil];
    self.clearsSelectionOnViewWillAppear = NO;
```

```
    self.contentSizeForViewInPopover = CGSizeMake(320.0,
        [self.languageCodes count] * 44.0);
}
```

This sets up the language arrays and also defines the size that this view will use if shown in a popover (which, as we know, it will be). Without defining the size, we would end up with a popover stretching vertically to fill nearly the whole screen, even with only four entries in it.

Farther down, we have a few methods generated by Xcode's template that don't contain particularly useful code—just a warning and some placeholder text. Let's replace those with something real:

```
- (NSInteger)numberOfSectionsInTableView:(UITableView *)tableView {
#warning Potentially incomplete method implementation.
    // Return the number of sections.
    return 0;
    return 1;
}


- (NSInteger)tableView:(UITableView *)tableView numberOfRowsInSection:(NSInteger)section
{
#warning Incomplete method implementation.
    // Return the number of rows in the section.
    return 0;
    return [self.languageCodes count];
}
```

Then add a line near the end of tableView:cellForRowAtIndexPath: to put a language name into a cell:

```
    // Configure the cell.
    cell.textLabel.text = [languageNames objectAtIndex:[indexPath row]];
    return cell;
```

Next, fix tableView:didSelectRowAtIndexPath: by eliminating the comment block it contains and adding this new code instead:

```
- (void)tableView:(UITableView *)tableView didSelectRowAtIndexPath:(NSIndexPath
*)indexPath {
    detailViewController.languageString = [self.languageCodes objectAtIndex:
        [indexPath row]];
}
```

Note that BIDDetailViewController doesn't actually have a languageString property. We'll take care of that in just a bit. But first, finish up BIDLanguageListController by making the following bookkeeping changes:

```
- (void)viewDidUnload {
    [super viewDidUnload];

    self.detailViewController = nil;
    self.languageNames = nil;
    self.languageCodes = nil;
}
```

Now, it's time to make the changes required for BIDDetailViewController to handle the popover, as well as generate the correct URL whenever the user either changes the display language or picks a different president. Start by making the following changes in *BIDDetailViewController.h*:

```
#import <UIKit/UIKit.h>

@interface BIDDetailViewController : UIViewController <UISplitViewControllerDelegate>

@property (strong, nonatomic) id detailItem;

@property (strong, nonatomic) IBOutlet UILabel *detailDescriptionLabel;
@property (weak, nonatomic) IBOutlet UIWebView *webView;

@property (strong, nonatomic) UIBarButtonItem *languageButton;
@property (strong, nonatomic) UIPopoverController *languagePopoverController;
@property (copy, nonatomic) NSString *languageString;
- (IBAction)touchLanguageButton;
@end
```

All we need to do now is fix *BIDDetailViewController.m* so that it can handle the language popover and the URL construction. Start by adding this import somewhere at the top:

```
#import "BIDLanguageListController.h"
```

Then synthesize the new properties just below the @implementation line:

```
@synthesize languageButton;
```

```
@synthesize languagePopoverController;
```

```
@synthesize languageString;
```

The next thing we're going to add is a function that takes as arguments a URL pointing to a Wikipedia page and a two-letter language code, and returns a URL that combines the two. We'll use this at appropriate spots in our controller code later. You can place this function just about anywhere, including within the class's implementation. The compiler is smart enough to always treat a function as just a function. Why don't you place it just after the last synthesize statement toward the top of the file?

```
static NSString * modifyUrlForLanguage(NSString *url, NSString *lang) {
    if (!lang) {
        return url;
    }

    // We're relying on a particular Wikipedia URL format here. This
    // is a bit fragile!
    NSRange languageCodeRange = NSMakeRange(7, 2);
    if ([[url substringWithRange:languageCodeRange] isEqualToString:lang]) {
        return url;
    } else {
        NSString *newUrl = [url stringByReplacingCharactersInRange:languageCodeRange
            withString:lang];
        return newUrl;
    }
}
```

Why make this a function instead of a method? There are a couple of reasons. First, instance methods in a class are typically meant to do something involving one or more instance variables. This function does not make use of any instance variables. It simply performs an operation on two strings and returns another. We could have made it a class method, but even that feels a bit wrong, since what the method does isn't really related specifically to our controller class. Sometimes, a function is just what you need.

Our next move is to update the setDetailItem: method. This method will use the function we just defined to combine the URL that's passed in with the chosen languageString to generate the correct URL. It also makes sure that our second popover, if present, disappears just like the first popover (the one that was defined for us) does.

```
- (void)setDetailItem:(id)newDetailItem {
    if (detailItem != newDetailItem) {
        _detailItem = newDetailItem;
        _detailItem = modifyUrlForLanguage(newDetailItem, languageString);

        // Update the view.
        [self configureView];
    }

    if (self.masterPopoverController != nil) {
        [self.masterPopoverController dismissPopoverAnimated:YES];
    }
}
```

Now, let's update the viewDidLoad method. Here, we're going to create a UIBarButtonItem and put it into the UINavigationItem at the top of the screen.

```
- (void)viewDidLoad
{
    [super viewDidLoad];
    // Do any additional setup after loading the view, typically from a nib.
    self.languageButton = [[UIBarButtonItem alloc] init];
    languageButton.title = @"Choose Language";
    languageButton.target = self;
    languageButton.action = @selector(touchLanguageButton);
    self.navigationItem.rightBarButtonItem = self.languageButton;

    [self configureView];
}
```

Next, we implement setLanguageString:. This also calls our modifyUrlForLanguage() function so that the URL can be regenerated (and the new page loaded) immediately. Add this method to the bottom of the file, just above the @end:

```
- (void)setLanguageString:(NSString *)newString {
    if (![newString isEqualToString:languageString]) {
        languageString = [newString copy];
        self.detailItem = modifyUrlForLanguage(_detailItem, languageString);
    }
    if (languagePopoverController != nil) {
        [languagePopoverController dismissPopoverAnimated:YES];
```

```
                self.languagePopoverController = nil;
        }
    }
```

Now, let's define what will happen when the user taps the *Choose Language* button. Simply put, we create a BIDLanguageListController, wrap it in a UIPopoverController, and display it. Place this method at the bottom of the file, just before the @end:

```
- (IBAction)touchLanguageButton {
    if (self.languagePopoverController == nil) {
        BIDLanguageListController *languageListController =
            [[BIDLanguageListController alloc] init];
        languageListController.detailViewController = self;
        UIPopoverController *poc = [[UIPopoverController alloc]
            initWithContentViewController:languageListController];
        [poc presentPopoverFromBarButtonItem:languageButton
                permittedArrowDirections:UIPopoverArrowDirectionAny
                               animated:YES];
        self.languagePopoverController = poc;
    } else {
        if (languagePopoverController != nil) {
            [languagePopoverController dismissPopoverAnimated:YES];
            self.languagePopoverController = nil;
        }
    }
}
```

The final change is to add these lines to the viewDidUnload method:

```
    self.languageButton = nil;
    self.languagePopoverController = nil;
```

And that's all! You should now be able to run the app in all its glory, switching willy-nilly between presidents and languages. Switching from one language to another should always leave the chosen president intact, and likewise switching from one president to another should leave the language intact.

iPad Wrap-Up

In this chapter, you learned about the main GUI components that are available only on the iPad: popovers and split views. You've also seen an example of how a complex iPad application with several interconnected view controllers can be configured entirely within Interface Builder. With this hard-won knowledge, you should be well on your way to building your first great iPad app. If you want to dig even further into the particulars of iPad development, you may want to take a look at *Beginning iPad Development for iPhone Developers* by David Mark, Jack Nutting, and Dave Wooldridge (Apress, 2010).

Next up, it's time to visit application settings and user defaults.

Chapter 12

Application Settings and User Defaults

All but the simplest computer programs today have a preferences window where the user can set application-specific options. On Mac OS X, the **Preferences…** menu item is usually found in the application menu. Selecting it brings up a window where the user can enter and change various options. The iPhone and other iOS devices have a dedicated application called Settings, which you no doubt have played with any number of times. In this chapter, we'll show you how to add settings for your application to the Settings application, and how to access those settings from within your application.

Getting to Know Your Settings Bundle

The Settings application lets the user enter and change preferences for any application that has a settings bundle. A **settings bundle** is a group of files built in to an application that tells the Settings application which preferences the application wishes to collect from the user.

Pick up your iOS device, and locate your Settings icon. By default, you'll find it on the home screen (see Figure 12–1).

Figure 12–1. *The Settings application icon is in the middle of the last column on this iPhone. It may be in a different spot on your device, but it's always available.*

When you touch the icon, the Settings application will launch. Ours is shown in Figure 12–2.

Figure 12–2. *The Settings application*

The Settings application acts as a common user interface for the iOS User Defaults mechanism. User Defaults is the part of the system that stores and retrieves preferences.

In an iOS application, User Defaults is implemented by the NSUserDefaults class. If you've done Cocoa programming on the Mac, you're probably already familiar with NSUserDefaults, because it is the same class that is used to store and read preferences on the Mac. Your applications will use NSUserDefaults to read and store preference data using a key value, just as you would access keyed data from an NSDictionary. The difference is that NSUserDefaults data is persisted to the file system, rather than stored in an object instance in memory.

In this chapter, we're going to create an application, add and configure a settings bundle, and then access and edit those preferences from within our application.

One nice thing about the Settings application is that it provides a solution so that you don't need to design your own user interface for your preferences. You create a property list defining your application's available settings, and the Settings application creates the interface for you.

Immersive applications, such as games, generally should provide their own preferences view so that the user doesn't need to quit in order to make a change. Even utility and productivity applications might, at times, have preferences that a user should be able to change without leaving the application. We'll also show you to how to collect preferences from the user directly in your application, and store those in iOS's User Defaults.

In addition, with the introduction of background processing in iOS 4, you can actually switch to the Settings application, change a preference, and then switch back to your application. We'll show you how to handle that situation at the end of this chapter.

The AppSettings Application

We're going to build a simple application in this chapter. First, we'll implement a settings bundle so that when the user launches the Settings application, there will be an entry for our application, AppSettings (see Figure 12–3).

Figure 12–3. *The Settings application showing an entry for our AppSettings application in the simulator*

If the user selects our application, Settings will drill down into a view that shows the preferences relevant to our application. As you can see from Figure 12–4, the Settings application uses text fields, secure text fields, switches, and sliders to coax values out of our intrepid user

Figure 12-4. *Our application's primary settings view*

Also notice the two items in the view that have disclosure indicators. The first one, *Protocol*, takes the user to another table view that displays the options available for that item. From that table view, the user can select a single value (see Figure 12-5).

The *More Settings* disclosure indicator allows the user to drill down to another set of preferences (see Figure 12-6). This child view can have the same kinds of controls as the main settings view, and can even have its own child views. You may have noticed that the Settings application uses a navigation controller, which it needs because it supports the building of hierarchical preference views.

Figure 12–5. *Selecting a single preference item from a list*

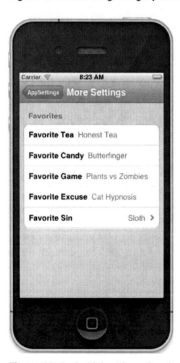

Figure 12–6. *A child settings view for our application*

When users launch our application, they will be presented with a list of the preferences gathered in the Settings application (see Figure 12–7).

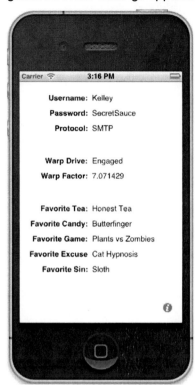

Figure 12–7. *Our application's main view*

In order to show how to update preferences from within our application, we also provide a little information button in the lower-right corner. This button takes users to another view where they can change additional preferences directly in our application (see Figure 12–8).

Figure 12–8. *Setting some preferences directly in our application*

Let's get started building AppSettings, shall we?

Creating the Project

In Xcode, press ⇧⌘N or select **File ➤ New ➤ New Project**.... When the new project assistant comes up, select *Application* from under the *iOS* heading in the left pane, click the *Utility Application* icon, and click *Next*. On the next screen, name your project *AppSettings*. Set *Device Family* to *iPhone*. Next, check that the *Use Storyboard* and *Use Automatic Reference Counting* checkboxes are checked, and that the *Use Core Data* and *Include Unit Tests* checkboxes are unchecked, and then click the *Next* button. Finally, choose a location for your project and click *Create*.

We haven't used this particular project template before, so let's take a quick look at the project before we proceed. The Utililty Application template creates an application similar to the multiview application we built in Chapter 6. The application has a main view and a secondary view called the **flipside view**. Tapping the information button on the main view takes you to the flipside view, and tapping the *Done* button on the flipside view takes you back to the main view.

It takes several controllers and views to implement this type of application. All of these are provided, as stubs, by the template. Expand the *AppSettings* folder, where you'll find the usual application delegate class, as well as two additional controller classes and a storyboard file to contain the GUI (see Figure 12–9).

Figure 12–9. *Our project created from the Utility Application template. Notice the application delegate, the storyboard, and the main and flipside view controllers.*

Working with the Settings Bundle

The Settings application bases the display of preferences for a given application on the contents of the settings bundle inside that application. Each settings bundle must have a property list, called *Root.plist*, which defines the root-level preferences view. This property list must follow a very precise format, which we'll talk about when we set up the property list for our application.

When the Settings application starts up, it checks each application for a settings bundle and adds a settings group for each application that includes a settings bundle. If we want our preferences to include any subviews, we need to add property lists to the bundle and add an entry to *Root.plist* for each child view. You'll see exactly how to do that in this chapter.

Adding a Settings Bundle to Our Project

In the project navigator, click the *AppSettings* folder, and then select File ➤ New File… or press ⌘N. In the left pane, select *Resource* under the *iOS* heading, and then select the *Settings Bundle* icon (see Figure 12–10). Click the *Next* button, leave the default name of *Settings.bundle*, and click *Create*.

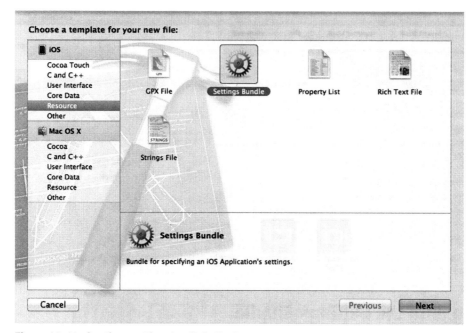

Figure 12–10. *Creating a settings bundle in Xcode*

You should now see a new item in the project window, called *Settings.bundle*. If it's not already opened, expand *Settings.bundle*, and you should see two subitems: a folder named *en.lproj*, containing a file named *Root.strings*, and an icon named *Root.plist*. We'll discuss *en.lproj* in Chapter 21 when we talk about localizing your application into other languages. Here, we'll concentrate on *Root.plist*.

Setting Up the Property List

Select *Root.plist*, and take a look at the editor pane. You're looking at Xcode's property list editor (see Figure 12–11). This editor functions in the same way as the Property List Editor application in */Developer/Applications/Utilities*.

Key	Type	Value
▶ PreferenceSpecifiers	Array	(4 items)
StringsTable	String	Root

Figure 12–11. *Root.plist in the property list editor pane. If your editing pane looks slightly different, don't panic. Control-click in the editing pane and select Show Raw Keys/Values from the contextual menu that appears.*

Notice the organization of the items in the property list. Property lists are essentially dictionaries, storing item types and values, and using a key to retrieve them, just as an NSDictionary does.

Several different types of nodes can be put into a property list. The *Boolean, Data, Date, Number, and String* node types are meant to hold individual pieces of data, but you also have a couple of ways to deal with whole collections of nodes as well. In addition to

Dictionary node types, which allow you to store other nodes under a key, there are *Array* nodes, which store an ordered list of other nodes similar to an NSArray. The *Dictionary* and *Array* types are the only property list node types that can contain other nodes.

> **NOTE:** Although you can use most kinds of objects as keys in an NSDictionary, keys in property list dictionary nodes must be strings. However, you are free to use any node type for the values.

When creating a settings property list, you need to follow a very specific format. Fortunately, *Root.plist*, the property list that came with the settings bundle you just added to your project, follows this format exactly. Let's take a look.

In the *Root.plist* editor pane, names of keys can either be displayed in their true, "raw" form or in a slightly more human-readable form. We're big fans of seeing things as they truly are whenever possible, so right-click anywhere in the editor and make sure the **Show Raw Keys/Values** option in the contextual menu is checked (see Figure 12–12). The rest of our discussion here uses the real names for all the keys we're going to talk about, so this step is important.

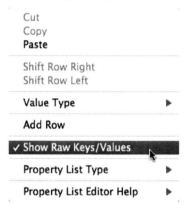

Figure 12–12. *Control-click anywhere in the property list editing pane and make sure the Show Raw Keys/Values item is checked. This will ensure that real names are used in the property list editor, which makes your editing experience more precise.*

> **CAUTION:** As of this writing, leaving the property list, either by editing a different file or by quitting Xcode, resets the **Show Raw Keys/Values** item to be unchecked. If your text suddenly looks a little different, take another look at that menu item and make sure it is checked.

One of the items in the dictionary is *StringsTable*. A strings table is used in translating your application into another language. We'll cover the strings table in Chapter 21, when we get into localization. Since the strings table is optional, you can delete that entry by clicking it and pressing the delete key. Or you can leave it there if you prefer, since it won't do any harm.

In addition to *StringsTable*, the property list contains a node named *PreferenceSpecifiers*, which is an array. This array node is designed to hold a set of dictionary nodes, each representing a single preference that the user can enter or a single child view that the user can drill down into.

You'll notice that Xcode's template kindly gave us four nodes (see Figure 12–13). Those nodes aren't likely to reflect our actual preferences, so delete *Item 1, Item 2*, and *Item 3* (select each one and press the delete key, one after another), leaving just *Item 0* in place.

> **NOTE:** To select an item in the property list, it is best to click on one side or another of the *Key* column, to avoid bringing up the *Key* column's drop-down menu.

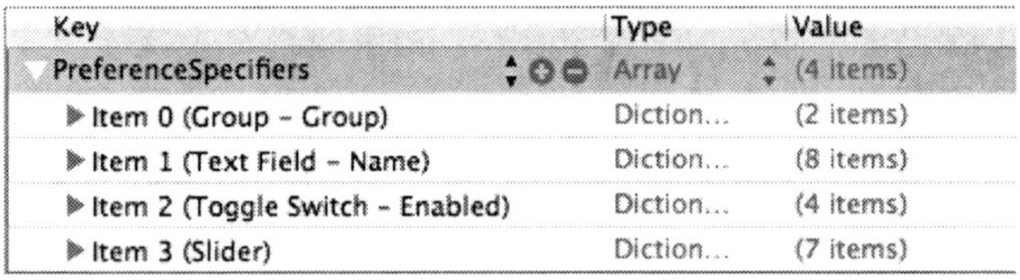

Figure 12–13. *Root.plist in the editor pane, this time with PreferenceSpecifiers expanded*

Single-click *Item 0* but don't expand it. Xcode's property list editor lets you add rows by pressing the return key. The current selection state—including which row is selected and whether or not it's expanded—determines where the new row will be inserted. When an unexpanded array or dictionary is selected, pressing return adds a sibling node after the selected row. In other words, it will add another node at the same level as the current selection. If you were to press return (but don't do that now), you would get a new row called *Item 1* immediately after *Item 0*. Figure 12–14 shows an example of hitting return to create a new row. Notice the drop-down menu that allows you to specify the kind of preference specifier this item represents—more on this in a bit.

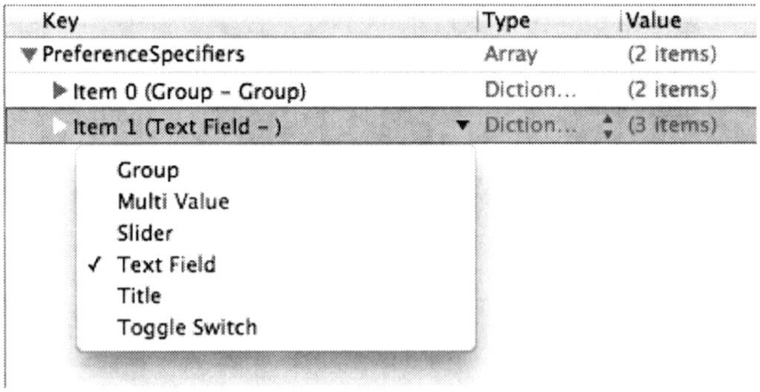

Figure 12–14. *We selected Item 0 and hit return to create a new sibling row. Note the drop-down menu that appears, allowing us to specify the kind of preference specifier this item represents.*

Now expand *Item 0*, and see what it contains (see Figure 12–15). The editor is now ready to add child nodes to the selected item. If you were to press return at this point (again, don't actually press it now), you would get a new first child row inside *Item 0*.

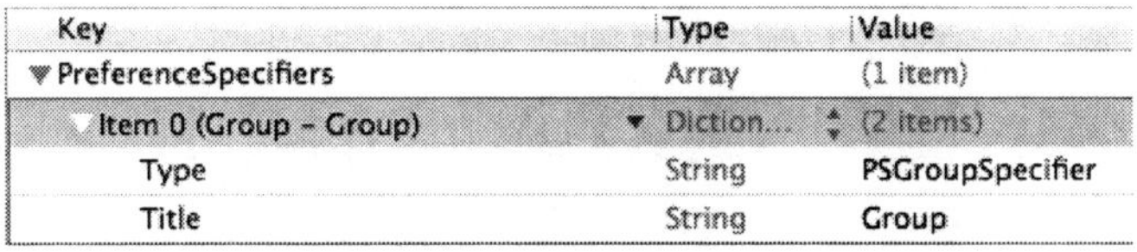

Key	Type	Value
▼ PreferenceSpecifiers	Array	(1 item)
▼ Item 0 (Group – Group)	▼ Diction... ↕	(2 items)
Type	String	PSGroupSpecifier
Title	String	Group

Figure 12–15. *When you expand Item 0, You'll find a row with a key of Type and a second row with a key of Title. This represents a group with a title of Group.*

One of the items inside *Item 0* has a key of *Type*. Every property list node in the *PreferenceSpecifiers* array must have an entry with this key. The *Type* key is typically the first entry, but order doesn't matter in a dictionary, so the *Type* key doesn't need to be first. The *Type* key tells the Settings application what type of data is associated with this item.

In *Item 0*, the *Type* item has a value of *PSGroupSpecifier*. This indicates that the item represents the start of a new group. Each item that follows will be part of this group, until the next item with a *Type* of *PSGroupSpecifier*.

If you look back at Figure 12–4, you'll see that the Settings application presents the application settings in a grouped table. *Item 0* in the *PreferenceSpecifiers* array in a settings bundle property list should always be a *PSGroupSpecifier* so the settings start in a new group, because you need at least one group in every Settings table.

The only other entry in *Item 0* has a key of *Title*, and this is used to set an optional header just above the group that is being started.

Now, take a closer look at the *Item 0* row itself, and you'll see that it's actually shown as *Item 0 (Group – Group)*. The values in parentheses represent the value of the *Type* item (the first *Group*) and the *Title* item (the second *Group*). This is a nice shortcut that Xcode gives you so that you can visually scan the contents of a settings bundle.

As shown back in Figure 12–4, we've called our first group *General Info*. Double-click the value next to *Title*, and change it from *Group to General Info* (see Figure 12–16). When you enter the new title, you may notice a slight change to *Item 0*. It's now shown as *Item 0 (Group – General Info)* to reflect the new title.

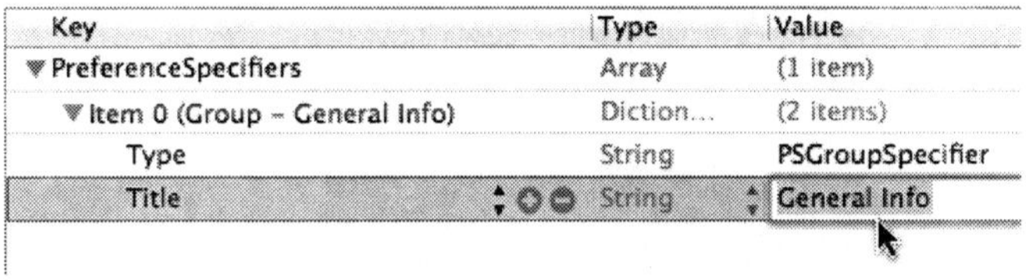

Key	Type	Value
▼ PreferenceSpecifiers	Array	(1 item)
▼ Item 0 (Group – General Info)	Diction...	(2 items)
Type	String	PSGroupSpecifier
Title ↕ ⊙ ⊖	String ↕	General Info

Figure 12–16. *We changed the title of the Item 0 group from Group to General Info.*

Adding a Text Field Setting

We now need to add a second item in this array, which will represent the first actual preference field. We're going to start with a simple text field.

If you single-click the *PreferenceSpecifiers* row in the editor pane (don't do this, just keep reading), and press return to add a child, the new row will be inserted at the beginning of the list, which is not what we want. We want to add a row at the end of the array.

To add the row, click the disclosure triangle to the left of *Item 0* to close it, and then select *Item 0* and press return, which will give you a new sibling row after the current row (see Figure 12–17). As usual, when the item is added, a drop-down menu appears, showing the default value of *Text Field*.

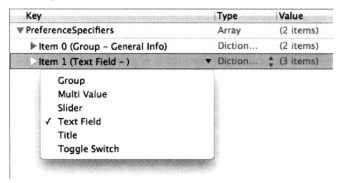

Figure 12–17. *Adding a new sibling row to Item 0*

Click somewhere outside the drop-down menu to make it go away, and then click the disclosure triangle next to *Item 1* to expand it. You'll see that it contains a *Type* row set to *PSTextFieldSpecifier*. This is the *Type* value used to tell the Settings application that we want the user to edit this setting in a text field. It also contains two empty rows for *Title* and *Key* (see Figure 12–18).

Key	Type	Value
▼ PreferenceSpecifiers	Array	(2 items)
▶ Item 0 (Group – General Info)	Diction...	(2 items)
▼ Item 1 (Text Field –)	Diction...	(3 items)
Type	String	PSTextFieldSpecifier
Title	String	
Key	String	

Figure 12–18. *Our text field item, expanded to show the type, title, and key*

Select the *Title* row, then double-click in the whitespace of the *Value* column. Type in *Username* to set the *Title* value. This is the text that will appear in the Settings app.

Now do the same for the *Key* row (no, that's not a misprint, you're really looking at a key called *Key*). For a value, type in *username* (note the lowercase first letter). Remember that user defaults work like a dictionary. This entry tells the Settings application which key to use when it stores the value entered in this text field.

Recall what we said about NSUserDefaults? It lets you store values using a key, similar to an NSDictionary. Well, the Settings application will do the same thing for each of the preferences it saves on your behalf. If you give it a key value of *foo*, then later in your application, you can request the value for *foo*, and it will give you the value the user entered for that preference. We will use this same key value later to retrieve this setting from the user defaults in our application.

NOTE: Notice that our *Title* has a value of *Username* and our *Key* has a value of *username*. This uppercase/lowercase difference will happen frequently. The *Title* is what appears on the screen, so the capital *U* makes sense. The *Key* is a text string we'll use to retrieve preferences from the user defaults, so all lowercase makes sense there. Could we use all lowercase for *Title*? You bet. Could we use all capitals for *Key*? Sure! As long as you capitalize it the same way when you save and when you retrieve, it doesn't matter which convention you use for your preference keys.

Now, select the last of the three *Item 1* rows (the one with a *Key* of *Key*) and press return to add another entry to our *Item 1* dictionary, giving this one a key of *AutocapitalizationType* and a value of *None*. This specifies that the text field shouldn't attempt to autocapitalize what the user types in this field. Note that as soon as you start typing *AutocapitalizationType*, Xcode presents you with a list of matching choices, so you can simply pick one from the list instead of typing the whole name.

Create one last new row, and give it a key of *AutocorrectionType* and a value of *No*. This will tell the Settings application not to try to autocorrect values entered into this text field. When you do want the text field to use autocorrection, change the value in this row to *Yes*. Again, Xcode presents you with a list of matching choices as you begin entering *AutocorrectionType*.

When you're finished, your property list should look like the one shown in Figure 12–19.

Key	Type	Value
▼ PreferenceSpecifiers	Array	(2 items)
▶ Item 0 (Group – General Info)	Diction...	(2 items)
▼ Item 1 (Text Field – Username)	Diction...	(5 items)
Type	String	PSTextFieldSpecifier
Title	String	Username
Key	String	username
AutocapitalizationType	String	None
AutocorrectionType	String	No

Figure 12–19. *The finished text field specified in Root.plist*

Adding an Application Icon

Before we try out our new setting, let's add an application icon to the project. You've done this before.

Save *Root.plist*, the property file you just edited. Then make your way into the source code archive and into the *12 – AppSettings* folder. Drag the file *icon.png* into your project's *AppSettings* folder and, when prompted, have Xcode copy the icon.

Next, open the *Supporting Files* folder, and click the file *AppSettings-info.plist*. When the property list editor appears, expand the *Icon files* row. Next, select the *Icon files* row, press return to create a new item inside it, and change the new item's value to *icon.png*.

That's it. Now compile and run the application by selecting **Product ➤ Run**. Press the home button, and then tap the icon for the Settings application. You will find an entry for our application, which uses the application icon we added earlier (see Figure 12–3). Click the *AppSettings* row, and you will be presented with a simple settings view with a single text field, as shown in Figure 12–20.

Figure 12–20. *Our root view in the Settings application after adding a group and a text field*

Quit the simulator, and go back to Xcode. We're not finished yet, but you should now have a sense of how easy it is to add preferences to your application. Let's add the rest of the fields for our root settings view. The first one we'll add is a secure text field for the user's password.

Adding a Secure Text Field Setting

Click *Root.plist* to return to your setting specifiers (don't forget to turn on *Show Raw Keys/Values*, assuming your friends at XcodeCorp have reset this). Collapse *Item 0* and *Item 1*. Now select *Item 1*. Press ⌘**C** to copy it to the clipboard, and then press ⌘**V** to paste it back. This will create a new *Item 2* that is identical to *Item 1*. Expand the new item, and change the *Title* to *Password* and the *Key* to *password* (one with a capital *P* and one with a lowercase *p*).

Next, add one more child to the new item. Remember that the order of items does not matter, so feel free to place it directly below the *Key* item you just edited. To do this, select the *Key/password* row, and then hit return.

Give the new item a *Key* of *IsSecure* (note the leading uppercase *I*), and change the *Type* to *Boolean*. Now change its *Value* from *NO* to *YES*, which tells the Settings application that this field needs to be a password field, rather than just an ordinary text field. Our finished *Item 2* is shown in Figure 12–21.

Key	Type	Value
▼ PreferenceSpecifiers	Array	(4 items)
▶ Item 0 (Group – General Info)	Diction...	(2 items)
▶ Item 1 (Text Field – Username)	Diction...	(5 items)
▼ Item 2 (Text Field – Password)	Diction...	(6 items)
Type	String	PSTextFieldSpecifier
Title	String	Password
Key	String	password
IsSecure	Boolean	YES
AutocapitalizationType	String	None
AutocorrectionType	String	No

Figure 12–21. *Our finished Item 2, a text field designed to accept a password*

Adding a Multivalue Field

The next item we're going to add is a multivalue field. This type of field will automatically generate a row with a disclosure indicator. Clicking it will take you down to another table where you can select one of several rows.

Collapse *Item 2*, select the row, and then press return to add *Item 3*. Use the popup attached to the *Key* field to select *Multi Value*, and expand *Item 3* by clicking the disclosure triangle.

The expanded *Item 3* already contains a few rows. One of them, the *Type* row, is set to *PSMultiValueSpecifier*. Look for the *Title* row and set its value to *Protocol*. Then find the *Key* row, and give it a value of *protocol*. The next part is a little tricky, so let's talk about it before we do it.

We're going to add two more children to *Item 3*, but they will be *Array* type nodes, not *String* type nodes, as follows:

- One array, called *Titles*, will hold a list of the values from which the user can select.

- The other array, called *Values*, will hold a list of the values that actually are stored in the user defaults.

So, if the user selects the first item in the list, which corresponds to the first item in the *Titles* array, the Settings application will actually store the first value from the *Values* array. This pairing of *Titles* and *Values* lets you present user-friendly text to the user but actually store something else, like a number, date, or different string.

Both of these arrays are required. If you want them both to be the same, you can create one array, copy it, paste it back in, and change the key so that you have two arrays with the same content but stored under different keys. We'll actually do just that.

Select *Item 3* (leave it open) and press return to add a new child. You'll see that once again, Xcode is aware of the type of file we're editing and seems to anticipate what we want to do, because the new child row already has its *Key* set to *Titles* and is configured to be an *Array*. Just what we wanted! Expand the *Titles* row and hit return to add a child node. Repeat this four more times, so you have a total of five child nodes. All five nodes should be *String* type and should contain the following values: *HTTP, SMTP, NNTP, IMAP*, and *POP3*.

Once you've entered all five nodes, collapse *Titles*, and select it. Then press ⌘C to copy it, and press ⌘V to paste it back. This will create a new item with a key of *Titles - 2*. Double-click *Titles - 2*, and change it to *Values*.

We're almost finished with our multivalue field. There's just one more required value in the dictionary, which is the default value. Multivalue fields must have one—and only one—row selected. So, we need to specify the default value to be used if none has yet been selected, and it needs to correspond to one of the items in the *Values* array (not the *Titles* array, if they are different). Xcode already added a *DefaultValue* row when we created this item, so all we need to do now is give it a value of *SMTP*. Figure 12–22 shows our version of *Item 3*.

Key	Type	Value
▼ PreferenceSpecifiers	Array	(4 items)
▶ Item 0 (Group - General Info)	Diction...	(2 items)
▶ Item 1 (Text Field - Username)	Diction...	(5 items)
▶ Item 2 (Text Field - Password)	Diction...	(6 items)
▼ Item 3 (Multi Value - Protocol)	Diction...	(6 items)
▼ Titles	Array	(5 items)
Item 0	String	HTTP
Item 1	String	SMTP
Item 2	String	NNTP
Item 3	String	IMAP
Item 4	String	POP3
▼ Values	Array	(5 items)
Item 0	String	HTTP
Item 1	String	SMTP
Item 2	String	NNTP
Item 3	String	IMAP
Item 4	String	POP3
Type	String	PSMultiValueSpecifier
Title	String	Protocol
Key	String	protocol
DefaultValue	String	SMTP

Figure 12–22. *Our finished Item 3, a multivalue field designed to let the user select from one of five possible values*

Let's check our work. Save the property list, and build and run the application again. When your application starts up, press the home button and launch the Settings application. When you select *AppSettings*, you should see three fields on your root-level view (see Figure 12–23). Go ahead and play with your creation, and then let's move on.

Figure 12–23. *Three fields down. Not too shabby!*

Adding a Toggle Switch Setting

The next item we need to get from the user is a Boolean value that indicates whether our warp engines are turned on. To capture a Boolean value in our preferences, we are going to tell the Settings application to use a UISwitch by adding another item to our *PreferenceSpecifiers* array with a type of *PSToggleSwitchSpecifier*.

Collapse *Item 3* if it's currently expanded, and then single-click it to select it. Press return to create *Item 4*. Use the drop-down menu to select *Toggle Switch*, and then click the disclosure triangle to expand *Item 4*. You'll see there's already a child row with a *Key* of *Type* and a *Value* of *PSToggleSwitchSpecifier*. Give the empty *Title* row a value of *Warp Drive*, and set the value of the *Key* row to *warp*.

We have one more required item in this dictionary, which is the default value. Just as with the *Multi Value* setup, here Xcode has already created a *DefaultValue* row for us. Let's turn on our warp engines by default by giving the *DefaultValue* row a value of *YES*. Figure 12–24 shows our completed *Item 4*.

Key	Type	Value
▼ PreferenceSpecifiers	Array	(5 items)
▶ Item 0 (Group – General Info)	Diction...	(2 items)
▶ Item 1 (Text Field – Username)	Diction...	(5 items)
▶ Item 2 (Text Field – Password)	Diction...	(6 items)
▶ Item 3 (Multi Value – Protocol)	Diction...	(6 items)
▼ Item 4 (Toggle Switch – Warp Drive)	Diction...	(4 items)
Type	String	PSToggleSwitchSpecifier
Title	String	Warp Drive
Key	String	warp
DefaultValue	Boolean	YES

Figure 12–24. *Our finished Item 4, a toggle switch to turn the warp engines on and off. Engage!*

Adding the Slider Setting

The next item we need to implement is a slider. In the Settings application, a slider can have a small image at each end, but it can't have a label. Let's put the slider in its own group with a header so that the user will know what the slider does.

Start by collapsing *Item 4*. Now, single-click *Item 4* and press return to create a new row. Use the popup to turn the new item into a *Group*, and then click the item's disclosure triangle to expand it. You'll see that *Type* is already set to *PSGroupSpecifier*. This will tell the Settings application to start a new group at this location. Double-click the value in the row labeled *Title,* and change the value to *Warp Factor*.

Collapse *Item 5* and select it, and then press return to add a new sibling row. Use the popup to change the new item into a *Slider*, which indicates to the Settings application that it should use a UISlider to get this information from the user. Expand *Item 6* and set the value of the *Key* row to *warpFactor* so that the Settings application knows which key to use when storing this value.

We're going to allow the user to enter a value from 1 to 10, and we'll set the default to *warp 5*. Sliders need to have a minimum value, a maximum value, and a starting (or default) value, and all of these need to be stored as numbers, not strings, in your property list. Fortunately, Xcode has already created rows for all these values. Give the *DefaultValue* row a value of *5*, the *MinimumValue* row a value of *1*, and the *MaximumValue* row a value of *10*.

If you want to test the slider, go ahead, but hurry back. We're going to do just a bit more customization.

As noted, sliders can have images. You can place a small 21 × 21-pixel image at each end of the slider. Let's provide little icons to indicate that moving the slider to the left slows us down, and moving it to the right speeds us up.

Adding Icons to the Settings Bundle

In the *12 - AppSettings* folder in the project archive that accompanies this book, you'll find two icons called *rabbit.png* and *turtle.png*. We need to add both of these to our settings bundle. Because these images need to be used by the Settings application, we can't just put them in our *AppSettings* folder; we need to put them in the settings bundle so the Settings application can access them.

To do that, find the *Settings.bundle* in the project navigator. We'll need to open this bundle in the Finder. Control-click the *Setting.bundle* icon in the project navigator. When the contextual menu appears, select **Show in Finder** (see Figure 12–25) to show the bundle in the Finder.

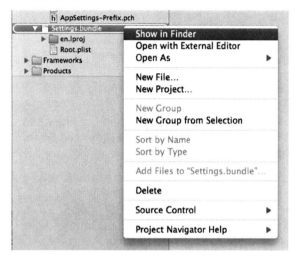

Figure 12–25. *The Settings.bundle contextual menu*

Remember that bundles look like files in the Finder, but they are really folders. When the Finder window opens to show the *Settings.bundle* file, control-click the file and select **Show Package Contents** from the contextual menu that appears. This will open the settings bundle in a new Finder window, and you should see the same two items that you see in *Settings.bundle* in Xcode. Copy the two icon files, *rabbit.png* and *turtle.png*, from the *12 - AppSettings* folder into the *Settings.bundle* package contents Finder window.

You can leave this window open in the Finder, as we'll need to copy another file here soon. Now, we'll return to Xcode and tell the slider to use these two images.

Back in Xcode, return to *Root.plist* and add two more child rows under *Item 6*. Give one a key of *MinimumValueImage* and a value of *turtle.png*. Give the other a key of *MaximumValueImage* and a value of *rabbit.png*. Our finished *Item 6* is shown in Figure 12–26.

Key	Type	Value
▼ PreferenceSpecifiers	Array	(9 items)
▶ Item 0 (Group – General Info)	Diction...	(2 items)
▶ Item 1 (Text Field – Username)	Diction...	(5 items)
▶ Item 2 (Text Field – Password)	Diction ..	(6 items)
▶ Item 3 (Multi Value – Protocol)	Diction...	(6 items)
▶ Item 4 (Toggle Switch – Warp Drive)	Diction...	(4 items)
▶ Item 5 (Group – Warp Factor)	Diction...	(2 items)
▼ Item 6 (Slider)	Diction...	(7 items)
Type	String	PSSliderSpecifier
Key	String	warpFactor
DefaultValue	Number	5
MinimumValue	Number	1
MinimumValueImage	String	turtle.png
MaximumValue	Number	10
MaximumValueImage	String	rabbit.png

Figure 12–26. *Our finished Item 6, a slider with turtle and rabbit icons to represent slow and fast*

Save your property list, and let's build and run our app to make sure everything is still hunky-dory. You should be able to navigate to the Settings application and find the slider waiting for you, with the sleepy turtle and the happy rabbit at each end (see Figure 12–27).

Figure 12–27. *We have text fields, multivalue fields, a toggle switch, and a slider. We're almost finished.*

Adding a Child Settings View

We're going to add another preference specifier to tell the Settings application that we want it to display a child settings view. This specifier will present a row with a disclosure indicator that, when tapped, will take the user down to a whole new view full of preferences. Let's get to it.

Since we don't want this new preference to be grouped with the slider, first we'll copy the group specifier in *Item 0* and paste it at the end of the *PreferenceSpecifiers* array to create a new group for our child settings view.

In *Root.plist*, collapse all open items, and then single-click *Item 0* to select it and press ⌘C to copy it to the clipboard. Next, select *Item 6*, and then press ⌘V to paste in a new *Item 7*. Expand Item 7, and double-click the *Value* column next to the key *Title*, changing it from *General Info* to *Additional Info*.

Now, collapse *Item 7* again. Select it, and press return to add *Item 8*, which will be our actual child view. Expand it by clicking the disclosure triangle. Find the *Type* row and give it a value of *PSChildPaneSpecifier*. Then set the value of the *Title* row to *More Settings*. You can ignore the *Key* row.

We need to add one final row to *Item 8*, which will tell the Settings application which property list to load for the *More Settings* view. Add another child row and give it a key of *File* and a value of *More* (see Figure 12–28). The file extension *.plist* is assumed and must not be included (if it is, the Settings application won't find the plist file).

Key	Type	Value
▼ PreferenceSpecifiers	Array	(9 items)
▶ Item 0 (Group – General Info)	Diction...	(2 items)
▶ Item 1 (Text Field – Username)	Diction...	(5 items)
▶ Item 2 (Text Field – Password)	Diction...	(6 items)
▶ Item 3 (Multi Value – Protocol)	Diction...	(6 items)
▶ Item 4 (Toggle Switch – Warp Drive)	Diction...	(4 items)
▶ Item 5 (Group – Warp Factor)	Diction...	(2 items)
▶ Item 6 (Slider)	Diction...	(7 items)
▼ Item 7 (Group – Additional Info)	Diction...	(2 items)
Type	String	PSGroupSpecifier
Title	String	Additional Info
▼ Item 8 (Child Pane – More Settings)	Diction...	(4 items)
Type	String	PSChildPaneSpecifier
Title	String	More Settings
Key	String	
File	String	More

Figure 12–28. *Our finished Items 7 and 8, setting up the new Additional Info settings group and providing the child pane link to the file More.plist*

We are adding a child view to our main preference view. The settings in that child view are specified in the *More.plist file*. We need to copy *More.plist* into the settings bundle. We can't add new files to the bundle in Xcode, and the Property List Editor's Save dialog will not let us save into a bundle. So, we need to create a new property list, save it somewhere else, and then drag it into the *Settings.bundle* window using the Finder.

You've now seen all the different types of preference fields that you can use in a settings bundle plist file. To save yourself some typing, you can grab *More.plist* out of the *12 - AppSettings* folder in the project archive that accompanies this book, and drag it into that *Settings.bundle* window we left open earlier.

> **TIP:** When you create your own child settings views, the easiest way is to make a copy of *Root.plist* and give it a new name. Then delete all of the existing preference specifiers except the first one, and add whatever preference specifiers you need for that new file.

We're finished with our settings bundle. Feel free to compile, run, and test the Settings application. You should be able to reach the child view and set values for all the other fields. Go ahead and play with it, and make changes to the property list if you want.

> **TIP:** We've covered almost every configuration option available (at least at the time of this writing). You can find the full documentation of the settings property list format in the document called *Settings Application Schema Reference* in the iOS Dev Center. You can get that document, along with a ton of other useful reference documents, from this page:
> `http://developer.apple.com/library/ios/navigation/`.

Before continuing, copy the *rabbit.png* and *turtle.png* icons from the *12 - AppSettings* folder in the project archive into your project's *AppSettings* folder. We'll use them in our application to show the value of the current settings.

You might have noticed that the two icons you just added are exactly the same ones you added to your settings bundle earlier, and you might be wondering why. Remember that iOS applications can't read files out of other applications' sandboxes. The settings bundle doesn't become part of our application's sandbox; it becomes part of the Settings application's sandbox. Since we also want to use those icons in our application, we need to add them separately to our *AppSettings* folder so they are copied into our application's sandbox as well.

Reading Settings in Our Application

We've now solved half of our problem. The user can get to our preferences, but how do we get to them? As it turns out, that's the easy part.

Retrieving User Settings

We'll take advantage of a class called NSUserDefaults to read in the user's settings. NSUserDefaults is implemented as a singleton, which means there is only one instance of NSUserDefaults running in your application. To get access to that one instance, we call the class method standardUserDefaults, like so:

```
NSUserDefaults *defaults = [NSUserDefaults standardUserDefaults];
```

Once we have a pointer to the standard user defaults, we use it just like an NSDictionary. To get a value from it, we can call objectForKey:, which will return an Objective-C object, such as an NSString, NSDate, or NSNumber. If we want to retrieve the value as a scalar—like an int, float, or BOOL—we can use another method, such as intForKey:, floatForKey:, or boolForKey:.

When you were creating the property list for this application, you added an array of *PreferenceSpecifiers*. Some of those specifiers were used to create groups. Others created interface objects that the user used to set the settings. Those are the specifiers we are really interested in, because they hold the real data. Every specifier that was tied to a user setting has a *Key* named *Key*. Take a minute to go back and check. For example, the *Key* for our slider has a value of *warpfactor*. The *Key* for our *Password* field is *password*. We'll use those keys to retrieve the user settings.

So that we have a place to display the settings, let's quickly set up our main view with a bunch of labels. Before going over to Interface Builder, let's create outlets for all the labels we'll need. Single-click *BIDMainViewController.h*, and make the following changes:

```
#import "BIDFlipsideViewController.h"
#define kUsernameKey        @"username"
#define kPasswordKey        @"password"
#define kProtocolKey        @"protocol"
#define kWarpDriveKey       @"warp"
#define kWarpFactorKey      @"warpFactor"
#define kFavoriteTeaKey     @"favoriteTea"
#define kFavoriteCandyKey   @"favoriteCandy"
#define kFavoriteGameKey    @"favoriteGame"
#define kFavoriteExcuseKey  @"favoriteExcuse"
#define kFavoriteSinKey     @"favoriteSin"

@interface BIDMainViewController : UIViewController
        <BIDFlipsideViewControllerDelegate>

@property (weak, nonatomic) IBOutlet UILabel *usernameLabel;
@property (weak, nonatomic) IBOutlet UILabel *passwordLabel;
@property (weak, nonatomic) IBOutlet UILabel *protocolLabel;
@property (weak, nonatomic) IBOutlet UILabel *warpDriveLabel;
@property (weak, nonatomic) IBOutlet UILabel *warpFactorLabel;
@property (weak, nonatomic) IBOutlet UILabel *favoriteTeaLabel;
@property (weak, nonatomic) IBOutlet UILabel *favoriteCandyLabel;
@property (weak, nonatomic) IBOutlet UILabel *favoriteGameLabel;
@property (weak, nonatomic) IBOutlet UILabel *favoriteExcuseLabel;
@property (weak, nonatomic) IBOutlet UILabel *favoriteSinLabel;

- (void)refreshFields;
@end
```

There's nothing new here. We declare a bunch of constants. These are the key values that we used in our plist file for the different preference fields. Then we declare ten outlets, all of them labels, and create properties for each of them. Finally, we declare a method that will read settings out of the user defaults and push those values into the various labels. We put this functionality in its own method, because we need to do this same task in more than one place.

Save your changes. Now that we have our outlets declared, let's head over to the storyboard file to create the GUI.

Creating the Main View

Select *MainStoryboard.storyboard* to edit it in Interface Builder. When it comes up, you'll see the main view on the left and the flipside view on the right, connected by a segue. Notice that the background of the main view is dark gray. Let's change it to white.

Single-click the *View* belonging to the *Main View Controller*, and bring up the attributes inspector. Use the color well labeled *Background* to change the background to white. Note that the color well also functions as a popup menu. If you prefer, use that menu to select *White Color*.

Put the layout area's dock in list mode by clicking the small triangle icon, if it's not already in that mode. In the dock, in the *Main View Controller Scene*, expand *Main View Controller*, and then within that, expand *View*. This reveals an item called *Button* (see Figure 12–29).

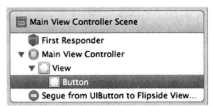

Figure 12–29. *In the dock, locate the Main View Controller Scene and expand Main View Controller, and then expand View and find the Button item.*

TIP: Got a complex Interface Builder list mode hierarchy that you want to open, all at once? Instead of expanding each of the items individually, you can expand the entire hierarchy by holding down the option key and clicking any of the list's disclosure triangles.

The *Button*, situated at the lower-right corner of the view, contains an icon that's mostly white, and is therefore hard to see against the white background. We're going to change this icon so it will look good on a white background. With the *Button* selected, bring up the attributes inspector. Change the button's *Type* from *Info Light* to *Info Dark*.

Now we're going to add a bunch of labels to the *View* so it looks like the one shown in Figure 12–30. We'll need a grand total of 20 labels. Half of them will be static labels that are right-aligned and **bold**; the other half will be used to display the actual values retrieved from the user defaults and will have outlets pointing to them.

Use Figure 12–30 as your guide to build this view. You don't need to match the appearance exactly, but you must have one label on the view for each of the outlets we declared. Go ahead and design the view. You don't need our help for this. When you're finished and have it looking the way you like, come back, and we'll continue. Just so you know, all our labels used 15-point System Font (or System Font Bold), but feel free to go wild with your own design.

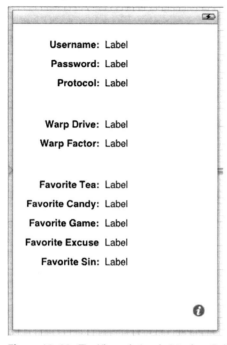

Figure 12–30. *The View window in Interface Builder showing the 20 labels we added*

The next thing we need to do is control-drag from the *Main View Controller* icon (which represents *File's Owner* in the storyboard) to each of the labels intended to display a settings value. You will control-drag a total of ten times, setting each label to a different outlet. Once you have all ten outlets connected to labels, save your changes.

Updating the Main View Controller

In Xcode, select *BIDMainViewController.m*, and add the following code at the beginning of the file:

```
#import "BIDMainViewController.h"

@implementation BIDMainViewController
@synthesize usernameLabel;
@synthesize passwordLabel;
@synthesize protocolLabel;
@synthesize warpDriveLabel;
@synthesize warpFactorLabel;
@synthesize favoriteTeaLabel;
@synthesize favoriteCandyLabel;
@synthesize favoriteGameLabel;
```

```
@synthesize favoriteExcuseLabel;
@synthesize favoriteSinLabel;

- (void)refreshFields {
    NSUserDefaults *defaults = [NSUserDefaults standardUserDefaults];
    usernameLabel.text = [defaults objectForKey:kUsernameKey];
    passwordLabel.text = [defaults objectForKey:kPasswordKey];
    protocolLabel.text = [defaults objectForKey:kProtocolKey];
    warpDriveLabel.text = [defaults boolForKey:kWarpDriveKey]
                            ? @"Engaged" : @"Disabled";
    warpFactorLabel.text = [[defaults objectForKey:kWarpFactorKey]
                            stringValue];
    favoriteTeaLabel.text = [defaults objectForKey:kFavoriteTeaKey];
    favoriteCandyLabel.text = [defaults objectForKey:kFavoriteCandyKey];
    favoriteGameLabel.text = [defaults objectForKey:kFavoriteGameKey];
    favoriteExcuseLabel.text = [defaults objectForKey:kFavoriteExcuseKey];
    favoriteSinLabel.text = [defaults objectForKey:kFavoriteSinKey];
}
.
.
.
- (void)viewDidAppear:(BOOL)animated {
    [super viewDidAppear:animated];
    [self refreshFields];
}
.
.
.
```

Also, let's be good memory citizens by inserting the following code into the existing viewDidUnload method:

```
- (void)viewDidUnload {
    [super viewDidUnload];
    // Release any retained subviews of the main view.
    // e.g. self.myOutlet = nil;
    self.usernameLabel = nil;
    self.passwordLabel = nil;
    self.protocolLabel = nil;
    self.warpDriveLabel = nil;
    self.warpFactorLabel = nil;
    self.favoriteTeaLabel = nil;
    self.favoriteCandyLabel = nil;
    self.favoriteGameLabel = nil;
    self.favoriteExcuseLabel = nil;
    self.favoriteSinLabel = nil;
}
```

When the user is finished using the flipside view where some preferences can be changed, our controller will be notified of the fact. When that happens, we need to make sure our labels are updated to show any changes. Add the following line of code to the existing flipsideViewControllerDidFinish: method:

```
- (void)flipsideViewControllerDidFinish:
        (BIDFlipsideViewController *)controller {
    [self refreshFields];
```

```
        [self dismissModalViewControllerAnimated:YES];
}
```

There's not really much here that should throw you. The new method, `refreshFields`, does nothing more than grab the standard user defaults, and sets the text property of all the labels to the appropriate object from the user defaults, using the key values that we put in our plist file. Notice that for `warpFactorLabel`, we're calling `stringValue` on the object returned. All of our other preferences are strings, which come back from the user defaults as `NSString` objects. The preference stored by the slider, however, comes back as an `NSNumber`, so we call `stringValue` on it to get a string representation of the value it holds.

After that, we fleshed out the `viewDidAppear:` method, where we call our `refreshFields` method. We call `refreshFields` again when we are notified that the flipside controller is being dismissed. This will cause our displayed fields to be set to the appropriate preference values when the view loads, and then to be refreshed when the flipside view is swapped out. Because the flipside view is handled modally, with the main view as its modal parent, the `BIDMainViewController`'s `viewDidAppear:` method will not be called when the flipside view is dismissed. Fortunately, the Utility Application template we chose has very kindly provided us with a delegate method we can use for exactly that purpose.

Registering Default Values

We've created a settings bundle, including some default settings for a few values, to give the Settings app access to our app's preferences. We've also set up our own app to access the same information, with a GUI to let the user see and edit it. However, one piece is missing: our app is completely unaware of the default values specified in the settings bundle. You can see this for yourself by deleting the AppSettings app from the iOS simulator or the device you're running on (thereby deleting the preferences stored for the app), and then running it from Xcode again. At the start of a fresh launch, the app will show you blank values for all the settings. Even the default values for the warp drive settings, which we defined in the settings bundle, are nowhere to be seen. If you then switch over to the Settings app, you'll see the default values, but unless you actually change the values there, you'll never see them back in our AppSettings app!

The reason our setting disappeared is that our app knows nothing about the settings bundle it contains. So, when it tries to read the value from `NSUserDefaults` for *warpFactor* and finds nothing saved under that key, it has nothing to show us. Fortunately, `NSUserDefaults` includes a method called `registerDefaults:` that lets us specify the default values that we should find if we try to look up a key/value that hasn't been set. To make this work throughout the app, it's best if this is called early during app startup. Select *BIDAppDelegate.m*, and include this header file somewhere at the top of the file, so we can access the key names we defined earlier:

```
#import "BIDMainViewController.h"
```

Then modify the `application:didFinishLaunchingWithOptions:` method as shown here:

```
- (BOOL)application:(UIApplication *)application
didFinishLaunchingWithOptions:(NSDictionary *)launchOptions
{
    // Override point for customization after application launch.
    NSDictionary *defaults = [NSDictionary dictionaryWithObjectsAndKeys:
                            [NSNumber numberWithBool:YES], kWarpDriveKey,
                            [NSNumber numberWithInt:5], kWarpFactorKey,
                            @"Greed", kFavoriteSinKey,
                            nil];
    [[NSUserDefaults standardUserDefaults] registerDefaults:defaults];
    return YES;
}
```

The first thing we do here is create a dictionary containing three key/value pairs, one for each of the keys available in Settings that requires a default value. We're using the same key names we defined earlier, to reduce the risk of mistyping a key name. Then we pass that entire dictionary to the standard `NSUserDefaults` instance. From that point on, `NSUserDefaults` will give us the values we specify here, as long as we haven't set different values either in our app or in the Settings app.

This class is complete. You should be able to compile and run your application. It will look something like Figure 12–7, except yours will be showing whatever values you entered in your Settings application, of course. Couldn't be much easier, could it?

Changing Defaults from Our Application

Now that we have the main view up and running, let's build the flipside view. As you can see in Figure 12–31, the flipside view features our warp drive switch, as well as the warp factor slider. We'll use the same controls that the Settings application uses for these two items: a switch and a slider. In addition to declaring our outlets, we'll also declare a method called `refreshFields`, just as we did in `BIDMainViewController`, and two action methods that will be triggered by the user touching the controls.

Figure 12–31. *Designing the flipside view in Interface Builder*

Select *BIDFlipsideViewController.h*, and make the following changes:

```
#import <UIKit/UIKit.h>

@class BIDFlipsideViewController;

@protocol BIDFlipsideViewControllerDelegate
- (void)flipsideViewControllerDidFinish:(BIDFlipsideViewController *)controller;
@end

@interface BIDFlipsideViewController : UIViewController

@property (weak, nonatomic) id <BIDFlipsideViewControllerDelegate> delegate;
@property (weak, nonatomic) IBOutlet UISwitch *engineSwitch;
@property (weak, nonatomic) IBOutlet UISlider *warpFactorSlider;

- (void)refreshFields;
- (IBAction)engineSwitchTapped;
- (IBAction)warpSliderTouched;
- (IBAction)done:(id)sender;

@end
```

> **NOTE:** Don't worry too much about the extra code here. As you saw before, the Utility Application template makes `BIDMainViewController` a delegate of the `BIDFlipsideViewController`. The extra code here that hasn't been in the other file templates we've used implements that delegate relationship.

Now, save your changes and select *MainStoryboard.storyboard* to edit the GUI in Interface Builder, this time focusing on the *Flipside View Controller Scene*. Hold down the option key and expand *Flipside View Controller* and everything below it. Next, double-click the flipside view title in the title bar and change it from *Title* to *Warp Settings*.

Next, select the *View* in the *Flipside View Controller Scene*, and then bring up the attributes inspector. First, change the background color by using the *Background* popup to select *Light Gray Color*. The default flipside view background color is too dark for black text to look good, but light enough that white text is hard to read.

Next, drag two *Labels* from the library and place them on the *View* window. Double-click one of them, and change it to read *Warp Engines:*. Double-click the other, and call it *Warp Factor:*. You can use Figure 12–31 as a placement guide.

Next, drag over a *Switch* from the library, and place it against the right side of the view, across from the label that reads *Warp Engines*. Control-drag from the *Flipside View Controller* icon to the new switch, and connect it to the *engineSwitch* outlet. Then control-drag from the switch back to the *Flipside View Controller* icon, and connect it to the *engineSwitchTapped* action.

Now drag over a *Slider* from the library, and place it below the label that reads *Warp Factor:*. Resize the slider so that it stretches from the blue guideline on the left margin to the one on the right, and then control-drag from the *Flipside View Controller* icon to the slider, and connect it to the *warpFactorSlider* outlet. Then control-drag from the slider to *Flipside View Controller*, and select the *warpSliderTouched* action.

Single-click the slider if it's not still selected, and bring up the attributes inspector. Set *Minimum* to *1.00*, *Maximum* to *10.00*, and *Current* to *5.00*. Next, select *turtle.png* for *Min Image* and *rabbit.png* for *Max Image* (you *did* drag them into the project, right?).

Now, let's finish the flipside view controller. Select *BIDFlipsideViewController.m*, and make the following changes:

```
#import "BIDFlipsideViewController.h"
#import "BIDMainViewController.h"

@implementation BIDFlipsideViewController

@synthesize delegate = _delegate;
@synthesize engineSwitch;
@synthesize warpFactorSlider;

    .
    .
    .


- (void)viewDidLoad {
    [super viewDidLoad];
    // Do any additional setup after loading the view, typically from a nib.

    [self refreshFields];
}

- (void)refreshFields {
    NSUserDefaults *defaults = [NSUserDefaults standardUserDefaults];
    engineSwitch.on = [defaults boolForKey:kWarpDriveKey];
    warpFactorSlider.value = [defaults floatForKey:kWarpFactorKey];

}

- (IBAction)engineSwitchTapped {
    NSUserDefaults *defaults = [NSUserDefaults standardUserDefaults];
    [defaults setBool:engineSwitch.on forKey:kWarpDriveKey];
}

- (IBAction)warpSliderTouched {
    NSUserDefaults *defaults = [NSUserDefaults standardUserDefaults];
    [defaults setFloat:warpFactorSlider.value forKey:kWarpFactorKey];
}
    .
    .
    .
```

Add the following lines of code to the existing viewDidUnload method:

```
- (void)viewDidUnload {
    [super viewDidUnload];
    // Release any retained subviews of the main view.
```

```
    // e.g. self.myOutlet = nil;
    self.engineSwitch = nil;
    self.warpFactorSlider = nil;
}
```

We added a call to our `refreshFields` method, whose three lines of code get a reference to the standard user defaults, and then use the outlets for the switch and slider to make them display the values stored in the user defaults.

```
- (void)refreshFields {
    NSUserDefaults *defaults = [NSUserDefaults standardUserDefaults];
    engineSwitch.on = [defaults boolForKey:kWarpDriveKey];
    warpFactorSlider.value = [defaults floatForKey:kWarpFactorKey];
}
```

We also implemented the `engineSwitchTapped` and `warpSliderTouched` action methods, so that we could stuff the values from our controls back into the user defaults when the user changes them.

Keeping It Real

Now you should be able to run your app, view the settings, and then press the home button and open the Settings app to tweak some values. Hit the home button again, launch your app again, and you may be in for a surprise. If you're running iOS 4.0 or later on your iOS device or simulator (and we bet you are), then when you go back to your app, you won't see the settings change! They'll remain as they are, showing the old values.

When you're using iOS 4, hitting the home button while an app is running doesn't actually quit the app. Instead, the operating system suspends the app in the background, leaving it ready to be quickly fired up again. This is great for switching back and forth between applications, since the amount of time it takes to reawaken a suspended app is much shorter than what it takes to launch it from scratch. However, in our case, we need to do a little more work so that when our app wakes up, it effectively gets a slap in the face, reloads the user preferences, and redisplays the values they contain.

You'll learn more about background applications in Chapter 15, but we'll give you a sneak peek at the basics of how to make your app notice that it has been brought back to life. To do this, we're going to sign up each of our controller classes to receive a notification that is sent by the application when it wakes up from its state of suspended execution.

A **notification** is a lightweight mechanism that objects can use to communicate with each other. Any object can define one or more notifications that it will publish to the application's **notification center**, which is a singleton object that exists only to pass these notifications between objects. Notifications are usually indications that some event occurred, and objects that publish notifications include a list of notifications in their documentation. The `UIApplication` class publishes a number of notifications (you can find them in the Xcode documentation viewer, toward the bottom of the *UIApplication* page). The purpose of most notifications is usually pretty obvious from their names, but the documentation contains further information if you find one whose purpose is unclear.

Our application needs to refresh its display when the application is about to come to the foreground, so we are interested in the notification called `UIApplicationWillEnterForegroundNotification`. When we write our `viewDidLoad` method, we will subscribe to that notification and tell the notification center to call this method when that notification happens. Add this method to both *BIDMainViewController.m* and *BIDFlipsideViewController.m*:

```
- (void)applicationWillEnterForeground:(NSNotification *)notification {
    NSUserDefaults *defaults = [NSUserDefaults standardUserDefaults];
    [defaults synchronize];
    [self refreshFields];
}
```

The method itself is quite simple. First, it gets a reference to the standard user defaults object, and calls its `synchronize` method, which forces the User Defaults system to save any unsaved changes and also reload any unmodified preferences from storage. In effect, we're forcing it to reread the stored preferences so that we can pick up the changes that were made in the Settings app. Then it calls the `refreshFields` method, which each class uses to update its display.

Now, we need to make each of our controllers subscribe to the notification we're interested in by adding the following lines to the bottom of the `viewDidLoad` method in both *BIDMainViewController.m* and *BIDFlipsideViewController.m*. Here's the version for *BIDMainViewController.m*:

```
- (void)viewDidLoad {
    [super viewDidLoad];
    // Do any additional setup after loading the view, typically from a nib.

    UIApplication *app = [UIApplication sharedApplication];
    [[NSNotificationCenter defaultCenter] addObserver:self
            selector:@selector(applicationWillEnterForeground:)
            name:UIApplicationWillEnterForegroundNotification
            object:app];
}
```

And here's the version for *BIDFlipsideViewController.m*:

```
- (void)viewDidLoad {
    [super viewDidLoad];
    // Do any additional setup after loading the view, typically from a nib.

    [self refreshFields];

    UIApplication *app = [UIApplication sharedApplication];
    [[NSNotificationCenter defaultCenter] addObserver:self
            selector:@selector(applicationWillEnterForeground:)
            name:UIApplicationWillEnterForegroundNotification
            object:app];
}
```

We start off by getting a reference to our application instance and use that to subscribe to the UIApplicationWillEnterForegroundNotification, using the default NSNotificationCenter instance and a method called addObserver:selector:name:object:. We then pass the following to this method:

- For an observer, we pass self, which means that our controller class (each of them individually, since this code is going into both of them) is the object that needs to be notified.

- For selector, we pass a selector to the applicationWillEnterForeground: method we just wrote, telling the notification center to call that method when the notification is posted.

- The third parameter, name:, is the name of the notification that we're interested in receiving.

- The final parameter, object:, is the object from which we're interested in getting the notification. If we passed nil for the final parameter, we would get notified any time any method posted the UIApplicationWillEnterForegroundNotification.

That takes care of updating the display, but we also need to consider what happens to the values that are put into the user defaults when the user manipulates the controls in our app. We need to make sure that they are saved to storage before control passes to another app. The easiest way to do that is to call synchronize as soon as the settings are changed, by adding one line to each of our new action methods in *BIDFlipsideViewController.m*:

```
- (IBAction)engineSwitchTapped {
    NSUserDefaults *defaults = [NSUserDefaults standardUserDefaults];
    [defaults setBool:engineSwitch.on forKey:kWarpDriveKey];
    [defaults synchronize];
}

- (IBAction)warpSliderTouched {
    NSUserDefaults *defaults = [NSUserDefaults standardUserDefaults];
    [defaults setFloat:warpFactorSlider.value forKey:kWarpFactorKey];
    [defaults synchronize];
}
```

> **NOTE:** Calling the synchronize method is a potentially expensive operation, since the entire contents of the user defaults in memory must be compared with what's in storage. When you're dealing with a whole lot of user defaults at once and want to make sure everything is in sync, it's best to try to minimize calls to synchronize so that this whole comparison isn't performed over and over again. However, calling it once in response to each user action, as we're doing here, won't cause any noticeable performance problems.

There's one more thing to take care of in order to make this work as cleanly as possible. You already know that you must clean up your memory by setting properties to nil when they're no longer in use, as well as performing other cleanup tasks. The notification system is another place where you need to clean up after yourself, by telling the default NSNotificationCenter that you don't want to listen to any more notifications. In our case, where we've registered each view controller to observe this notification in its viewDidLoad method, we should unregister in the matching viewDidUnload method. So, in both *BIDMainViewController.m* and *BIDFlipsideViewController.m*, put the following line at the top of the viewDidUnload method:

```
- (void)viewDidUnload {
    [[NSNotificationCenter defaultCenter] removeObserver:self];
    .
    .
    .
}
```

Note that it's possible to unregister for specific notifications using the removeObserver:name:object: method, by passing in the same values that were used to register your observer in the first place. But the preceding line is a handy way to make sure that the notification center forgets about our observer completely, no matter how many notifications it was registered for.

With that in place, it's time to build and run the app, and see what happens when you switch between your app and the Settings app. Changes you make in the Settings app should now be immediately reflected in your app when you switch back to it.

Beam Me Up, Scotty

At this point, you should have a very solid grasp on both the Settings application and the User Defaults mechanism. You know how to add a settings bundle to your application and how to build a hierarchy of views for your application's preferences. You also learned how to read and write preferences using NSUserDefaults, and how to let the user change preferences from within your application. You even got a chance to use a new project template in Xcode. There really shouldn't be much in the way of application preferences that you are not equipped to handle now.

In the next chapter, we're going to show you how to keep your application's data around after your application quits. Ready? Let's go!

Chapter 13

Basic Data Persistence

So far, we've focused on the controller and view aspects of the MVC paradigm. Although several of our applications have read data out of the application bundle, none of them has saved data to any form of persistent storage—nonvolatile storage that survives a restart of the computer or device. With the exception of Application Settings (in Chapter 12), so far, every sample application either did not store data or used volatile or nonpersistent storage. Every time one of our sample applications launched, it appeared with exactly the same data it had the first time you launched it.

This approach has worked for us up to this point. But in the real world, your applications will need to persist data. When users make changes, they usually like to find those changes when they launch the program again.

A number of different mechanisms are available for persisting data on an iOS device. If you've programmed in Cocoa for Mac OS X, you've likely used some or all of these techniques.

In this chapter, we're going to look at four different mechanisms for persisting data to the iOS file system:

- Property lists
- Object archives (or archiving)
- SQLite3 (iOS's embedded relational database)
- Core Data (Apple's provided persistence tool)

We will write example applications that use all four approaches.

> **NOTE:** Property lists, object archives, SQLite3, and Core Data are not the only ways you can persist data on iOS. They are just the most common and easiest. You always have the option of using traditional C I/O calls like `fopen()` to read and write data. You can also use Cocoa's low-level file-management tools. In almost every case, doing so will result in a lot more coding effort and is rarely necessary, but those tools are there if you need them.

Your Application's Sandbox

All four of this chapter's data-persistence mechanisms share an important common element: your application's */Documents* folder. Every application gets its own */Documents* folder, and applications are allowed to read and write from only their own */Documents* directory.

To give you some context, let's take a look at how applications are organized in iOS, by examining the folder layout used by the iPhone simulator. In order to see this, you'll need to look inside the *Library* directory contained in your home directory. On Mac OS X 10.6 and earlier, this is no problem, but starting with 10.7, Apple decided to make the *Library* folder hidden by default, so there's a small extra hoop to jump through. Open a Finder window, and navigate to your home directory. If you can see your *Library* folder, that's great. If not, select **Go ➤ Go to Folder…** to open a small sheet that prompts you for the name of a directory. Type *Library* and press enter, and the Finder will take you there.

Within the *Library* folder, drill down into *Application Support/iPhone Simulator/*. Within that directory, you'll see a subdirectory for each version of iOS supported by your current Xcode installation. For example, you might see one directory named *4.3* and another named *5.0*. Drill down into the directory representing the latest version of iOS supported by your version of Xcode. At this point, you should see four subfolders, including one named *Applications* (see Figure 13–1).

> **NOTE:** If you've installed multiple versions of the SDK, you may see a few additional folders inside the *iPhone Simulator* directory, with names indicating the iOS version number they represent. That's perfectly normal.

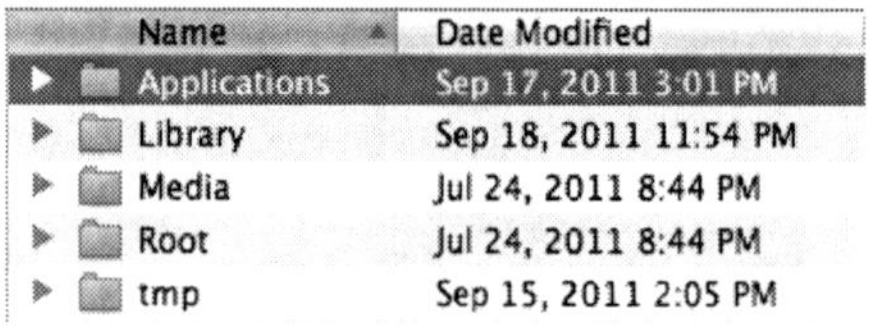

Name	▲	Date Modified
▶ 📁 Applications		Sep 17, 2011 3:01 PM
▶ 📁 Library		Sep 18, 2011 11:54 PM
▶ 📁 Media		Jul 24, 2011 8:44 PM
▶ 📁 Root		Jul 24, 2011 8:44 PM
▶ 📁 tmp		Sep 15, 2011 2:05 PM

Figure 13–1. *The layout of one user's Library/Application Support/iPhone Simulator/5.0/ directory showing the Applications folder*

Although this listing represents the simulator, the file structure is similar to what's on the actual device. As is probably obvious, the *Applications* folder is where iOS stores its applications. If you open the *Applications* folder, you'll see a bunch of folders and files with names that are long strings of characters. These names are globally unique identifiers (GUIDs) and are generated automatically by Xcode. Each of these folders contains one application and its supporting folders.

If you open one of the application subdirectories, you should see something that looks familiar. You'll find one of the iOS applications you've built, along with three support folders:

- ▨ **Documents**: Your application stores its data in *Documents*, with the exception of NSUserDefaults-based preference settings.

- ▨ **Library**: NSUserDefaults-based preference settings are stored in the *Library/Preferences* folder.

- ▨ **tmp**: The *tmp* directory offers a place where your application can store temporary files. Files written into *tmp* will not be backed up by iTunes when your iOS device syncs, but your application does need to take responsibility for deleting the files in *tmp* once they are no longer needed, to avoid filling up the file system.

Getting the Documents Directory

Since our application is in a folder with a seemingly random name, how do we retrieve the full path to the *Documents* directory so that we can read and write our files? It's actually quite easy. The C function NSSearchPathForDirectoriesInDomain() will locate various directories for you. This is a Foundation function, so it is shared with Cocoa for Mac OS X. Many of its available options are designed for Mac OS X and won't return any values on iOS, either because those locations don't exist on iOS (such as the *Downloads* folder) or because your application doesn't have rights to access the location due to iOS's sandboxing mechanism.

Here's some code to retrieve the path to the *Documents* directory:

```
NSArray *paths = NSSearchPathForDirectoriesInDomains(NSDocumentDirectory,
    NSUserDomainMask, YES);
NSString *documentsDirectory = [paths objectAtIndex:0];
```

The constant NSDocumentDirectory says we are looking for the path to the *Documents* directory. The second constant, NSUserDomainMask, indicates that we want to restrict our search to our application's sandbox. In Mac OS X, this same constant is used to indicate that we want the function to look in the user's home directory, which explains its somewhat odd name.

Though an array of matching paths is returned, we can count on our *Documents* directory residing at index 0 in the array. Why? We know that only one directory meets the criteria we've specified, since each application has only one *Documents* directory.

We can create a file name by appending another string onto the end of the path we just retrieved. We'll use an NSString method designed for just that purpose called stringByAppendingPathComponent:.

```
NSString *filename = [documentsDirectory
    stringByAppendingPathComponent:@"theFile.txt"];
```

After this call, filename would contain the full path to a file called *theFile.txt* in our application's *Documents* directory, and we can use filename to create, read, and write from that file.

Getting the tmp Directory

Getting a reference to your application's temporary directory is even easier than getting a reference to the *Documents* directory. The Foundation function called NSTemporaryDirectory() will return a string containing the full path to your application's temporary directory. To create a file name for a file that will be stored in the temporary directory, first find the temporary directory:

```
NSString *tempPath = NSTemporaryDirectory();
```

Then create a path to a file in that directory by appending a file name to that path, like this:

```
NSString *tempFile = [tempPath
    stringByAppendingPathComponent:@"tempFile.txt"];
```

File-Saving Strategies

All four approaches we're going to look at in this chapter make use of the iOS file system. In the case of SQLite3, you'll create a single SQLite3 database file and let SQLite3 worry about storing and retrieving your data. In its simplest form, Core Data takes care of all the file system management for you. With the other two persistence mechanisms—property lists and archiving—you need to put some thought into whether you are going to store your data in a single file or in multiple files.

Single-File Persistence

Using a single file for data storage is the easiest approach, and with many applications, it is a perfectly acceptable one. You start off by creating a root object, usually an NSArray or NSDictionary (your root object can also be based on a custom class when using archiving). Next, you populate your root object with all the program data that needs to be persisted. Whenever you need to save, your code rewrites the entire contents of that root object to a single file. When your application launches, it reads the entire contents of that file into memory. When it quits, it writes out the entire contents. This is the approach we'll use in this chapter.

The downside of using a single file is that you need to load all of your application's data into memory, and you must write all of it to the file system for even the smallest changes. But if your application isn't likely to manage more than a few megabytes of data, this approach is probably fine, and its simplicity will certainly make your life easier.

Multiple-File Persistence

Using multiple files for persistence is an alternative approach. For example, an e-mail application might store each e-mail message in its own file.

There are obvious advantages to this method. It allows the application to load only data that the user has requested (another form of lazy loading), and when the user makes a change, only the files that changed need to be saved. This method also gives you the opportunity to free up memory when you receive a low-memory notification. Any memory that is being used to store data that the user is not currently viewing can be flushed, and then simply reloaded from the file system the next time it's needed.

The downside of multiple-file persistence is that it adds a fair amount of complexity to your application. For now, we'll stick with single-file persistence.

Next, we'll get into the specifics of each of our persistence methods: property lists, object archives, SQLite3, and Core Data. We'll explore each of these in turn and build an application that uses each mechanism to save some data to the device's file system. We'll start with property lists.

Using Property Lists

Several of our sample applications have made use of property lists, most recently when we used a property list to specify our application preferences. Property lists are convenient. They can be edited manually using Xcode or the Property List Editor application. Also, both NSDictionary and NSArray instances can be written to and created from property lists, as long as the dictionary or array contains only specific serializable objects.

Property List Serialization

A **serialized object** is one that has been converted into a stream of bytes so it can be stored in a file or transferred over a network. Although any object can be made serializable, only certain objects can be placed into a collection class, such as an NSDictionary or NSArray, and then stored to a property list using the collection class's writeToFile:atomically: method. The following Objective-C classes can be serialized this way:

- NSArray
- NSMutableArray
- NSDictionary
- NSMutableDictionary
- NSData
- NSMutableData

- NSString
- NSMutableString
- NSNumber
- NSDate

If you can build your data model from just these objects, you can use property lists to save and load your data.

If you're going to use property lists to persist your application data, you'll use either an NSArray or an NSDictionary to hold the data that needs to be persisted. Assuming that all of the objects that you put into the NSArray or NSDictionary are serializable objects from the preceding list, you can write a property list by calling the writeToFile:atomically: method on the dictionary or array instance, like so:

```
[myArray writeToFile:@"/some/file/location/output.plist" atomically:YES];
```

> **NOTE:** In case you were wondering, the atomically parameter tells the method to write the data to an auxiliary file, not to the specified location. Once it has successfully written the file, it will then copy that auxiliary file to the location specified by the first parameter. This is a safer way to write a file, because if the application crashes during the save, the existing file (if there was one) will not be corrupted. It adds a bit of overhead, but in most situations, it's worth the cost.

One problem with the property list approach is that custom objects cannot be serialized into property lists. You also can't use other delivered classes from Cocoa Touch that aren't specified in the previous list of serializable objects, which means that classes like NSURL, UIImage, and UIColor cannot be used directly.

Apart from the serialization issue, keeping all your model data in the form of property lists means that you can't easily create derived or calculated properties (such as a property that is the sum of two other properties), and some of your code that really should be contained in model classes must be moved to your controller classes. Again, these restrictions are OK for simple data models and simple applications. Most of the time, however, your application will be much easier to maintain if you create dedicated model classes.

Simple property lists can still be useful in complex applications. They are a great way to include static data in your application. For example, when your application has a picker, often the best way to include the list of items for it is to create a plist file and place that file in your project's *Resources* folder, which will cause it to be compiled into your application.

Let's a build a simple application that uses property lists to store its data.

The First Version of the Persistence Application

We're going to build a program that lets you enter data into four text fields, saves those fields to a plist file when the application quits, and then reloads the data back from that plist file the next time the application launches (see Figure 13–2).

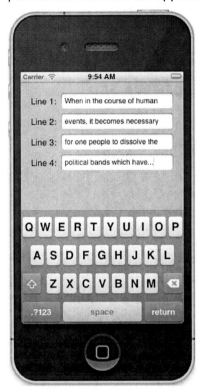

Figure 13–2. *The Persistence application*

> **NOTE:** In this chapter's applications, we won't be taking the time to set up all the user interface niceties that we have added in previous examples. Tapping the return key, for example, will neither dismiss the keyboard nor take you to the next field. If you want to add such polish to the application, doing so would be good practice, so we encourage you to do that on your own.

Creating the Persistence Project

In Xcode, create a new project using the *Single View Application* template, name it *Persistence*, and make sure to turn off the *Use Storyboard* option. This project contains all the files that we'll need to build our application, so we can dive right in.

Before we build the view with the four text fields, let's create the outlets we need. Expand the *Classes* folder. Then single-click the *BIDViewController.h* file, and make the following changes:

```
#import <UIKit/UIKit.h>

@interface BIDViewController : UIViewController

@property (weak, nonatomic) IBOutlet UITextField *field1;
@property (weak, nonatomic) IBOutlet UITextField *field2;
@property (weak, nonatomic) IBOutlet UITextField *field3;
@property (weak, nonatomic) IBOutlet UITextField *field4;
- (NSString *)dataFilePath;
- (void)applicationWillResignActive:(NSNotification *)notification;
@end
```

In addition to defining four text field outlets, we've also defined two additional methods. One method, dataFilePath, will create and return the full pathname to our data file by concatenating a file name onto the path for the *Documents* directory. The other method, applicationWillResignActive: will be called when our application quits and will save data to the plist file. We'll discuss these methods when we edit the persistence classes.

Next, select *BIDViewController.xib* to edit the GUI.

Designing the Persistence Application View

Once Xcode switches over to Interface Builder mode, click the *View* icon to open the *View* window in the nib editing pane. Drag a *Text Field* from the library, and place it against the top and right blue guidelines. Bring up the attributes inspector. Make sure the box labeled *Clear When Editing Begins* is unchecked.

Now, drag a *Label* to the window, and place it to the left of the text field using the left blue guideline, and use the horizontal centering blue guideline to line up the label with the text field. Double-click the label and change it to say *Line 1:*. Finally, resize the text field using the left resize handle to bring it close to the label. Use Figure 13–3 as a guide.

Next, select the label and text field, hold down the option key, and drag down to make a copy below the first set. Use the blue guidelines to guide your placement. Now, select both labels and both text fields, hold down the option key, and drag down again. You should have four labels next to four text fields. Double-click each of the remaining labels and change their names to *Line 2:*, *Line 3:*, and *Line 4:*. Again, compare your results with Figure 13–3.

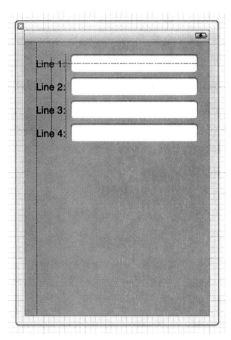

Figure 13–3. *Designing the Persistence application's view*

Once you have all four text fields and labels placed, control-drag from the *File's Owner* icon to each of the four text fields. Connect the topmost text field to the outlet called *field1*, the next one to *field2*, the third to *field3*, and the bottom one to *field4*. When you have all four text fields connected to outlets, save the changes you made to *BIDViewController.xib*.

Editing the Persistence Classes

In the project navigator, select *BIDViewController.m*, and add the following code at the beginning of the file:

```
#import "BIDViewController.h"

#define kFilename        @"data.plist"

@implementation BIDViewController
@synthesize field1;
@synthesize field2;
@synthesize field3;
@synthesize field4;

- (NSString *)dataFilePath {
    NSArray *paths = NSSearchPathForDirectoriesInDomains(
        NSDocumentDirectory, NSUserDomainMask, YES);
    NSString *documentsDirectory = [paths objectAtIndex:0];
    return [documentsDirectory stringByAppendingPathComponent:kFilename];
```

```
}
.
.
.
```

Then go down a bit to find the viewDidLoad and viewDidUnload methods, and fill in their contents like this:

```objc
- (void)viewDidLoad {
    [super viewDidLoad];
    // Do any additional setup after loading the view, typically from a nib.
    NSString *filePath = [self dataFilePath];
    if ([[NSFileManager defaultManager] fileExistsAtPath:filePath]) {
        NSArray *array = [[NSArray alloc] initWithContentsOfFile:filePath];
        field1.text = [array objectAtIndex:0];
        field2.text = [array objectAtIndex:1];
        field3.text = [array objectAtIndex:2];
        field4.text = [array objectAtIndex:3];
    }

    UIApplication *app = [UIApplication sharedApplication];
    [[NSNotificationCenter defaultCenter] addObserver:self
        selector:@selector(applicationWillResignActive:)
        name:UIApplicationWillResignActiveNotification
        object:app];
}

- (void)viewDidUnload {
    [super viewDidUnload];
    // Release any retained subviews of the main view.
    // e.g. self.myOutlet = nil;
    self.field1 = nil;
    self.field2 = nil;
    self.field3 = nil;
    self.field4 = nil;
}
```

Finally, add the following new method at the bottom of the file, just before @end:

```objc
- (void)applicationWillResignActive:(NSNotification *)notification {
    NSMutableArray *array = [[NSMutableArray alloc] init];
    [array addObject:field1.text];
    [array addObject:field2.text];
    [array addObject:field3.text];
    [array addObject:field4.text];
    [array writeToFile:[self dataFilePath] atomically:YES];
}
```

The first method we added, dataFilePath, returns the full pathname of our data file by finding the *Documents* directory and appending kFilename to it. This method will be called from any code that needs to load or save data.

```objc
- (NSString *)dataFilePath {
    NSArray *paths = NSSearchPathForDirectoriesInDomains(
        NSDocumentDirectory, NSUserDomainMask, YES);
    NSString *documentsDirectory = [paths objectAtIndex:0];
```

```
    return [documentsDirectory stringByAppendingPathComponent:kFilename];
}
```

In the `viewDidLoad` method, we do a few more things. First, we check to see if a data file already exists. If there isn't one, we don't want to bother trying to load it. If the file does exist, we instantiate an array with the contents of that file, and then copy the objects from that array to our four text fields. Because arrays are ordered lists, by copying them in the same order as we saved them, we are always sure to get the correct values in the correct fields.

```
- (void)viewDidLoad {
    [super viewDidLoad];
    NSString *filePath = [self dataFilePath];
    if ([[NSFileManager defaultManager] fileExistsAtPath:filePath]) {
        NSArray *array = [[NSArray alloc] initWithContentsOfFile:filePath];
        field1.text = [array objectAtIndex:0];
        field2.text = [array objectAtIndex:1];
        field3.text = [array objectAtIndex:2];
        field4.text = [array objectAtIndex:3];
    }
}
```

After we load the data from the property list, we get a reference to our application instance and use that to subscribe to `UIApplicationWillResignActiveNotification`, using the default `NSNotificationCenter` instance and a method called `addObserver:selector:name:object:`. We pass an observer of `self`, specifying that our `BIDViewController` instance should be notified. For `selector`, we pass a selector to the `applicationWillResignActive:` method, telling the notification center to call that method when the notification is posted. The third parameter, `name:`, is the name of the notification that we're interested in receiving. The final parameter, `object:`, is the object we're interested in getting the notification from.

```
    UIApplication *app = [UIApplication sharedApplication];
    [[NSNotificationCenter defaultCenter] addObserver:self
            selector:@selector(applicationWillResignActive:)
            name:UIApplicationWillResignActiveNotification
        object:app];
```

The final new method is called `applicationWillResignActive:`. Notice that it takes a pointer to an `NSNotification` as an argument. You probably recognize this pattern from Chapter 12. `applicationWillResignActive:` is a notification method, and all notifications take a single `NSNotification` instance as their argument.

Our application needs to save its data before the application is terminated or sent to the background, so we are interested in the notification called `UIApplicationWillResignActiveNotification`. This notification is posted whenever an app is no longer the one with which the user is interacting. This includes when the user quits the application and (in iOS 4 and later) when the application is pushed to the background, perhaps to later be brought back to the foreground. Earlier, in the `viewDidLoad` method, we used the notification center to subscribe to that particular notification. This method is called when that notification happens:

```
- (void)applicationWillResignActive:(NSNotification *)notification {
    NSMutableArray *array = [[NSMutableArray alloc] init];
```

```
        [array addObject:field1.text];
        [array addObject:field2.text];
        [array addObject:field3.text];
        [array addObject:field4.text];
        [array writeToFile:[self dataFilePath] atomically:YES];
}
```

This method is pretty simple. We create a mutable array, add the text from each of the four fields to the array, and then write the contents of that array out to a plist file. That's all there is to saving our data using property lists.

That wasn't too bad, was it? When our main view is finished loading, we look for a plist file. If it exists, we copy data from it into our text fields. Next, we register to be notified when the application becomes inactive (either by being quit or pushed to the background). When that happens, we gather the values from our four text fields, stick them in a mutable array, and write that mutable array to a property list.

Why don't you compile and run the application? It should build and then launch in the simulator. Once it comes up, you should be able to type into any of the four text fields. When you've typed something in them, press the home button (the circular button with the rounded square in it at the bottom of the simulator window). It's very important that you press the home button. If you just exit the simulator, that's the equivalent of forcibly quitting your application. In that case, you will never receive the notification that the application is terminating, and your data will not be saved.

> **NOTE:** Starting in iOS 4, pressing the home button doesn't typically quit the app—at least not at first. The app is put into a background state, ready to be instantly reactivated in case the user switches back to it. We'll dig into the details of these states and their implications for running and quitting apps in Chapter 15. In the meantime, if you want to verify that the data really was saved, you can quit the iPhone simulator entirely, and then restart your app from Xcode. Quitting the simulator is basically the equivalent of rebooting an iPhone, so when it starts up again, your app will have a fresh relaunch experience.

Property list serialization is pretty cool and easy to use. However, it's a little limiting, since only a small selection of objects can be stored in property lists. Let's look at a somewhat more robust approach.

Archiving Model Objects

In the last part of Chapter 9, when we built the Presidents data model object, you saw an example of the process of loading archived data using NSCoder. In the Cocoa world, the term **archiving** refers to another form of serialization, but it's a more generic type that any object can implement. Any model object specifically written to hold data should support archiving. The technique of archiving model objects lets you easily write complex objects to a file and then read them back in.

As long as every property you implement in your class is either a scalar, like `int` or `float`, or an instance of a class that conforms to the `NSCoding` protocol, you can archive your objects completely. Since most Foundation and Cocoa Touch classes capable of storing data do conform to `NSCoding` (though there are a few noteworthy exceptions, such as `UIImage`), archiving is actually relatively easy to implement for most classes.

Although not strictly required to make archiving work, another protocol should be implemented along with `NSCoding`: the `NSCopying` protocol, which is a protocol that allows your object to be copied. Being able to copy an object gives you a lot more flexibility when using data model objects. For example, in the Presidents application in Chapter 9, instead of that complex code we wrote to store changes the user made so we could handle both the *Cancel* and *Save* buttons, we could have made a copy of the president object and stored the changes in that copy. If the user tapped *Save*, we would just copy the changed version over to replace the original version.

Conforming to NSCoding

The `NSCoding` protocol declares two methods, which are both required. One encodes your object into an archive; the other one creates a new object by decoding an archive. Both methods are passed an instance of `NSCoder`, which you work with in very much the same way as `NSUserDefaults`, introduced in the previous chapter. You can encode and decode both objects and native datatypes like `int` and `float` values using key-value coding.

A method to encode an object might look like this:

```
- (void)encodeWithCoder:(NSCoder *)encoder {
    [encoder encodeObject:foo forKey:kFooKey];
    [encoder encodeObject:bar forKey:kBarKey];
    [encoder encodeInt:someInt forKey:kSomeIntKey];
    [encoder encodeFloat:someFloat forKey:kSomeFloatKey]
}
```

To support archiving in our object, we need to encode each of our instance variables into encoder using the appropriate encoding method. We need to implement a method that initializes an object from an `NSCoder`, allowing us to restore an object that was previously archived. If you are subclassing a class that also conforms to `NSCoding`, you need to make sure you call encodeWithCoder: on your superclass, meaning your method would look like this instead:

```
- (void)encodeWithCoder:(NSCoder *)encoder {
    [super encodeWithCoder:encoder];
    [encoder encodeObject:foo forKey:kFooKey];
    [encoder encodeObject:bar forKey:kBarKey];
    [encoder encodeInt:someInt forKey:kSomeIntKey];
    [encoder encodeFloat:someFloat forKey:kSomeFloatKey]
}
```

Implementing the initWithCoder: method is slightly more complex than implementing encodeWithcoder:. If you are subclassing NSObject directly, or subclassing some other class that doesn't conform to NSCoding, your method would look something like the following:

```
- (id)initWithCoder:(NSCoder *)decoder {
    if (self = [super init]) {
        foo = [decoder decodeObjectForKey:kFooKey];
        bar = [decoder decodeObjectForKey:kBarKey];
        someInt = [decoder decodeIntForKey:kSomeIntKey];
        someFloat = [decoder decodeFloatForKey:kAgeKey];
    }
    return self;
}
```

The method initializes an object instance using [super init]. If that's successful, it sets its properties by decoding values from the passed-in instance of NSCoder. When implementing NSCoding for a class with a superclass that also conforms to NSCoding, the initWithCoder: method needs to look slightly different. Instead of calling init on super, it needs to call initWithCoder:, like so:

```
- (id)initWithCoder:(NSCoder *)decoder {
    if (self = [super initWithCoder:decoder]) {
        foo = [decoder decodeObjectForKey:kFooKey];
        bar = [decoder decodeObjectForKey:kBarKey];
        someInt = [decoder decodeIntForKey:kSomeIntKey];
        someFloat = [decoder decodeFloatForKey:kAgeKey];
    }
    return self;
}
```

And that's basically it. As long as you implement these two methods to encode and decode all of your object's properties, your object is archivable and can be written to and read from archives.

Implementing NSCopying

As we mentioned earlier, conforming to NSCopying is a very good idea for any data model objects. NSCopying has one method, called copyWithZone:, which allows objects to be copied. Implementing NSCopying is similar to implementing initWithCoder:. You just need to create a new instance of the same class, and then set all of that new instance's properties to the same values as this object's properties. Here's what a copyWithZone: method might look like:

```
- (id)copyWithZone:(NSZone *)zone {
    MyClass *copy = [[[self class] allocWithZone:zone] init];
    copy.foo = [self.foo copyWithZone:zone];
    copy.bar = [self.bar copyWithZone:zone];
    copy.someInt = self.someInt;
    copy.someFloat = self.someFloat;
    return copy;
}
```

> **NOTE:** Don't worry too much about the NSZone parameter. This pointer is to a struct that is used by the system to manage memory. Only in rare circumstances did developers ever need to worry about zones or create their own, and nowadays, it's almost unheard of to have multiple zones. Calling copy on an object is the same as calling copyWithZone: using the default zone, which is almost always what you want.

Archiving and Unarchiving Data Objects

Creating an archive from an object or objects that conforms to NSCoding is relatively easy. First, we create an instance of NSMutableData to hold the encoded data, and then we create an NSKeyedArchiver instance to archive objects into that NSMutableData instance:

```
NSMutableData *data = [[NSMutableData alloc] init];
NSKeyedArchiver *archiver = [[NSKeyedArchiver alloc]
    initForWritingWithMutableData:data];
```

After creating both of those, we then use key-value coding to archive any objects we wish to include in the archive, like this:

```
[archiver encodeObject:myObject forKey:@"keyValueString"];
```

Once we've encoded all the objects we want to include, we just tell the archiver we're finished, and write the NSMutableData instance to the file system:

```
[archiver finishEncoding];
BOOL success = [data writeToFile:@"/path/to/archive" atomically:YES];
```

If anything went wrong while writing the file, success will be set to NO. If success is YES, the data was successfully written to the specified file. Any objects created from this archive will be exact copies of the objects that were last written into the file.

To reconstitute objects from the archive, we go through a similar process. We create an NSData instance from the archive file and create an NSKeyedUnarchiver to decode the data:

```
NSData *data = [[NSData alloc] initWithContentsOfFile:path];
NSKeyedUnarchiver *unarchiver = [[NSKeyedUnarchiver alloc]
    initForReadingWithData:data];
```

After that, we read our objects from the unarchiver using the same key that we used to archive the object:

```
self.object = [unarchiver decodeObjectForKey:@"keyValueString"];
```

Finally, we tell the archiver we are finished:

```
[unarchiver finishDecoding];
```

If you're feeling a little overwhelmed by archiving, don't worry. It's actually fairly straightforward. We're going to retrofit our Persistence application to use archiving, so you'll get to see it in action. Once you've done it a few times, archiving will become second nature, as all you're really doing is storing and retrieving your object's properties using key-value coding.

The Archiving Application

Let's redo the Persistence application so it uses archiving instead of property lists. We're going to be making some fairly significant changes to the Persistence source code, so you might want to make a copy of your project before continuing.

Implementing the BIDFourLines Class

Once you're ready to proceed and have a copy of your *Persistence* project open in Xcode, select the *Persistence* folder and press ⌘N or select File ➤ New ➤ New File…. When the new file assistant comes up, select *Cocoa Touch*, select *Objective-C class*, and click *Next*. On the next screen, name the class *BIDFourLines*, and select *NSObject* in the *Subclass of* control. Click *Next* again. Then choose the *Persistence* folder to save the files, and click *Create*. This class is going to be our data model. It will hold the data that we're currently storing in a dictionary in the property list application.

Single-click *BIDFourLines.h*, and make the following changes:

```
#import <Foundation/Foundation.h>

@interface BIDFourLines : NSObject
@interface BIDFourLines : NSObject <NSCoding, NSCopying>

@property (copy, nonatomic) NSString *field1;
@property (copy, nonatomic) NSString *field2;
@property (copy, nonatomic) NSString *field3;
@property (copy, nonatomic) NSString *field4;
@end
```

This is a very straightforward data model class with four string properties. Notice that we've conformed the class to the NSCoding and NSCopying protocols. Now, switch over to *BIDFourLines.m*, and add the following code:

```
#import "BIDFourLines.h"

#define     kField1Key     @"Field1"
#define     kField2Key     @"Field2"
#define     kField3Key     @"Field3"
#define     kField4Key     @"Field4"

@implementation BIDFourLines
@synthesize field1;
@synthesize field2;
@synthesize field3;
@synthesize field4;

#pragma mark NSCoding
- (void)encodeWithCoder:(NSCoder *)encoder {
    [encoder encodeObject:field1 forKey:kField1Key];
```

```
    [encoder encodeObject:field2 forKey:kField2Key];
    [encoder encodeObject:field3 forKey:kField3Key];
    [encoder encodeObject:field4 forKey:kField4Key];
}

- (id)initWithCoder:(NSCoder *)decoder {
    if (self = [super init]) {
        field1 = [decoder decodeObjectForKey:kField1Key];
        field2 = [decoder decodeObjectForKey:kField2Key];
        field3 = [decoder decodeObjectForKey:kField3Key];
        field4 = [decoder decodeObjectForKey:kField4Key];
    }
    return self;
}

#pragma mark -
#pragma mark NSCopying
- (id)copyWithZone:(NSZone *)zone {
    BIDFourLines *copy = [[[self class] allocWithZone:zone] init];
    copy.field1 = [self.field1 copyWithZone:zone];
    copy.field2 = [self.field2 copyWithZone:zone];
    copy.field3 = [self.field3 copyWithZone:zone];
    copy.field4 = [self.field4 copyWithZone:zone];
    return copy;
}
@end
```

We just implemented all the methods necessary to conform to NSCoding and NSCopying. We encode all four of our properties in encodeWithCoder: and decode all four of them using the same four key values in initWithCoder:. In copyWithZone:, we create a new BIDFourLines object and copy all four strings to it. See? It's not hard at all.

Implementing the BIDViewController Class

Now that we have an archivable data object, let's use it to persist our application data. Select *BIDViewController.m*, and make the following changes:

```
#import "BIDViewController.h"
#import "BIDFourLines.h"

#define kFilename         @"data.plist"
#define kFilename         @"archive"
#define kDataKey          @"Data"

@implementation BIDViewController
@synthesize field1;
@synthesize field2;
@synthesize field3;
@synthesize field4;

- (NSString *)dataFilePath {
    NSArray *paths = NSSearchPathForDirectoriesInDomains(
        NSDocumentDirectory, NSUserDomainMask, YES);
    NSString *documentsDirectory = [paths objectAtIndex:0];
```

```
        return [documentsDirectory stringByAppendingPathComponent:kFilename];
}

#pragma mark -
- (void)viewDidLoad {
    [super viewDidLoad];
    // Do any additional setup after loading the view, typically from a nib.
    NSString *filePath = [self dataFilePath];
    if ([[NSFileManager defaultManager] fileExistsAtPath:filePath]) {
        NSArray *array =[[NSArray alloc] initWithContentsOfFile:filePath];
        field1.text = [array objectAtIndex:0];
        field2.text = [array objectAtIndex:1];
        field3.text = [array objectAtIndex:2];
        field4.text = [array objectAtIndex:3];

        NSData *data = [[NSMutableData alloc]
            initWithContentsOfFile:[self dataFilePath]];
        NSKeyedUnarchiver *unarchiver = [[NSKeyedUnarchiver alloc]
            initForReadingWithData:data];
        BIDFourLines *fourLines = [unarchiver decodeObjectForKey:kDataKey];
        [unarchiver finishDecoding];

        field1.text = fourLines.field1;
        field2.text = fourLines.field2;
        field3.text = fourLines.field3;
        field4.text = fourLines.field4;
    }

    UIApplication *app = [UIApplication sharedApplication];
    [[NSNotificationCenter defaultCenter] addObserver:self
            selector:@selector(applicationWillResignActive:)
            name:UIApplicationWillResignActiveNotification
            object:app];
}
.
.
.
- (void)applicationWillResignActive:(NSNotification *)notification {
    NSMutableArray *array = [[NSMutableArray alloc] init];
    [array addObject:field1.text];
    [array addObject:field2.text];
    [array addObject:field3.text];
    [array addObject:field4.text];
    [array writeToFile:[self dataFilePath] atomically:YES];

    BIDFourLines *fourLines = [[BIDFourLines alloc] init];
    fourLines.field1 = field1.text;
    fourLines.field2 = field2.text;
    fourLines.field3 = field3.text;
    fourLines.field4 = field4.text;

    NSMutableData *data = [[NSMutableData alloc] init];
    NSKeyedArchiver *archiver = [[NSKeyedArchiver alloc]
        initForWritingWithMutableData:data];
    [archiver encodeObject:fourLines forKey:kDataKey];
    [archiver finishEncoding];
```

```
    [data writeToFile:[self dataFilePath] atomically:YES];
}
...
```

Save your changes and take this version of Persistence for a spin.

Not very much has changed, really. We started off by specifying a new file name so that our program doesn't try to load the old property list as an archive. We also defined a new constant that will be the key value we use to encode and decode our object. Then we redefined the loading and saving, using BIDFourLines to hold the data, and using the NSCoding methods to do the actual loading and saving. The GUI is identical to the previous version.

This new version takes several more lines of code to implement than property list serialization, so you might be wondering if there really is an advantage to using archiving over just serializing property lists. For this application, the answer is simple: no, there really isn't any advantage. But think back to the last example in Chapter 9, where we allowed the user to edit a list of presidents, and each president had four different fields that could be edited. To handle archiving that list of presidents with a property list would involve iterating through the list of presidents, creating an NSDictionary instance for each president, copying the value from each of their fields over to the NSDictionary instance, and adding that instance to another array, which could then be written to a plist file. And that's assuming that we restricted ourselves to using only serializable properties. If we didn't, using property list serialization wouldn't even be an option without doing a lot of conversion work.

On the other hand, if we had an array of archivable objects, such as the BIDFourLines class that we just built, we could archive the entire array by archiving the array instance itself. Collection classes like NSArray, when archived, archive all of the objects they contain. As long as every object you put into an array or dictionary conforms to NSCoding, you can archive the array or dictionary and restore it, so that all the objects that were in it when you archived it will be in the restored array or dictionary.

In other words, this approach scales beautifully (in terms of code size, at least). No matter how many objects you add, the work to write those objects to disk (assuming you're using single-file persistence) is exactly the same. With property lists, the amount of work increases with every object you add.

Using iOS's Embedded SQLite3

The third persistence option we're going to discuss is using iOS's embedded SQL database, called SQLite3. SQLite3 is very efficient at storing and retrieving large amounts of data. It's also capable of doing complex aggregations on your data, with much faster results than you would get doing the same thing using objects.

For example, if your application needs to calculate the sum of a particular field across all the objects in your application, or if you need the sum from just the objects that meet certain criteria, SQLite3 allows you to do that without loading every object into memory. Getting aggregations from SQLite3 is several orders of magnitude faster than loading all the objects into memory and summing their values. Being a full-fledged embedded database, SQLite3 contains tools to make it even faster by, for example, creating table indexes that can speed up your queries.

> **NOTE:** There are several schools of thought about the pronunciation of "SQL" and "SQLite." Most official documentation says to pronounce "SQL" as "Ess-Queue-Ell" and "SQLite" as "Ess-Queue-Ell-Light." Many people pronounce them, respectively, as "Sequel" and "Sequel Light." A small cadre of hardened rebels prefer "Squeal" and "Squeal Light." Pick whatever works best for you (and be prepared to be mocked and shunned by the infidels if you choose to join the "Squeal" movement).

SQLite3 uses the Structured Query Language (SQL). SQL is the standard language used to interact with relational databases. Whole books have been written on the syntax of SQL (hundreds of them, in fact), as well as on SQLite itself. So, if you don't already know SQL and you want to use SQLite3 in your application, you have a little work ahead of you. We'll show you how to set up and interact with the SQLite database from your iOS applications, and you'll see some of the basics of the syntax in this chapter. But to really make the most of SQLite3, you'll need to do some additional research and exploration. A couple of good starting points are "An Introduction to the SQLite3 C/C++ Interface" (http://www.sqlite.org/cintro.html) and "SQL As Understood by SQLite" (http://www.sqlite.org/lang.html).

Relational databases, including SQLite3, and object-oriented programming languages use fundamentally different approaches to storing and organizing data. The approaches are different enough that numerous techniques and many libraries and tools for converting between the two have been developed. These different techniques are collectively called **object-relational mapping** (ORM). There are currently several ORM tools available for Cocoa Touch. In fact, we'll look at one ORM solution provided by Apple, called Core Data, later in the chapter.

In this chapter, we're going to focus on the SQLite3 basics, including setting it up, creating a table to hold your data, and using the database in an application. Obviously, in the real world, such a simple application as the one we're working on wouldn't warrant the investment in SQLite3. But this application's simplicity is exactly what makes it a good learning example.

Creating or Opening the Database

Before you can use SQLite3, you must open the database. The function that's used to do that, sqlite3_open(), will open an existing database, or if none exists at the specified location, it will create a new one. Here's what the code to open a new database might look like:

```
sqlite3 *database;
int result = sqlite3_open("/path/to/database/file", &database);
```

If `result` is equal to the constant `SQLITE_OK`, then the database was successfully opened. Note that the path to the database file must be passed in as a C string, not as an `NSString`. SQLite3 was written in portable C, not Objective-C, and it has no idea what an `NSString` is. Fortunately, there is an `NSString` method that generates a C string from an `NSString` instance:

```
const char *stringPath = [pathString UTF8String];
```

When you're finished with an SQLite3 database, close it:

```
sqlite3_close(database);
```

Databases store all their data in tables. You can create a new table by crafting an SQL `CREATE` statement and passing it in to an open database using the function `sqlite3_exec`, like so:

```
char *errorMsg;
const char *createSQL = "CREATE TABLE IF NOT EXISTS PEOPLE ↩
    (ID INTEGER PRIMARY KEY AUTOINCREMENT, FIELD_DATA TEXT)";
int result = sqlite3_exec(database, createSQL, NULL, NULL, &errorMsg);
```

As you did before, you need to verify that `result` is equal to `SQLITE_OK` to make sure your command ran successfully. If it didn't, `errorMsg` will contain a description of the problem that occurred.

The function `sqlite3_exec` is used to run any command against SQLite3 that doesn't return data, including updates, inserts, and deletes. Retrieving data from the database is little more involved. You first need to prepare the statement by feeding it your SQL `SELECT` command:

```
NSString *query = @"SELECT ID, FIELD_DATA FROM FIELDS ORDER BY ROW";
sqlite3_stmt *statement;
int result = sqlite3_prepare_v2(database, [query UTF8String],
    -1, &statement, nil);
```

> **NOTE:** All of the SQLite3 functions that take strings require an old-fashioned C string. In the example, we created and passed a C string. We created an `NSString` and derived a C string by using one of `NSString`'s methods called `UTF8String`. Either method is acceptable. If you need to do manipulation on the string, using `NSString` or `NSMutableString` will be easier, but converting from `NSString` to a C string incurs a bit of extra overhead.

If `result` equals `SQLITE_OK`, your statement was successfully prepared, and you can start stepping through the result set. Here is an example of stepping through a result set and retrieving an `int` and an `NSString` from the database:

```
while (sqlite3_step(statement) == SQLITE_ROW) {
    int rowNum = sqlite3_column_int(statement, 0);
    char *rowData = (char *)sqlite3_column_text(statement, 1);
    NSString *fieldValue = [[NSString alloc] initWithUTF8String:rowData];
    // Do something with the data here
}
sqlite3_finalize(statement);
```

Using Bind Variables

Although it's possible to construct SQL strings to insert values, it is common practice to use something called **bind variables** for this purpose. Handling strings correctly—making sure they don't have invalid characters and that quotes are inserted properly—can be quite a chore. With bind variables, those issues are taken care of for us.

To insert a value using a bind variable, you create your SQL statement as normal, but put a question mark (?) into the SQL string. Each question mark represents one variable that must be bound before the statement can be executed. Then you prepare the SQL statement, bind a value to each of the variables, and execute the command.

Here's an example that prepares an SQL statement with two bind variables, binds an int to the first variable and a string to the second variable, and then executes and finalizes the statement:

```
char *sql = "insert into foo values (?, ?);";
sqlite3_stmt *stmt;
if (sqlite3_prepare_v2(database, sql, -1, &stmt, nil) == SQLITE_OK) {
    sqlite3_bind_int(stmt, 1, 235);
    sqlite3_bind_text(stmt, 2, "Bar", -1, NULL);
}
if (sqlite3_step(stmt) != SQLITE_DONE)
    NSLog(@"This should be real error checking!");
sqlite3_finalize(stmt);
```

There are multiple bind statements available depending on the datatype you wish to use. Most bind functions take only three parameters:

- The first parameter to any bind function, regardless of the datatype, is a pointer to the `sqlite3_stmt` used previously in the `sqlite3_prepare_v2()` call.

- The second parameter is the index of the variable to which you're binding. This is a one-indexed value, meaning that the first question mark in the SQL statement has index 1, and each one after it is one higher than the one to its left.

- The third parameter is always the value that should be substituted for the question mark.

A few bind functions, such as those for binding text and binary data, have two additional parameters:

- The first additional parameter is the length of the data being passed in the third parameter. In the case of C strings, you can pass -1 instead of the string's length, and the function will use the entire string. In all other cases, you need to tell it the length of the data being passed in.

- The final parameter is an optional function callback in case you need to do any memory cleanup after the statement is executed. Typically, such a function would be used to free memory allocated using `malloc()`.

The syntax that follows the bind statements may seem a little odd, since we're doing an insert. When using bind variables, the same syntax is used for both queries and updates. If the SQL string had an SQL query, rather than an update, we would need to call `sqlite3_step()` multiple times until it returned `SQLITE_DONE`. Since this was an update, we call it only once.

The SQLite3 Application

We've covered the basics, so let's see how this would work in practice. We're going to retrofit our Persistence application again, this time storing its data using SQLite3. We'll use a single table and store the field values in four different rows of that table. We'll give each row a row number that corresponds to its field, so for example, the value from `field1` will get stored in the table with a row number of 1. Let's get started.

Linking to the SQLite3 Library

SQLite 3 is accessed through a procedural API that provides interfaces to a number of C function calls. To use this API, we'll need to link our application to a dynamic library called *libsqlite3.dylib*, located in */usr/lib* on both Mac OS X and iOS. The process of linking a dynamic library into your project is exactly the same as that of linking in a framework.

Use the Finder to make a copy of your last *Persistence* project directory, and then open the new copy's *.xcodeproj* file. Select the *Persistence* item at the very top of the project navigator's list (leftmost pane), and then select *Persistence* from the *TARGETS* section in the main area (middle pane; see Figure 13–4). Be careful that you have selected *Persistence* from the *TARGETS* section, and not from the *PROJECT* section.

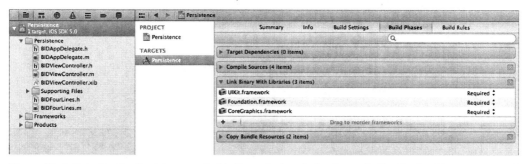

Figure 13–4. *Selecting the Persistence project in the project navigator, then selecting the Persistence target, and finally, selecting the Build Phases tab*

With the *Persistence* target selected, click the *Build Phases* tab in the rightmost pane. You'll see a list of items, initially all collapsed, which represent the various steps Xcode goes through to build the application. Expand the item labeled *Link Binary With Libraries*. You'll see the standard frameworks that our application is set up to link with by default: *UIKit.framework*, *Foundation.framework*, and *CoreGraphics.framework*.

Now, let's add the SQLite3 library to our project. Click the + button at the bottom of the linked frameworks list, and you'll be presented with a sheet that lists all available frameworks and libraries. Find *libsqlite3.dylib* in the list (or use the handy search field), and click the *Add* button. Note that there may be several other entries in that directory that start with *libsqlite3*. Be sure you select *libsqlite3.dylib*. It is an alias that always points to the latest version of the SQLite3 library. Adding this to the project puts it at the top level of the project navigator. For the sake of keeping things organized, you may want to drag it to the project's *Frameworks* folder.

Modifying the Persistence View Controller

Now, it's time to change things around again. This time, we'll replace the NSCoding code with its SQLite equivalent. Once again, we'll change the file name so that we won't be using the same file that we used in the previous version, and the file name properly reflects the type of data it holds. Then we're going to change the methods that save and load the data.

Select *BIDViewController.m*, and make the following changes:

```
#import "BIDViewController.h"
#import "BIDFourLines.h"
#import <sqlite3.h>

#define kFilename    @"archive.plist"
#define kDataKey     @"Data"
#define kFilename    @"data.sqlite3"

@implementation BIDViewController
@synthesize field1;
@synthesize field2;
@synthesize field3;
@synthesize field4;

- (NSString *)dataFilePath {
    NSArray *paths = NSSearchPathForDirectoriesInDomains(
        NSDocumentDirectory, NSUserDomainMask, YES);
    NSString *documentsDirectory = [paths objectAtIndex:0];
    return [documentsDirectory stringByAppendingPathComponent:kFilename];
}

#pragma mark -
- (void)viewDidLoad {
    [super viewDidLoad];
    // Do any additional setup after loading the view, typically from a nib.
    NSString *filePath = [self dataFilePath];
    if ([[NSFileManager defaultManager] fileExistsAtPath:filePath])
    {
        NSData *data = [[NSMutableData alloc]
            initWithContentsOfFile:[self dataFilePath]];
        NSKeyedUnarchiver *unarchiver =
            [[NSKeyedUnarchiver alloc] initForReadingWithData:data];
        BIDFourLines *fourLines = [unarchiver decodeObjectForKey:kDataKey];
        [unarchiver finishDecoding];
```

```
        field1.text = fourLines.field1;
        field2.text = fourLines.field2;
        field3.text = fourLines.field3;
        field4.text = fourLines.field4;
    }
    sqlite3 *database;
    if (sqlite3_open([[self dataFilePath] UTF8String], &database)
            != SQLITE_OK) {
        sqlite3_close(database);
        NSAssert(0, @"Failed to open database");
    }

    // Useful C trivia: If two inline strings are separated by nothing
    // but whitespace (including line breaks), they are concatenated into
    // a single string:
    NSString *createSQL = @"CREATE TABLE IF NOT EXISTS FIELDS "
                          "(ROW INTEGER PRIMARY KEY, FIELD_DATA TEXT);";
    char *errorMsg;
    if (sqlite3_exec (database, [createSQL UTF8String],
        NULL, NULL, &errorMsg) != SQLITE_OK) {
        sqlite3_close(database);
        NSAssert(0, @"Error creating table: %s", errorMsg);
    }
    NSString *query = @"SELECT ROW, FIELD_DATA FROM FIELDS ORDER BY ROW";
    sqlite3_stmt *statement;
    if (sqlite3_prepare_v2(database, [query UTF8String],
        -1, &statement, nil) == SQLITE_OK) {
        while (sqlite3_step(statement) == SQLITE_ROW) {
            int row = sqlite3_column_int(statement, 0);
            char *rowData = (char *)sqlite3_column_text(statement, 1);

            NSString *fieldName = [[NSString alloc]
                initWithFormat:@"field%d", row];
            NSString *fieldValue = [[NSString alloc]
                initWithUTF8String:rowData];
            UITextField *field = [self valueForKey:fieldName];
            field.text = fieldValue;
            }
        sqlite3_finalize(statement);
    }
    sqlite3_close(database);

    UIApplication *app = [UIApplication sharedApplication];
    [[NSNotificationCenter defaultCenter] addObserver:self
            selector:@selector(applicationWillResignActive:)
            name:UIApplicationWillResignActiveNotification
            object:app];
}

- (void)applicationWillResignActive:(NSNotification *)notification {
    BIDFourLines *fourLines = [[BIDFourLines alloc] init];
    fourLines.field1 = field1.text;
    fourLines.field2 = field2.text;
    fourLines.field3 = field3.text;
    fourLines.field4 = field4.text;
```

```
    NSMutableData *data = [[NSMutableData alloc] init];
    NSKeyedArchiver *archiver = [[NSKeyedArchiver alloc]
            initForWritingWithMutableData:data];
    [archiver encodeObject:fourLines forKey:kDataKey];
    [archiver finishEncoding];
    [data writeToFile:[self dataFilePath] atomically:YES];

    sqlite3 *database;
    if (sqlite3_open([[self dataFilePath] UTF8String], &database)
        != SQLITE_OK) {
        sqlite3_close(database);
        NSAssert(0, @"Failed to open database");
    }
    for (int i = 1; i <= 4; i++) {
        NSString *fieldName = [[NSString alloc]
                            initWithFormat:@"field%d", i];
        UITextField *field = [self valueForKey:fieldName];

        // Once again, inline string concatenation to the rescue:
        char *update = "INSERT OR REPLACE INTO FIELDS (ROW, FIELD_DATA) "
                    "VALUES (?, ?);";
        char *errorMsg;
        sqlite3_stmt *stmt;
        if (sqlite3_prepare_v2(database, update, -1, &stmt, nil)
                == SQLITE_OK) {
            sqlite3_bind_int(stmt, 1, i);
            sqlite3_bind_text(stmt, 2, [field.text UTF8String], -1, NULL);
        }
        if (sqlite3_step(stmt) != SQLITE_DONE)
            NSAssert(0, @"Error updating table: %s", errorMsg);
        sqlite3_finalize(stmt);

    }
    sqlite3_close(database);
}
.
.
.
```

The first new code is in the viewDidLoad method. We begin by opening the database. If
we hit a problem with opening the database, we close it and raise an assertion.

```
    sqlite3 *database;
    if (sqlite3_open([[self dataFilePath] UTF8String], &database)
        != SQLITE_OK) {
        sqlite3_close(database);
        NSAssert(0, @"Failed to open database");
    }
```

Next, we need to make sure that we have a table to hold our data. We can use SQL
CREATE TABLE to do that. By specifying IF NOT EXISTS, we prevent the database from
overwriting existing data. If there is already a table with the same name, this command
quietly exits without doing anything, so it's safe to call every time our application
launches without explicitly checking to see if a table exists.

```
NSString *createSQL = @"CREATE TABLE IF NOT EXISTS FIELDS "
                       "(ROW INTEGER PRIMARY KEY, FIELD_DATA TEXT);";
char *errorMsg;
if (sqlite3_exec (database, [createSQL UTF8String], NULL, NULL,
    &errorMsg) != SQLITE_OK) {
    sqlite3_close(database);
    NSAssert1(0, @"Error creating table: %s", errorMsg);
}
```

Finally, we need to load our data. We do this using an SQL SELECT statement. In this simple example, we create an SQL SELECT that requests all the rows from the database and ask SQLite3 to prepare our SELECT. We also tell SQLite3 to order the rows by the row number, so that we always get them back in the same order. Absent this, SQLite3 will return the rows in the order in which they are stored internally.

```
NSString *query = @"SELECT ROW, FIELD_DATA FROM FIELDS ORDER BY ROW";
sqlite3_stmt *statement;
if (sqlite3_prepare_v2( database, [query UTF8String],
    -1, &statement, nil) == SQLITE_OK) {
```

Then we step through each of the returned rows:

```
while (sqlite3_step(statement) == SQLITE_ROW) {
```

We grab the row number and store it in an int, and then we grab the field data as a C string.

```
int row = sqlite3_column_int(statement, 0);
char *rowData = (char *)sqlite3_column_text(statement, 1);
```

Next, we create a field name based on the row number (such as field1 for row 1), convert the C string to an NSString, and use that to set the appropriate field with the value retrieved from the database.

```
NSString *fieldName = [[NSString alloc]
    initWithFormat:@"field%d", row];
NSString *fieldValue = [[NSString alloc]
    initWithUTF8String:rowData];
UITextField *field = [self valueForKey:fieldName];
field.text = fieldValue;
```

Finally, we close the database connection, and we're all finished.

```
    }
    sqlite3_finalize(statement);
}
sqlite3_close(database);
```

Note that we close the database connection as soon as we're finished creating the table and loading any data it contains, rather than keeping it open the entire time the application is running. It's the simplest way of managing the connection, and in this little app, we can just open the connection those few times we need it. In a more database-intensive app, you might want to keep the connection open all the time,

The other changes we made are in the applicationWillResignActive: method, where we need to save our application data. Because the data in the database is stored in a table, our application's data will look something like Table 13–1 when stored.

Table 13–1. *Data Stored in the FIELDS Table of the Database*

ROW	FIELD_DATA
1	When in the course of human
2	events, it becomes necessary
3	for one people to dissolve the
4	political bands which have...

The `applicationWillResignActive:` method starts off by once again opening the database.

```
sqlite3 *database;
if (sqlite3_open([[self dataFilePath] UTF8String], &database)
    != SQLITE_OK) {
    sqlite3_close(database);
    NSAssert(0, @"Failed to open database");
}
```

To save the data, we loop through all four fields and issue a separate command to update each row of the database.

```
for (int i = 1; i <= 4; i++) {
    NSString *fieldName = [[NSString alloc]
        initWithFormat:@"field%d", i];
    UITextField *field = [self valueForKey:fieldName];
```

The first thing we do in the loop is craft a field name so we can retrieve the correct text field outlet. Remember that `valueForKey:` allows you to retrieve a property based on its name. We also declare a pointer to be used for the error message if we encounter an error.

We craft an `INSERT OR REPLACE` SQL statement with two bind variables. The first represents the row that's being stored; the second is for the actual string value to be stored. By using `INSERT OR REPLACE` instead of the more standard `INSERT`, we don't need to worry about whether a row already exists.

```
char *update = "INSERT OR REPLACE INTO FIELDS (ROW, FIELD_DATA) "
               "VALUES (?, ?);";
```

Next, we declare a pointer to a statement, prepare our statement with the bind variables, and bind values to both of the bind variables.

```
sqlite3_stmt *stmt;
if (sqlite3_prepare_v2(database, update, -1, &stmt, nil) == SQLITE_OK) {
    sqlite3_bind_int(stmt, 1, i);
    sqlite3_bind_text(stmt, 2, [field.text UTF8String], -1, NULL);
}
```

Then we call `sqlite3_step` to execute the update, check to make sure it worked, and finalize the statement, ending the loop.

```
        if (sqlite3_step(stmt) != SQLITE_DONE) {
            NSAssert(0, @"Error updating table.");
        }
        sqlite3_finalize(stmt);
    }
```

Notice that we used an assertion here to check for an error condition. We use assertions rather than exceptions or manual error checking because this condition should happen only if we, the developers, make a mistake. Using this assertion macro will help us debug our code, and it can be stripped out of our final application. If an error condition is one that a user might reasonably experience, you should probably use some other form of error checking.

> **NOTE:** There is one condition that could cause an error to occur in the preceding SQLite code that is not a programmer error. If the device's storage is completely full—to the extent that SQLite can't save its changes to the database—then an error will occur here as well. However, this condition is fairly rare, and will probably result in deeper problems for the user, outside the scope of our app's data. Probably our app wouldn't even launch successfully if the system were in that state. So we're going to just sidestep the issue entirely.

Once we're finished with the loop, we close the database.

```
    sqlite3_close(database);
```

Why don't you compile and run the app? Enter some data, and then press the iPhone simulator's home button. Quit the simulator (to force the app to actually quit), and then relaunch the Persistence application. That data should be right where you left it. As far as the user is concerned, there's absolutely no difference between the various versions of this application, but each version uses a very different persistence mechanism.

Using Core Data

The final technique we're going to demonstrate in this chapter is how to implement persistence using Apple's Core Data framework. Core Data is a robust, full-featured persistence tool. Here, we will show you how to use Core Data to re-create the same persistence you've seen in our Persistence application so far.

> **NOTE:** For more comprehensive coverage of Core Data, check out *More iOS 5 Development* by Alex Horovitz and Kevin Kim (Apress, 2011). That book devotes several chapters to Core Data.

In Xcode, create a new project. This time, select the *Empty Application* template and click *Next*. Name the product *Core Data Persistence*, and select *iPhone* from the *Device Family* popup, but don't click the *Next* button just yet. If you look just below the *Device Family* popup, you should see a checkbox labeled *Use Core Data*. There's a certain amount of complexity involved in adding Core Data to an existing project, so Apple has kindly provided an option with some application project templates to do much of the work for you.

Check the *Use Core Data* checkbox (see Figure 13–5), and then click the *Next* button. When prompted, choose a directory to store your project, and click *Create*.

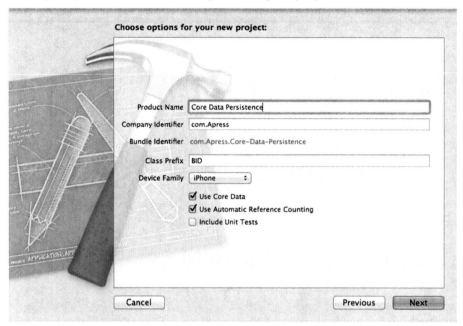

Figure 13–5. *Some project templates, including Empty Application, offer the option to use Core Data for persistence.*

Before we move on to our code, let's take a look at the project window, which contains some new stuff. Expand the *Core Date Persistence* and *Supporting Files* folders (see Figure 13–6).

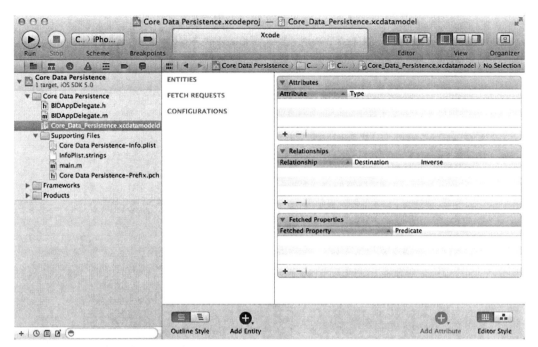

Figure 13-6. *Our project template with the files needed for Core Data. The Core Data model is selected, and the data model editor is shown in the editing pane.*

Entities and Managed Objects

Most of what you see in the project navigator should be familiar: the application delegate and the various files in the *Supporting Files* folder. In addition, you'll find a file called *Core_Data_Persistence.xcdatamodeld*, which contains our data model. Within Xcode, Core Data lets us design our data models visually, without writing code, and stores that data model in the *.xcdatamodeld* file.

Single-click the *.xcdatamodeld* file now, and you will be presented with the **data model editor** (see the right side of Figure 13-6). The data model editor gives you two distinct views into your data model, depending on the setting of the control in the lower-right corner of the project window. In Table mode, the mode shown in Figure 13-6, the elements that make up your data model will be shown in a series of editable tables. In Graph mode, you'll see a graphical depiction of the same elements. At the moment, both views reflect the same, empty data model.

Before Core Data, the traditional way to create data models was to create subclasses of NSObject and conform them to NSCoding and NSCopying so that they could be archived, as we did earlier in this chapter. Core Data uses a fundamentally different approach. Instead of classes, you create **entities** here in the data model editor, and then in your code, you create **managed objects** from those entities.

> **NOTE:** The terms *entity* and *managed object* can be a little confusing, since both refer to data model objects. *Entity* refers to the description of an object. *Managed object* refers to actual concrete instances of that entity created at runtime. So, in the data model editor, you create entities, but in your code, you create and retrieve managed objects. The distinction between entities and managed objects is similar to the distinction between a class and instances of that class.

An entity is made up of properties. There are three types of properties:

- **Attributes**: An attribute serves the same function in a Core Data entity as an instance variable does in an Objective-C class. They both hold the data.

- **Relationships**: As the name implies, a relationship defines the relationship between entities. For example, to create a Person entity, you might start by defining a few attributes, like hairColor, eyeColor, height, and weight. You might define address attributes, like state and ZIP code, or you might embed those in a separate, HomeAddress entity. Using the latter approach, you would then create a relationship between a Person and a HomeAddress. Relationships can be **to-one** and **to-many**. The relationship from Person to HomeAddress is probably to-one, since most people have only a single home address. The relationship from HomeAddress to Person might be to-many, since there may be more than one Person living at that HomeAddress.

- **Fetched properties**: A fetched property is an alternative to a relationship. Fetched properties allow you to create a query that is evaluated at fetch time to see which objects belong to the relationship. To extend our earlier example, a Person object could have a fetched property called Neighbors that finds all HomeAddress objects in the data store that have the same ZIP code as the Person's own HomeAddress. Due to the nature of how fetched properties are constructed and used, they are always one-way relationships. Fetched properties are also the only kind of relationship that lets you traverse multiple data stores.

Typically, attributes, relationships, and fetched properties are defined using Xcode's data model editor. In our Core Data Persistence application, we'll build a simple entity so you can get a sense of how this all works together.

Key-Value Coding

In your code, instead of using accessors and mutators, you will use **key-value coding** to set properties or retrieve their existing values. Key-value coding may sound intimidating, but you've already used it quite a bit in this book. Every time we used NSDictionary, for example, we were using key-value coding, because every object in a dictionary is stored under a unique key value. The key-value coding used by Core Data is a bit more complex than that used by NSDictionary, but the basic concept is the same.

When working with a managed object, the key you will use to set or retrieve a property's value is the name of the attribute you wish to set. So, here's how to retrieve the value stored in the attribute called name from a managed object:

```
NSString *name = [myManagedObject valueForKey:@"name"];
```

Similarly, to set a new value for a managed object's property, do this:

```
[myManagedObject setValue:@"Gregor Overlander" forKey:@"name"];
```

Putting It All in Context

So, where do these managed objects live? They live in something called a **persistent store**, also referred to as a **backing store**. Persistent stores can take several different forms. By default, a Core Data application implements a backing store as an SQLite database stored in the application's *Documents* directory. Even though your data is stored via SQLite, classes in the Core Data framework do all the work associated with loading and saving your data. If you use Core Data, you don't need to write any SQL statements. You just work with objects, and Core Data figures out what it needs to do behind the scenes.

SQLite isn't the only option Core Data has for storage. Backing stores can also be implemented as binary flat files, or even stored in an XML format. Another option is to create an in-memory store, which you might use if you're writing a caching mechanism, but it doesn't save data beyond the end of the current session. In almost all situations, you should just leave it as the default and use SQLite as your persistent store.

Although most applications will have only one persistent store, it is possible to have multiple persistent stores within the same application. If you're curious about how the backing store is created and configured, take a look at the file *BIDAppDelegate.m* in your Xcode project. The Xcode project template we chose provided us with all the code needed to set up a single persistent store for our application.

Other than creating it (which is handled for you in your application delegate), you generally won't work with your persistent store directly, but rather will use something called a **managed object context**, often referred to as just a **context**. The context manages access to the persistent store and maintains information about which properties have changed since the last time an object was saved. The context also registers all changes with the undo manager, meaning that you always have the ability to undo a single change or roll back all the way to the last time data was saved.

> **NOTE:** You can have multiple contexts pointing to the same persistent store, though most iOS applications will use only one. You can find out more about using multiple contexts and the undo manager in the Apress book *More iOS 5 Development*.

Many Core Data method calls require an NSManagedObjectContext as a parameter or must be executed against a context. With the exception of very complicated, multithreaded iOS applications, you can just use the managedObjectContext property provided by your application delegate, which is a default context that is created for you automatically, also courtesy of the Xcode project template.

You may notice that, in addition to a managed object context and a persistent store coordinator, the provided application delegate also contains an instance of NSManagedObjectModel. This class is responsible for loading and representing, at runtime, the data model you will create using the data model editor in Xcode. You generally won't need to interact directly with this class. It's used behind the scenes by the other Core Data classes so they can identify which entities and properties you've defined in your data model. As long as you create your data model using the provided file, there's no need to worry about this class at all.

Creating New Managed Objects

Creating a new instance of a managed object is pretty easy, though not quite as straightforward as creating a normal object instance using alloc and init. Instead, you use the insertNewObjectForEntityForName:inManagedObjectContext: factory method in a class called NSEntityDescription. NSEntityDescription's job is to keep track of all the entities defined in the app's data model. This method returns an instance representing a single entity in memory. It returns either an instance of NSManagedObject that is set up with the correct properties for that particular entity or, if you've configured your entity to be implemented with a specific subclass of NSManagedObject, an instance of that class. Remember that entities are like classes. An entity is a description of an object and defines which properties a particular entity has.

To create a new object, do this:

```
theLine = [NSEntityDescription
    insertNewObjectForEntityForName:@"EntityName"
             inManagedObjectContext:context];
```

The method is called insertNewObjectForEntityForName:inManagedObjectContext: because, in addition to creating the object, it inserts the newly created object into the context and then returns that object. After this call, the object exists in the context but is not yet part of the persistent store. The object will be added to the persistent store the next time the managed object context's save: method is called.

Retrieving Managed Objects

To retrieve managed objects from the persistent store, you'll make use of a **fetch request**, which is Core Data's way of handling a predefined query. For example, you might say, "Give me every Person whose eyeColor is blue."

After first creating a fetch request, you provide it with an NSEntityDescription that specifies the entity of the object or objects you wish to retrieve. Here is an example that creates a fetch request:

```
NSFetchRequest *request = [[NSFetchRequest alloc] init];
NSEntityDescription *entityDescr = [NSEntityDescription
    entityForName:@"EntityName" inManagedObjectContext:context];
[request setEntity:entityDescr];
```

Optionally, you can also specify criteria for a fetch request using the NSPredicate class. A **predicate** is similar to the SQL WHERE clause and allows you to define the criteria used to determine the results of your fetch request. Here is a simple example of a predicate:

```
NSPredicate *pred = [NSPredicate predicateWithFormat:@"(name = %@)", nameString];
[request setPredicate: pred];
```

The predicate created by the first line of code tells a fetch request that, instead of retrieving all managed objects for the specified entity, retrieve just those where the name property is set to the value currently stored in the nameString variable. So, if nameString is an NSString that holds the value @"Bob", we are telling the fetch request to bring back only managed objects that have a name property set to "Bob". This is a simple example, but predicates can be considerably more complex and can use Boolean logic to specify the precise criteria you might need in most any situation.

> **NOTE:** *Learn Objective-C on the Mac* by Mark Dalrymple and Scott Knaster (Apress, 2009) has an entire chapter devoted to the use of NSPredicate.

After you've created your fetch request, provided it with an entity description, and optionally given it a predicate, you **execute** the fetch request using an instance method on NSManagedObjectContext:

```
NSError *error;
NSArray *objects = [context executeFetchRequest:request error:&error];
if (objects == nil) {
    // handle error
}
```

executeFetchRequest:error: will load the specified objects from the persistent store and return them in an array. If an error is encountered, you will get a nil array, and the error pointer you provided will point to an NSError object that describes the specific problem. Otherwise, you will get a valid array, though it may not have any objects in it, since it is possible that none meet the specified criteria. From this point on, any changes you make to the managed objects returned in that array will be tracked by the managed object context you executed the request against, and saved when you send that context a save: message.

The Core Data Application

Let's take Core Data for a spin now. First, we'll return our attention to Xcode and create our data model.

Designing the Data Model

Select *Core_Data_Persistence.xcdatamodel* to open Xcode's data model editor. The data model editing pane shows all the entities, fetch requests, and configurations that are contained within your data model.

> **NOTE:** The Core Data concept of **configurations** lets you define one or more named subsets of the entities contained in your data model, which can be useful in certain situations. For example, if you want to create a suite of apps that share the same data model, but some apps shouldn't have access to everything (perhaps there's one app for normal users and another for sysadmins), this approach lets you do that. You can also use multiple configurations within a single app as it switches between different modes of operation. In this book, we're not going to deal with configurations at all, but since the list of configurations (including the single default configuration that contains everything in your model) is right there, staring you in the face beneath the entities and fetch requests, we thought it was worth a mention here.

As shown in Figure 13–6, those lists are empty now, because we haven't created anything yet. Remedy that by clicking the plus icon labeled *Add Entity* in the lower-left corner of the entity pane. This will create a brand-new entity with the name *Entity* (see Figure 13–7).

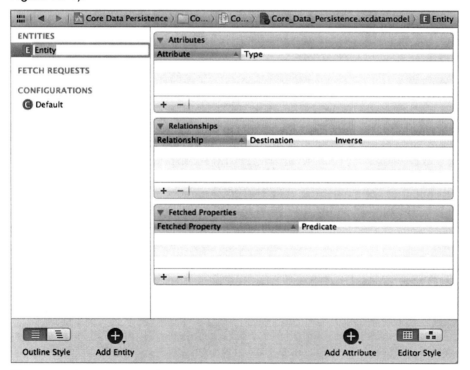

Figure 13–7. *The data model editor, showing our newly added entity*

As you build your data model, you'll probably find yourself switching between Table view and Graph view using the control at the bottom right of the editing area. Switch to Graph view now. Graph view presents a little box representing our entity, which itself contains sections for showing the entity's attributes and relationships, also currently empty (see Figure 13–8). Graph view is really useful if your model contains multiple entities, since it shows a graphic representation of all the relationships between your entities.

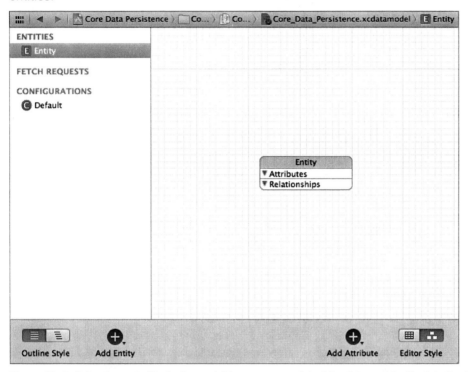

Figure 13–8. *Using the control in the lower-right corner, we switched the data model editor into Graph mode. Note that Graph mode shows the same entities as Table mode, just in a graphic form. This is useful if you have multiple entities with relationships between them.*

> **NOTE:** If you prefer working graphically, you can actually build your entire model in Graph view. We're going to stick with Table view in this chapter because it's easier to explain. When you're creating your own data models, feel free to work in Graph view if that approach suits you better.

Whether you're using Table view or Graph view for designing your data model, you'll almost always want to bring up the Core Data data model inspector. This inspector lets you view and edit relevant details for whatever item is selected in the data model editor—whether it's an entity, attribute, relationship, or anything else. You can browse an existing model without the data model inspector, but to really work on a model, you'll invariably need to use this inspector, much like you frequently use the attributes inspector when editing nib files.

Press **Style** for menu shortcut to open the data model inspector. At the moment, the inspector shows information about the entity we just added. Change the *Name* field from *Entity* to *Line* (see Figure 13–9).

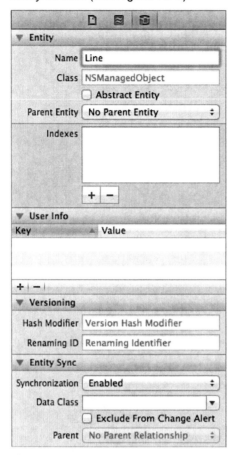

Figure 13–9. *Using the data model inspector to change our entity's name to Line*

If you're currently in Graph view, switch to Table view now. Table view shows more details for each piece of the entity we're working on, so it's usually more useful than Graph view when creating a new entity. In Table view, most of the data model editor is taken up by the table showing the entity's attributes, relationships, and fetched properties. This is where we'll set up our entity.

Notice that at the lower right of the editing area, there's an icon containing a plus sign, similar to the one at the lower left, which you used to create the entity. If you select your entity, and then click the plus sign and hold down the mouse button, a popup menu will appear, allowing you to add an attribute, relationship, or fetched property to your entity (see Figure 13–10).

NOTE: Notice that you don't need to press and hold to add an attribute. You'll get the same result if you click the plus icon. Shortcut!

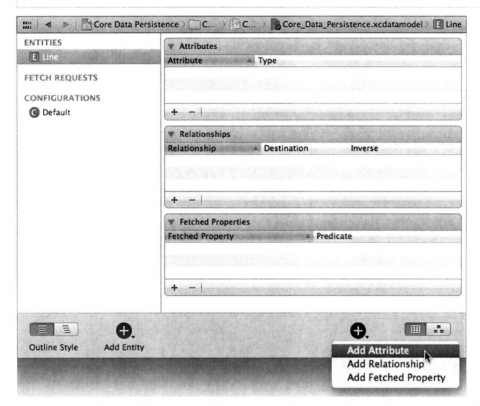

Figure 13–10. *With an entity selected, press and hold the right plus sign icon to add an attribute, relationship, or fetched property to your entity.*

Go ahead and use this technique to add an attribute to your *Line* entity. A new attribute, creatively named *attribute*, is added to the *Attributes* section of the table and selected. In the table, you'll see that not only is the row selected, but the attribute's name is selected as well. This means that immediately after clicking the plus sign, you can start typing the name of the new attribute without further clicking.

Change the new attribute's name from *attribute* to *lineNum,* and click the popup next to the name to change its *Type* from *Undefined* to *Int16*, which turns this attribute into one that will hold an integer value. We will be using this attribute to identify for which of the four fields the managed object holds data. Since we have only four options, we selected the smallest integer type available.

Now direct your attention to the data model inspector, where additional details can be configured. The checkbox below the *Name* field on the right, *Optional*, is selected by default. Click it to deselect it. We don't want this attribute to be optional—a line that doesn't correspond to a label on our interface is useless.

Selecting the *Transient* checkbox creates a transient attribute, which is used to specify a value that is held by managed objects while the app is running, but never saved to the data store. Since we do want the line number saved to the data store, leave the *Transient* checkbox unchecked.

Selecting the *Indexed* checkbox will cause an index in the underlying SQL database to be created on the column that holds this attribute's data. Leave the *Indexed* checkbox unchecked. Since the amount of data is small, and we won't provide the user with a search capability, there's no need for an index.

Beneath that are more settings, allowing you to do some simple data validation by specifying minimum and maximum values for the integer, a default value, and more. We won't be using any of these settings in this example.

Now, make sure the *Line* entity is selected and click the plus sign to add a second attribute. Change the name of your new attribute to *lineText* and change its *Type* to *String*. This attribute will hold the actual data from the text field. Leave the *Optional* checkbox checked for this one; it is altogether possible that the user won't enter a value for a given field.

> **NOTE:** When you change the *Type* to *String*, you'll notice that the inspector shows a slightly different set of options for setting a default value or limiting the length of the string. Although we won't be using any of those options for this application, it's nice to know they're there.

Guess what? Your data model is complete. That's all there is to it. Core Data lets you point and click your way to an application data model. Let's finish building the application so you can see how to use our data model from our code.

Creating the Persistence View and Controller

Because we selected the *Empty Application* template, we weren't provided with a view controller. Go back to the project navigator, single-click the *Core Data Persistence* folder, and press ⌘N or select File ➤ New ➤ New File… to bring up the new file assistant. Select *UIViewController subclass* from the *Cocoa Touch* heading, and click *Next*. On the next sheet, name the class *BIDViewController,* select a *Subclass of UIViewController*, make sure the box labeled *Targeted for iPad* is unchecked, and check the box that says *With XIB for user interface* to have Xcode create a nib file automatically. Click *Next*, and choose the directory in which to save the file. When you're finished, *BIDViewController.h, BIDViewController.m,* and *BIDViewController.xib* will be placed in your *Core Data Persistence* folder.

Select *BIDViewController.h*, and make the following changes, which should look very familiar to you:

```
#import <UIKit/UIKit.h>

@interface BIDViewController : UIViewController

@property (weak, nonatomic) IBOutlet UITextField *line1;
@property (weak, nonatomic) IBOutlet UITextField *line2;
@property (weak, nonatomic) IBOutlet UITextField *line3;
@property (weak, nonatomic) IBOutlet UITextField *line4;
@end
```

Save this file. Next, select *BIDViewController.xib* to edit the GUI in Interface Builder. Design the view, and connect the outlets by following the instructions in the "Designing the Persistence Application View" section earlier in this chapter. Once your design is complete, select the view, bring up the attributes inspector, and select *Light Gray Color* from the *Background* popup menu. You might also find it useful to refer back to Figure 13–3. Once you've created the view, save the nib file.

In *BIDViewController.m*, insert the following code at the top of the file:

```
#import "BIDViewController.h"
#import "BIDAppDelegate.h"

@implementation BIDViewController
@synthesize line1;
@synthesize line2;
@synthesize line3;
@synthesize line4;
```

Then insert the following code into the existing viewDidLoad and viewDidUnload methods:

```
- (void)viewDidLoad {
    [super viewDidLoad];
    // Do any additional setup after loading the view from its nib.
    BIDAppDelegate *appDelegate =
        [[UIApplication sharedApplication] delegate];
    NSManagedObjectContext *context = [appDelegate managedObjectContext];
    NSEntityDescription *entityDescription = [NSEntityDescription
                 entityForName:@"Line"
        inManagedObjectContext:context];
    NSFetchRequest *request = [[NSFetchRequest alloc] init];
    [request setEntity:entityDescription];

    NSError *error;
    NSArray *objects = [context executeFetchRequest:request error:&error];
    if (objects == nil) {
        NSLog(@"There was an error!");
        // Do whatever error handling is appropriate
    }

    for (NSManagedObject *oneObject in objects) {
```

```
            NSNumber *lineNum = [oneObject valueForKey:@"lineNum"];
            NSString *lineText = [oneObject valueForKey:@"lineText"];

            NSString *fieldName = [NSString
                stringWithFormat:@"line%d", [lineNum integerValue]];
            UITextField *theField = [self valueForKey:fieldName];
            theField.text = lineText;
        }

    UIApplication *app = [UIApplication sharedApplication];
    [[NSNotificationCenter defaultCenter] addObserver:self
        selector:@selector(applicationWillResignActive:)
        name:UIApplicationWillResignActiveNotification
        object:app];
}

- (void)viewDidUnload {
    [super viewDidUnload];
    // Release any retained subviews of the main view.
    // e.g. self.myOutlet = nil;
    self.line1 = nil;
    self.line2 = nil;
    self.line3 = nil;
    self.line4 = nil;
}
```

Then add the following new method down at the bottom, just before the @end marker:

```
- (void)applicationWillResignActive:(NSNotification *)notification {
    BIDAppDelegate *appDelegate = [[UIApplication sharedApplication] delegate];
    NSManagedObjectContext *context = [appDelegate managedObjectContext];
    NSError *error;
    for (int i = 1; i <= 4; i++) {
        NSString *fieldName = [NSString stringWithFormat:@"line%d", i];
        UITextField *theField = [self valueForKey:fieldName];

        NSFetchRequest *request = [[NSFetchRequest alloc] init];

        NSEntityDescription *entityDescription = [NSEntityDescription
            entityForName:@"Line"
            inManagedObjectContext:context];
        [request setEntity:entityDescription];
        NSPredicate *pred = [NSPredicate
            predicateWithFormat:@"(lineNum = %d)", i];
        [request setPredicate:pred];

        NSManagedObject *theLine = nil;

        NSArray *objects = [context executeFetchRequest:request
            error:&error];

        if (objects == nil) {
            NSLog(@"There was an error!");
            // Do whatever error handling is appropriate
        }
```

```
        if ([objects count] > 0)
            theLine = [objects objectAtIndex:0];
        else
            theLine = [NSEntityDescription
                    insertNewObjectForEntityForName:@"Line"
                            inManagedObjectContext:context];

        [theLine setValue:[NSNumber numberWithInt:i] forKey:@"lineNum"];
        [theLine setValue:theField.text forKey:@"lineText"];

    }
    [context save:&error];
}
.
.
.
```

Now, let's look at the viewDidLoad method, which needs to check if there is any existing data in the persistent store. If there is, it should load the data and populate the fields with it. The first thing we do in that method is to get a reference to our application delegate, which we then use to get the managed object context that was created for us.

```
BIDAppDelegate *appDelegate = UIApplication sharedApplication] delegate];
NSManagedObjectContext *context = [appDelegate managedObjectContext];
```

Next, we create an entity description that describes our entity.

```
NSEntityDescription *entityDescription = [NSEntityDescription
            entityForName:@"Line"
    inManagedObjectContext:context];
```

The next order of business is to create a fetch request and pass it the entity description so it knows which type of objects to retrieve.

```
NSFetchRequest *request = [[NSFetchRequest alloc] init];
[request setEntity:entityDescription];
```

Since we want to retrieve all Line objects in the persistent store, we do not create a predicate. By executing a request without a predicate, we're telling the context to give us every Line object in the store.

```
NSError *error;
NSArray *objects = [context executeFetchRequest:request error:&error];
```

We make sure we got back a valid array, and log it if we didn't.

```
if (objects == nil) {
    NSLog(@"There was an error!");
    // Do whatever error handling is appropriate
}
```

Next, we use fast enumeration to loop through the array of retrieved managed objects, pull the lineNum and lineText values from it, and use that information to update one of the text fields on our user interface.

```
for (NSManagedObject *oneObject in objects) {
    NSNumber *lineNum = [oneObject valueForKey:@"lineNum"];
    NSString *lineText = [oneObject valueForKey:@"lineText"];
```

```
    NSString *fieldName = [NSString stringWithFormat:@"line%@",
        lineNum];
    UITextField *theField = [self valueForKey:fieldName];
    theField.text = lineText;
}
```

Then, just as with all the other applications in this chapter, we register to be notified when the application is about to move out of the active state (either being shuffled to the background or exited completely), so we can save any changes the user has made to the data.

```
UIApplication *app = [UIApplication sharedApplication];
[[NSNotificationCenter defaultCenter] addObserver:self
    selector:@selector(applicationWillResignActive:)
        name:UIApplicationWillResignActiveNotification
      object:app];
[super viewDidLoad];
```

Let's look at applicationWillResignActive: next. We start out the same way as the previous method, by getting a reference to the application delegate and using that to get a pointer to our application's default context.

```
BIDAppDelegate *appDelegate = [[UIApplication sharedApplication] delegate];
NSManagedObjectContext *context = [appDelegate managedObjectContext];
```

After that, we go into a loop that executes four times, one time for each label.

```
for (int i = 1; i <= 4; i++) {
```

We construct the name of one of the four fields by appending i to the word line and use that to get a reference to the correct field using valueForKey:.

```
    NSString *fieldName = [NSString stringWithFormat:@"line%d", i];
    UITextField *theField = [self valueForKey:fieldName];
```

Next, we create our fetch request:

```
    NSFetchRequest *request = [[NSFetchRequest alloc] init];
```

After that, we create an entity description that describes the *Line* entity we designed earlier in the data model editor and that uses the context we retrieved from the application delegate. Once we create the description, we feed it to the fetch request, so the request knows which type of entity to look for.

```
    NSEntityDescription *entityDescription = [NSEntityDescription
                entityForName:@"Line"
        inManagedObjectContext:context];
    [request setEntity:entityDescription];
```

Next, we need to find out if there's already a managed object in the persistent store that corresponds to this field, so we create a predicate that identifies the correct object for the field.

```
    NSPredicate *pred = [NSPredicate
        predicateWithFormat:@"(lineNum = %d)", i];
    [request setPredicate:pred];
```

After that, we declare a pointer to an NSManagedObject and set it to nil. We do this because we don't know yet if we're going to load a managed object from the persistent store or create a new one. We also declare an NSError that the system will use to notify us of the specific nature of the problem if we get back a nil array.

```
NSManagedObject *theLine = nil;
NSError *error;
```

Next, we execute the fetch request against the context.

```
NSArray *objects = [context executeFetchRequest:request
    error:&error];
```

Then we check to make sure that objects is not nil. If it is nil, then there was an error, and we should do whatever error checking is appropriate for our application. For this simple application, we're just logging the error and moving on.

```
if (objects == nil) {
    NSLog(@"There was an error!");
    // Do whatever error handling is appropriate
}
```

After that, we check if an object that matched our criteria was returned. If there is one, we load it. If there isn't one, we create a new managed object to hold this field's text.

```
if ([objects count] > 0)
    theLine = [objects objectAtIndex:0];
else
    theLine = [NSEntityDescription
        insertNewObjectForEntityForName:@"Line"
                inManagedObjectContext:context];
```

Then we use key-value coding to set the line number and text for this managed object.

```
    [theLine setValue:[NSNumber numberWithInt:i] forKey:@"lineNum"];
    [theLine setValue:theField.text forKey:@"lineText"];
}
```

Finally, once we're finished looping, we tell the context to save its changes.

```
    [context save:&error];
}
```

Making the Persistence View Controller the Application's Root Controller

Because we used the Empty Application template instead of the Single View Application template, we have one more step to take before our fancy new Core Data application will work. We need to create an instance of BIDViewController to act as our application's root controller and add its view as a subview of our application's main window. Let's do that now.

First, the application delegate needs a property to point to our view controller. Select *BIDAppDelegate.h*, and make the following changes to declare that property:

```
#import <UIKit/UIKit.h>

@class BIDViewController;

@interface BIDAppDelegate : UIResponder <UIApplicationDelegate>

@property (strong, nonatomic) IBOutlet UIWindow *window;

@property (readonly, strong, nonatomic) NSManagedObjectContext
    *managedObjectContext;
@property (readonly, strong, nonatomic) NSManagedObjectModel
    *managedObjectModel;
@property (readonly, strong, nonatomic) NSPersistentStoreCoordinator
    *persistentStoreCoordinator;
@property (strong, nonatomic) BIDViewController *rootController;

- (void)saveContext;
- (NSURL *)applicationDocumentsDirectory;

@end
```

To make the root controller's view a subview of the application's window so that the user can interact with it, switch to *BIDAppDelegate.m*, and make the following changes at the top of that file:

```
#import "BIDAppDelegate.h"
#import "BIDViewController.h"

@implementation BIDAppDelegate

@synthesize window = _window;
@synthesize managedObjectContext = __managedObjectContext;
@synthesize managedObjectModel = __managedObjectModel;
@synthesize persistentStoreCoordinator = __persistentStoreCoordinator;
@synthesize rootController;

- (BOOL)application:(UIApplication *)application
    didFinishLaunchingWithOptions:(NSDictionary *)launchOptions {
    self.window = [[UIWindow alloc] initWithFrame:[[UIScreen mainScreen] bounds]];
    // Override point for customization after application launch.
    self.rootController = [[BIDViewController alloc]
                          initWithNibName:@"BIDViewController" bundle:nil];

    UIView *rootView = self.rootController.view;
    CGRect viewFrame = rootView.frame;
    viewFrame.origin.y += [UIApplication
        sharedApplication].statusBarFrame.size.height;
    rootView.frame = viewFrame;
    [self.window addSubview:rootView];
    self.window.backgroundColor = [UIColor whiteColor];
    [self.window makeKeyAndVisible];
    return YES;
}
    .
    .
    .
```

That's it! Build and run the app to make sure it works. The Core Data version of your application should behave exactly the same as the previous versions.

It may seem that Core Data entails a lot of work and, for a simple application like this, doesn't offer much of an advantage. But in more complex applications, Core Data can substantially decrease the amount of time you spend designing and writing your data model.

Persistence Rewarded

You should now have a solid handle on four different ways of preserving your application data between sessions—five ways if you include the user defaults that you learned how to use in the previous chapter. We built an application that persisted data using property lists and modified the application to save its data using object archives. We then made a change and used the iOS's built-in SQLite3 mechanism to save the application data. Finally, we rebuilt the same application using Core Data. These mechanisms are the basic building blocks for saving and loading data in almost all iOS applications.

Ready for more? In the next chapter, we're going to continue talking about saving and loading data, and introduce you to iOS 5's document system. This system not only provides a nice abstraction for dealing with saving and loading documents in files stored on your device, but it also lets you save your documents to Apple's iCloud, one of the biggest new features in iOS 5.

Chapter 14

Hey! You! Get onto iCloud!

One of the biggest new features touted with the announcement of iOS 5 is Apple's new iCloud service, which provides cloud storage services for iOS 5 devices, as well as for computers running Mac OS X and Microsoft Windows. Most iOS users will probably encounter the iCloud device backup option immediately when setting up a new device or upgrading an old device to iOS 5, and will quickly discover the advantages of automatic backup that doesn't even require the use of a computer.

Computerless backup is a great feature, but it only scratches the surface of what iCloud can do. What may be even a bigger feature of iCloud is that it provides app developers with a mechanism for transparently saving data to Apple's cloud servers with very little effort. You can make your apps save data to iCloud, and have that data automatically transfer to any other devices that are registered to the same iCloud user. Users may create a document on their iPad, and later view the same document on their iPhone without any intervening steps; the document just appears.

A system process takes care of making sure the user has a valid iCloud login and manages the file transfers, so you don't need to worry about networks or authentication. Apart from a small amount of app configuration, just a few small changes to your methods for saving files and locating available files will get you well on your way to having an iCloud-backed app.

One key component of the iCloud filing system is the UIDocument class, which is also new to iOS 5. UIDocument takes a portion of the work out of creating a document-based app, by handling some of the common aspects of reading and writing files. That way, you can spend more of your time focusing on the unique features of your app, instead of building the same plumbing for every app you create.

Whether you're using iCloud or not, UIDocument provides some powerful tools for managing document files in iOS. To demonstrate these features, the first portion of this chapter is dedicated to creating TinyPix, a simple document-based app that saves files to local storage. This is an approach that can work well for all kinds of iOS-based apps.

Later in this chapter, we'll show you how to iCloud-enable TinyPix. For that to work, you'll need to have one or more iCloud-connected iOS devices at hand. You'll also need a paid iOS developer account so that you can install on devices, because apps running in the simulator don't have access to iCloud services.

Managing Document Storage with UIDocument

Anyone who has used a desktop computer for anything besides just surfing the Web has probably worked with a document-based application. From TextEdit to Microsoft Word to GarageBand to Xcode, any piece of software that lets you deal with multiple collections of data, saving each collection to a separate file, could be considered a document-based application. Often, there's a one-to-one correspondence between an on-screen window and the document it contains, but sometimes (such as in Xcode), a single window can display multiple files that are all related in some way.

On iOS devices, we don't have the luxury of multiple windows, but plenty of apps can still benefit from a document-based approach. Now iOS developers have a little boost in making it work thanks to the UIDocument class, which takes care of the most common aspects of document file storage. You won't need to deal with files directly (just URLs), and all the necessary reading and writing happen on a background thread, so your app can remain responsive even while file access is occurring. It also automatically saves edited documents periodically and whenever the app is suspended (such as when the device is shut down, the home button is pressed, and so on), so there's no need for any sort of save button. All of this helps make your apps behave the way users expect their iOS apps to behave.

Building TinyPix

We're going to build an app called TinyPix that lets you edit simple 8 × 8 images, in glorious 1-bit color (see Figure 14–1)! For the user's convenience, each picture is blown up to the full-screen size for editing. And, of course, we'll be using UIDocument to represent the data for each image.

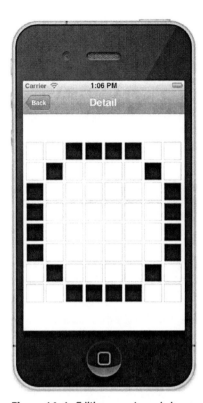

Figure 14-1. *Editing an extremely low-resolution icon in TinyPix*

Start off by creating a new project in Xcode. From the *iOS Application* section, select the *Master-Detail Application* template, and then click *Next*. Name this new app *TinyPix*, set the *Device Family* popup to *iPhone*, and make sure the *Use Storyboard* checkbox is checked. Then click *Next* again, and choose the location to save your project.

In Xcode's project navigator, you'll see that your project contains files for BIDAppDelegate, BIDMasterViewController, and BIDDetailViewController, as well as the *MainStoryboard.storyboard* file. We'll make changes to most of these files to some extent, and we will create a few new classes along the way as well.

Creating BIDTinyPixDocument

The first new class we're going to create is the document class that will contain the data for each TinyPix image that's loaded from file storage. Select the *TinyPix* folder in Xcode, and press ⌘N to create a new file. From the *iOS Cocoa Touch* section, select *Objective-C class*, and click *Next*. Enter *BIDTinyPixDocument* in the *Class* field, enter *UIDocument* in the *Subclass of* field, and click *Next*. Then click *Create* to create the files.

Let's think about the public API of this class before we get into its implementation details. This class is going to represent an 8 × 8 grid of pixels, where each pixel consists of a single on or off value. So, let's give it a method that takes a pair of grid and column indexes and returns a BOOL value. Let's also provide a method to set a specific state at a specified grid and column and, as a convenience, another method that simply toggles the state at a particular place.

Select *BIDTinyPixDocument.h* to edit the new class's header. Add the following bold lines:

```
#import <UIKit/UIKit.h>

@interface BIDTinyPixDocument : UIDocument

// row and column range from 0 to 7
- (BOOL)stateAtRow:(NSUInteger)row column:(NSUInteger)column;
- (void)setState:(BOOL)state atRow:(NSUInteger)row column:(NSUInteger)column;
- (void)toggleStateAtRow:(NSUInteger)row column:(NSUInteger)column;

@end
```

Now switch over to *BIDTinyPixDocument.m*, where we'll implement storage for our 8 × 8 grid, the methods defined in our public API, and the required UIDocument methods that will enable loading and saving our documents.

Let's start by defining the storage for our 8 × 8 bitmap data. We'll hold this data in an instance of NSMutableData, which lets us work directly with an array of byte data that is still contained inside an object, so that the usual Cocoa memory management will take care of freeing the memory when we're finished with it. Add this class extension and property synthesis to make it happen:

```
#import "BIDTinyPixDocument.h"

@interface BIDTinyPixDocument ()
@property (strong, nonatomic) NSMutableData *bitmap;
@end

@implementation BIDTinyPixDocument
@synthesize bitmap;

@end
```

The UIDocument class has a designated initializer that all subclasses should use. This is where we'll create our initial bitmap. In true bitmap style, we're going to minimize memory usage by using a single byte to contain each row. Each bit in the byte represents the on/off value of a column index within that row. In total, our document contains just 8 bytes.

> **NOTE:** This section contains a small amount of bitwise operations, as well as some C pointer and array manipulation. This is all pretty mundane for C developers, but if you don't have much C experience, it may seem puzzling or even impenetrable. In that case, feel free to simply copy and use the code provided (it works just fine). If you want to really understand what's going on, you may want to dig deeper into C itself, perhaps by adding a copy of *Learn C on the Mac* by Dave Mark (Apress, 2009) to your bookshelf.

Add this method to our document's implementation, placing it directly above the @end at the bottom of the file:

```
- (id)initWithFileURL:(NSURL *)url {
    self = [super initWithFileURL:url];
    if (self) {
        unsigned char startPattern[] = {
            0x01,
            0x02,
            0x04,
            0x08,
            0x10,
            0x20,
            0x40,
            0x80
        };

        self.bitmap = [NSMutableData dataWithBytes:startPattern length:8];
    }
    return self;
}
```

This starts off each bitmap with a simple diagonal pattern stretching from one corner to another.

Now, it's time to implement the methods that make up the public API we defined in the header. Let's tackle the method for reading the state of a single bit first. This simply grabs the relevant byte from our array of bytes, then does a bit shift and an AND operation to determine if the specified bit was set, and returns YES or NO accordingly. Add this method above the @end:

```
- (BOOL)stateAtRow:(NSUInteger)row column:(NSUInteger)column {
    const char *bitmapBytes = [bitmap bytes];
    char rowByte = bitmapBytes[row];
    char result = (1 << column) & rowByte;
    if (result != 0)
        return YES;
    else
        return NO;
}
```

Next comes the inverse: a method that sets the value specified at a given row and column. Here, we once again grab a pointer to the relevant byte for the specified row and do a bit shift. But this time, instead of using the shifted bit to examine the contents of the row, we use it to either set or unset a bit in the row. Add this method above the @end:

```
- (void)setState:(BOOL)state atRow:(NSUInteger)row column:(NSUInteger)column {
    char *bitmapBytes = [bitmap mutableBytes];
    char *rowByte = &bitmapBytes[row];

    if (state)
        *rowByte = *rowByte | (1 << column);
    else
        *rowByte = *rowByte & ~(1 << column);
}
```

Now, let's add the convenience method, which lets outside code simply toggle a single cell. Add this method above the @end:

```
- (void)toggleStateAtRow:(NSUInteger)row column:(NSUInteger)column {
    BOOL state = [self stateAtRow:row column:column];
    [self setState:!state atRow:row column:column];
}
```

Our document class requires two final pieces before it fits into the puzzle of a document-based app: methods for reading and writing. As we mentioned earlier, you don't need to deal with files directly. You don't even need to worry about the URL that was passed into the `initWithFileURL:` method earlier. All that'you need to do is implement one method that transforms the document's data structure into an NSData object, ready for saving, and another that takes a freshly loaded NSData object and pulls the object's data structure out of it. Since our document's internal structure is already contained in an NSMutableData object, which is a subclass of NSData, these implementations are pleasingly simple. Add these two methods above the @end:

```
- (id)contentsForType:(NSString *)typeName error:(NSError **)outError {
    NSLog(@"saving document to URL %@", self.fileURL);
    return [bitmap copy];
}
```

```
- (BOOL)loadFromContents:(id)contents ofType:(NSString *)typeName
        error:(NSError **)outError {
    NSLog(@"loading document from URL %@", self.fileURL);
    self.bitmap = [contents mutableCopy];
    return true;
}
```

The first of these methods, `contentsForType:error:`, is called whenever our document is about to be saved to storage. It simply returns an immutable copy of our bitmap data, which the system will take care of storing later.

The second method, `loadFromContents:ofType:error:`, is called whenever the system has just loaded data from storage and wants to provide this data to an instance of our document class. Here, we just grab a mutable copy of the data that has been passed in. We've included some logging statements, just so you can see what's happening in the Xcode log later on.

Each of these methods allows you to do some things that we're ignoring in this app. They both provide a typeName parameter, which you could use to distinguish between different types of data storage that your document can load from or save to. They also have an outError parameter, which you could use to specify that an error occurred while copying data to or from your document's in-memory data structure. In our case, however, what we're doing is so simple that these aren't important concerns.

That's all we need for our document class. Sticking to MVC principles, our document sits squarely in the model camp, knowing nothing about how it's displayed. And thanks to the UIDocument superclass, the document is even shielded from most of the details about how it's stored.

Code Master

Now that we have our document class ready to go, it's time to address the first view that a user sees when running our app: the list of existing TinyPix documents, which is taken care of by the BIDMasterViewController class. We need to let this class know how to grab the list of available documents, create and name a new document, and let the user choose an existing document. When a document is created or chosen, it's then passed along to the detail controller for display.

Start off by selecting *BIDMasterViewController.h*, where we'll make a few changes. We're going to use an alert panel later on to let the user name a new document, so we want to declare that this class implements the relevant delegate protocol.

We'll also include a segmented control in our GUI, which will allow the user to choose the color that will be used to display the TinyPix pixels. Though this is not a particularly useful feature in and of itself, it will help demonstrate the iCloud mechanism, as the highlight color setting makes its way from the device on which you set it to another of your connected devices running the same app. The first version of the app will use the color as a per-device setting. Later in the chapter, we'll add the code to make the color setting propagate through iCloud to the user's other devices.

To implement the color segmented control, we'll add an outlet and an action to our code as well. Make these changes to *BIDMasterViewController.h*:

```
#import <UIKit/UIKit.h>

@interface BIDMasterViewController : UITableViewController
@property (weak, nonatomic) IBOutlet UISegmentedControl *colorControl;
- (IBAction)chooseColor:(id)sender;

@end
```

Now, switch over to *BIDMasterViewController.m*. We're going to start by importing the header for our document class, adding some private properties and methods (for later use) in a class extension, and synthesizing accessors for the new properties we just added.

```
#import "BIDMasterViewController.h"
#import "BIDTinyPixDocument.h"
```

```
@interface BIDMasterViewController () <UIAlertViewDelegate>
@property (strong, nonatomic) NSArray *documentFilenames;
@property (strong, nonatomic) BIDTinyPixDocument *chosenDocument;
- (NSURL *)urlForFilename:(NSString *)filename;
- (void)reloadFiles;
@end

@implementation BIDMasterViewController
@synthesize colorControl;
@synthesize documentFilenames;
@synthesize chosenDocument;
    .
    .
    .
```

Let's take care of those private methods right away. The first of these takes a file name, combines it with the file path of the app's *Documents* directory, and returns a URL pointing to that specific file. The *Documents* directory is a special location that iOS sets aside, one for each app installed on an iOS device. You can use it to store documents created by your app, and rest assured that those documents will be automatically included whenever users back up their iOS device, whether it's to iTunes or iCloud.

Add this method to our implementation, placing it directly above the @end at the bottom of the file:

```
- (NSURL *)urlForFilename:(NSString *)filename {
    NSArray *paths = NSSearchPathForDirectoriesInDomains(NSDocumentDirectory,
        NSUserDomainMask, YES);
    NSString *documentDirectory = [paths objectAtIndex:0];
    NSString *filePath = [documentDirectory stringByAppendingPathComponent:filename];
    NSURL *url = [NSURL fileURLWithPath:filePath];
    return url;
}
```

The second private method is a bit longer. It also uses the *Documents* directory, this time to search for files representing existing documents. The method takes the files it finds and sorts them by creation date, so that the user will see the list of documents sorted "blog-style," with the newest items first. The document file names are stashed away in the documentFilenames property, and then the table view (which we admittedly haven't yet dealt with) is reloaded. Add this method above the @end:

```
- (void)reloadFiles {
    NSArray *paths = NSSearchPathForDirectoriesInDomains(NSDocumentDirectory,
        NSUserDomainMask, YES);
    NSString *path = [paths objectAtIndex:0];
    NSFileManager *fm = [NSFileManager defaultManager];

    NSError *dirError;
    NSArray *files = [fm contentsOfDirectoryAtPath:path error:&dirError];
    if (!files) {
        NSLog(@"Encountered error while trying to list files in directory %@: %@",
            path, dirError);
    }
```

```
    NSLog(@"found files: %@", files);

    files = [files sortedArrayUsingComparator:
            ^NSComparisonResult(id filename1, id filename2) {
        NSDictionary *attr1 = [fm attributesOfItemAtPath:
                                [path stringByAppendingPathComponent:filename1]
                                            error:nil];
        NSDictionary *attr2 = [fm attributesOfItemAtPath:
                                [path stringByAppendingPathComponent:filename2]
                                            error:nil];
        return [[attr2 objectForKey:NSFileCreationDate] compare:
                [attr1 objectForKey:NSFileCreationDate]];
    }];
    self.documentFilenames = files;
    [self.tableView reloadData];
}
```

Now, let's deal with our dear old friends, the table view data source methods. These should be pretty familiar to you by now. Add the following three methods above the @end:

```
- (NSInteger)numberOfSectionsInTableView:(UITableView *)tableView {
    return 1;
}

- (NSInteger)tableView:(UITableView *)tableView
        numberOfRowsInSection:(NSInteger)section {
    return [self.documentFilenames count];
}

- (UITableViewCell *)tableView:(UITableView *)tableView
        cellForRowAtIndexPath:(NSIndexPath *)indexPath {
    UITableViewCell *cell = [tableView dequeueReusableCellWithIdentifier:@"FileCell"];

    NSString *path = [self.documentFilenames objectAtIndex:indexPath.row];
    cell.textLabel.text = path.lastPathComponent.stringByDeletingPathExtension;
    return cell;
}
```

These methods are based on the contents of the array stored in the documentFilenames property. The tableView:cellForForAtIndexPath: method relies on the existence of a cell attached to the table view with "FileCell" set as its identifier, so we must be sure to set that up in the storyboard a little later.

If not for the fact that we haven't touched our storyboard yet, the code we have now would almost be something we could run and see in action, but with no preexisting TinyPix documents, we would have nothing to display in our table view. And so far, we don't have any way to create new documents either. Also, we have not yet dealt with the color-selection control we're going to add. So, let's do a bit more work before we try to run our app.

The user's choice of highlight color will be stored in NSUserDefaults for later retrieval. Here's the action method that will do that by passing along the segmented control's chosen index. Add this method above the @end:

```
- (IBAction)chooseColor:(id)sender {
    NSInteger selectedColorIndex = [(UISegmentedControl *)sender selectedSegmentIndex];
    NSUserDefaults *prefs = [NSUserDefaults standardUserDefaults];
    [prefs setInteger:selectedColorIndex forKey:@"selectedColorIndex"];
}
```

We realize that we haven't yet set this up in the storyboard, but we'll get there!

We also need to add the following few lines to the viewWillAppear: method, to make
sure that the segmented control in our app's GUI will show the current value from
NSUserDefaults as soon as it's about to be displayed:

```
- (void)viewWillAppear:(BOOL)animated
{
    [super viewWillAppear:animated];
    NSUserDefaults *prefs = [NSUserDefaults standardUserDefaults];
    NSInteger selectedColorIndex = [prefs integerForKey:@"selectedColorIndex"];
    self.colorControl.selectedSegmentIndex = selectedColorIndex;
}
```

Now, let's set up a few things in our viewDidLoad method. We'll start off by adding a
button to the right side of the navigation bar. The user will press this button to create a
new TinyPix document. We finish off by calling the reloadFiles method that we
implemented earlier. Make this change to viewDidLoad:

```
- (void)viewDidLoad
{
    [super viewDidLoad];
    // Do any additional setup after loading the view, typically from a nib.

    UIBarButtonItem *addButton = [[UIBarButtonItem alloc]
        initWithBarButtonSystemItem:UIBarButtonSystemItemAdd
        target:self
        action:@selector(insertNewObject)];
    self.navigationItem.rightBarButtonItem = addButton;

    [self reloadFiles];
}
```

You may have noticed that when we created the UIBarButtonItem in this method, we
told it to call the insertNewObject method when it's pressed. We haven't written that
method yet, so let's do so now. Add this method above the @end:

```
- (void)insertNewObject {
    // get the name
    UIAlertView *alert =
    [[UIAlertView alloc] initWithTitle:@"Filename"
                               message:@"Enter a name for your new TinyPix document."
                              delegate:self
                     cancelButtonTitle:@"Cancel"
                     otherButtonTitles:@"Create", nil];
    alert.alertViewStyle = UIAlertViewStylePlainTextInput;
    [alert show];
}
```

This method creates an alert panel that includes a text-input field and displays it. The responsibility of creating a new item instead falls to the delegate method that the alert view calls when it's finished, which we'll also address now. Add this method above the @end:

```
- (void)alertView:(UIAlertView *)alertView
        didDismissWithButtonIndex:(NSInteger)buttonIndex {
    if (buttonIndex == 1) {
        NSString *filename = [NSString stringWithFormat:@"%@.tinypix",
                            [alertView textFieldAtIndex:0].text];
        NSURL *saveUrl = [self urlForFilename:filename];
        self.chosenDocument = [[BIDTinyPixDocument alloc] initWithFileURL:saveUrl];
        [chosenDocument saveToURL:saveUrl
                forSaveOperation:UIDocumentSaveForCreating
                completionHandler:^(BOOL success) {
            if (success) {
                NSLog(@"save OK");
                [self reloadFiles];
                [self performSegueWithIdentifier:@"masterToDetail" sender:self];
            } else {
                NSLog(@"failed to save!");
            }
        }];
    }
}
```

This method starts out simply enough. It checks the value of buttonIndex to see which button was pressed (a 0 indicates that the user pressed the *Cancel* button). It then creates a file name based on the user's entry, a URL based on that file name (using the urlForFilename: method we wrote earlier), and a new BIDTinyPixDocument instance using that URL.

What comes next is a little more subtle. It's important to understand here that just creating a new document with a given URL doesn't create the file. In fact, at the time that the initWithFileURL: is called, the document doesn't yet know if the given URL refers to an existing file or to a new file that needs to be created. We need to tell it what to do. In this case, we tell it to save a new file at the given URL with this code:

```
[chosenDocument saveToURL:saveUrl
        forSaveOperation:UIDocumentSaveForCreating
        completionHandler:^(BOOL success) {
.
.
.
}];
```

Of interest is the purpose and usage of the block that is passed in as the last argument. This method, which we're calling saveToURL:forSaveOperation:completionHandler:, doesn't have a return value to tell us how it all worked out. In fact, the method returns immediately after it's called, long before the file is actually saved. Instead, it starts the file-saving work, and later, when it's done, calls the block that we gave it, using the success parameter to let us know whether it succeeded. To make it all work as smoothly as possible, the file-saving work is actually performed on a background thread. The block we pass in, however, is called on the main thread, so we can safely use any facilities that require the main thread, such as UIKit. With that in mind, take a look again at what happens inside that block:

```
if (success) {
    NSLog(@"save OK");
    [self reloadFiles];
    [self performSegueWithIdentifier:@"masterToDetail" sender:self];
} else {
    NSLog(@"failed to save!");
}
```

This is the content of the block we passed in to the file-saving method, and it's called later after the file operation is completed. We check to see if it succeeded; if so, we do an immediate file reload, and then initiate a segue to another view controller. This is an aspect of segues that we didn't cover in Chapter 10, but it's pretty straightforward.

The idea is that a segue in a storyboard file can have an identifier, just like a table view cell, and you can use that identifier to trigger a segue programmatically. In this case, we'll just need to remember to configure that segue in the storyboard when we get to it. But before we do that, let's add the last method this class needs, to take care of that segue. Insert this method above the @end:

```
- (void)prepareForSegue:(UIStoryboardSegue *)segue sender:(id)sender {
    if (sender == self) {
        // if sender == self, a new document has just been created,
        // and chosenDocument is already set.

        UIViewController *destination = segue.destinationViewController;
        if ([destination respondsToSelector:@selector(setDetailItem:)]) {
            [destination setValue:self.chosenDocument forKey:@"detailItem"];
        }
    } else {
        // find the chosen document from the tableview
        NSIndexPath *indexPath = [self.tableView indexPathForSelectedRow];
        NSString *filename = [documentFilenames objectAtIndex:indexPath.row];
        NSURL *docUrl = [self urlForFilename:filename];
        self.chosenDocument = [[BIDTinyPixDocument alloc] initWithFileURL:docUrl];
        [self.chosenDocument openWithCompletionHandler:^(BOOL success) {
            if (success) {
                NSLog(@"load OK");
                UIViewController *destination = segue.destinationViewController;
                if ([destination respondsToSelector:@selector(setDetailItem:)]) {
                    [destination setValue:self.chosenDocument forKey:@"detailItem"];
                }
            } else {
                NSLog(@"failed to load!");
            }
        }];
    }
}
```

This method has two clear paths of execution, determined by the condition at the top. Remember from our discussion of storyboards in Chapter 10 that this method is called on a view controller whenever a new controller is about to be pushed onto the navigation stack. The `sender` parameter points out the object that initiated the segue, and we use that to figure out just what to do here. If the segue is initiated by the programmatic method call we performed in the alert view delegate method, then `sender` will be equal to `self`. In that case, we know that the `chosenDocument` property is already set, and we simply pass its value off to the destination view controller.

Otherwise, we know we're responding to the user touching a row in the table view, and that's where things get a little more complicated. That's the time to construct a URL (much as we did when creating a document), create a new instance of our document class, and try to open the file. You'll see that the method we call to open the file, `openWithCompletionHandler:`, works similarly to the save method we used earlier. We pass it a block that it will save for later execution. Just as with the file-saving method, the loading occurs in the background, and this block will be executed on the main thread when it's complete. At that point, if the loading succeeded, we pass the document along to the detail view controller.

Note that both of these methods use the key-value coding technique that we've used a few times before (such as in Chapter 10), letting us set the `detailItem` property of the segue's destination controller, even though we don't include its header. This will work out just fine for us, since `BIDDetailViewController`—the detail view controller class created as part of the Xcode project—happens to include a property called `detailItem` right out of the box.

With the amount of code we now have in place, it's high time we configure the storyboard so that we can run our app and make something happen. Save your code and continue.

Initial Storyboarding

Select *MainStoryboard.storyboard* in the Xcode project navigator, and take a look at what's already there. You'll find scenes for the navigation controller, the master view controller, and the detail view controller (see Figure 14–2). You can ignore the navigation controller entirely, since all our work will be with the other two.

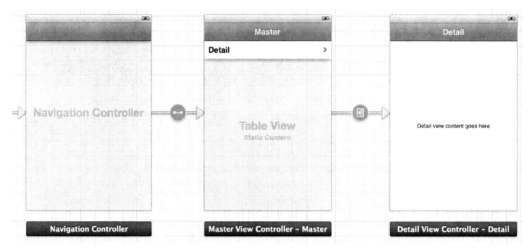

Figure 14–2. *The TinyPix storyboard, showing the navigation controller, master view controller, and detail view controller*

Let's start by dealing with the master view controller scene. This is where the table view showing the list of all our TinyPix documents is configured. By default, this scene's table view is configured to use static cells instead of dynamic cells (see Chapter 10 if you need a refresher on the difference between these two cell types). We want our table view to get its contents from the data source methods we implemented, so select the table view. You can do this in the dock by first finding the item named *Master View Controller – Master*, and then opening its disclosure triangle and selecting the *Table View* item just below it. With the table view selected, open the attributes inspector and set the *Content* popup to *Dynamic Prototypes*.

This change actually deletes the preexisting segue that connected a table view cell to the detail view controller. Re-create that segue by first selecting the *Table View Cell* within the *Table View*. Next, control-drag from the cell to the detail view controller, and select *Push* from the storyboard segues menu that pops up.

Now, select that same prototype table view cell you just dragged from, and use the attributes inspector to set its *Style* to *Basic* and its *Identifier* to *FileCell*. This will let the data source code we wrote earlier access the table view cell.

We also need to create the segue that we're triggering in our code. Do this by control-dragging from the master detail view controller's icon (an orange circle at the bottom of its scene or the *Master View Controller - Master* icon in the dock) over to the detail view controller, and selecting *Push* from the storyboard segues menu.

You'll now see two segues that seem to connect the two scenes. By selecting each of them, you can tell where they're coming from. Selecting one segue highlights the whole master scene; selecting the second one highlights just the table view cell. Select the segue that highlights the whole scene, and use the attributes inspector to set its *Identifier* to *masterToDetail*.

The final touch needed for the master view controller scene is to let the user pick which color will be used to represent an "on" point in the detail view. Instead of implementing some kind of comprehensive color picker, we're just going to add a segmented control that will let the user pick from a set of predefined colors.

Find a *Segmented Control* in the object library, drag it out, and place it in the navigation bar at the top of the master view (see Figure 14–3).

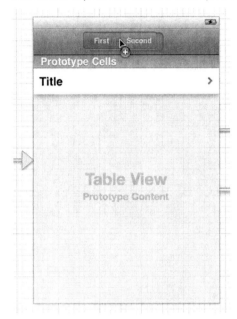

Figure 14–3. *The TinyPix storyboard, showing the master view controller with a segmented control being dropped on the controller's navigation bar*

Make sure the segmented control is selected, and open the attributes inspector. In the *Segmented Control* section at the top of the inspector, use the stepper control to change the number of *Segments* from *2* to *3*. Then double-click the title of each segment in turn, changing them to *Black*, *Red*, and *Green*, respectively.

Next, control-drag from the segmented control to the icon representing the master controller (the orange circle below the controller or the dock icon labeled *Master View Controller – Master*), and select the *chooseColor:* method. Then control-drag from the master controller back to the segmented control, and select the *colorControl* outlet.

We've finally reached a point where we can run the app and see all our hard work brought to life! Run your app. You'll see it start up and display an empty table view with a segmented control at the top and a plus button in the upper-right corner (see Figure 14–4).

Hit the plus button, and the app will ask you to name the new document. Give it a name, tap *Create*, and you'll see the app transition to the detail display, which is, well, under construction right now. All the default implementation of the detail view controller does is display the description of its detailItem in a label. Of course, there's more information in the console pane. It's not much, but it's something!

Figure 14–4. *The TinyPix app when it first appears. Click the plus icon to add a new document. You'll be prompted to name your new TinyPix document. At the moment, all the detail view does is display the document name in a label.*

Tap the back button to return to the master list, where you'll see the item you added. Go ahead and create one or two more items to see that they're correctly added to the list. Then head back to Xcode, because we've got more work to do!

Creating BIDTinyPixView

Our next order of business is the creation of a view class to display our grid and let the user edit it. Select the *TinyPix* folder in the project navigator, and press ⌘N to create a new file. In the *iOS Cocoa Touch* section, select *Objective-C class,* and click *Next.* Name the new class *BIDTinyPixView*, and choose *UIView* in the *Subclass of* popup. Click *Next,* verify that the save location is OK, and click *Create.*

> **NOTE:** The implemention of our view class includes some drawing and touch handling that we
> haven't covered yet. Rather than bog down this chapter with too many details about these topics,
> we're just going to quickly show you the code. We'll cover details about drawing in Chapter 16
> and responding to touches and drags in Chapter 17.

Select *BIDTinyPixView.h*, and make the following changes:

```
#import <UIKit/UIKit.h>
#import "BIDTinyPixDocument.h"

@interface BIDTinyPixView : UIView

@property (strong, nonatomic) BIDTinyPixDocument *document;
@property (strong, nonatomic) UIColor *highlightColor;

@end
```

All we're doing here is adding a couple of properties so that the controller can pass
along the document, as well as set the color used for the "on" squares in our grid.

Now, switch over to *BIDTinyPixView.m*, where we have some more substantial work
ahead of us. Start by adding this class extension and synthesizing all our properties at
the top of the file:

```
#import "BIDTinyPixView.h"

typedef struct {
    NSUInteger row;
    NSUInteger column;
} GridIndex;

@interface BIDTinyPixView ()
@property (assign, nonatomic) CGSize blockSize;
@property (assign, nonatomic) CGSize gapSize;
@property (assign, nonatomic) GridIndex selectedBlockIndex;
- (void)initProperties;
- (void)drawBlockAtRow:(NSUInteger)row column:(NSUInteger)column;
- (GridIndex)touchedGridIndexFromTouches:(NSSet *)touches;
- (void)toggleSelectedBlock;
@end

@implementation BIDTinyPixView

@synthesize document;
@synthesize highlightColor;

@synthesize blockSize;
@synthesize gapSize;
@synthesize selectedBlockIndex;

    .
    .
    .
```

Here, we defined a C struct called GridIndex as a handy way to deal with row/column pairs. We also defined a class extension with some properties and private methods that we'll need to use later, and synthesized all our properties inside the implementation.

The default empty UIView subclass contains an initWithFrame: method, which is really the default initializer for the UIView class. However, since this class is going to be loaded from a storyboard, it will instead be initialized using the initWithCoder: method. We'll implement both of these, making each call a third method that initializes our properties. Make this change to initWithFrame: and add the code just below it:

```
- (id)initWithFrame:(CGRect)frame
{
    self = [super initWithFrame:frame];
    if (self) {
        // Initialization code
        [self initProperties];
    }
    return self;
}

- (id)initWithCoder:(NSCoder *)aDecoder {
    self = [super initWithCoder:aDecoder];
    if (self) {
        [self initProperties];
    }
    return self;
}

- (void)initProperties {
    blockSize = CGSizeMake(34, 34);
    gapSize = CGSizeMake(5, 5);
    selectedBlockIndex.row = NSNotFound;
    selectedBlockIndex.column = NSNotFound;
    highlightColor = [UIColor blackColor];
}
```

The blockSize and gapSize values are specifically tuned to a view that's 310 pixels across. If we wanted to be extra clever here, we could have defined them dynamically based on the view's actual frame, but this is the simplest approach that works for our case, so we're sticking with it!

Now, let's take a look at the drawing routines. We override the standard UIView drawRect: method, and use that to simply walk through all the blocks in our grid, and call another method for each block. Add the following bold code, and don't forget to remove the comment marks around the drawRect: method:

```
/*
// Only override drawRect: if you perform custom drawing.
// An empty implementation adversely affects performance during animation.
- (void)drawRect:(CGRect)rect
{
    // Drawing code
    if (!document) return;
```

```
        for (NSUInteger row = 0; row < 8; row++) {
            for (NSUInteger column = 0; column < 8; column++) {
                [self drawBlockAtRow:row column:column];
            }
        }
    }
*/

- (void)drawBlockAtRow:(NSUInteger)row column:(NSUInteger)column {
    CGFloat startX = (blockSize.width + gapSize.width) * (7 - column) + 1;
    CGFloat startY = (blockSize.height + gapSize.height) * row + 1;
    CGRect blockFrame = CGRectMake(startX, startY, blockSize.width, blockSize.height);
    UIColor *color = [document stateAtRow:row column:column] ?
        self.highlightColor : [UIColor whiteColor];
    [color setFill];
    [[UIColor lightGrayColor] setStroke];
    UIBezierPath *path = [UIBezierPath bezierPathWithRect:blockFrame];
    [path fill];
    [path stroke];
}
```

Finally, we add a set of methods that respond to touch events by the user. Both
touchesBegan:withEvent: and touchesMoved:withEvent: are standard methods that
every UIView subclass can implement in order to capture touch events that happen
within the view's frame. These two methods make use of other methods we defined in
the class extension to calculate a grid location based on a touch location, and to toggle
a specific value in the document. Add these four methods at the bottom of the file, just
above the @end:

```
- (GridIndex)touchedGridIndexFromTouches:(NSSet *)touches {
    GridIndex result;
    UITouch *touch = [touches anyObject];
    CGPoint location = [touch locationInView:self];
    result.column = 8 - (location.x * 8.0 / self.bounds.size.width);
    result.row = location.y * 8.0 / self.bounds.size.height;
    return result;
}

- (void)toggleSelectedBlock {
    [document toggleStateAtRow:selectedBlockIndex.row column:selectedBlockIndex.column];
    [[document.undoManager prepareWithInvocationTarget:document]
     toggleStateAtRow:selectedBlockIndex.row column:selectedBlockIndex.column];
    [self setNeedsDisplay];
}

- (void)touchesBegan:(NSSet *)touches withEvent:(UIEvent *)event {
    self.selectedBlockIndex = [self touchedGridIndexFromTouches:touches];
    [self toggleSelectedBlock];
}

- (void)touchesMoved:(NSSet *)touches withEvent:(UIEvent *)event {
    GridIndex touched = [self touchedGridIndexFromTouches:touches];
    if (touched.row != selectedBlockIndex.row
        || touched.column != selectedBlockIndex.column) {
```

```
            selectedBlockIndex = touched;
            [self toggleSelectedBlock];
        }
    }
```

Sharp-eyed readers may have noticed that the toggleSelectedBlock method does something a bit special. After calling the document's toggleStateAtRow:column: method, it does something more. Let's take another look:

```
- (void)toggleSelectedBlock {
    [document toggleStateAtRow:selectedBlockIndex.row column:selectedBlockIndex.column];
    [[document.undoManager prepareWithInvocationTarget:document]
        toggleStateAtRow:selectedBlockIndex.row column:selectedBlockIndex.column];
    [self setNeedsDisplay];
}
```

The call to document.undoManager returns an instance of NSUndoManager. We haven't dealt with this directly anywhere else in this book, but NSUndoManager is the structural underpinning for the undo/redo functionality in both iOS and Mac OS X. The idea is that anytime the user performs an action in the GUI, you use NSUndoManager to leave a sort of breadcrumb by "recording" a method call that will undo what the user just did. NSUndoManager will store that method call on a special undo stack, which can be used to backtrack through a document's state whenever the user activates the system's undo functionality.

The way it works is that the prepareWithInvocationTarget: method returns a special kind of proxy object that you can send any message to, and the message will be packed up with the target and pushed onto the undo stack. So, while it may look like you're calling toggleStateAtRow:column: twice in a row, the second time it's not being called, but instead is just being queued up for later potential use. This kind of spectacularly dynamic behavior is an area where Objective-C really stands out in comparison to static languages such as C++, where techniques such as letting one object act as a proxy to another or packing up a method invocation for later use have no language support and are nearly impossible (and therefore many tasks such as building undo support can be quite tedious).

So, why are we doing this? We haven't been giving any thought to undo/redo issues up to this point, so why now? The reason is that registering an undoable action with the document's undoManager marks the document as "dirty," and ensures that it will be saved automatically at some point in the next few seconds. The fact that the user's actions are also undoable is just icing on the cake, at least in this application. In an app with a more complex document structure, allowing document-wide undo support can be hugely beneficial.

Save your changes. Now that our view class is ready to go, let's head back to the storyboard to configure the GUI for the detail view.

Storyboard Detailing

Select *MainStoryboard.storyboard*, find the detail scene, and take a look at what's there right now.

All the GUI contains is a label ("Detail view content goes here"), which is the one that contained the document's description when you ran the app earlier. That label isn't particularly useful, so select the label in the detail view controller and press the delete key to remove it.

Use the object library to find a *View*, and drag it into the detail view. Interface Builder will help you line it up so that it fills the entire area. After dropping it there, use the size inspector to set both its width and height to 310. Finally, drag the view and use the guidelines to center it in its container (see Figure 14–5).

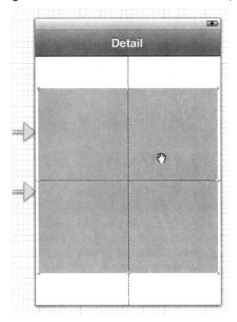

Figure 14–5. *We replaced the label in the detail view with another view, 310 × 310 pixels, centered in its containing view.*

Switch over to the identity inspector so we can change this UIView instance into an instance of our custom class. In the *Custom Class* section at the top of the inspector, select the *Class* popup list and choose *BIDTinyPixView*.

Now, we need to wire up the custom view to our detail view controller. We haven't prepared an outlet for our custom view yet, but that's OK, since Xcode 4's drag-to-code will do that for us.

Activate the assistant editor. A text editor should slide into place alongside the GUI editor, displaying the contents of *BIDDetailViewController.h*. If it's showing you anything else, use the jump bar at the top of the text editor to make *BIDDetailViewController.h* come into view.

To make the connection, control-drag from the *Tiny Pix View* to the code, releasing the drag just above the @end line. In the popup window that appears, make sure that *Connection* is set to *Outlet*, name the new outlet *pixView*, and click the *Connect* button.

You should see that making those connections has added this line to *BIDDetailViewController.h*:

```
@property (weak, nonatomic) IBOutlet BIDTinyPixView *pixView;
```

One thing it didn't add, however, is a header import for our custom view. Let's take care of that by adding this line toward the top of *BIDDetailViewController.h*:

```
#import <UIKit/UIKit.h>
#import "BIDTinyPixView.h"
@interface BIDDetailViewController : UIViewController
.
.
.
```

Next, switch over to *BIDDetailViewController.m*, and you'll see that Xcode also added this method synthesizer toward the top of the file:

```
@synthesize pixView = _pixView;
```

Xcode also added a line to the top of viewDidUnload:

```
- (void)viewDidUnload
{
    [self setPixView:nil];
    [super viewDidUnload];
    // Release any retained subviews of the main view.
    // e.g. self.myOutlet = nil;
}
```

From this start, this class already has a class extension that declares a private method. Let's add a property to *BIDDetailViewController.m*, to keep track of which color the user has chosen:

```
.
.
.
@interface BIDDetailViewController ()
@property (assign, nonatomic) NSUInteger selectedColorIndex;
- (void)configureView;
@end
.
.
.
```

Next, we'll implement the getter and setter for this property. We'll use the synthesized getter, but write our own setter so that whenever this value is set (which will happen when the user taps the segmented control), we set the highlight color in our custom view.

```
.
.
.
@synthesize pixView = _pixView;
@synthesize selectedColorIndex;

- (void)setSelectedColorIndex:(NSUInteger)i {
    if (selectedColorIndex == i) return;

    selectedColorIndex = i;
    switch (selectedColorIndex) {
        case 0:
            self.pixView.highlightColor = [UIColor blackColor];
            break;
        case 1:
            self.pixView.highlightColor = [UIColor redColor];
            break;
        case 2:
            self.pixView.highlightColor = [UIColor greenColor];
            break;
        default:
            break;
    }
    [self.pixView setNeedsDisplay];
}
.
.
.
```

Now, let's modify the configureView method. This isn't a standard UIViewController method. It's just a private method that the project template included in this class as a convenient spot to put code that needs to update the view after anything changes. Since we're not using the description label, we delete the line that sets that. Then we add a bit of code to pass the chosen document along to our custom view, and tell it to redraw itself by calling setNeedsDisplay.

```
- (void)configureView
{
    // Update the user interface for the detail item.

    if (self.detailItem) {
        self.detailDescriptionLabel.text = [self.detailItem description];
        self.pixView.document = self.detailItem;
        [self.pixView setNeedsDisplay];
    }
    NSUserDefaults *prefs = [NSUserDefaults standardUserDefaults];
    self.selectedColorIndex = [prefs integerForKey:@"selectedColorIndex"];
}
```

We ended this method by pulling the color choice out of NSUserDefaults. We set our own selectedColorIndex property. That, in turn, will call the setter we defined earlier, which will pass the chosen color to our custom view.

While we're on the subject of dealing with color choices, scroll down to the bottom and fill in the implementation for the chooseColor: method:

```
- (IBAction)chooseColor:(id)sender {
    NSInteger selectedColorIndex = [(UISegmentedControl *)sender
                                    selectedSegmentIndex];
    NSUserDefaults *prefs = [NSUserDefaults standardUserDefaults];
    [prefs setInteger:selectedColorIndex forKey:@"selectedColorIndex"];
}
```

We're nearly finished with this class, but we need to make one more change. Remember when we mentioned the autosaving that takes place when a document is notified that some editing has occurred, triggered by registering an undoable action? The save normally happens within about ten seconds after the edit occurs. Like the other saving and loading procedures we described earlier in this chapter, it happens in a background thread, so that normally the user won't even notice. However, that works only as long as the document is still around.

With our current setup, there's a risk that when the user hits the back button to go back to the master list, the document instance will be deallocated without any save operation occurring, and the user's latest changes will be lost. To make sure this doesn't happen, we need to add some code to the viewWillDisappear: method to close the document as soon as the user navigates away from the detail view. Closing a document causes it to be automatically saved, and again, the saving occurs on a background thread. In this particular case, we don't need to do anything when the save is done, so we pass in nil instead of a block.

Make this change to viewWillDisappear::

```
- (void)viewWillDisappear:(BOOL)animated
{
    [super viewWillDisappear:animated];
    UIDocument *doc = self.detailItem;
    [doc closeWithCompletionHandler:nil];
}
```

And with that, this version of our first truly document-based app is ready to try out! Fire it up, and bask in the glory. You can create new documents, edit them, flip back to the list, and then select another document (or the same document), and it all just works. If you open the Xcode console while doing this, you'll see some output each time a document is loaded or saved. Using the autosaving system, you don't have direct control over just when saves occur (except for when closing a document), but it can be interesting to watch the logs just to get a feel for when they happen.

Adding iCloud Support

You now have a fully working document-based app, but we're not going to stop here. We promised you iCloud support in this chapter, and it's time to deliver!

Modifying TinyPix to work with iCloud is pretty straightforward. Considering all that's happening behind the scenes, this requires a surprisingly small number of changes.

We'll need to make some revisions to the method that loads the list of available files and the method that specifies the URL for loading a new file, but that's about it.

Apart from the code changes, we will also need to deal with some additional administrative details. Apple allows an app to save to iCloud only if it contains an embedded provisioning profile that is configured to allow iCloud usage. This means that to add the iCloud support to our app, you must have a paid iOS developer membership and have installed your developer certificate. It also works only with actual devices, not the simulator, so you'll need to have at least one iOS device registered with iCloud to run the new iCloud-backed TinyPix. With two devices, you'll have even more fun, as you can see how changes made on one device propagate to the other.

Creating a Provisioning Profile

First, you need to create an iCloud-enabled provisioning profile for TinyPix. Go to http://developer.apple.com and log in to your developer account. Then find your way to the iOS provisioning portal. Apple changes the layout of the developer areas now and then, so we're not going into great detail about how the web site looks. Instead, we'll just describe the basic steps you'll need to take.

Go to the *App IDs* section, and create a new app ID based on the identifier you used when creating TinyPix. You can see this identifier by selecting the top-level *TinyPix* item in Xcode's project navigator, selecting the *Summary* tab, and looking in the *iOS Application Target* section in the *Identifier* field. If you've been using *com.apress* as the base for your application identifiers, the identifier for TinyPix will be *com.apress.TinyPix*. You get the idea.

In the current version of the portal, we set the *common name* to *TinyPix AppID*, left the popup menu set to *Use Team ID*, and entered *com.apress.TinyPix* as the *Bundle Identifier*. Then we clicked *Submit*.

After creating your app ID, you'll see it appear in a table showing various characteristics of each app ID. Under *iCloud*, you should see a yellow dot and the word *Configurable*. Click the *Configure* link next to that, and on the next page, click to enable the *Enable for iCloud* checkbox. Then click *Done* to go back to your list of app IDs.

Now, switch to the provisioning section and create a new provisioning profile specific to that app ID. The *Provisioning* section is just below the *App IDs* section. Click the *New Profile* button, and enter a *Profile Name* of *TinyPixAppPP*. If you don't already have one, you'll need to create a development certificate. In that case, click the *Development Certificate* link, and follow the instructions on that page. Once you have your development certificate set up, select *TinyPix AppID* as your *App ID*. Finally, select the devices on which you want your app to run.

When you're ready, download the new provisioning profile to your Mac, and double-click it to install it in Xcode. In the TinyPix project window, select the top-level *TinyPix* object, select the *TinyPix* project itself (as opposed to the TinyPix build target), and then select the *Build Settings* tab. Scroll down to the *Code Signing* section, where you'll find an item called *Code Signing Identity*. That contains an item called *Debug*, in which you'll find an item labeled *Any iOS SDK*. Click the light-green popup in that row, and choose the developer certificate name listed under *TinyPixAppPP*.

> **NOTE**: Dealing with certificates and provisioning profiles is a pain in the neck, but it seems to be a necessary evil in our little corner of the programming world. If it's any consolation, as random as the current tool work flow seems, it's actually a lot better than it was just a couple of years ago. If this trend continues, perhaps in a few years, we'll reach a point where configuring provisioning profiles will be as straightforward as creating a new project. We're not holding our collective breath, but we're keeping our fingers crossed. You might find it useful to work through the provisioning specifics with a friend who's been through the process before, and take copious notes!

The other bit of new configuration is, thankfully, quite a bit simpler. We need to enable iCloud entitlements for this project, so that it can make use of the iCloud capability baked into the provisioning profile.

Enabling iCloud Entitlements

With the top-level *TinyPix* item selected in the project navigator, select the *TinyPix* target from the list of projects and targets shown just to the right of the navigator. Switch to the *Summary* tab, and scroll down to the *Entitlements* section, which is currently empty. Click the *Enable Entitlements* checkbox at the top of this section, and you'll see that Xcode populates the remaining fields for you. It specifies an *Entitlements File* named *TinyPix*, and fills in your application identifier in the three other sections.

You're finished! Your app now has the necessary permissions to access iCloud from your code. The rest is a simple matter of programming.

How to Query

Select *BIDMasterViewController.m* so we can start making changes for iCloud. The biggest change is going to be the way we look for available documents. In the first version of TinyPix, we used `NSFileManager` to see what's available on the local file system. This time, we're going to do things a little differently. Here, we will fire up a special sort of query to look for documents.

Start by adding a pair of properties: one to hold a pointer to an ongoing query and the other to hold the list of all the documents the query finds.

```
@interface BIDMasterViewController ()
@property (strong, nonatomic) NSArray *documentFilenames;
@property (strong, nonatomic) BIDTinyPixDocument *chosenDocument;
@property (strong, nonatomic) NSMetadataQuery *query;
@property (strong, nonatomic) NSMutableArray *documentURLs;
- (NSURL *)urlForFilename:(NSString *)filename;
- (void)reloadFiles;
@end
```

```
@implementation BIDMasterViewController
@synthesize documentFilenames;
@synthesize chosenDocument;
@synthesize query;
@synthesize documentURLs;
```

Now for the new file-listing method. Remove the entire reloadFiles method, and replace it with this:

```
- (void)reloadFiles {
    NSFileManager *fileManager = [NSFileManager defaultManager];
    // passing nil is OK here, matches first entitlement
    NSURL *cloudURL = [fileManager URLForUbiquityContainerIdentifier:nil];
    NSLog(@"got cloudURL %@", cloudURL);  // returns nil in simulator

    self.query = [[NSMetadataQuery alloc] init];
    query.predicate = [NSPredicate predicateWithFormat:@"%K like '*.tinypix'",
                    NSMetadataItemFSNameKey];
    query.searchScopes = [NSArray arrayWithObject:
                            NSMetadataQueryUbiquitousDocumentsScope];
    [[NSNotificationCenter defaultCenter]
     addObserver:self
     selector:@selector(updateUbiquitousDocuments:)
     name:NSMetadataQueryDidFinishGatheringNotification
     object:nil];
    [[NSNotificationCenter defaultCenter]
     addObserver:self
     selector:@selector(updateUbiquitousDocuments:)
     name:NSMetadataQueryDidUpdateNotification
     object:nil];
    [query startQuery];
}
```

There are some new things here that are definitely worth mentioning. The first is seen in this line:

```
    NSURL *cloudURL = [fileManager URLForUbiquityContainerIdentifier:nil];
```

That's a mouthful, for sure. Ubiquity? What are we talking about here? When it comes to iCloud, a lot of Apple's terminology for identifying resources in iCloud storage includes words like "ubiquity" and "ubiquitous," to indicate that something is omnipresent — accessible from any device using the same iCloud login credentials.

In this case, we're asking the file manager to give us a base URL that will let us access the iCloud directory associated with a particular container identifier. A container identifier is normally a string containing your company's unique bundle seed ID and the application identifier, and is used to pick one of the iCloud entitlements contained within your app. Passing nil here is a shortcut that just means "give me the first one in the list." Since our app contains only one item in that list (created in the previous section), that shortcut suits our needs perfectly.

After that, we create and configure an instance of NSMetadataQuery:

```
self.query = [[NSMetadataQuery alloc] init];
query.predicate = [NSPredicate predicateWithFormat:@"%K like '*.tinypix'",
                 NSMetadataItemFSNameKey];
query.searchScopes = [NSArray arrayWithObject:
                   NSMetadataQueryUbiquitousDocumentsScope];
```

This class was originally written for use with the Spotlight search facility on Mac OS X, but it's now doing extra duty as a way to let iOS apps search iCloud directories. We give the query a predicate, which limits its search results to include only those with the correct sort of file name, and we give it a search scope that limits it to look just within the *Documents* folder in the app's iCloud storage. Then we set up some notifications to let us know when the query is complete, and fire up the query.

Now, we need to implement the method that those notifications call when the query is done. Add this method just below the reloadFiles method:

```
- (void)updateUbiquitousDocuments:(NSNotification *)notification {
    self.documentURLs = [NSMutableArray array];
    self.documentFilenames = [NSMutableArray array];

    NSLog(@"updateUbiquitousDocuments, results = %@", self.query.results);
    NSArray *results = [self.query.results sortedArrayUsingComparator:
        ^NSComparisonResult(id obj1, id obj2) {
        NSMetadataItem *item1 = obj1;
        NSMetadataItem *item2 = obj2;
        return [[item2 valueForAttribute:NSMetadataItemFSCreationDateKey] compare:
                [item1 valueForAttribute:NSMetadataItemFSCreationDateKey]];
    }];

    for (NSMetadataItem *item in results) {
        NSURL *url = [item valueForAttribute:NSMetadataItemURLKey];
        [self.documentURLs addObject:url];
        [(NSMutableArray *)documentFilenames addObject:[url lastPathComponent]];
    }

    [self.tableView reloadData];
}
```

The query's results contain a list of NSMetadataItem objects, from which we can get items like file URLs and creation dates. We use this to sort the items by date, and then grab all the URLs for later use.

Save Where?

The next change is to the urlForFilename: method, which once again is completely different. Here, we're using a ubiquitous URL to create a full path URL for a given file name. We insert "Documents" in the generated path as well, to make sure we're using the app's *Documents* directory. Delete the old method, and replace it with this new one:

```
- (NSURL *)urlForFilename:(NSString *)filename {
    // be sure to insert "Documents" into the path
```

```
        NSURL *baseURL = [[NSFileManager defaultManager]
                            URLForUbiquityContainerIdentifier:nil];
        NSURL *pathURL = [baseURL URLByAppendingPathComponent:@"Documents"];
        NSURL *destinationURL = [pathURL URLByAppendingPathComponent:filename];
        return destinationURL;
}
```

Now, build and run your app on an actual iOS device (not the simulator). If you've run the previous version of the app on that device, you'll find that any TinyPix masterpieces you created earlier are now nowhere to be seen. This new version ignores the local *Documents* directory for the app, and relies completely on iCloud. However, you should be able to create new documents, and find that they stick around after quitting and restarting the app. Moreover, you can even delete the TinyPix app from your device entirely, run it again from Xcode, and find that all your iCloud-saved documents are available at once. If you have an additional iOS device configured with the same iCloud user, use Xcode to run the app on that device, and you'll see all the same documents appear there as well! It's pretty sweet.

Storing Preferences on iCloud

We can "cloudify" one more piece of functionality with just a bit of effort. iOS 5 includes a new class called NSUbiquitousKeyValueStore, which works a lot like an NSDictionary (or NSUserDefaults, for that matter), but whose keys and values are stored in the cloud. This is great for application preferences, login tokens, and anything else that doesn't belong in a document but could be useful when shared among all of a user's devices.

In TinyPix, we'll use this feature to store the user's preferred highlight color. That way, instead of needing to be configured on each device, the user sets the color once, and it shows up everywhere.

Select *BIDMasterViewController.m* so we can make a couple of small changes. First, find chooseColor:, and make the following changes:

```
- (IBAction)chooseColor:(id)sender {
    NSInteger selectedColorIndex = [(UISegmentedControl *)sender selectedSegmentIndex];
    NSUserDefaults *prefs = [NSUserDefaults standardUserDefaults];
    [prefs setInteger:selectedColorIndex forKey:@"selectedColorIndex"];
    NSUbiquitousKeyValueStore *prefs = [NSUbiquitousKeyValueStore defaultStore];
    [prefs setLongLong:selectedColorIndex forKey:@"selectedColorIndex"];
}
```

Here, we grab a slightly different object instead of NSUserDefaults. This new class doesn't have a setInteger: method, so we use setLongLong: instead, which will do the same thing.

Then find the viewWillAppear: method, and change it as shown here:

```
- (void)viewWillAppear:(BOOL)animated
{
    [super viewWillAppear:animated];
    NSUserDefaults *prefs = [NSUserDefaults standardUserDefaults];
    NSInteger selectedColorIndex = [prefs integerForKey:@"selectedColorIndex"];
```

```
        NSUbiquitousKeyValueStore *prefs = [NSUbiquitousKeyValueStore defaultStore];
        NSInteger selectedColorIndex = [prefs longLongForKey:@"selectedColorIndex"];
        self.colorControl.selectedSegmentIndex = selectedColorIndex;
}
```

We also need to make a change to the detail display, so that it will pick up the color from the correct place. Select *BIDDetailViewController.m*, find the configureView method, and change its last few lines as shown here:

```
        NSUserDefaults *prefs = [NSUserDefaults standardUserDefaults];
        self.selectedColorIndex = [prefs integerForKey:@"selectedColorIndex"];
        NSUbiquitousKeyValueStore *prefs = [NSUbiquitousKeyValueStore defaultStore];
        self.selectedColorIndex = [prefs longLongForKey:@"selectedColorIndex"];
```

That's it! You can now run the app on multiple devices configured for the same iCloud user, and will see that setting the color on one device results in the new color appearing on the other device the next time a document is opened there. Piece of cake!

What We Didn't Cover

We now have the basics of an iCloud-enabled, document-based application up and running, but there are a few more issues that you may want to consider. We're not going to cover these topics in this book, but if you're serious about making a great iCloud-based app, you'll want to think about these areas:

- Documents stored in iCloud are prone to conflicts. What happens if you edit the same TinyPix file on several devices at once? Fortunately, Apple has already thought of this, and provides some ways to deal with these conflicts in your app. It's up to you to decide if you want to ignore conflicts, try to fix them automatically, or ask the user to help sort out the problem. For full details, search for "resolving document version conflicts" in the Xcode documentation viewer.

- Apple recommends that you design your application to work in a completely offline mode in case the user isn't using iCloud for some reason. It also recommends that you provide a way for a user to move files between iCloud storage and local storage. Sadly, Apple doesn't provide or suggest any standard GUI for helping a user manage this, and current apps that provide this functionality, such as Apple's iWork apps, don't seem to handle it in a particularly user-friendly way. See Apple's "Managing the Life Cycle of a Document" in the Xcode documentation for more on this.

- Apple supports using iCloud for Core Data storage, and even provides a class called UIManagedDocument that you can subclass if you want to make that work. See the UIManagedDocument class reference for more information, or take a look at *More iOS 5 Development: Further Explorations of the iOS SDK*, (http://apress.com/book/view/1430232528) by Dave Mark, Alex Horowitz, Kevin Kim, and Jeff

LaMarche (Apress, 2012) for a hands-on guide to building an iCloud-backed Core Data app.

What's up next? In Chapter 15, we'll take you through the process of having your apps work properly in a multithreaded, multitasking environment.

Grand Central Dispatch, Background Processing, and You

If you've ever tried your hand at multithreaded programming, in any environment, chances are you've come away from the experience with a feeling of dread, terror, or worse. Fortunately, technology marches on, and lately Apple has come up with a new approach that makes multithreaded programming much easier. This approach is called **Grand Central Dispatch**, and we'll get you started using it in this chapter. We'll also dig into the multitasking capabilities of iOS, showing you how to adjust your applications to play nicely in this new world, as well as using the new capabilities to make your apps work even better than before.

Grand Central Dispatch

One of the biggest challenges facing developers today is to write software that can perform complex actions in response to user input while remaining responsive so that the user isn't constantly kept waiting while the processor does some behind-the-scenes task. If you think about it, that challenge has been with us all along, and in spite of the advances in computing technology that bring us faster CPUs, the problem persists. If you want evidence, you need look no further than your nearest computer screen. Chances are that the last time you sat down to work at your computer, at some point, your work flow was interrupted by a spinning mouse cursor of some kind or another.

So why does this continue to vex us, given all the advances in system architecture? One part of the problem is the way that software is typically written: as a sequence of events to be performed in order. Such software can scale up as CPU speeds increase, but only to a certain point. As soon as the program gets stuck waiting for an external resource, such as a file or a network connection, the entire sequence of events is effectively paused. All modern operating systems now allow the use of multiple threads of execution within a program, so that even if a single thread is stuck waiting for a specific event, the other threads can keep going. Even so, many developers see multithreaded programming as something of a black art and shy away from it.

Fortunately, Apple has some good news for anyone who wants to break up their code into simultaneous chunks without too much hands-on intimacy with the system's threading layer. This good news is called Grand Central Dispatch (GCD). It provides an entirely new API for splitting up the work your application needs to do into smaller chunks that can be spread across multiple threads and, with the right hardware, multiple CPUs.

Much of this new API is accessed using **blocks**, another Apple innovation that adds a sort of anonymous in-line function capability to C and Objective-C. Blocks have a lot in common with similar features in languages such as Ruby and Lisp, and they can provide interesting new ways to structure interactions between different objects while keeping related code closer together in your methods.

Introducing SlowWorker

As a platform for demonstrating how GCD works, we'll create an application called SlowWorker, which consists of a simple interface driven by a single button and a text view. Click the button, and a synchronous task is immediately started, locking up the app for about ten seconds. Once the task completes, some text appears in the text view (see Figure 15–1).

Figure 15–1. *The SlowWorker application hides its interface behind a single button. Click the button, and the interface hangs for about ten seconds while the application does its work.*

Start by using the *Single View Application* template to make a new application in Xcode, as you've done many times before. Name this one *SlowWorker*, set *Device Family* to *iPhone*, and turn off the *Use Storyboard* option. Make the following additions to *BIDViewController.h*:

```
#import <UIKit/UIKit.h>

@interface BIDViewController : UIViewController

@property (strong, nonatomic) IBOutlet UIButton *startButton;
@property (strong, nonatomic) IBOutlet UITextView *resultsTextView;

- (IBAction)doWork:(id)sender;

@end
```

This simply defines a couple of outlets to the two objects visible in our GUI and an action method to be triggered by the button.

Now, enter the following code near the top of *BIDViewController.m*:

```
#import "BIDViewController.h"

@implementation BIDViewController

@synthesize startButton, resultsTextView;

- (NSString *)fetchSomethingFromServer {
    [NSThread sleepForTimeInterval:1];
    return @"Hi there";
}

- (NSString *)processData:(NSString *)data {
    [NSThread sleepForTimeInterval:2];
    return [data uppercaseString];
}

- (NSString *)calculateFirstResult:(NSString *)data {
    [NSThread sleepForTimeInterval:3];
    return [NSString stringWithFormat:@"Number of chars: %d",
            [data length]];
}

- (NSString *)calculateSecondResult:(NSString *)data {
    [NSThread sleepForTimeInterval:4];
    return [data stringByReplacingOccurrencesOfString:@"E"
                                           withString:@"e"];
}

- (IBAction)doWork:(id)sender {
    NSDate *startTime = [NSDate date];
    NSString *fetchedData = [self fetchSomethingFromServer];
    NSString *processedData = [self processData:fetchedData];
    NSString *firstResult = [self calculateFirstResult:processedData];
    NSString *secondResult = [self calculateSecondResult:processedData];
```

```
        NSString *resultsSummary = [NSString stringWithFormat:
                                @"First: [%@]\nSecond: [%@]", firstResult,
                                secondResult];
        resultsTextView.text = resultsSummary;
        NSDate *endTime = [NSDate date];
        NSLog(@"Completed in %f seconds",
              [endTime timeIntervalSinceDate:startTime]);
}
.
.
.
```

Next, add the usual cleanup code in viewDidUnload:

```
- (void)viewDidUnload {
    [self viewDidUnload];
    // Release any retained subviews of the main view.
    // e.g. self.myOutlet = nil;
    self.startButton = nil;
    self.resultsTextView = nil;
}
```

As you can see, the work of this class (such as it is) is split up into a number of small
chunks. This code is just meant to simulate some slow activities, and none of those
methods really do anything time-consuming at all. To make things interesting, each
method contains a call to the sleepForTimeInterval: class method in NSThread, which
simply makes the program (specifically, the thread from which the method is called)
effectively pause and do nothing at all for the given number of seconds. The doWork:
method also contains code at the beginning and end to calculate the amount of time it
took for all the work to be done.

Now, open *BIDViewController.xib*, and drag a *Round Rect Button* and a *Text View* into
the empty *View* window, laying things out as shown in Figure 15–2. Control-drag from
File's Owner to connect the view controller's two outlets to the button and the text view.

Next, select the button, and go to the connections inspector to connect the button's
Touch Up Inside event to *File's Owner*, selecting the view controller's doWork: method.
Finally, select the text view, use the attributes inspector to uncheck the *Editable*
checkbox (it's in the upper-right corner), and delete the default text from the text view.

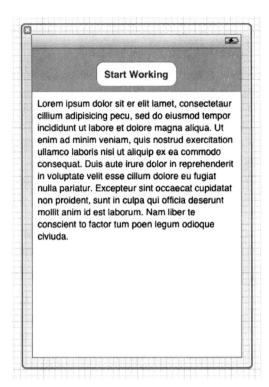

Figure 15–2. *The SlowWorker interface consists of a round rect button and a text view. Be sure to uncheck the Editable checkbox for the text view and delete all of its text.*

Save your work. Then select *Run*. Your app should start up, and pressing the button will make it work for about ten seconds (the sum of all those sleep amounts) before showing you the results. During your wait, you'll see that the *Start Working!* button remains dark blue the entire time, never turning back to its normal color until the "work" is done. Also, until the work is complete, the application's view is unresponsive. Tapping anywhere on the screen has no effect. In fact, the only way you can interact with your application during this time is by tapping the home button to switch away from it. This is exactly the state of affairs we want to avoid!

In this particular case, the wait is not too bad, since the application appears to be hung for just a few seconds, but if your app regularly hangs this way for much longer, using it will be a frustrating experience. In the worst of cases, the operating system may actually kill your app if it's unresponsive for too long. In any case, you'll end up with some unhappy users—and maybe even some ex-users!

Threading Basics

Before we start implementing solutions, let's go over some of the basics involved in concurrency. This is far from a complete description of threading in iOS or threading in general. We just want to explain enough for you to understand what we're doing in this chapter.

Most modern operating systems (including, of course, iOS) support the notion of threads of execution. Each process can contain multiple threads, which all run concurrently. If there's just one processor core, the operating system will switch between all executing threads, much like it switches between all executing processes. If more than one core is available, the threads will be distributed among them, just as processes are.

All threads in a process share the same executable program code and the same global data. Each thread can also have some data that is exclusive to the thread. Threads can make use of a special structure called a **mutex** (short for mutual exclusion) or a lock, which can ensure that a particular chunk of code can't be run by multiple threads at once. This is useful for ensuring correct outcomes when multiple threads access the same data simultaneously, by locking out other threads when one thread is updating a value (in what's called a **critical section** of your code).

A common concern when dealing with threads is the idea of code being **thread-safe**. Some software libraries are written with thread concurrency in mind and have all their critical sections properly protected with mutexes. Some code libraries aren't thread-safe.

For example, in Cocoa Touch, the Foundation framework (containing basic classes appropriate for all sorts of Objective-C programming, such as NSString, NSArray, and so on) is generally considered to be thread-safe. However, the UIKit framework (containing the classes specific to building GUI applications, such as UIApplication, UIView and all its subclasses, and so on) is for the most part not thread-safe. This means that in a running iOS application, all method calls that deal with any UIKit objects should be executed from within the same thread, which is commonly known as the **main thread**. If you access UIKit objects from another thread, all bets are off! You are likely to encounter seemingly inexplicable bugs (or, even worse, you won't experience any problems, but some of your users will be affected by them after you ship your app).

By default, the main thread is where all the action of your iOS app (such as dealing with actions triggered by user events) occurs, so for simple applications, it's nothing you need to worry about. Action methods triggered by a user are already running in the main thread. Up to this point in the book, our code has been running exclusively on the main thread, but that's about to change.

> **TIP:** A lot has been written about thread-safety, and it's well worth your time to dig in and try to digest as much of it as you can. One great place to start is Apple's own documentation. Take a few minutes and read through this page (it will definitely help):
>
> http://developer.apple.com/library/ios/#documentation/Cocoa/Conceptual/M
> ultithreading/ThreadSafetySummary/ThreadSafetySummary.html

Units of Work

The problem with the threading model described earlier is that for the average programmer, writing error-free, multithreaded code is nearly impossible. This is not meant as a critique of our industry or of the average programmer's abilities; it's simply an observation. The complex interactions you must account for in your code when synchronizing data and actions across multiple threads are really just too much for most people to tackle. Imagine that 5% of all people have the capacity to write software at all. Only a small fraction of those 5% are really up to the task of writing heavy-duty multithreaded applications. Even people who have done it successfully will often advise others to not follow their example!

Fortunately, all hope is not lost. It is possible to implement some concurrency without too much low-level thread-twisting. Just as we have the ability to display data on the screen without directly poking bits into video RAM, and to read data from disk without interfacing directly with disk controllers, software abstractions exist that let us run our code on multiple threads without requiring us to do much directly with the threads.

The solutions that Apple encourages us to use are centered around the ideas of splitting up long-running tasks into units of work and putting those units into queues for execution. The system manages the queues for us, executing units of work on multiple threads. We don't need to start and manage the background threads directly and are freed from much of the bookkeeping that's usually involved in implementing concurrent applications; the system takes care of that for us.

GCD: Low-Level Queueing

This idea of putting units of work into queues that can be executed in the background, with the system managing the threads for you, is really powerful and greatly simplifies many development situations where concurrency is needed. In the 10.6 release of Mac OS X, GCD made its debut, providing the infrastructure to do just that. With the release of iOS 4.0, this technology came to the iOS platform as well. This technology works not only with Objective-C, but also with C and C++.

GCD puts some great concepts—units of work, painless background processing, and automatic thread management—into a C interface that can be used from all of the C-based languages. To top things off, Apple has made its implementation of GCD open source so that it can be ported to other Unix-like operating systems as well.

One of the key concepts of GCD is the queue. The system provides a number of predefined queues, including a queue that's guaranteed to always do its work on the main thread. It's perfect for non-thread-safe UIKit! You can also create your own queues—as many as you like. GCD queues are strictly first-in, first-out (FIFO). Units of work added to a GCD queue will always be started in the order they were placed in the queue. That being said, they may not always finish in the same order, since a GCD queue will automatically distribute its work among multiple threads, if possible.

Each queue has access to a pool of threads that are reused throughout the lifetime of the application. GCD will try to maintain a pool of threads that's appropriate for the machine's architecture, automatically taking advantage of a more powerful machine by utilizing more processor cores when it has work to do. Until recently, iOS devices were all single-core, so this wasn't much of an issue. But now that Apple's current crop of iOS devices, starting with the iPad 2 and the iPhone 4S, feature dual-core processors, GCD is becoming truly useful.

Becoming a Blockhead

Along with GCD, Apple has added a bit of new syntax to the C language itself (and, by extension, Objective-C and C++) to implement a language feature called **blocks** (also known as **closures** or **lambdas** in some other languages), which are really important for getting the most out of GCD. The idea behind a block is to let a particular chunk of code be treated like any other C-language type. A block can be assigned to a variable, passed as an argument to a function or method, and (unlike most other types) executed. In this way, blocks can be used as an alternative to the delegate pattern in Objective-C or to callback functions in C.

Much like a method or function, a block can take one or more parameters and specify a return value. To declare a block variable, you use the caret (^) symbol along with some additional parenthesized bits to declare parameters and return types. To define the block itself, you do roughly the same, but follow it up with the actual code defining the block wrapped in curly braces.

```
// Declare a block variable "loggerBlock" with no parameters and no return value.
void (^loggerBlock)(void);

// Assign a block to the variable declared above.  A block without parameters
// and with no return value, like this one, needs no "decorations" like the use
// of void in the preceding variable declaration.
loggerBlock = ^{ NSLog(@"I'm just glad they didn't call it a lambda"); };

// Execute the block, just like calling a function.
loggerBlock();  // this produces some output in the console
```

If you've done much C programming, you may recognize that this is similar to the concept of a function pointer in C. However, there are a few critical differences. Perhaps the biggest difference—the one that's the most striking when you first see it—is that blocks can be defined in-line in your code. You can define a block right at the point where it's going to be passed to another method or function. Another big difference is that a block can access variables available in the scope where it's created. By default, the block makes a copy of any variable you access this way, leaving the original intact, but you can make an outside variable "read/write" by prepending the storage qualifier __block before its declaration. Note that there are two underscores before block, not just one.

```
// define a variable that can be changed by a block
__block int a = 0;

// define a block that tries to modify a variable in its scope
void (^sillyBlock)(void) = ^{ a = 47; };

// check the value of our variable before calling the block
NSLog(@"a == %d", a); // outputs "a == 0"

// execute the block
sillyBlock();

// check the values of our variable again, after calling the block
NSLog(@"a == %d", a); // outputs "a == 47"
```

As we mentioned, blocks really shine when used with GCD, which lets you take a block and add it to a queue in a single step. When you do this with a block that you define immediately at that point, rather than a block stored in a variable, you have the added advantage of being able to see the relevant code directly in the context where it's being used.

Improving SlowWorker

To see how blocks work, let's revisit SlowWorker's doWork: method. It currently looks like this:

```
- (IBAction)doWork:(id)sender {
    NSDate *startTime = [NSDate date];
    NSString *fetchedData = [self fetchSomethingFromServer];
    NSString *processedData = [self processData:fetchedData];
    NSString *firstResult = [self calculateFirstResult:processedData];
    NSString *secondResult = [self calculateSecondResult:processedData];
    NSString *resultsSummary = [NSString stringWithFormat:
                        @"First: [%@]\nSecond: [%@]", firstResult,
                        secondResult];
    resultsTextView.text = resultsSummary;
    NSDate *endTime = [NSDate date];
    NSLog(@"Completed in %f seconds",
        [endTime timeIntervalSinceDate:startTime]);
}
```

We can make this method run entirely in the background by wrapping all the code in a block and passing it to a GCD function called dispatch_async. This function takes two parameters: a GCD queue and a block to assign to the queue. Make these two changes to your copy of doWork:. Be sure to add the closing brace and parenthesis at the end of the method.

```
- (IBAction)doWork:(id)sender {
    NSDate *startTime = [NSDate date];
    dispatch_async(dispatch_get_global_queue(0, 0), ^{
        NSString *fetchedData = [self fetchSomethingFromServer];
        NSString *processedData = [self processData:fetchedData];
        NSString *firstResult = [self calculateFirstResult:processedData];
        NSString *secondResult = [self calculateSecondResult:processedData];
        NSString *resultsSummary = [NSString stringWithFormat:
                                    @"First: [%@]\nSecond: [%@]", firstResult,
                                    secondResult];
        resultsTextView.text = resultsSummary;
        NSDate *endTime = [NSDate date];
        NSLog(@"Completed in %f seconds",
            [endTime timeIntervalSinceDate:startTime]);
    });
}
```

The first line grabs a preexisting global queue that's always available, using the dispatch_get_global_queue() function. That function takes two arguments: the first lets you specify a priority, and the second is currently unused and should always be 0. If you specify a different priority in the first argument, such as DISPATCH_QUEUE_PRIORITY_HIGH or DISPATCH_QUEUE_PRIORITY_LOW (passing a 0 is the same as passing DISPATCH_QUEUE_PRIORITY_DEFAULT), you will actually get a different global queue, which the system will prioritize differently. For now, we'll stick with the default global queue.

The queue is then passed to the dispatch_async() function, along with the block of code that comes after. GCD takes that entire block and passes it to a background thread, where it will be executed one step at a time, just as when it was running in the main thread.

Note that we define a variable called startTime just before the block is created, and then use its value at the end of the block. Intuitively, this doesn't seem to make sense, since by the time the block is executed, the doWork: method has exited, so the NSDate instance that the startTime variable is pointing to should already be released! This is a crucial point of block usage: if a block accesses any variables from "the outside" during its execution, then some special setup happens when the block is created, allowing the block access to those variables. The values contained by such variables will either be duplicated (if they are plain C types such as int or float) or retained (if they are pointers to objects) so that the values they contain can be used inside the block. When dispatch_async is called in the second line of doWork:, and the block shown in the code is created, startTime is actually sent a retain message, whose return value is assigned to what is essentially a new static variable with the same name (startTime) inside the block.

The startTime variable needs to be static inside the block so that code inside the block can't accidentally mess with a variable that's defined outside the block. If that were allowed all the time, it would just be confusing for everyone. Sometimes, however, you actually do want to let a block write to a value defined on the outside, and that's where the __block storage qualifier (which we mentioned a couple of pages ago) comes in handy. If __block is used to define a variable, then it is directly available to any and all blocks that are defined within the same scope. An interesting side effect of this is that __block-qualified variables are not duplicated or retained when used inside a block.

Don't Forget That Main Thread

Getting back to the project at hand, there's one problem here: UIKit thread-safety. Remember that messaging any GUI object, including our resultsTextView, from a background thread is a no-no. Fortunately, GCD provides a way to deal with this, too. Inside the block, we can call another dispatching function, passing work back to the main thread! We do this by once again calling dispatch_async(), this time passing in the queue returned by the dispatch_get_main_queue() function. This always gives us the special queue that lives on the main thread, ready to execute blocks that require the use of the main thread. Make one more change to your version of doWork::

```
- (IBAction)doWork:(id)sender {
    NSDate *startTime = [NSDate date];
    dispatch_async(dispatch_get_global_queue(0, 0), ^{
        NSString *fetchedData = [self fetchSomethingFromServer];
        NSString *processedData = [self processData:fetchedData];
        NSString *firstResult = [self calculateFirstResult:processedData];
        NSString *secondResult = [self calculateSecondResult:processedData];
        NSString *resultsSummary = [NSString stringWithFormat:
                                    @"First: [%@]\nSecond: [%@]", firstResult,
                                    secondResult];
        dispatch_async(dispatch_get_main_queue(), ^{
            resultsTextView.text = resultsSummary;
        });
        NSDate *endTime = [NSDate date];
        NSLog(@"Completed in %f seconds",
              [endTime timeIntervalSinceDate:startTime]);
    });
}
```

Giving Some Feedback

If you build and run your app at this point, you'll see that it now seems to work a bit more smoothly, at least in some sense. The button no longer gets stuck in a highlighted position after you touch it, which perhaps leads you to tap again, and again, and so on. If you look in Xcode's console log, you'll see the result of each of those taps, but only the results of the last tap will be shown in the text view.

What we really want to do is enhance the GUI so that after the user presses the button, the display is immediately updated in a way that indicates that an action is underway, and that the button is disabled while the work is in progress. We'll do this by adding a UIActivityIndicatorView to our display. This class provides the sort of spinner seen in many applications and web sites. Start by declaring it in *BIDViewController.h*:

```
@interface BIDViewController : UIViewController

@property (strong, nonatomic) IBOutlet UIButton *startButton;
```

```
@property (strong, nonatomic) IBOutlet UITextView *resultsTextView;
@property (strong, nonatomic) IBOutlet UIActivityIndicatorView *spinner;
  .
  .
  .
```

Then open *BIDViewController.xib*, locate an *Activity Indicator View* in the library, and drag it into our view, next to the button (see Figure 15–3).

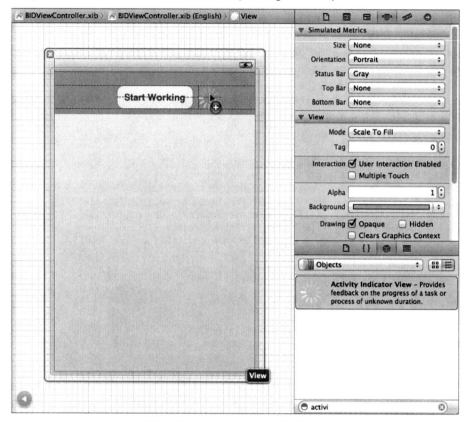

Figure 15–3. *Dragging an activity indicator view into our main view in Interface Builder*

With the activity indicator spinner selected, use the attributes inspector to check the *Hide When Stopped* checkbox so that our spinner will appear only when we tell it to start spinning (no one wants an unspinning spinner in their GUI).

Next, control-drag from the *File's Owner* icon to the spinner, and connect the *spinner* outlet. Save your changes.

Now, open *BIDViewController.m*. Here, we'll first add the usual code for handling an outlet:

```
@implementation BIDViewController

@synthesize startButton, resultsTextView;
@synthesize spinner;
```

.
.
.

```
- (void)viewDidUnload {
    [super viewDidUnload];
    // Release any retained subviews of the main view.
    // e.g. self.myOutlet = nil;
    self.startButton = nil;
    self.resultsTextView = nil;
    self.spinner = nil;
}
```

Let's next work on the doWork: method a bit, adding a few lines to manage the appearance of the button and the spinner when the user clicks and when the work is done. We'll first set the button's enabled property to NO, which prevents it from registering any taps but doesn't give any visual cue. To let the user see that the button is disabled, we'll set its alpha value to 0.5. You can think of the alpha value as a transparency setting, where 0.0 is fully transparent (that is, invisible) and 1.0 is not transparent at all. We'll talk more about alpha values in Chapter 16.

```
- (IBAction)doWork:(id)sender {
    startButton.enabled = NO;
    startButton.alpha = 0.5;
    [spinner startAnimating];
    NSDate *startTime = [NSDate date];
    dispatch_async(dispatch_get_global_queue(0, 0), ^{
        NSString *fetchedData = [self fetchSomethingFromServer];
        NSString *processedData = [self processData:fetchedData];
        NSString *firstResult = [self calculateFirstResult:processedData];
        NSString *secondResult = [self calculateSecondResult:processedData];
        NSString *resultsSummary = [NSString stringWithFormat:
                                    @"First: [%@]\nSecond: [%@]", firstResult,
                                    secondResult];
        dispatch_async(dispatch_get_main_queue(), ^{
            startButton.enabled = YES;
            startButton.alpha = 1.0;
            [spinner stopAnimating];
            resultsTextView.text = resultsSummary;
        });
        NSDate *endTime = [NSDate date];
        NSLog(@"Completed in %f seconds",
              [endTime timeIntervalSinceDate:startTime]);
    });
}
```

Build and run the app, and press the button. That's more like it, eh? Even though the work being done takes a few seconds, the user isn't just left hanging. The button is disabled and looks the part as well. Also, the animated spinner lets the user know that the app hasn't actually hung and can be expected to return to normal at some point.

Concurrent Blocks

So far, so good, but we're not quite finished yet! The sharp-eyed among you will notice that after going through these motions, we still haven't really changed the basic sequential layout of our algorithm (if you can even call this simple list of steps an algorithm). All that we're doing is moving a chunk of this method to a background thread and then finishing up in the main thread. The Xcode console output proves it: this work takes ten seconds to run, just as it did at the outset. The 900-pound gorilla in the room is that calculateFirstResult: and calculateSecondResult: don't need to be performed in sequence, and doing them concurrently could give us a substantial speedup.

Fortunately, GCD has a way to accomplish this by using what's called a **dispatch group**. All blocks that are dispatched asynchronously within the context of a group, via the dispatch_group_async() function, are set loose to execute as fast as they can, including being distributed to multiple threads for concurrent execution, if possible. We can also use dispatch_group_notify() to specify an additional block that will be executed when all the blocks in the group have been run to completion.

Make the following changes to your copy of doWork:. Again, make sure you get that trailing bit of curly brace and parenthesis.

```objc
- (IBAction)doWork:(id)sender {
    NSDate *startTime = [NSDate date];
    dispatch_async(dispatch_get_global_queue(0, 0), ^{
        NSString *fetchedData = [self fetchSomethingFromServer];
        NSString *processedData = [self processData:fetchedData];
        NSString *firstResult = [self calculateFirstResult:processedData];
        NSString *secondResult = [self calculateSecondResult:processedData];
        __block NSString *firstResult;
        __block NSString *secondResult;
        dispatch_group_t group = dispatch_group_create();
        dispatch_group_async(group, dispatch_get_global_queue(0, 0), ^{
            firstResult = [self calculateFirstResult:processedData];
        });
        dispatch_group_async(group, dispatch_get_global_queue(0, 0), ^{
            secondResult = [self calculateSecondResult:processedData];
        });
        dispatch_group_notify(group, dispatch_get_global_queue(0, 0), ^{
            NSString *resultsSummary = [NSString stringWithFormat:
                                        @"First: [%@]\nSecond: [%@]", firstResult,
                                        secondResult];
            dispatch_async(dispatch_get_main_queue(), ^{
                startButton.enabled = YES;
                startButton.alpha = 1.0;
                [spinner stopAnimating];
                resultsTextView.text = resultsSummary;
            });
            NSDate *endTime = [NSDate date];
            NSLog(@"Completed in %f seconds",
                  [endTime timeIntervalSinceDate:startTime]);
        });
    });
}
```

One complication here is that each of the `calculate` methods returns a value that we want to grab, so we must first create the variables using the `__block` storage modifier. This ensures the values set inside the blocks are made available to the code that runs later.

With this in place, build and run the app again. You'll see that your efforts have paid off. What was once a ten-second operation now takes just seven seconds, thanks to the fact that we're running both of the calculations simultaneously.

Obviously, our contrived example gets the maximum effect because these two "calculations" don't actually do anything but cause the thread they're running on to sleep. In a real application, the speedup would depend on what sort of work is being done and which resources are available. The performance of CPU-intensive calculations is helped by this technique only if multiple CPU cores are available. At the time of this writing, only the latest iOS devices—the iPhone 4S and iPad 2—have more than one CPU core. Other uses, such as fetching data from multiple network connections at once, would see a speed increase even with just one CPU.

As you can see, GCD is not a panacea. Using GCD won't automatically speed up every application. But by carefully applying these techniques at those points in your app where speed is essential, or where you find that your application feels like it's lagging in its responses to the user, you can easily provide a better user experience, even in situations where you can't improve the real performance.

Background Processing

Another important addition that arrived with iOS 4 is the introduction of background processing. This allows your apps to run in the background—in some circumstances, even after the user has pressed the home button.

This functionality should not be confused with the true multitasking that modern desktop operating systems now feature, where all the programs you launch remain resident in the system RAM until you explicitly quit them. iOS devices are still too low on RAM to be able to pull that off very well. Instead, this background processing is meant to allow applications that require specific kinds of system functionality to continue to run in a constrained manner. For instance, if you have an app that plays an audio stream from an Internet radio station, iOS will let that app continue to run, even if the user switches to another app. Beyond that, it will even provide standard pause and volume controls in the iOS system taskbar (the bar that appears at the bottom when you double-tap the home button) while your app is playing audio.

NOTE: The background processing features are available only on devices that meet a certain minimum hardware standard. At the time of this writing, this includes the iPhone 3GS and beyond, the third- and fourth-generation iPod touch, and the iPad. Basically, if you have any iPhone or iPod touch that was available before mid-2009, your device isn't welcome on the multitasking playground. Sorry!

Specifically, if you're creating an app that plays audio, that wants continuous location updates, or that implements Voice over IP (VoIP) to let users send and receive phone calls on the Internet, you can declare this situation in your app's *Info.plist* file, and the system will treat your app in a special way. This usage, while interesting, is probably not something that most readers of this book will be tackling, so we're not going to delve into it here.

Besides running apps in the background, iOS also includes the ability to put an app into a suspended state after the user presses the home button. This state of suspended execution is conceptually similar to putting your Mac into sleep mode. The entire working memory of the application is held in RAM; it just isn't executed while suspended. As a result, switching back to such an application is lightning-fast. This isn't limited to special applications. In fact, it is the default behavior of any app you compile with the iOS 5 SDK (though this can be disabled by another setting in the *Info.plist* file). To see this in action, open your device's Mail application and drill down into a message. Then press the home button, open the Notes application, and select a note. Now double-tap the home button and switch back to Mail. You'll see that there's no perceptible lag; it just slides into place as if it had been running all along.

For most applications, this sort of automatic suspending and resuming is all you're likely to need. However, in some situations, your app may need to know when it's about to be suspended and when it has just been awakened. The system provides ways of notifying an app about changes to its execution state via the UIApplication class, which has a number of delegate methods and notifications for just this purpose. We'll show you how to use them later in this chapter.

When your application is about to be suspended, one thing it can do, regardless of whether it's one of the special backgroundable application types, is request a bit of additional time to run in the background. The idea is to make sure your app has enough time to close any open files, network resources, and so on. We'll give you an example of this in a bit.

Application Life Cycle

Before we get into the specifics of how to deal with changes to your app's execution state, let's talk a bit about the various states:

- **Not Running**: This is the state that all apps are in on a freshly rebooted device. An application that has been launched at any point after the device is turned on will return to this state only under specific conditions:

  - If its *Info.plist* includes the UIApplicationExitsOnSuspend key (with its value set to YES)

  - If it was previously Suspended and the system needs to clear out some memory

  - If it crashes while running

- **Active**: This is the normal running state of an application when it's displayed on the screen. It can receive user input and update the display.

- **Background**: In this state, an app is given some time to execute some code but can't directly access the screen or get any user input. All apps enter this state briefly when the user presses the home button; most of them quickly move on to the Suspended state. Apps that want to run in the background stay here until they're made Active again.

- **Suspended**: A Suspended app is frozen. This is what happens to normal apps after their brief stint in the Background state. All the memory the app was using while it was active is held just as it was. If the user brings the app back to the Active state, it will pick up right where it left off. On the other hand, if the system needs more memory for whichever app is currently Active, any Suspended apps may be terminated (and placed back into the Not Running state) and their memory freed for other use.

- **Inactive**: An app enters the Inactive state only as a temporary rest stop between two other states. The only way an app can stay Inactive for any length of time is if the user is dealing with a system prompt (such as those shown for an incoming call or SMS message) or if the user has locked the screen. This state is basically a sort of limbo.

State-Change Notifications

To manage changes between these states, UIApplication defines a number of methods that its delegate can implement. In addition to the delegate methods, UIApplication also defines a matching set of notification names (see Table 15–1). This allows other objects besides the app delegate to register for notifications when the application's state changes.

Table 15–1. *Delegate Methods for Tracking Your Application's Execution State and Their Corresponding Notification Names*

Delegate Method	Notification Name
application:didFinishLaunchingWithOptions:	UIApplicationDidFinishLaunchingNotification
applicationWillResignActive:	UIApplicationWillResignActiveNotification
applicationDidBecomeActive:	UIApplicationDidBecomeActiveNotification
applicationDidEnterBackground:	UIApplicationDidEnterBackgroundNotification
applicationWillEnterForeground:	UIApplicationWillEnterForegroundNotification
applicationWillTerminate:	UIApplicationWillTerminateNotification

Note that each of these methods is directly related to one of the running states: Active, Inactive, and Background. Each delegate method is called (and each notification posted) in only one of those states. The most important state transitions are between Active and other states. Some transitions, like from Background to Suspended, occur without any notice whatsoever. Let's go through these methods and discuss how they're meant to be used.

The first of these, application:didFinishLaunchingWithOptions:, is one you've already seen many times in this book. It's the primary way of doing application-level coding directly after the app has launched.

The next two methods, applicationWillResignActive: and applicationDidBecomeActive:, are both used in a number of circumstances. If the user presses the home button, applicationWillResignActive: will be called. If the user later brings the app back to the foreground, applicationDidBecomeActive: will be called. The same sequence of events occurs if the user receives a phone call. To top it all off, applicationDidBecomeActive: is also called when the application launches for the first time! In general, this pair of methods brackets the movement of an application from the Active state to the Inactive state. They are good places to enable and disable any animations, in-app audio, or other items that deal with the app's presentation to the user. Because of the multiple situations where applicationDidBecomeActive: is used, you may want to put some of your app initialization code there instead of in application:didFinishLaunchingWithOptions:. Note that you should not assume in applicationWillResignActive: that the application is about to be sent to the background, because it may just be a temporary change that ends up with a move back to the Active state.

After those methods come applicationDidEnterBackground: and applicationWillEnterForeground:, which have a slightly different usage area: dealing with an app that is definitely being sent to the background.

applicationDidEnterBackground: is where your app should free all resources that can be re-created later, save all user data, close network connections, and so on. This is also the spot where you can request more time to run in the background if you need to, as we'll demonstrate shortly. If you spend too much time doing things in applicationDidEnterBackground:—more than about five seconds—the system will decide that your app is misbehaving and terminate it. You should implement applicationWillEnterForeground: to re-create whatever was torn down in applicationDidEnterBackground:, such as reloading user data, reestablishing network connections, and so on. Note that when applicationDidEnterBackground: is called, you can safely assume that applicationWillResignActive: has also been recently called. Likewise, when applicationWillEnterForeground: is called, you can assume that applicationDidBecomeActive: will soon be called as well.

Last in the list is applicationWillTerminate:, which you'll probably use seldom, if ever. Prior to iOS 4, this was the method you would implement to save user data and so on, but now that applicationDidEnterBackground: exists, we don't need the old method. It is called only if your application is already in the background and the system decides to skip suspension for some reason and simply terminate the app.

Now, you should have a basic theoretical understanding of the states an application transitions between. Let's put this knowledge to the test with a simple app that does nothing more than write a message to Xcode's console log each time one of these methods is called. Then we'll manipulate the running app in a variety of ways, just as a user might, and see which transitions occur.

Creating State Lab

In Xcode, create a new project based on the *Single View Application* template, and name it *State Lab*. This app won't display anything but the default gray screen it's born with. All the output it's going to generate will end up in the Xcode console instead. The *BIDAppDelegate.m* file already contains all the methods we're interested in. We just need to add some logging, as shown in bold. Note that we've also removed the comments from these methods, just for the sake of brevity.

```
- (BOOL)application:(UIApplication *)application didFinishLaunchingWithOptions:
    (NSDictionary *)launchOptions
{
    self.window = [[UIWindow alloc] initWithFrame:[[UIScreen mainScreen] bounds]];
    // Override point for customization after application launch.
    NSLog(@"%@", NSStringFromSelector(_cmd));

    self.viewController = [[BIDViewController alloc]
        initWithNibName:@"BIDViewController" bundle:nil];
    self.window.rootViewController = self.viewController;
    [self.window makeKeyAndVisible];
    return YES;
}

- (void)applicationWillResignActive:(UIApplication *)application
{
```

```
    NSLog(@"%@", NSStringFromSelector(_cmd));
}

- (void)applicationDidEnterBackground:(UIApplication *)application
{
    NSLog(@"%@", NSStringFromSelector(_cmd));
}

- (void)applicationWillEnterForeground:(UIApplication *)application
{
    NSLog(@"%@", NSStringFromSelector(_cmd));
}

- (void)applicationDidBecomeActive:(UIApplication *)application
{
    NSLog(@"%@", NSStringFromSelector(_cmd));
}

- (void)applicationWillTerminate:(UIApplication *)application
{
    NSLog(@"%@", NSStringFromSelector(_cmd));
}
```

You may be wondering about that NSLog call we're using in all these methods. Objective-C provides a handy built-in variable called _cmd that always contains the selector of the current method. A **selector**, in case you need a refresher, is simply Objective-C's way of referring to a method. The NSStringFromSelector function returns an NSString representation of a given selector. Our usage here simply gives us a shortcut for outputting the current method name without needing to retype it or copy and paste it.

Exploring Execution States

Now, build and run the app. The simulator will appear and launch our application. Switch back to Xcode and take a look at the console (**View** ➤ **Debug Area** ➤ **Activate Console**), where you should see something like this:

```
2011-10-31 11:56:52.674 State Lab[83116:f803] application:didFinishLaunchingWithOptions:
2011-10-31 11:56:52.677 State Lab[83116:f803] applicationDidBecomeActive:
```

Here, you can see that the application has successfully launched and been moved into the Active state. Now, go back to the simulator and press the home button, and you should see the following in the console:

```
2011-10-31 11:56:55.874 State Lab[83116:f803] applicationWillResignActive:
2011-10-31 11:56:55.875 State Lab[83116:f803] applicationDidEnterBackground:
```

These two lines show the app actually transitioning between two states: it first becomes Inactive and then goes to Background. What you can't see here is that the app also switches to a third state: Suspended. Remember that you do not get any notification that this has happened; it's completely outside your control. Note that the app is still live in some sense, and Xcode is still connected to it, even though it's not actually getting any CPU time. Verify this by tapping the app's icon to relaunch it, which should produce this output:

```
2011-10-31 11:57:00.886 State Lab[83116:f803] applicationWillEnterForeground:
2011-10-31 11:57:00.888 State Lab[83116:f803] applicationDidBecomeActive:
```

There you are, back in business. The app was previously Suspended, is woken up to Inactive, and then ends up Active again. So, what happens when the app is really terminated? Tap the home button again, and you'll see this:

```
2011-10-31 11:57:03.569 State Lab[83116:f803] applicationWillResignActive:
2011-10-31 11:57:03.570 State Lab[83116:f803] applicationDidEnterBackground:
```

Now double-tap the home button. The shelf of apps should appear. Press and hold the State Lab icon until the little "kill" icon (the minus in a red circle) comes up. Press the kill icon to terminate State Lab. What happens? You may be surprised to see that none of our NSLog calls print anything to the console. Instead, Xcode itself prints a somewhat cryptic line like sharedlibrary apply-load-rules all before leaving you with a (gdb) prompt. At this point, State Lab is truly and completely terminated.

As it turns out, the applicationWillTerminate: method isn't normally called when the system is moving an app from Suspended to Not Running state. When an app is Suspended, whether the system decides to dump it to reclaim memory or you manually force-quit it, the app simply vanishes and doesn't get a chance to do anything. The applicationWillTerminate: method is called only if the app being terminated is in the Background state. This can occur, for instance, if your app is actively running in the Background state, using system resources in one of the predefined ways (audio playback, GPS usage, and so on) and is force-quit either by the user or by the system. In the case we just explored with State Lab, the app was in the Suspended state, not Background, and was therefore terminated immediately without any notification.

There's one more interesting interaction to examine here. It's what happens when the system shows an alert dialog, temporarily taking over the input stream from the app and putting it into an Inactive state. This state can be readily triggered only when running on a real device instead of the simulator, using the built-in Messages app. Messages, like many other apps, can receive messages from the outside and display them in several ways.

To see how these are set up, run the Settings app on your device, choose Notifications from the list at the upper left, and then select the Messages app from the list of apps on the right. The hot new way to show messages in iOS 5 is called Banners. This works by showing a small banner overlaid at the top of the screen, which doesn't need to interrupt whatever app is currently being run. What we want to show is the bad old Alerts method, which makes a modal panel appear in front of the current app, requiring a user action. Select that, so that the Messages app turns back into the kind of pushy jerk that users of iOS 4 and below always had to deal with.

Now back to your computer. In Xcode, use the popup at the upper left to switch from the simulator to your device, and then hit the *Run* button to build and run the app on your device. Now, all you need to do is send a message to your device from the outside. If your device is an iPhone, you can send it an SMS message from another phone. If it's an iPod touch or an iPad, you're limited to Apple's own iMessage communication, which works on all iOS 5 devices (including the iPhone). Figure out what works for your setup, and have someone else send your device a message via SMS or iMessage. When your device displays the system alert showing the incoming message, this will appear in the Xcode console:

```
2011-10-31 12:05:15.391 State Lab[1069:307] applicationWillResignActive:
```

Note that our app didn't get sent to the background. It's in the Inactive state and can still be seen behind the system alert. If this app were a game or had any video, audio, or animations running, this is where we would probably want to pause them.

Press the Close button on the alert, and you'll get this:

```
2011-10-31 12:05:24.808 State Lab[1069:307] applicationDidBecomeActive:
```

Now let's see what happens if you decide to reply to the message instead. Have someone send you another message, generating this:

```
2011-10-31 12:11:04.154 State Lab[1069:307] applicationWillResignActive:
```

This time hit Reply, which switches you over to the Messages app, and you should see the following flurry of activity:

```
2011-10-31 12:11:07.826 State Lab[1069:307] applicationDidBecomeActive:
2011-10-31 12:11:07.966 State Lab[1069:307] applicationWillResignActive:
2011-10-31 12:11:07.984 State Lab[1069:307] applicationDidEnterBackground:
```

Interesting! Our app quickly becomes Active again, then becomes Inactive again, and finally goes to Background (and then, silently, Suspended).

Making Use of Execution State Changes

So, what should we make of all this? Based on what we've just demonstrated, it seems like there's a clear strategy to follow when dealing with these state changes:

Active ➤ Inactive

Use applicationWillResignActive:/UIApplicationWillResignActiveNotification to "pause" your app's display. If your app is a game, you probably already have the ability to pause the gameplay in some way. For other kinds of apps, make sure no time-critical demands for user input are in the works, because your app won't be getting any user input for a while.

Inactive ➤ Background

Use applicationDidEnterBackground:/UIApplicationDidEnterBackgroundNotification to release any resources that don't need to be kept around when the app is backgrounded (such as cached images or other easily reloadable data) or that wouldn't survive backgrounding anyway (such as active network connections). Getting rid of excess memory usage here will make your app's eventual Suspended snapshot smaller, thereby decreasing the risk that your app will be purged from RAM entirely. You should also use this opportunity to save any application data that will help your users pick up where they left off the next time your app is relaunched. If your app comes back to the

Active state, normally this won't matter, but in case it's purged and must be relaunched, your users will appreciate starting off in the same place.

Background ➤ Inactive

Use applicationWillEnterForeground:/UIApplicationWillEnterForeground to undo anything you did when switching from Inactive to Background. For example, here you can reestablish persistent network connections.

Inactive ➤ Active

Use applicationDidBecomeActive:/UIApplicationDidBecomeActive to undo anything you did when switching from Active to Inactive. Note that if your app is a game, this probably does not mean dropping out of pause straight to the game; you should let your users do that on their own. Also keep in mind that this method and notification are used when an app is freshly launched, so anything you do here must work in that context as well.

There is one special consideration for the **Inactive ➤ Background** transition. Not only does it have the longest description in the previous list, but it's also probably the most code- and time-intensive transition in applications because of the amount of bookkeeping you may want your app to do. When this transition is underway, the system won't give you the benefit of an unlimited amount of time to save your changes here. It gives you about five seconds. If your app takes longer than that to return from the delegate method (and handle any notifications you've registered for), then your app will be summarily purged from memory and pushed into the Not Running state! If this seems unfair, don't worry, because there is a reprieve available. While handling that delegate method or notification, you can ask the system to perform some additional work for you in a background queue, which buys you some extra time. We'll demonstrate that technique in the next section.

Handling the Inactive State

The simplest state change your app is likely to encounter is from Active to Inactive and then back to Active. You may recall that this is what happens if your iPhone receives an SMS message while your app is running and displays it for the user. In this section, we're going to make State Lab do something visually interesting so that you can see what happens if you ignore that state change, and then we'll show you how to fix it.

We'll add a UILabel to our display and make it move using Core Animation, which is a really nice way of animating objects in iOS.

Start by adding a UILabel as an instance variable and property in *BIDViewController.h*:

```
@interface BIDViewController : UIViewController

@property (strong, nonatomic) UILabel *label;
@end
```

Then do the usual memory-management work for this property in *BIDViewController.m*:

```
@implementation BIDViewController
@synthesize label;
    .
    .
    .
- (void)viewDidUnload {
    [super viewDidUnload];
    // Release any retained subviews of the main view.
    // e.g. self.myOutlet = nil;
    self.label = nil;
}
```

Now, let's set up the label when the view loads. Add the bold lines shown here to the viewDidLoad method:

```
- (void)viewDidLoad {
    [super viewDidLoad];
    // Do any additional setup after loading the view, typically from a nib.

    CGRect bounds = self.view.bounds;
    CGRect labelFrame = CGRectMake(bounds.origin.x, CGRectGetMidY(bounds) - 50,
                                   bounds.size.width, 100);
    self.label = [[UILabel alloc] initWithFrame:labelFrame];
    label.font = [UIFont fontWithName:@"Helvetica" size:70];
    label.text = @"Bazinga!";
    label.textAlignment = UITextAlignmentCenter;
    label.backgroundColor = [UIColor clearColor];
    [self.view addSubview:label];
}
```

It's time to set up some animation. We'll define two methods: one to rotate the label to an upside-down position and one to rotate it back to normal. Let's declare these methods in an class extension at the top of the file, just before the class's @implementation begins:

```
@interface BIDViewController ()
- (void)rotateLabelUp;
- (void)rotateLabelDown;
@end
```

The method definitions themselves can then be inserted anywhere within the @implementation block:

```
- (void)rotateLabelDown {
    [UIView animateWithDuration:0.5
                     animations:^{
                         label.transform = CGAffineTransformMakeRotation(M_PI);
                     }
                     completion:^(BOOL finished){
                         [self rotateLabelUp];
                     }];
}

- (void)rotateLabelUp {
    [UIView animateWithDuration:0.5
                     animations:^{
```

```
            label.transform = CGAffineTransformMakeRotation(0);
        }
        completion:^(BOOL finished){
            [self rotateLabelDown];
        }];
}
```

This deserves a bit of explanation. UIView defines a class method called animateWithDuration:animations:completion:, which sets up an animation. Any animatable attributes that we set within the animations block don't have an immediate effect on the receiver. Instead, Core Animation will smoothly transition that attribute from its current value to the new value we specify. This is what's called an **implicit animation** and is one of the main features of Core Animation. The final completion block lets us specify what will happen after the animation is complete.

So, each of these methods sets the label's transform property to a particular rotation, specified in radians. Each also sets up a completion block to just call the other method, so the text will continue to animate back and forth forever.

Finally, we need to set up a way to kick-start the animation. For now, we'll do this by adding this line at the end of viewDidLoad (but we'll change this later, for reasons we'll describe at that time):

```
    [self rotateLabelDown];
```

Now, build and run the app. You should see the Bazinga! label rotate back and forth (see Figure 15–4).

To test the Active ➤ Inactive transition, you really need to once again run this on an actual iPhone and send an SMS message to it from elsewhere. Unfortunately, there's no way to simulate this behavior in any version of the iOS simulator that Apple has released so far. If you don't yet have the ability to build and install on a device or don't have an iPhone, you won't be able to try this for yourself, but please follow along as best you can!

Build and run the app on an iPhone, and see that the animation is running along. Now, send an SMS message to the device. When the system alert comes up to show the message, you'll see that the animation keeps on running! That may be slightly comical, but it's probably irritating for a user. We will use transition notifications to stop our animation when this occurs.

Figure 15–4. *The State Lab application doing its label rotating magic*

Our controller class will need to have some internal state to keep track of whether it should be animating at any given time. For this purpose, let's add an ivar to *BIDViewController.m*. Because this simple BOOL doesn't need to be accessed by any outside classes, we skip the header and add it to the class extension we created earlier.

```
@interface BIDViewController ()
@property (assign, nonatomic) BOOL animate;
- (void)rotateLabelUp;
- (void)rotateLabelDown;
@end

@implementation BIDViewController
@synthesize label;
@synthesize animate;
```

Since our class isn't the application delegate, we can't just implement the delegate methods and expect them to work. Instead, we sign up to receive notifications from the application when the execution state changes. Do this by adding a few lines at the top of the viewDidLoad method in BIDViewController.m:

```
- (void)viewDidLoad {
    [super viewDidLoad];
    // Do any additional setup after loading the view, typically from a nib.
```

```
    [[NSNotificationCenter defaultCenter] addObserver:self
                            selector:@selector(applicationWillResignActive)
                                name:UIApplicationWillResignActiveNotification
                              object:[UIApplication sharedApplication]];
    [[NSNotificationCenter defaultCenter] addObserver:self
                            selector:@selector(applicationDidBecomeActive)
                                name:UIApplicationDidBecomeActiveNotification
                              object:[UIApplication sharedApplication]];
    CGRect bounds = self.view.bounds;
.
.
.
```

This sets up these two notifications to each call a method in our class at the appropriate time. Define these methods anywhere you like inside the @implementation block:

```
- (void)applicationWillResignActive {
    NSLog(@"VC: %@", NSStringFromSelector(_cmd));
    animate = NO;
}

- (void)applicationDidBecomeActive {
    NSLog(@"VC: %@", NSStringFromSelector(_cmd));
    animate = YES;
    [self rotateLabelDown];
}
```

Here, we've included the same method logging as before, just so you can see where the methods occur in the Xcode console. We added the preface "VC: " to distinguish this call from the NSLog() calls in the delegate (VC is for view controller). The first of these methods just turns off the animate flag. The second turns the flag back on, and then actually starts up the animations again. For that first method to have any effect, we need to add some code to check the animate flag and keep on animating only if it's enabled.

```
- (void)rotateLabelUp {
    [UIView animateWithDuration:0.5
                    animations:^{
                        label.transform = CGAffineTransformMakeRotation(0);
                    }
                    completion:^(BOOL finished){
                        if (animate) {
                            [self rotateLabelDown];
                        }
                    }];
}
```

We added this to the completion block of rotateLabelUp, and only there, so that our animation will stop only when the text is right-side up.

Now, build and run the app again, and see what happens. Chances are, you'll see some flickery madness, with the label rapidly flipping up and down, not even rotating! The reason for this is simple but perhaps not obvious (though we did hint at it earlier).

Remember that we started up the animations at the end of viewDidLoad by calling rotateLabelDown? Well, we're now calling rotateLabelDown in `applicationDidBecomeActive` as well. And remember that `applicationDidBecomeActive` will be called not only when we switch from Inactive back to Active, but also when the app launches and becomes Active in the first place! That means we're starting our animations twice, and Core Animation doesn't seem to deal well with multiple animations trying to change the same attributes at the same time. The solution is simply to delete the line you previously added at the end of `viewDidLoad`:

~~[self rotateLabelDown];~~

Now build and run the app again, and you should see that it's animating properly. Once again, send an SMS message to your iPhone. This time when the system alert appears, you'll see that the animation in the background stops as soon as the text is right-side up. Tap the Close button, and the animation starts back up.

Now you've seen what to do for the simple case of switching from Active to Inactive and back. The bigger task, and perhaps the more important one, is dealing with a switch to the background and then back to foreground.

Handling the Background State

As we mentioned earlier, switching to the Background state is pretty important to ensure the best possible user experience. This is the spot where you'll want to discard any resources that can easily be reacquired (or will be lost anyway when your app goes silent) and save information about your app's current state, all without occupying the main thread for more than five seconds.

To demonstrate some of these behaviors, we're going to extend State Lab in a few ways. First, we're going to add an image to the display so that we can later show you how to get rid of the in-memory image. Then we're going to show you how to save some information about the app's state so we can easily restore it later. Finally, we'll show you how to make sure these activities aren't taking up too much main thread time, by putting all this work into a background queue.

Removing Resources When Entering the Background

Start by adding *smiley.png* from the book's source archive to your project's *State Lab* folder. Be sure to enable the checkbox that tells Xcode to copy the file to your project directory.

Now, let's add properties for both an image and an image view to *BIDViewController.h*:

```
@interface BIDViewController : UIViewController

@property (strong, nonatomic) UILabel *label;
@property (strong, nonatomic) UIImage *smiley;
@property (strong, nonatomic) UIImageView *smileyView;
@end
```

Then switch to the *.m* file again, and add the usual memory-management code:

```
@implementation BIDViewController
```

```
@synthesize label;
@synthesize animate;
@synthesize smiley, smileyView;
.
.
.
- (void)viewDidUnload {
    // Release any retained subviews of the main view.
    // e.g. self.myOutlet = nil;
    self.label = nil;
    self.smiley = nil;
    self.smileyView = nil;
    [super viewDidUnload];
}
```

Now, let's set up the image view and put it on the screen by modifying the viewDidLoad method as shown here:

```
- (void)viewDidLoad {
    [super viewDidLoad];
    [[NSNotificationCenter defaultCenter] addObserver:self
                             selector:@selector(applicationWillResignActive)
                                 name:UIApplicationWillResignActiveNotification
                               object:[UIApplication sharedApplication]];
    [[NSNotificationCenter defaultCenter] addObserver:self
                             selector:@selector(applicationDidBecomeActive)
                                 name:UIApplicationDidBecomeActiveNotification
                               object:[UIApplication sharedApplication]];
    CGRect bounds = self.view.bounds;
    CGRect labelFrame = CGRectMake(bounds.origin.x, CGRectGetMidY(bounds) - 50,
                            bounds.size.width, 100);
    self.label = [[UILabel alloc] initWithFrame:labelFrame];
    label.font = [UIFont fontWithName:@"Helvetica" size:70];
    label.text = @"Bazinga!";
    label.textAlignment = UITextAlignmentCenter;
    label.backgroundColor = [UIColor clearColor];

    // smiley.png is 84 x 84
    CGRect smileyFrame = CGRectMake(CGRectGetMidX(bounds) - 42,
                            CGRectGetMidY(bounds)/2 - 42,
                            84, 84);
    self.smileyView = [[UIImageView alloc] initWithFrame:smileyFrame];
    self.smileyView.contentMode = UIViewContentModeCenter;
    NSString *smileyPath = [[NSBundle mainBundle] pathForResource:@"smiley"
                                                    ofType:@"png"];
    self.smiley = [UIImage imageWithContentsOfFile:smileyPath];
    self.smileyView.image = self.smiley;

    [self.view addSubview:smileyView];
    [self.view addSubview:label];
}
```

Build and run the app, and you'll see the incredibly happy-looking smiley face toward the top of your screen (see Figure 15–5).

Figure 15–5. *The State Lab application doing its label-rotating magic with the addition of a smiley icon*

Next, press the home button to switch your app to the background, and then tap its icon to launch it again. You'll see that the app starts up right where it left off. That's good for the user, but we're not optimizing system resources as well as we could.

Remember that the fewer resources we use while our app is Suspended, the lower the risk that iOS will terminate our app entirely. By clearing any easily re-created resources from memory when we can, we increase the chance that our app will stick around and therefore relaunch super-quickly.

Let's see what we can do about that smiley face. We would really like to free up that image when going to the Background state and re-create it when coming back from the Background state. To do that, we'll need to add two more notification registrations toward the top of viewDidLoad, just after [super viewDidLoad]:

```
[[NSNotificationCenter defaultCenter] addObserver:self
                           selector:@selector(applicationDidEnterBackground)
                               name:UIApplicationDidEnterBackgroundNotification
                             object:[UIApplication sharedApplication]];
[[NSNotificationCenter defaultCenter] addObserver:self
                           selector:@selector(applicationWillEnterForeground)
                               name:UIApplicationWillEnterForegroundNotification
                             object:[UIApplication sharedApplication]];
```

And we want to implement the two new methods:

```
- (void)applicationDidEnterBackground {
    NSLog(@"VC: %@", NSStringFromSelector(_cmd));
    self.smiley = nil;
    self.smileyView.image = nil;
}

- (void)applicationWillEnterForeground {
    NSLog(@"VC: %@", NSStringFromSelector(_cmd));
    NSString *smileyPath = [[NSBundle mainBundle] pathForResource:@"smiley"
                                                           ofType:@"png"];
```

```
    self.smiley = [UIImage imageWithContentsOfFile:smileyPath];
    self.smileyView.image = self.smiley;
}
```

Build and run the app, and repeat the same steps of backgrounding your app and switching back to it. You should see that from the user's standpoint, the behavior appears to be about the same. If you want to verify for yourself that this is really happening, comment out the contents of the applicationWillEnterForeground method, and build and run the app again. You'll see that the image really does disappear.

Saving State When Entering the Background

Now that you've seen an example of how to free up some resources when entering the Background state, it's time to think about saving state. Remember that the idea is to save all information relevant to what the user is doing, so that in case your application is later dumped from memory, the next time users return, they can still pick up right where they left off.

The kind of state we're talking about here is really application-specific. You might want to keep track of which document users were looking at, their cursor location in a text field, which application view was open, and so on. In our case, we're just going to keep track of the selection in a segmented control.

Start by adding a new instance variable and property in *BIDViewController.h*:

```
@interface BIDViewController : UIViewController

@property (strong, nonatomic) UILabel *label;
@property (strong, nonatomic) UIImage *smiley;
@property (strong, nonatomic) UIImageView *smileyView;
@property (strong, nonatomic) UISegmentedControl *segmentedControl;
@end
```

Then implement the usual boilerplate code for accessors and memory management in *BIDViewController.m*:

```
    .
    .
    .
@implementation BIDViewController

@synthesize label;
@synthesize smiley, smileyView;
@synthesize segmentedControl;
    .
    .
    .
- (void)viewDidUnload {
    [super viewDidUnload];
    // Release any retained subviews of the main view.
    // e.g. self.myOutlet = nil;
    self.label = nil;
    self.smiley = nil;
    self.smileyView = nil;
```

```
        self.segmentedControl = nil;
}
```

Now, move to the end of the viewDidLoad method, where we'll create the segmented control and add it to the view:

```
    .
    .
    .

    self.smileyView.image = self.smiley;

    self.segmentedControl = [[UISegmentedControl alloc] initWithItems:
                                [NSArray arrayWithObjects:
                                    @"One", @"Two", @"Three", @"Four", nil]] ;
    self.segmentedControl.frame = CGRectMake(bounds.origin.x + 20,
                                        CGRectGetMaxY(bounds) - 50,
                                        bounds.size.width - 40, 30);

    [self.view addSubview:segmentedControl];
    [self.view addSubview:smileyView];
    [self.view addSubview:label];
}
```

Build and run the app. You should see the segmented control and be able to click its segments to select them one at a time.

Here's where we should mention a slight backward-compatibility issue, since something subtle but important has changed between iOS 4 and iOS 5. To see the difference, we're going to blast into the past, to the dark days of iOS 4.3, and see what happens. Don't worry; you don't need to downgrade a device. We'll do it in Xcode and the iOS simulator.

A Brief Journey to Yesteryear

In the project navigator, select the topmost item, the one that represents your project, to examine your project details. You've seen this view before, which shows you various settings for the project and the application target. Select the *State Lab* item in the *TARGETS* section, and make sure the *Summary* tab is selected at the top of the detail view. The uppermost section, *iOS Application Target*, contains a *Deployment Target* popup, currently set to the latest version of iOS that your copy of Xcode knows about. Click the control and choose *4.3*. This tells Xcode not only that it should build the app with iOS 4.3 in mind, but also run the iOS simulator using iOS 4.3. Next, click the scheme/device popup control near the upper left of the window, and choose *iPhone 4.3 Simulator* from the popup. You're now ready to take a trip in the way-back machine!

If you build and run your app at this point, you'll see one glaring problem: the segmented control doesn't seem to work! You can tap those segments all you like, and nothing will happen. The problem actually lies with the animation. By default, the Core Animation method we used to set up animation actually prevents some amount of user input from being collected while animations are running (presumably this is a performance optimization). The key difference here between iOS 4 and iOS 5 is that while iOS 5 turns off user interaction for the views that are currently animated, iOS 4 turns off user interaction for the entire application!

Fortunately, there is an optional way to enable user interaction, which requires us to use a longer method name in each of our rotate methods. Modify them as shown here:

```
- (void)rotateLabelDown {
    [UIView animateWithDuration:0.5
                    delay:0
                  options:UIViewAnimationOptionAllowUserInteraction
               animations:^{
                   label.transform = CGAffineTransformMakeRotation(M_PI);
               }
               completion:^(BOOL finished){
                   [self rotateLabelUp]; }];
}

- (void)rotateLabelUp {
    [UIView animateWithDuration:0.5
                    delay:0
                  options:UIViewAnimationOptionAllowUserInteraction
               animations:^{
                   label.transform = CGAffineTransformMakeRotation(0);
               }
               completion:^(BOOL finished){
                   if (animate) {
                       [self rotateLabelDown];
                   }
               }];
}
```

Build and run the app again, and see what happens. That's more like it, eh? As we said, this difference between iOS 4 and iOS 5 is subtle, but quite important if your apps use Core Animation and you need to support iOS 4. Though you could strip that code back out, since we're about to return to the iOS 5 simulator, there's really no harm in leaving it in and allowing the code to work under iOS 4 as well.

Back to the Background

Let's return to the present. Select **iPhone 5.0 Simulator** from the popup menu in the upper-left part of the project window. Now, touch any one of the four segments, and then go through the now-familiar sequence of backgrounding your app and bringing it back up. You'll see that the segment you chose (bet it was Three) is still selected—no surprise there. Background your app again by clicking the home button, bring up the taskbar (by double-clicking the home button) and kill your app, and then relaunch it. You'll find yourself back at square one, with no segment selected. That's what we need to fix next.

> **CAUTION:** When you kill your app in the simulator, it's possible (depending on which version of Xcode you are running) that upon relaunching your app, you'll find yourself back in Xcode as the result of a SIGKILL signal. This is perfectly normal. If this happens, click the stop button at the top left of the project window, and then rerun your project to bring your project back to life in the simulator.

Saving the selection is simple enough. We just need to add a few lines to the end of the applicationDidEnterBackground method in *BIDViewController.m*:

```
- (void)applicationDidEnterBackground {
    NSLog(@"VC: %@", NSStringFromSelector(_cmd));
    self.smiley = nil;
    self.smileyView.image = nil;

    NSInteger selectedIndex = self.segmentedControl.selectedSegmentIndex;
    [[NSUserDefaults standardUserDefaults] setInteger:selectedIndex
                                forKey:@"selectedIndex"];
}
```

But where should we restore this selection index and use it to configure the segmented control? The inverse of this method, applicationWillEnterForeground, isn't what we want. When that method is called, the app has already been running, and the setting is still intact. Instead, we need to access this when things are being set up after a new launch, which brings us back to the viewDidLoad method. Add the bold lines shown here at the end of the method:

```
    .
    .
    .

    [self.view addSubview:label];

    NSNumber *indexNumber;
    if (indexNumber = [[NSUserDefaults standardUserDefaults]
                    objectForKey:@"selectedIndex"]) {
        NSInteger selectedIndex = [indexNumber intValue];
        self.segmentedControl.selectedSegmentIndex = selectedIndex;
    }
}
```

We needed to include a little sanity check here to see whether there's a value stored for the selectedIndex key, to cover cases such as the first app launch, where nothing has been selected.

Now build and run the app, touch a segment, and then do the full background-kill-restart dance. There it is—your selection is intact!

Obviously, what we've shown here is pretty minimal, but the concept can be extended to all kinds of application state. It's up to you to decide how far you want to take it in order to maintain the illusion for the users that your app was always there, just waiting for them to come back!

Requesting More Backgrounding Time

Earlier, we mentioned the possibility of your app being dumped from memory if moving to the Background state takes too much time. For example, your app may be in the middle of doing a file transfer that it would really be a shame not to finish, but trying to hijack the applicationDidEnterBackground method to make it complete the work there, before the application is really backgrounded, isn't really an option. Instead, you should use applicationDidEnterBackground as a platform for telling the system that you have some extra work you would like to do, and then start up a block to actually do it. Assuming that the system has enough available RAM to keep your app in memory while the user does something else, the system will oblige you and keep your app running for a while.

We'll demonstrate this not with an actual file transfer, but with a simple sleep call. Once again, we'll be using our new acquaintances GCD and blocks to make the contents of our applicationDidEnterBackground method run in a separate queue.

In *BIDViewController.m*, modify the applicationDidEnterBackground method as follows:

```
- (void)applicationDidEnterBackground {
    NSLog(@"VC: %@", NSStringFromSelector(_cmd));
    UIApplication *app = [UIApplication sharedApplication];

    __block UIBackgroundTaskIdentifier taskId;
    taskId = [app beginBackgroundTaskWithExpirationHandler:^{
        NSLog(@"Background task ran out of time and was terminated.");
        [app endBackgroundTask:taskId];
    }];

    if (taskId == UIBackgroundTaskInvalid) {
        NSLog(@"Failed to start background task!");
        return;
    }

    dispatch_async(dispatch_get_global_queue(0, 0), ^{
        NSLog(@"Starting background task with %f seconds remaining",
                app.backgroundTimeRemaining);
        self.smiley = nil;
        self.smileyView.image = nil;

        NSInteger selectedIndex = self.segmentedControl.selectedSegmentIndex;
        [[NSUserDefaults standardUserDefaults] setInteger:selectedIndex
                                        forKey:@"selectedIndex"];
        // simulate a lengthy (25 seconds) procedure
        [NSThread sleepForTimeInterval:25];

        NSLog(@"Finishing background task with %f seconds remaining",
                app.backgroundTimeRemaining);
        [app endBackgroundTask:taskId];
    });
}
```

Let's look through this code piece by piece. First, we grab the shared UIApplication instance, since we'll be using it several times in this method. Then comes this:

```
__block UIBackgroundTaskIdentifier taskId;
taskId = [app beginBackgroundTaskWithExpirationHandler:^{
    NSLog(@"Background task ran out of time and was terminated.");
    [app endBackgroundTask:taskId];
}];
```

The call to beginBackgroundTaskWithExpirationHandler: returns an identifier that we'll need to keep track of for later use. We've declared the taskId variable it's stored in with the __block storage qualifier, since we want to be sure the identifier returned by the method is shared among any blocks we create in this method.

With the call to beginBackgroundTaskWithExpirationHandler:, we're basically telling the system that we need more time to accomplish something, and we promise to let it know when we're finished. The block we give as a parameter may be called if the system decides that we've been going way too long anyway and decides to stop running.

Note that the block we gave ended with a call to endBackgroundTask:, passing along taskId. That tells the system that we're finished with the work for which we previously requested extra time. It's important to balance each call to beginBackgroundTaskWithExpirationHandler: with a matching call to endBackgroundTask: so that the system knows when we've completed the work.

> **NOTE:** Depending on your computing background, the use of the word *task* here may evoke associations with what we usually call a *process*, consisting of a running program that may contain multiple threads, and so on. In this case, try to put that out of your mind. The use of *task* in this context really just means "something that needs to get done." Any task you create here is running within your still-executing app.

Next, we do this:

```
if (taskId == UIBackgroundTaskInvalid) {
    NSLog(@"Failed to start background task!");
    return;
}
```

If our earlier call to beginBackgroundTaskWithExpirationHandler: returned the special value UIBackgroundTaskInvalid, that means the system is refusing to grant us any additional time. In that case, you could try to do the quickest part of whatever needs doing anyway and hope that it completes quickly enough that your app won't be terminated before it's finished. This is mostly likely to be an issue when running on older devices, such as the iPhone 3G, that let you run iOS 4 but don't support multitasking. In this example, however, we're just letting it slide.

Next comes the interesting part where the work itself is actually done:

```
dispatch_async(dispatch_get_global_queue(0, 0), ^{
    NSLog(@"Starting background task with %f seconds remaining",
        app.backgroundTimeRemaining);
```

```
        self.smiley = nil;
        self.smileyView.image = nil;

        NSInteger selectedIndex = self.segmentedControl.selectedSegmentIndex;
        [[NSUserDefaults standardUserDefaults] setInteger:selectedIndex
                                        forKey:@"selectedIndex"];
        // simulate a lengthy (25 seconds) procedure
        [NSThread sleepForTimeInterval:25];

        NSLog(@"Finishing background task with %f seconds remaining",
            app.backgroundTimeRemaining);
        [app endBackgroundTask:taskId];
    });
```

All this does is take the same work our method was doing in the first place and place it in a background queue. At the end of that block, we call endBackgroundTask: to let the system know that we're finished.

With that in place, build and run the app, and then background your app by pressing the home button. Watch the Xcode console as well as the status bar at the bottom of the Xcode window. You'll see that this time, your app stays running (you don't get the "Debugging terminated" message in the status bar), and after 25 seconds, you will see the final log in your output. A complete run of the app up to this point should give you console output along these lines:

```
2011-10-30 22:35:28.608 State Lab[7449:207] application:didFinishLaunchingWithOptions:
2011-10-30 22:35:28.616 State Lab[7449:207] applicationDidBecomeActive:
2011-10-30 22:35:28.617 State Lab[7449:207] VC: applicationDidBecomeActive
2011-10-30 22:35:31.869 State Lab[7449:207] applicationWillResignActive:
2011-10-30 22:35:31.870 State Lab[7449:207] VC: applicationWillResignActive
2011-10-30 22:35:31.871 State Lab[7449:207] applicationDidEnterBackground:
2011-10-30 22:35:31.873 State Lab[7449:207] VC: applicationDidEnterBackground
2011-10-30 22:35:31.874 State Lab[7449:1903] Starting background task with 599.995069
seconds remaining
2011-10-30 22:35:56.877 State Lab[7449:1903] Finishing background task with 574.993956
seconds remaining
```

As you can see, the system is much more generous with time when doing things in the background than in the main thread of your app, so following this procedure can really help you out if you have any ongoing tasks to deal with.

Note that we asked for a single background task identifier, but in practice, you can ask for as many as you need. For example, if you have multiple network transfers happening at Background time and you need to complete them, you can ask for an identifier for each and allow them to continue running in a background queue. So, you can easily allow multiple operations to run in parallel during the available time. Also consider that the task identifier you receive is a normal C-language value (not an object), and apart from being stored in a local __block variable, it can just as well be stored as an instance variable if that better suits your class design.

Grand Central Dispatch, Over and Out

This has been a pretty heavy chapter, with a lot of new concepts thrown your way. Not only have you learned about a complete new feature set Apple added to the C language, but you've also discovered a new conceptual paradigm for dealing with concurrency without worrying about threads. We also demonstrated some techniques for making sure your apps play nicely in the multitasking world of iOS. Now that we've gotten some of this heavy stuff out of the way, let's move on to the next chapter, which focuses on drawing. Pencils out, let's draw!

Chapter 16

Drawing with Quartz and OpenGL

Every application we've built so far has been constructed from views and controls that are part of the UIKit framework. You can do a lot with the UIKit stock components, and a great many applications can be constructed using only these objects. Some applications, however, can't be fully realized without going beyond what the UIKit stock components offer.

Sometimes, an application needs to be able to do custom drawing. Fortunately, we have not one, but two separate libraries we can call on for our drawing needs:

- Quartz 2D, which is part of the Core Graphics framework
- OpenGL ES, which is a cross-platform graphics library

OpenGL ES is a slimmed-down version of another cross-platform graphic library called OpenGL. OpenGL ES is a subset of OpenGL, designed specifically for embedded systems (hence the letters *ES*), such as the iPhone, iPad, and iPod touch.

In this chapter, we'll explore these powerful graphics environments. We'll build sample applications in both, and give you a sense of which environment to use and when.

Two Views of a Graphical World

Although Quartz 2D and OpenGL ES overlap a lot in terms of what you can accomplish with them, there are distinct differences between them.

Quartz 2D is a set of functions, datatypes, and objects designed to let you draw directly into a view or an image in memory. Quartz 2D treats the view or image that is being drawn into as a virtual canvas. It follows what's called a **painter's model**, which is just a fancy way of saying that the drawing commands are applied in much the same way as paint is applied to a canvas.

If a painter paints an entire canvas red, and then paints the bottom half of the canvas blue, the canvas will be half red and half either blue or purple (blue if the paint is opaque; purple if the paint is semitransparent). Quartz 2D's virtual canvas works the same way. If you paint the whole view red, and then paint the bottom half of the view blue, you'll have a view that's half red and half either blue or purple, depending on whether the second drawing action was fully opaque or partially transparent. Each drawing action is applied to the canvas on top of any previous drawing actions.

On the other hand, OpenGL ES is implemented as a **state machine**. This concept is somewhat more difficult to grasp, because it doesn't resolve to a simple metaphor like painting on a virtual canvas. Instead of letting you take actions that directly impact a view, window, or image, OpenGL ES maintains a virtual three-dimensional world. As you add objects to that world, OpenGL ES keeps track of the state of all objects.

Instead of a virtual canvas, OpenGL ES gives you a virtual window into its world. You add objects to the world and define the location of your virtual window with respect to the world. OpenGL ES then draws what you can see through that window based on the way it is configured and where the various objects are in relation to each other. This concept is a bit abstract, but it will make more sense when we build our OpenGL ES drawing application later in this chapter.

Quartz 2D provides a variety of line, shape, and image drawing functions. Though easy to use, Quartz 2D is limited to two-dimensional drawing. Although many Quartz 2D functions do result in drawing that takes advantage of hardware acceleration, there is no guarantee that any particular action you take in Quartz 2D will be accelerated.

OpenGL ES, though considerably more complex and conceptually more difficult, offers a lot more power than Quartz 2D. It has tools for both two-dimensional and three-dimensional drawing, and is specifically designed to take full advantage of hardware acceleration. OpenGL ES is also extremely well suited to writing games and other complex, graphically intensive programs.

Now that you have a general idea of the two drawing libraries, let's try them out. We'll start with the basics of how Quartz 2D works, and then build a simple drawing application with it. Then we'll re-create the same application using OpenGL ES.

The Quartz 2D Approach to Drawing

When using Quartz 2D (Quartz for short), you'll usually add the drawing code to the view doing the drawing. For example, you might create a subclass of UIView and add Quartz function calls to that class's drawRect: method. The drawRect: method is part of the UIView class definition and is called every time a view needs to redraw itself. If you insert your Quartz code in drawRect:, that code will be called, and then the view will redraw itself.

Quartz 2D's Graphics Contexts

In Quartz, as in the rest of Core Graphics, drawing happens in a **graphics context**, usually referred to just as a **context**. Every view has an associated context. You retrieve the current context, use that context to make various Quartz drawing calls, and let the context worry about rendering your drawing onto the view.

This line of code retrieves the current context:

```
CGContextRef context = UIGraphicsGetCurrentContext();
```

> **NOTE:** Notice that we're using Core Graphics C functions, rather than Objective-C objects, to do our drawing. Both Core Graphics and OpenGL are C-based APIs, so most of the code we write in this part of the chapter will consist of C function calls.

Once you've defined your graphics context, you can draw into it by passing the context to a variety of Core Graphics drawing functions. For example, this sequence will create a **path** consisting of a 4-pixel-wide line in the context, and then draw that line:

```
CGContextSetLineWidth(context, 4.0);
CGContextSetStrokeColorWithColor(context, [UIColor redColor].CGColor);
CGContextMoveToPoint(context, 10.0f, 10.0f);
CGContextAddLineToPoint(context, 20.0f, 20.0f);
CGContextStrokePath(context);
```

The first call specifies that lines used to create the current path should be drawn 4 pixels wide. Think of this as selecting the size of the brush you're about to paint with. Until you call this function again with a different number, all lines will have a width of four lines when drawn. We then specify that the stroke color should be red. In Core Graphics, two colors are associated with drawing actions:

- The **stroke color** is used in drawing lines and for the outline of shapes.

- The **fill color** is used to fill in shapes.

A context has a sort of invisible pen associated with it that does the line drawing. As drawing commands are executed, the movements of this pen form a path. When you call CGContextMoveToPoint(), you move the end point of the current path to that location, without actually drawing anything. Whatever operation comes next, it will do its work relative to the point to which you moved the pen. In the earlier example, for instance, we first moved the pen to (10, 10). The next function call drew a line from the current pen location (10, 10) to the specified location (20, 20), which became the new pen location.

When you draw in Core Graphics, you're not drawing anything you can actually see. You're creating a path, which can be a shape, a line, or some other object, but it contains no color or other features to make it visible. It's like writing in invisible ink. Until you do something to make it visible, your path can't be seen. So, the next step is to tell

Quartz to draw the line using `CGContextStrokePath()`. This function will use the line width and the stroke color we set earlier to actually color (or "paint") the path and make it visible.

The Coordinate System

In the previous chunk of code, we passed a pair of floating-point numbers as parameters to `CGContextMoveToPoint()` and `CGContextLineToPoint()`. These numbers represent positions in the Core Graphics coordinate system. Locations in this coordinate system are denoted by their x and y coordinates, which we usually represent as (x, y). The upper-left corner of the context is (0, 0). As you move down, y increases. As you move to the right, x increases.

In the previous code snippet, we drew a diagonal line from (10, 10) to (20, 20), which would look like the one shown in Figure 16–1.

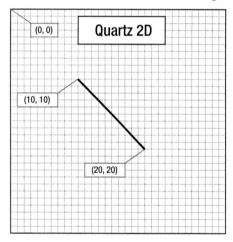

Figure 16–1. *Drawing a line using Quartz 2D's coordinate system*

The coordinate system is one of the gotchas in drawing with Quartz, because Quartz's coordinate system is flipped from what many graphics libraries use and from the traditional Cartesian coordinate system (introduced by René Descartes in the seventeenth century). In OpenGL ES, for example, (0, 0) is in the lower-left corner, and as the y coordinate increases, you move toward the top of the context or view, as shown in Figure 16–2. When working with OpenGL, you must translate the position from the view's coordinate system to OpenGL's coordinate system. That's easy enough to do, as you'll see when we work with OpenGL ES later in the chapter.

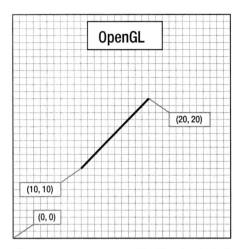

Figure 16–2. *In many graphics libraries, including OpenGL, drawing from (10, 10) to (20, 20) would produce a line that looks like this instead of the line in Figure 16–1.*

To specify a point in the coordinate system, some Quartz functions require two floating-point numbers as parameters. Other Quartz functions ask for the point to be embedded in a CGPoint, a struct that holds two floating-point values: x and y. To describe the size of a view or other object, Quartz uses CGSize, a struct that also holds two floating-point values: width and height. Quartz also declares a datatype called CGRect, which is used to define a rectangle in the coordinate system. A CGRect contains two elements: a CGPoint called origin, which identifies the top left of the rectangle, and a CGSize called size, which identifies the width and height of the rectangle.

Specifying Colors

An important part of drawing is color, so understanding the way colors work on iOS is critical. This is one of the areas where the UIKit does provide an Objective-C class: UIColor. You can't use a UIColor object directly in Core Graphic calls, but since UIColor is just a wrapper around CGColor (which is what the Core Graphic functions require), you can retrieve a CGColor reference from a UIColor instance by using its CGColor property, as we showed earlier, in this code snippet:

```
CGContextSetStrokeColorWithColor(context, [UIColor redColor].CGColor);
```

We created a UIColor instance using a convenience method called redColor, and then retrieved its CGColor property and passed that into the function.

A Bit of Color Theory for Your iOS Device's Display

In modern computer graphics, a common way to represent colors is to use four components: red, green, blue, and alpha. In Quartz, each of these values is represented as CGFloat (which is a 4-byte floating-point value, the same as float). These values should always contain a value between 0.0 and 1.0.

> **NOTE:** A floating-point value that is expected to be in the range 0.0 to 1.0 is often referred to as a **clamped floating-point variable**, or sometimes just a **clamp**.

The red, green, and blue components are fairly easy to understand, as they represent the **additive primary colors**, or the **RGB color model** (see Figure 16–3). If you add together light of these three colors in equal proportions, the result will appear to the eye as either white or a shade of gray, depending on the intensity of the light mixed. Combining the three additive primaries in different proportions gives you a range of different colors, referred to as a **gamut**.

In grade school, you probably learned that the primary colors are red, yellow, and blue. These primaries, which are known as the **historical subtractive primaries**, or the **RYB color model**, have little application in modern color theory and are almost never used in computer graphics. The color gamut of the RYB color model is much more limited than the RGB color model, and it also doesn't lend itself easily to mathematical definition. As much as we hate to tell you that your wonderful third-grade art teacher, Mrs. Smedlee, was wrong about anything, well, in the context of computer graphics, she was. For our purposes, the primary colors are red, green, and blue, not red, yellow, and blue.

Figure 16–3. *A simple representation of the additive primary colors that make up the RGB color model*

In addition to red, green, and blue, both Quartz and OpenGL ES use another color component, called **alpha**, which represents how transparent a color is. When drawing one color on top of another color, alpha is used to determine the final color that is drawn. With an alpha of 1.0, the drawn color is 100% opaque and obscures any colors beneath it. With any value less than 1.0, the colors below will show through and mix with the color above. When an alpha component is used, the color model is sometimes referred to as the **RGBA color model,** although technically speaking, the alpha isn't really part of the color; it just defines how the color will interact with other colors when it is drawn.

Other Color Models

Although the RGB model is the most commonly used in computer graphics, it is not the only color model. Several others are in use, including the following:

- Hue, saturation, value (HSV)
- Hue, saturation, lightness (HSL)
- Cyan, magenta, yellow, key (CMYK), which is used in four-color offset printing
- Grayscale

To make matters even more confusing, there are different versions of some of these models, including several variants of the RGB color space.

Fortunately, for most operations, we don't need to worry about the color model that is being used. We can just pass CGColor from our UIColor object, and in most cases, Core Graphics will handle any necessary conversions. If you use UIColor or CGColor when working with OpenGL ES, it's important to keep in mind that they support other color models, because OpenGL ES requires colors to be specified in RGBA.

Color Convenience Methods

UIColor has a large number of convenience methods that return UIColor objects initialized to a specific color. In our previous code sample, we used the redColor method to initialize a color to red.

Fortunately, the UIColor instances created by most of these convenience methods all use the RGBA color model. The only exceptions are the predefined UIColors that represent grayscale values—such as blackColor, whiteColor, and darkGrayColor— which are defined only in terms of white level and alpha. In our examples here, we're not using those, so we can assume RGBA for now.

If you need more control over color, instead of using one of those convenience methods based on the name of the color, you can create a color by specifying all four of the components. Here's an example:

```
return [UIColor colorWithRed:1.0f green:0.0f blue:0.0f alpha:1.0f];
```

Drawing Images in Context

Quartz allows you to draw images directly into a context. This is another example of an Objective-C class (UIImage) that you can use as an alternative to working with a Core Graphics data structure (CGImage). The UIImage class contains methods to draw its image into the current context. You'll need to identify where the image should appear in the context using either of the following techniques:

- By specifying a CGPoint to identify the image's upper-left corner
- By specifying a CGRect to frame the image, resized to fit the frame, if necessary

You can draw a UIImage into the current context like so:

```
CGPoint drawPoint = CGPointMake(100.0f, 100.0f);
[image drawAtPoint:drawPoint];
```

Drawing Shapes: Polygons, Lines, and Curves

Quartz provides a number of functions to make it easier to create complex shapes. To draw a rectangle or a polygon, you don't need to calculate angles, draw lines, or do any math at all. You can just call a Quartz function to do the work for you. For example, to draw an ellipse, you define the rectangle into which the ellipse needs to fit and let Core Graphics do the work:

```
CGRect theRect = CGMakeRect(0,0,100,100);
CGContextAddEllipseInRect(context, theRect);
CGContextDrawPath(context, kCGPathFillStroke);
```

You use similar methods for rectangles. Quartz also provides methods that let you create more complex shapes, such as arcs and Bezier paths.

> **NOTE:** We won't be working with complex shapes in this chapter's examples. To learn more about arcs and Bezier paths in Quartz, check out the *Quartz 2D Programming Guide* in the iOS Dev Center at
> http://developer.apple.com/documentation/GraphicsImaging/Conceptua l/drawingwithquartz2d/ or in Xcode's online documentation.

Quartz 2D Tool Sampler: Patterns, Gradients, and Dash Patterns

Although not as expansive as OpenGL ES, Quartz does offer quite an impressive array of tools. For example, Quartz supports filling polygons with gradients, not just solid colors, and an assortment of dash patterns in addition to solid lines. Take a look at the screenshots in Figure 16–4, which are from Apple's QuartzDemo sample code, to see a sampling of what Quartz can do for you.

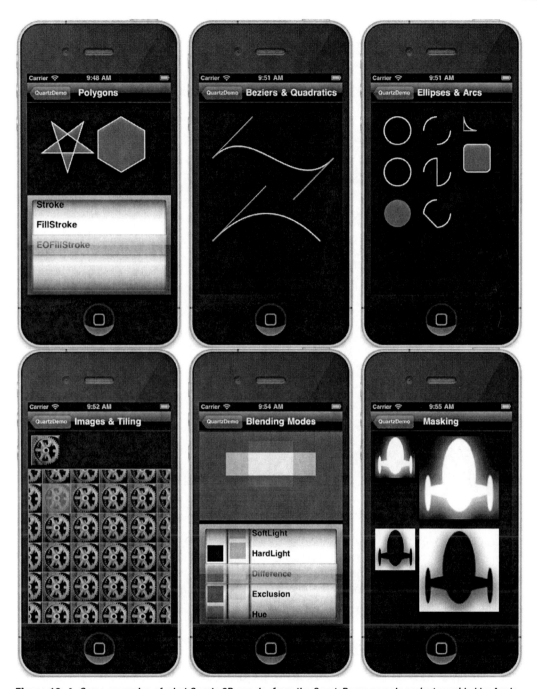

Figure 16–4. *Some examples of what Quartz 2D can do, from the QuartzDemo sample project provided by Apple*

Now that you have a basic understanding of how Quartz works and what it is capable of doing, let's try it out.

The QuartzFun Application

Our next application is a simple drawing program (see Figure 16–5). We're going to build this application twice: now using Quartz, and later using OpenGL ES. This will give you a real feel for the difference between the two environments.

Figure 16–5. *Our chapter's simple drawing application in action*

The application features a bar across the top and one across the bottom, each with a segmented control. The control at the top lets you change the drawing color, and the one at the bottom lets you change the shape to be drawn. When you touch and drag, the selected shape will be drawn in the selected color. To minimize the application's complexity, only one shape will be drawn at a time.

Setting Up the QuartzFun Application

In Xcode, create a new iPhone project using the *Single View Application* template (with ARC but not using storyboards) and call it *QuartzFun*. The template has already provided us with an application delegate and a view controller. We're going to be executing our custom drawing in a custom view, so we need to also create a subclass of UIView where we'll do the drawing by overriding the drawRect: method.

With the *QuartzFun* folder selected (the folder that currently contains the app delegate and view controller files), press ⌘N to bring up the new file assistant, and then select *Objective-C class* from the *Cocoa Touch* section. Name the new class *BIDQuartzFunView* and make it a subclass of *UIView*.

We're going to define some constants, as we've done in previous projects, but this time, our constants will be needed by more than one class. We'll create a header file just for the constants.

Select the *QuartzFun* group again and press ⌘N to bring up the new file assistant. Select the *Header File* template from the *C and C++* heading, and name the file *BIDConstants.h*.

We have two more files to go. If you look at Figure 16–5, you can see that we offer an option to select a random color. UIColor doesn't have a method to return a random color, so we'll need to write code to do that. We could put that code into our controller class, but because we're savvy Objective-C programmers, we'll put it into a category on UIColor.

Again, select the *QuartzFun* folder and press ⌘N to bring up the new file assistant. Select the *Objective-C category* from the *Cocoa Touch* heading and hit *Next*. When prompted, name the category *BIDRandom* and make it a *Category on UIColor*. Click *Next*, and save the file into your project folder.

You should now have a new pair of files named *UIColor+BIDRandom.h* and *UIColor+BIDRandom.m* for your category.

Creating a Random Color

Let's tackle the category first. Add the following line to *UIColor+BIDRandom.h*:

```
#import <UIKit/UIKit.h>

@interface UIColor (BIDRandom)
+ (UIColor *)randomColor;
@end
```

Now, switch over to *UIColor+BIDRandom.m*, and add this code:

```
#import "UIColor+BIDRandom.h"

@implementation UIColor (BIDRandom)
+ (UIColor *)randomColor {
    static BOOL seeded = NO;
    if (!seeded) {
        seeded = YES;
        srandom(time(NULL));
    }
    CGFloat red = (CGFloat)random() / (CGFloat)RAND_MAX;
    CGFloat blue = (CGFloat)random() / (CGFloat)RAND_MAX;
    CGFloat green = (CGFloat)random() / (CGFloat)RAND_MAX;
    return [UIColor colorWithRed:red green:green blue:blue alpha:1.0f];
}
@end
```

This is fairly straightforward. We declare a static variable that tells us if this is the first time through the method. The first time this method is called during an application's run, we will seed the random-number generator. Doing this here means we don't need to rely on the application doing it somewhere else, and as a result, we can reuse this category in other iOS projects.

Once we've made sure the random-number generator is seeded, we generate three random CGFloats with a value between 0.0 and 1.0, and use those three values to create a new color. We set alpha to 1.0 so that all generated colors will be opaque.

Defining Application Constants

Next, we'll define constants for each of the options that the user can select using the segmented controllers. Single-click *BIDConstants.h*, and add the following code:

```
#ifndef QuartzFun_BIDConstants_h
#define QuartzFun_BIDConstants_h

typedef enum {
    kLineShape = 0,
    kRectShape,
    kEllipseShape,
    kImageShape
} ShapeType;

typedef enum {
    kRedColorTab = 0,
    kBlueColorTab,
    kYellowColorTab,
    kGreenColorTab,
    kRandomColorTab
} ColorTabIndex;

#define degreesToRadian(x) (M_PI * (x) / 180.0)

#endif
```

To make our code more readable, we've declared two enumeration types using typedef. One will represent the shape options available in our application; the other will represent the various color options available. The values these constants hold will correspond to segments on the two segmented controllers we'll create in our application.

> **NOTE:** Just in case you haven't seen this form before, the purpose of the #ifndef compiler directive is to first test if QuartzFun_BIDConstants_h is defined and, if not, to define it. Why not just put in the #define? This way, if a .*h* file is included more than once, either directly or via other .*h* files, the directive won't be duplicated.

Implementing the QuartzFunView Skeleton

Since we're going to do our drawing in a subclass of UIView, let's set up that class with everything it needs except for the actual code to do the drawing, which we'll add later. Single-click *BIDQuartzFunView.h*, and add the following code:

```
#import <UIKit/UIKit.h>
#import "BIDConstants.h"

@interface BIDQuartzFunView : UIView
@property (nonatomic) CGPoint firstTouch;
@property (nonatomic) CGPoint lastTouch;
@property (strong, nonatomic) UIColor *currentColor;
@property (nonatomic) ShapeType shapeType;
@property (nonatomic, strong) UIImage *drawImage;
@property (nonatomic) BOOL useRandomColor;
@end
```

First, we import the *BIDConstants.h* header we just created so we can make use of our enumeration values. We then declare our properties. The first two will track the user's finger as it drags across the screen. We'll store the location where the user first touches the screen in firstTouch. We'll store the location of the user's finger while dragging and when the drag ends in lastTouch. Our drawing code will use these two variables to determine where to draw the requested shape.

Next, we define a color to hold the user's color selection and a ShapeType to keep track of the shape the user wants to draw. After that is a UIImage property that will hold the image to be drawn on the screen when the user selects the rightmost toolbar item on the bottom toolbar (see Figure 16–6). The last property is a Boolean that will be used to keep track of whether the user is requesting a random color.

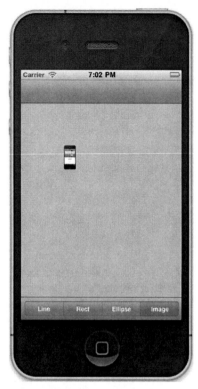

Figure 16–6. *When drawing a UIImage to the screen, notice that the color control disappears. Can you tell which app is running on the tiny iPhone?*

Switch to *BIDQuartzFunView.m*. We have several changes we need to make in this file. First, we're going to need access to the randomColor category method we wrote earlier in the chapter, at the top of the file. We'll also need to synthesize our properties. So add these lines directly below the existing import statement:

```
#import "UIColor+BIDRandom.h"
```

and add this one right after the @implementation **declaration:**

```
@synthesize firstTouch, lastTouch, currentColor, drawImage, useRandomColor, shapeType;
```

The template gave us a method called initWithFrame:, but we won't be using that. Keep in mind that object instances in nibs are stored as archived objects, which is the same mechanism we used in Chapter 13 to archive and load our objects to disk. As a result, when an object instance is loaded from a nib, neither init nor initWithFrame: is ever called. Instead, initWithCoder: is used, so this is where we need to add any initialization code. In our case, we'll set the initial color value to red, initialize useRandomColor to NO, and load the image file that we're going to draw later in the chapter. Delete the existing stub implementation of initWithFrame:, and replace it with the following method:

```
-   (id)initWithCoder:(NSCoder*)coder {

if (self = [super initWithCoder:coder]) {
    currentColor = [UIColor redColor];
    useRandomColor = NO;
    self.drawImage = [UIImage imageNamed:@"iphone.png"] ;
}
    return self;
}
```

After initWithCoder:, we need to add a few more methods to respond to the user's touches. Insert the following three methods after initWithCoder:.

```
#pragma mark - Touch Handling

- (void)touchesBegan:(NSSet *)touches withEvent:(UIEvent *)event {
    if (useRandomColor) {
        self.currentColor = [UIColor randomColor];
    }
    UITouch *touch = [touches anyObject];
    firstTouch = [touch locationInView:self];
    lastTouch = [touch locationInView:self];
    [self setNeedsDisplay];
}

- (void)touchesEnded:(NSSet *)touches withEvent:(UIEvent *)event {
    UITouch *touch = [touches anyObject];
    lastTouch = [touch locationInView:self];

    [self setNeedsDisplay];
}

- (void)touchesMoved:(NSSet *)touches withEvent:(UIEvent *)event {
    UITouch *touch = [touches anyObject];
    lastTouch = [touch locationInView:self];

    [self setNeedsDisplay];
}
```

These three methods are inherited from UIView (but actually declared in UIView's parent UIResponder). They can be overridden to find out where the user is touching the screen. They work as follows:

- touchesBegan:withEvent: is called when the user's finger first touches the screen. In that method, we change the color if the user has selected a random color using the new randomColor method we added to UIColor earlier. After that, we store the current location so that we know where the user first touched the screen, and we indicate that our view needs to be redrawn by calling setNeedsDisplay on self.

- touchesMoved:withEvent: is continuously called while the user is dragging a finger on the screen. All we do here is store the new location in lastTouch and indicate that the screen needs to be redrawn.

- touchesEnded:withEvent: is called when the user lifts that finger off the screen. Just as in the touchesMoved:withEvent: method, all we do is store the final location in the lastTouch variable and indicate that the view needs to be redrawn.

Don't worry if you don't fully grok the rest of the code here. We'll get into the details of working with touches and the specifics of the touchesBegan:withEvent:, touchesMoved:withEvent:, and touchesEnded:withEvent: methods in Chapter 17.

We'll come back to this class once we have our application skeleton up and running. That drawRect: method, which is currently commented out, is where we will do this application's real work, and we haven't written that yet. Let's finish setting up the application before we add our drawing code.

Creating and Connecting Outlets and Actions

Before we can start drawing, we need to add the segmented controls to our nib, and then hook up the actions and outlets. Single-click *BIDViewController.xib* to edit the file.

The first order of business is to change the class of the view. Single-click the *View* icon in the dock and press ⌥⌘3 to bring up the identity inspector. Change the class from *UIView* to *BIDQuartzFunView*.

Select the newly renamed *QuartzFunView* icon and look for a *Navigation Bar* in the library. Make sure you are grabbing a *Navigation Bar*, not a *Navigation Controller*. We want the bar that goes at the top of the view. Place the navigation bar snugly against the top of the view, just beneath the status bar.

Next, look for a *Segmented Control* in the library, and drag that directly on top of the navigation bar. Drop it in the center of the navigation bar, not on the left or right side (see Figure 16–7).

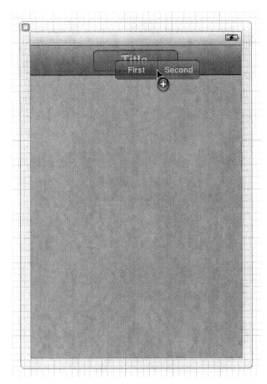

Figure 16–7. *Dragging out a segmented control, being sure to drop it on top of the navigation bar*

Once you drop the control, it should stay selected. Grab one of the resize dots on either side of the segmented control and resize it so that it takes up the entire width of the navigation bar. You don't get any blue guidelines, but Interface Builder won't let you make the bar any bigger than you want it in this case, so just drag until it won't expand any farther.

With the segmented control still selected, bring up the attributes inspector, and change the number of segments from *2* to *5*. Double-click each segment in turn, changing its label to (from left to right) *Red*, *Blue*, *Yellow*, *Green*, and *Random*, in that order. At this point, your view should look like Figure 16–8.

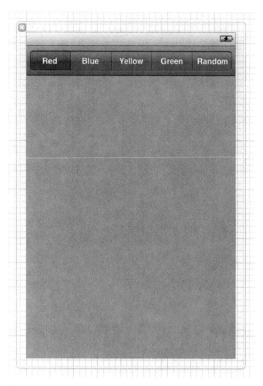

Figure 16–8. *The completed navigation bar*

Bring up the assistant editor, if it's not already open, and select *BIDViewController.h* from the jump bar. Now, control-drag from the segmented control in the dock to the *BIDViewController.h* file on the right. When your cursor is between the @interface and @end declarations, release the mouse to create a new outlet. Name the new outlet *colorControl*, and leave all the other options at their default values. Make sure you are dragging from the segmented control, not from the navigation bar or navigation item.

Next, let's add an action. Control-drag once again from the same segmented control over to the header file, directly above the @end declaration. This time, insert an action called changeColor:. The popup should default to using the *Value Changed* event, which is what we want.

Now, look for a *Toolbar* in the library (*not* a *Navigation Bar*), and drag one of those over, snug to the bottom of the view window. The toolbar from the library has a button on it that we don't need, so select the button and press the delete key on your keyboard. The button should disappear, leaving a blank toolbar in its stead.

With the toolbar in place, grab another *Segmented Control*, and drop it onto the toolbar (see Figure 16–9).

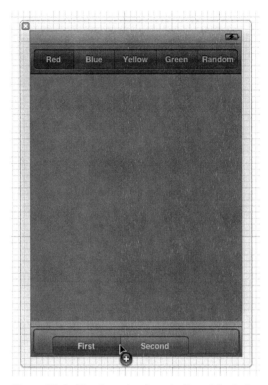

Figure 16–9. *The view, showing a toolbar at the bottom of the window with a segmented control dropped onto the toolbar*

As it turns out, segmented controls are a bit harder to center in a toolbar than in a navigation bar, so we'll bring in a little help. Drag a *Flexible Space Bar Button Item* from the library onto the toolbar, to the left of our segmented control. Next, drag a second *Flexible Space Bar Button Item* onto the toolbar, to the right of our segmented control (see Figure 16–10). These items will keep the segmented control centered in the toolbar as we resize it.

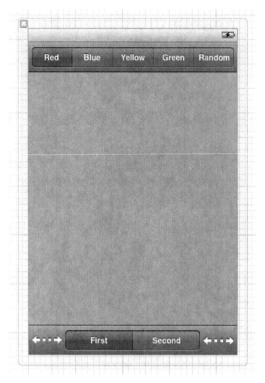

Figure 16–10. *The segmented control after we dropped the Flexible Space Bar Button Item on either side. Note that we have not yet resized the segmented control to fill the toolbar.*

It's time to resize the segmented control. In the dock, select the middle of the three *Bar Button Item*s, the one with the *Segmented Control* as a subitem. A resize handle should appear on the left side of the segmented control in the editing area. Drag that handle to resize the segmented control, and resize it so it fills the toolbar, leaving just a bit of space to the left and right. Interface Builder won't give you guidelines or prevent you from making the segmented control wider than the toolbar, as it did with the navigation bar, so you'll need to be a little careful to resize the segmented control to the correct size.

Next, select the *Segmented Control* in the dock, bring up the attributes inspector, and change the number of segments from *2* to *4*. Then double-click each segment and change the titles of the four segments to *Line*, *Rect*, *Ellipse*, and *Image*, in that order.

Once you've done that, be sure the *Segmented Control* is selected in the dock, and then control-drag from the segmented control over to *BIDViewController.h* to create another action. Change the connection type to *Action*, and name this new action *changeShape:*.

Our next task is to implement our action methods.

Implementing the Action Methods

Save the nib and feel free to close the assistant editor. Now, single-click *BIDViewController.m*. The first thing we need to do is to import our constants file so that we have access to our enumeration values. We'll also be interacting with our custom view, so we need to import its header as well. At the top of the file, immediately below the existing `import` statement, add the following lines of code:

```
#import "BIDConstants.h"
#import "BIDQuartzFunView.h"
```

Next, look for the stub implementation of `changeColor:` that Xcode created for you, and add the following code to it:

```
- (IBAction)changeColor:(id)sender {
    UISegmentedControl *control = sender;
    NSInteger index = [control selectedSegmentIndex];

    BIDQuartzFunView *quartzView = (BIDQuartzFunView *)self.view;

    switch (index) {
        case kRedColorTab:
            quartzView.currentColor = [UIColor redColor];
            quartzView.useRandomColor = NO;
            break;
        case kBlueColorTab:
            quartzView.currentColor = [UIColor blueColor];
            quartzView.useRandomColor = NO;
            break;
        case kYellowColorTab:
            quartzView.currentColor = [UIColor yellowColor];
            quartzView.useRandomColor = NO;
            break;
        case kGreenColorTab:
            quartzView.currentColor = [UIColor greenColor];
            quartzView.useRandomColor = NO;
            break;
        case kRandomColorTab:
            quartzView.useRandomColor = YES;
            break;
        default:
            break;
    }
}
```

This is pretty straightforward. We simply look at which segment was selected and create a new color based on that selection to serve as our current drawing color. In order to keep the compiler happy, we cast view, which is declared as an instance of UIView in our superclass, to QuartzFunView. After that, we set the currentColor property so that our class knows which color to use when drawing, except when a random color is selected. When a random color is chosen, it will look at the useRandomColor property, so we also set that to the appropriate value for each selection. Since all the drawing code will be in the view itself, we don't need to do anything else in this method.

Next, look for the existing implementation of changeShape:, and add the following code to it:

```
- (IBAction)changeShape:(id)sender {
    UISegmentedControl *control = sender;
    [(BIDQuartzFunView *)self.view setShapeType:[control
                                        selectedSegmentIndex]];

    if ([control selectedSegmentIndex] == kImageShape)
        colorControl.hidden = YES;
    else
        colorControl.hidden = NO;
}
```

In this method, all we do is set the shape type based on the selected segment of the control. Do you recall the ShapeType enum? The four elements of the enum correspond to the four toolbar segments at the bottom of the application view. We set the shape to be the same as the currently selected segment, and we hide and unhide the colorControl based on whether the *Image* segment was selected

> **NOTE:** You may have wondered why we put a navigation bar at the top of the view and a toolbar at the bottom of the view. According to the *Human Interface Guidelines* published by Apple, navigation bars were specifically designed to be placed at the top of the screen and toolbars are designed for the bottom. If you read the descriptions of the *Toolbar* and *Navigation Bar* in Interface Builder's library window, you'll see this design intention spelled out.

Make sure that everything is in order by compiling and running your app. You won't be able to draw shapes on the screen yet, but the segmented controls should work, and when you tap the *Image* segment in the bottom control, the color controls should disappear.

Now that we have everything working, let's do some drawing.

Adding Quartz 2D Drawing Code

We're ready to add the code that does the drawing. We'll draw a line, some shapes, and an image. We're going to work incrementally, adding a small amount of code, and then running the app to see what that code does.

Drawing the Line

Let's do the simplest drawing option first: drawing a single line. Select *BIDQuartzFunView.m*, and replace the commented-out drawRect: method with this one:

```
- (void)drawRect:(CGRect)rect {
    CGContextRef context = UIGraphicsGetCurrentContext();

    CGContextSetLineWidth(context, 2.0);
    CGContextSetStrokeColorWithColor(context, currentColor.CGColor);

    switch (shapeType) {
        case kLineShape:
            CGContextMoveToPoint(context, firstTouch.x, firstTouch.y);
            CGContextAddLineToPoint(context, lastTouch.x, lastTouch.y);
            CGContextStrokePath(context);
            break;
        case kRectShape:
            break;
        case kEllipseShape:
            break;
        case kImageShape:
            break;
        default:
            break;
    }
}
```

We start things off by retrieving a reference to the current context so we know where to draw.

```
CGContextRef context = UIGraphicsGetCurrentContext();
```

Next, we set the line width to 2.0, which means that any line that we stroke will be 2 pixels wide.

```
CGContextSetLineWidth(context, 2.0);
```

After that, we set the color for stroking lines. Since UIColor has a CGColor property, which is what this method needs, we use that property of our currentColor property to pass the correct color on to this function.

```
CGContextSetStrokeColorWithColor(context, currentColor.CGColor);
```

We use a switch to jump to the appropriate code for each shape type. As we mentioned earlier, we'll start off with the code to handle kLineShape, get that working, and then we'll add code for each shape in turn as we make our way through this example.

```
switch (shapeType) {
    case kLineShape:
```

To draw a line, we tell the graphics context to create a path starting at the first place the user touched. Remember that we stored that value in the touchesBegan: method, so it will always reflect the starting point of the most recent touch or drag.

```
CGContextMoveToPoint(context, firstTouch.x, firstTouch.y);
```

Next, we draw a line from that spot to the last spot the user touched. If the user's finger is still in contact with the screen, lastTouch contains Mr. Finger's current location. If the user is no longer touching the screen, lastTouch contains the location of the user's finger when it was lifted off the screen.

```
    CGContextAddLineToPoint(context, lastTouch.x, lastTouch.y);
```

Then we stroke the path. This function will stroke the line we just drew using the color and width we set earlier:

```
    CGContextStrokePath(context);
```

After that, we finish the switch statement.

```
        break;
    case kRectShape:
        break;
    case kEllipseShape:
        break;
    case kImageShape:
        break;
    default:
        break;
    }
```

And that's it for now. At this point, you should be able to compile and run the app once more. The *Rect*, *Ellipse*, and *Shape* options won't work, but you should be able to draw lines just fine using any of the color choices (see Figure 16–11).

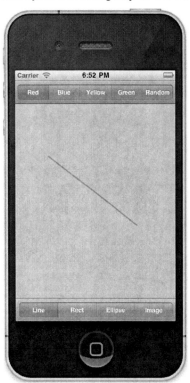

Figure 16–11. *The line-drawing part of our application is now complete. Here, we are drawing using the color red.*

Drawing the Rectangle and Ellipse

Let's write the code to draw the rectangle and the ellipse at the same time, since Quartz implements both of these objects in basically the same way. Add the following bold code to your existing drawRect: method:

```
- (void)drawRect:(CGRect)rect {
    CGContextRef context = UIGraphicsGetCurrentContext();

    CGContextSetLineWidth(context, 2.0);
    CGContextSetStrokeColorWithColor(context, currentColor.CGColor);

    CGContextSetFillColorWithColor(context, currentColor.CGColor);
    CGRect currentRect = CGRectMake(firstTouch.x,
                                    firstTouch.y,
                                    lastTouch.x - firstTouch.x,
                                    lastTouch.y - firstTouch.y);

    switch (shapeType) {
        case kLineShape:
            CGContextMoveToPoint(context, firstTouch.x, firstTouch.y);
            CGContextAddLineToPoint(context, lastTouch.x, lastTouch.y);
            CGContextStrokePath(context);
            break;
        case kRectShape:
            CGContextAddRect(context, currentRect);
            CGContextDrawPath(context, kCGPathFillStroke);
            break;
        case kEllipseShape:
            CGContextAddEllipseInRect(context, currentRect);
            CGContextDrawPath(context, kCGPathFillStroke);
            break;
        case kImageShape:
            break;
        default:
            break;
    }
}
```

Because we want to paint both the ellipse and the rectangle in a solid color, we add a call to set the fill color using currentColor.

```
CGContextSetFillColorWithColor(context, currentColor.CGColor);
```

Next, we declare a CGRect variable. We do this here because both the rectangle and ellipse are drawn based on a rect. We'll use currentRect to hold the rectangle described by the user's drag. Remember that a CGRect has two members: size and origin. A function called CGRectMake() lets us create a CGRect by specifying the x, y, width, and height values, so we use that to make our rectangle.

The code to create the rectangle is pretty straightforward. We use the point stored in firstTouch to create the origin. Then we figure out the size by getting the difference between the two x values and the two y values. Note that depending on the direction of the drag, one or both size values may end up with negative numbers, but that's OK. A

CGRect with a negative size will simply be rendered in the opposite direction of its origin point (to the left for a negative width; upward for a negative height).

```
CGRect currentRect = CGRectMake(firstTouch.x,
                                firstTouch.y,
                                lastTouch.x - firstTouch.x,
                                lastTouch.y - firstTouch.y);
```

Once we have this rectangle defined, drawing either a rectangle or an ellipse is as easy as calling two functions: one to draw the rectangle or ellipse in the CGRect we defined, and the other to stroke and fill it.

```
case kRectShape:
    CGContextAddRect(context, currentRect);
    CGContextDrawPath(context, kCGPathFillStroke);
    break;
case kEllipseShape:
    CGContextAddEllipseInRect(context, currentRect);
    CGContextDrawPath(context, kCGPathFillStroke);
    break;
```

Compile and run your application. Try out the *Rect* and *Ellipse* tools to see how you like them. Don't forget to change colors, including using a random color.

Drawing the Image

For our last trick, let's draw an image. The *16 - QuartzFun* folder contains an image named *iphone.png* that you can add to your *Supporting Files* folder, or you can use any *.png* file you prefer, as long as you remember to change the file name in your code to point to that image.

Add the following code to your drawRect: method:

```
- (void)drawRect:(CGRect)rect {

    CGContextRef context = UIGraphicsGetCurrentContext();

    CGContextSetLineWidth(context, 2.0);
    CGContextSetStrokeColorWithColor(context, currentColor.CGColor);

    CGContextSetFillColorWithColor(context, currentColor.CGColor);
    CGRect currentRect = CGRectMake(firstTouch.x,
                                    firstTouch.y,
                                    lastTouch.x - firstTouch.x,
                                    lastTouch.y - firstTouch.y);

    switch (shapeType) {
        case kLineShape:
            CGContextMoveToPoint(context, firstTouch.x, firstTouch.y);
            CGContextAddLineToPoint(context, lastTouch.x, lastTouch.y);
            CGContextStrokePath(context);
            break;
        case kRectShape:
            CGContextAddRect(context, currentRect);
            CGContextDrawPath(context, kCGPathFillStroke);
            break;
```

```
        case kEllipseShape:
            CGContextAddEllipseInRect(context, currentRect);
            CGContextDrawPath(context, kCGPathFillStroke);
            break;
        case kImageShape: {
            CGFloat horizontalOffset = drawImage.size.width / 2;
            CGFloat verticalOffset = drawImage.size.height / 2;
            CGPoint drawPoint = CGPointMake(lastTouch.x - horizontalOffset,
                                            lastTouch.y - verticalOffset);
            [drawImage drawAtPoint:drawPoint];
            break;
        }
        default:
            break;
    }
}
```

> **NOTE:** Notice that in the switch statement, we added curly braces around the code under case
> kImageShape:. The compiler has a problem with variables declared in the first line after a case
> statement. These curly braces are our way of telling the compiler to stop complaining. We could
> also have declared horizontalOffset before the switch statement, but this approach keeps
> the related code together.

First, we calculate the center of the image, since we want the image drawn centered on the point where the user last touched. Without this adjustment, the image would be drawn with the upper-left corner at the user's finger, also a valid option. We then make a new CGPoint by subtracting these offsets from the x and y values in lastTouch.

```
CGFloat horizontalOffset = drawImage.size.width / 2;
CGFloat verticalOffset = drawImage.size.height / 2;
CGPoint drawPoint = CGPointMake(lastTouch.x - horizontalOffset,
                                lastTouch.y - verticalOffset);
```

Now, we tell the image to draw itself. This line of code will do the trick:

```
[drawImage drawAtPoint:drawPoint];
```

Optimizing the QuartzFun Application

Our application does what we want, but we should consider a bit of optimization. In our little application, you won't notice a slowdown, but in a more complex application, running on a slower processor, you might see some lag.

The problem occurs in *BIDQuartzFunView.m*, in the methods touchesMoved: and touchesEnded:. Both methods include this line of code:

```
[self setNeedsDisplay];
```

Obviously, this is how we tell our view that something has changed and it needs to redraw itself. This code works, but it causes the entire view to be erased and redrawn, even if only a tiny bit changed. We do want to erase the screen when we get ready to drag out a new shape, but we don't want to clear the screen several times a second as we drag out our shape.

Rather than forcing the entire view to be redrawn many times during our drag, we can use setNeedsDisplayInRect: instead. setNeedsDisplayInRect: is a UIView method that marks just one rectangular portion of a view's region as needing redisplay. By using this method, we can be more efficient by marking only the part of the view that is affected by the current drawing operation as needing to be redrawn.

We need to redraw not just the rectangle between firstTouch and lastTouch, but any part of the screen encompassed by the current drag. If the user touched the screen and then scribbled all over, but we redrew only the section between firstTouch and lastTouch, we would leave a lot of stuff drawn on the screen that we don't want to remain.

The solution is to keep track of the entire area that has been affected by a particular drag in a CGRect instance variable. In touchesBegan:, we reset that instance variable to just the point where the user touched. Then in touchesMoved: and touchesEnded:, we use a Core Graphics function to get the union of the current rectangle and the stored rectangle, and we store the resulting rectangle. We also use it to specify which part of the view needs to be redrawn. This approach gives us a running total of the area impacted by the current drag.

Now, we'll calculate the current rectangle in the drawRect: method for use in drawing the ellipse and rectangle shapes. We'll move that calculation into a new method so that it can be used in all three places without repeating code. Ready? Let's do it.

Make the following changes to *BIDQuartzFunView.h*:

```
#import <UIKit/UIKit.h>
#import "BIDConstants.h"

@interface BIDQuartzFunView : UIView
@property (nonatomic) CGPoint firstTouch;
@property (nonatomic) CGPoint lastTouch;
@property (nonatomic, strong) UIColor *currentColor;
@property (nonatomic) ShapeType shapeType;
@property (nonatomic, strong) UIImage *drawImage;
@property (nonatomic) BOOL useRandomColor;
@property (readonly) CGRect currentRect;
@property CGRect redrawRect;
@end
```

We declare a CGRect called redrawRect that we will use to keep track of the area that needs to be redrawn. We also declare a read-only property called currentRect, which will return that rectangle that we were previously calculating in drawRect:.

Switch over to *BIDQuartzFunView.m,* and insert the following code at the top of the file, after the existing @synthesize statement:

```
@synthesize redrawRect, currentRect;

- (CGRect)currentRect {
    return CGRectMake (firstTouch.x,
```

```
                    firstTouch.y,
                    lastTouch.x - firstTouch.x,
                    lastTouch.y - firstTouch.y);
}
```

Now, in the drawRect: method, change all references to currentRect to self.currentRect so that the code uses that new accessor we just created. Then delete the lines of code where we calculated currentRect.

```
- (void)drawRect:(CGRect)rect {
    CGContextRef context = UIGraphicsGetCurrentContext();

    CGContextSetLineWidth(context, 2.0);
    CGContextSetStrokeColorWithColor(context, currentColor.CGColor);

    CGContextSetFillColorWithColor(context, currentColor.CGColor);
    CGRect currentRect = CGRectMake(firstTouch.x,
                                    firstTouch.y,
                                    lastTouch.x - firstTouch.x,
                                    lastTouch.y - firstTouch.y);

    switch (shapeType) {
        case kLineShape:
            CGContextMoveToPoint(context, firstTouch.x, firstTouch.y);
            CGContextAddLineToPoint(context, lastTouch.x, lastTouch.y);
            CGContextStrokePath(context);
            break;
        case kRectShape:
            CGContextAddRect(context, self.currentRect);
            CGContextDrawPath(context, kCGPathFillStroke);
            break;
        case kEllipseShape:
            CGContextAddEllipseInRect(context, self.currentRect);
            CGContextDrawPath(context, kCGPathFillStroke);
            break;
        case kImageShape:{
            CGFloat horizontalOffset = drawImage.size.width / 2;
            CGFloat verticalOffset = drawImage.size.height / 2;
            CGPoint drawPoint = CGPointMake(lastTouch.x - horizontalOffset,
                                            lastTouch.y - verticalOffset);
            [drawImage drawAtPoint:drawPoint];
            break;
        }
        default:
            break;
    }
}
```

We also need to make some changes to touchesEnded:withEvent: and touchesMoved:withEvent:. We will recalculate the space impacted by the current operation and use that to indicate that only a portion of our view needs to be redrawn. Replace the existing touchesEnded: and touchesMoved: methods with these new versions:

```
- (void)touchesEnded:(NSSet *)touches withEvent:(UIEvent *)event {
    UITouch *touch = [touches anyObject];
    lastTouch = [touch locationInView:self];
```

```
    if (shapeType == kImageShape) {
        CGFloat horizontalOffset = drawImage.size.width / 2;
        CGFloat verticalOffset = drawImage.size.height / 2;
        redrawRect = CGRectUnion(redrawRect,
                                 CGRectMake(lastTouch.x - horizontalOffset,
                                            lastTouch.y - verticalOffset,
                                            drawImage.size.width,
                                            drawImage.size.height));
    }
    else
        redrawRect = CGRectUnion(redrawRect, self.currentRect);
    redrawRect = CGRectInset(redrawRect, -2.0, -2.0);
    [self setNeedsDisplayInRect:redrawRect];
}

- (void)touchesMoved:(NSSet *)touches withEvent:(UIEvent *)event {
    UITouch *touch = [touches anyObject];
    lastTouch = [touch locationInView:self];

    if (shapeType == kImageShape) {
        CGFloat horizontalOffset = drawImage.size.width / 2;
        CGFloat verticalOffset = drawImage.size.height / 2;
        redrawRect = CGRectUnion(redrawRect,
                                 CGRectMake(lastTouch.x - horizontalOffset,
                                            lastTouch.y - verticalOffset,
                                            drawImage.size.width,
                                            drawImage.size.height));
    }
    redrawRect = CGRectUnion(redrawRect, self.currentRect);
    [self setNeedsDisplayInRect:redrawRect];
}
```

With only a few additional lines of code, we reduced the amount of work necessary to redraw our view by getting rid of the need to erase and redraw any portion of the view that hasn't been affected by the current drag. Being kind to your iOS device's precious processor cycles like this can make a big difference in the performance of your applications, especially as they get more complex.

> **NOTE:** If you're interested in a more in-depth exploration of Quartz 2D topics, you might want to take a look at *Beginning iPad Development for iPhone Developers: Mastering the iPad SDK* by Jack Nutting, Dave Wooldridge, and David Mark (Apress, 2010), which covers a lot of Quartz 2D drawing. All the drawing code and explanations in that book apply to the iPhone as well as the iPad.

The GLFun Application

As explained earlier in the chapter, OpenGL ES and Quartz take fundamentally different approaches to drawing. A detailed introduction to OpenGL ES would be a book in and of itself, so we're not going to attempt that here. Instead, we'll re-create our Quartz application using OpenGL ES, just to give you a sense of the basics and some sample code you can use to kick-start your own OpenGL ES applications.

Let's get started with our application.

> **TIP:** If you want to create a full-screen OpenGL ES application, you don't need to build it manually. Xcode has a template you can use. It sets up the screen and the buffers for you, and even puts some sample drawing and animation code into the class, so you can see where to put your code. If you want to try this out after you finish up GLFun, create a new *iOS application*, and choose the *OpenGL ES Application* template.

Setting Up the GLFun Application

Almost everything we do in this new app is identical to QuartzFun, with the exception of the drawing code. Since the drawing code is contained in a single class (*BIDQuartzFunView*), we can just copy the existing application and replace that view with a new view. That will save us from needing to redo all the work necessary to get the application up and running.

Close the *QuartzFun* Xcode project, and in the Finder, make a copy of the project folder. Rename the copy *GLFun* (do *not* rename the project in the Finder—just rename it in the enclosing folder), and then double-click the folder to open it and double-click *QuartzFun.xcodeproj* to open that file.

Once *QuartzFun.xcodeproj* is opened, you'll notice that the top of the project navigator still says *QuartzFun*. Single-click directly on the name *QuartzFun* at the very top of the project navigator, wait about a second, and then single-click it again. When the name becomes editable, change it to *GLFun*, and then hit return to commit the change. The timing on the project rename is a little tricky, so it may take you more than one try to get this to work.

Once you change the project name, you'll be prompted with a sheet asking you if you want to rename its contents. Click the *Rename* button, and it will go through and rename all the component parts of your project to match the new name. It will prompt you to take a **snapshot** before doing so. Snapshots are like backups of a project at a given point in time, and Xcode will often ask you to make one before doing something that's potentially disruptive.

Now we've renamed everything except for the group called *QuartzFun* that contains all of our source code files. You can rename that using the same trick. Single-click, pause, single-click again, and you can change *QuartzFun* to *GLFun*.

Before we proceed, you'll need to add a few more files to your project. In the *16 - GLFun* folder, you'll find four files named *Texture2D.h*, *Texture2D.m*, *OpenGLES2DView.h*, and *OpenGLES2DView.m*. The code in the first two files was written by Apple to make drawing images in OpenGL ES much easier than it otherwise would be. The second pair of files is a class we've provided based on sample code from Apple that configures

OpenGL ES to do two-dimensional drawing (in other words, we've done the necessary configuration for you). You can feel free to use any of these files in your own programs if you wish. Add all four files to your project.

Lastly, we no longer need the class that draws using Quartz. Select *BIDQuartzFunView.h* and *BIDQuartzFunView.m*, and press the delete button on your keyboard to delete them. When asked if you want to remove references or delete, select *Delete*.

Creating BIDGLFunView

Create a new file in your project by selecting the *GLFun* folder and pressing ⌘N. Select the *Objective-C class* from the *Cocoa Touch Class* heading and hit *Next*. When prompted, name the class *BIDGLFunView* and make it a subclass of *OpenGLES2DView*.

OpenGLES2DView is a subclass of UIView that uses OpenGL ES to do two-dimensional drawing. We set up this view so that the coordinate systems of OpenGL ES and the coordinate system of the view are mapped on a one-to-one basis. To use the OpenGLES2DView class, we just need to subclass it, and then implement the draw method to do our actual drawing.

Hit *Next*, and save the file into your project folder. This should create a new pair of files called *BIDGLFunView.h* and *BIDGLFunView.m*, which is where we'll do the OpenGL ES-based drawing.

Single-click the newly created *BIDGLFunView.h* file, and replace its contents with the following:

```
#import "BIDConstants.h"
#import "OpenGLES2DView.h"

@class Texture2D;

@interface BIDGLFunView : OpenGLES2DView
@property CGPoint firstTouch;
@property CGPoint lastTouch;
@property (nonatomic, strong) UIColor *currentColor;
@property BOOL useRandomColor;
@property ShapeType shapeType;
@property (nonatomic, strong) Texture2D *sprite;
@end
```

OpenGL ES doesn't have sprites or images, per se; it has one kind of image called a **texture**. Textures must be drawn onto a shape or object. The way you draw an image in OpenGL ES is to draw a square (technically speaking, it's two triangles), and then map a texture onto that square so that it exactly matches the square's size. Texture2D is a class provided by Apple that encapsulates that relatively complex process into a single, easy-to-use class.

Note that despite the one-to-one relationship between the view and the OpenGL context, the y coordinates are still flipped. We need to translate the y coordinate from the view coordinate system, where increases in y represent moving down, to the OpenGL coordinate system, where increases in y represent moving up.

Switch over to *BIDGLFunView.m*, and add the following code:

```
#import "BIDGLFunView.h"
#import "UIColor+BIDRandom.h"
#import "Texture2D.h"

@implementation BIDGLFunView
@synthesize firstTouch;
@synthesize lastTouch;
@synthesize currentColor;
@synthesize useRandomColor;
@synthesize shapeType;
@synthesize sprite;

- (id)initWithCoder:(NSCoder*)coder {
    if (self = [super initWithCoder:coder]) {
        self.currentColor = [UIColor redColor];
        useRandomColor = NO;
        sprite = [[Texture2D alloc] initWithImage:[UIImage
                                            imageNamed:@"iphone.png"]];
        glBindTexture(GL_TEXTURE_2D, sprite.name);
    }
    return self;
}

- (void)draw  {
    glLoadIdentity();

    glClearColor(0.78f, 0.78f, 0.78f, 1.0f);
    glClear(GL_COLOR_BUFFER_BIT);

    CGColorRef color = currentColor.CGColor;
    const CGFloat *components = CGColorGetComponents(color);
    CGFloat red = components[0];
    CGFloat green = components[1];
    CGFloat blue = components[2];

    glColor4f(red,green, blue, 1.0);
    switch (shapeType) {
        case kLineShape: {
            glDisable(GL_TEXTURE_2D);
            GLfloat vertices[4];

            // Convert coordinates
            vertices[0] = firstTouch.x;
            vertices[1] = self.frame.size.height - firstTouch.y;
            vertices[2] = lastTouch.x;
            vertices[3] = self.frame.size.height - lastTouch.y;
            glLineWidth(2.0);
            glVertexPointer(2, GL_FLOAT, 0, vertices);
            glDrawArraysGL_LINES, 0, 2);
            break;
        }
        case kRectShape: {
```

```
            glDisable(GL_TEXTURE_2D);
            // Calculate bounding rect and store in vertices
            GLfloat vertices[8];
            GLfloat minX = (firstTouch.x > lastTouch.x) ?
            lastTouch.x : firstTouch.x;
            GLfloat minY = (self.frame.size.height - firstTouch.y >
                            self.frame.size.height - lastTouch.y) ?
            self.frame.size.height - lastTouch.y :
            self.frame.size.height - firstTouch.y;
            GLfloat maxX = (firstTouch.x > lastTouch.x) ?
            firstTouch.x : lastTouch.x;
            GLfloat maxY = (self.frame.size.height - firstTouch.y >
                            self.frame.size.height - lastTouch.y) ?
            self.frame.size.height - firstTouch.y :
            self.frame.size.height - lastTouch.y;

            vertices[0] = maxX;
            vertices[1] = maxY;
            vertices[2] = minX;
            vertices[3] = maxY;
            vertices[4] = minX;
            vertices[5] = minY;
            vertices[6] = maxX;
            vertices[7] = minY;

            glVertexPointer(2, GL_FLOAT , 0, vertices);
            glDrawArrays(GL_TRIANGLE_FAN, 0, 4);
            break;
        }
        case kEllipseShape: {
            glDisable(GL_TEXTURE_2D);
            GLfloat vertices[720];

            GLfloat xradius = fabsf((firstTouch.x - lastTouch.x) / 2);
            GLfloat yradius = fabsf((firstTouch.y - lastTouch.y) / 2);
            for (int i = 0; i <= 720; i += 2) {
                GLfloat xOffset = (firstTouch.x > lastTouch.x) ?
                lastTouch.x + xradius : firstTouch.x + xradius;
                GLfloat yOffset = (firstTouch.y < lastTouch.y) ?
                self.frame.size.height - lastTouch.y + yradius :
                self.frame.size.height - firstTouch.y + yradius;
                vertices[i] = (cos(degreesToRadian(i / 2)) * xradius) + xOffset;
                vertices[i+1] = (sin(degreesToRadian(i / 2)) * yradius) +
                yOffset;
            }

            glVertexPointer(2, GL_FLOAT , 0, vertices);
            glDrawArrays(GL_TRIANGLE_FAN, 0, 360);
            break;
        }
        case kImageShape:
            glEnable(GL_TEXTURE_2D);
            [sprite drawAtPoint:CGPointMake(lastTouch.x,
                                        self.frame.size.height - lastTouch.y)];
```

```
            break;
        default:
            break;
    }

    glBindRenderbufferOES(GL_RENDERBUFFER_OES, viewRenderbuffer);
    [context presentRenderbuffer:GL_RENDERBUFFER_OES];
}

- (void)touchesBegan:(NSSet *)touches withEvent:(UIEvent *)event {
    if (useRandomColor)
        self.currentColor = [UIColor randomColor];

    UITouch* touch = [[event touchesForView:self] anyObject];
    firstTouch = [touch locationInView:self];
    lastTouch = [touch locationInView:self];
    [self draw];
}

- (void)touchesMoved:(NSSet *)touches withEvent:(UIEvent *)event {

    UITouch *touch = [touches anyObject];
    lastTouch = [touch locationInView:self];

    [self draw];
}

- (void)touchesEnded:(NSSet *)touches withEvent:(UIEvent *)event {
    UITouch *touch = [touches anyObject];
    lastTouch = [touch locationInView:self];

    [self draw];
}
@end
```

You can see that using OpenGL ES isn't by any means easier or more concise than using Quartz. Although OpenGL ES is more powerful than Quartz, you're also closer to the metal, so to speak. OpenGL ES can be daunting at times.

Because this view is being loaded from a nib, we added an initWithCoder: method, and in it, we create and assign a UIColor to currentColor. We also defaulted useRandomColor to NO and created our Texture2D object, just as we did in *QuartzFunView* earlier.

After the initWithCoder: method, we have our draw method, which is where you can really see the difference between the two libraries.

Let's take a look at process of drawing a line. Here's how we drew the line in the Quartz version (we've removed the code that's not directly relevant to drawing):

```
CGContextRef context = UIGraphicsGetCurrentContext();
CGContextSetLineWidth(context, 2.0);
CGContextSetStrokeColorWithColor(context, currentColor.CGColor);
CGContextMoveToPoint(context, firstTouch.x, firstTouch.y);
CGContextAddLineToPoint(context, lastTouch.x, lastTouch.y);
CGContextStrokePath(context);
```

In OpenGL ES, we needed to take a few more steps to draw that same line. First, we reset the virtual world so that any rotations, translations, or other transforms that might have been applied to it are gone.

```
glLoadIdentity();
```

Next, we clear the background to the same shade of gray that was used in the Quartz version of the application.

```
glClearColor(0.78, 0.78f, 0.78f, 1.0f);
glClear(GL_COLOR_BUFFER_BIT);
```

After that, we need to set the OpenGL ES drawing color by dissecting a UIColor and pulling the individual RGB components out of it. Fortunately, because we used the convenience class methods, we don't need to worry about which color model the UIColor uses. We can safely assume it will use the RGBA color space.

```
CGColorRef color = currentColor.CGColor;
const CGFloat *components = CGColorGetComponents(color);
CGFloat red = components[0];
CGFloat green = components[1];
CGFloat blue = components[2];
glColor4f(red,green, blue, 1.0);
```

Next, we turn off OpenGL ES's ability to map textures.

```
glDisable(GL_TEXTURE_2D);
```

Any drawing code that fires from the time we make this call until there's a call to glEnable(GL_TEXTURE_2D) will be drawn without a texture, which is what we want. If we allowed a texture to be used, the color we just set wouldn't show.

To draw a line, we need two vertices, which means we need an array with four elements. As we've discussed, a point in two-dimensional space is represented by two values: x and y. In Quartz, we used a CGPoint struct to hold these. In OpenGL ES, points are not embedded in structs. Instead, we pack an array with all the points that make up the shape we need to draw. So, to draw a line from point (100, 150) to point (200, 250) in OpenGL ES, we need to create a vertex array that looks like this:

```
vertex[0] = 100;
vertex[1] = 150;
vertex[2] = 200;
vertex[3] = 250;
```

Our array has the format {x1, y1, x2, y2, x3, y3}. The next code in this method converts two CGPoint structs into a vertex array.

```
GLfloat vertices[4];
vertices[0] = firstTouch.x;
vertices[1] = self.frame.size.height - firstTouch.y;
vertices[2] = lastTouch.x;
vertices[3] = self.frame.size.height - lastTouch.y;
```

Once we've defined the vertex array that describes what we want to draw (in this example, a line), we specify the line width, pass the array into OpenGL ES using the method glVertexPointer(), and tell OpenGL ES to draw the arrays.

```
glLineWidth(2.0);
glVertexPointer (2, GL_FLOAT , 0, vertices);
glDrawArrays (GL_LINES, 0, 2);
```

Whenever we finish drawing in OpenGL ES, we need to instruct it to render its buffer, and tell our view's context to show the newly rendered buffer.

```
glBindRenderbufferOES(GL_RENDERBUFFER_OES, viewRenderbuffer);
[context presentRenderbuffer:GL_RENDERBUFFER_OES];
```

To clarify, the process of drawing in OpenGL ES consists of three steps:

1. Draw in the context.

2. After all of your drawing is complete, render the context into the buffer.

3. Present your render buffer, which is when the pixels are actually drawn onto the screen.

As you can see, the OpenGL ES example is considerably longer.

The difference between Quartz and OpenGL ES becomes even more dramatic when we look at the process of drawing an ellipse. OpenGL ES doesn't know how to draw an ellipse. OpenGL, the big brother and predecessor to OpenGL ES, has a number of convenience functions for generating common two- and three-dimensional shapes, but those convenience functions are some of the functionality that was stripped out of OpenGL ES to make it more streamlined and suitable for use in embedded devices like the iPhone. As a result, a lot more responsibility falls into the developer's lap.

As a reminder, here is how we drew the ellipse using Quartz:

```
CGContextRef context = UIGraphicsGetCurrentContext();
CGContextSetLineWidth(context, 2.0);
CGContextSetStrokeColorWithColor(context, currentColor.CGColor);
CGContextSetFillColorWithColor(context, currentColor.CGColor);
CGRect currentRect;
CGContextAddEllipseInRect(context, self.currentRect);
CGContextDrawPath(context, kCGPathFillStroke);
```

For the OpenGL ES version, we start off with the same steps as before, resetting any movement or rotations, clearing the background to white, and setting the draw color based on currentColor.

```
glLoadIdentity();
glClearColor(1.0f, 1.0f, 1.0f, 1.0f);
glClear(GL_COLOR_BUFFER_BIT);
glDisable(GL_TEXTURE_2D);
CGColorRef color = currentColor.CGColor;
const CGFloat *components = CGColorGetComponents(color);
CGFloat red = components[0];
CGFloat green = components[1];
CGFloat blue = components[2];
glColor4f(red,green, blue, 1.0);
```

Since OpenGL ES doesn't know how to draw an ellipse, we need to roll our own, which means dredging up painful memories of Ms. Picklebaum's geometry class. We'll define a vertex array that holds 720 GLfloats, which will hold an x and a y position for 360 points, one for each degree around the circle. We could change the number of points to increase or decrease the smoothness of the circle. This approach looks good on any view that will fit on the iPhone screen, but probably does require more processing than strictly necessary if you're just drawing smaller circles.

```
GLfloat vertices[720];
```

Next, we figure out the horizontal and vertical radii of the ellipse based on the two points stored in firstTouch and lastTouch.

```
GLfloat xradius = fabsf((firstTouch.x - lastTouch.x)/2);
GLfloat yradius = fabsf((firstTouch.y - lastTouch.y)/2);
```

Then we loop around the circle, calculating the correct points around the circle.

```
for (int i = 0; i <= 720; i+=2) {
    GLfloat xOffset = (firstTouch.x > lastTouch.x) ?
        lastTouch.x + xradius : firstTouch.x + xradius;
    GLfloat yOffset = (firstTouch.y < lastTouch.y) ?
        self.frame.size.height - lastTouch.y + yradius :
        self.frame.size.height - firstTouch.y + yradius;
    vertices[i] = (cos(degreesToRadian(i / 2))*xradius) + xOffset;
    vertices[i+1] = (sin(degreesToRadian(i / 2))*yradius) +
        yOffset;
}
```

Finally, we feed the vertex array to OpenGL ES, tell it to draw it and render it, and then tell our context to present the newly rendered image.

```
glVertexPointer (2, GL_FLOAT , 0, vertices);
glDrawArrays (GL_TRIANGLE_FAN, 0, 360);
...
glBindRenderbufferOES(GL_RENDERBUFFER_OES, viewRenderbuffer);
[context presentRenderbuffer:GL_RENDERBUFFER_OES];
```

We won't review the rectangle method, because it uses the same basic technique. We define a vertex array with the four vertices to define the rectangle, and then we render and present it.

There's also not much to talk about with the image drawing, since that lovely Texture2D class from Apple makes drawing a sprite just as easy as it is in Quartz. There is one important item to notice there, though:

```
glEnable(GL_TEXTURE_2D);
```

Since it is possible that the ability to draw textures was previously disabled, we must make sure it's enabled before we attempt to use the Texture2D class.

After the draw method, we have the same touch-related methods as the previous version. The only difference is that instead of telling the view that it needs to be displayed, we just call the draw method. We don't need to tell OpenGL ES which parts of the screen will be updated; it will figure that out and leverage hardware acceleration to draw in the most efficient manner.

Updating BIDViewController

We need to make a few minor changes in *BIDViewController.m*. One is to change all references to *BIDQuartzFunView* to *BIDGLFunView*. First, replace the line

```
#import "BIDQuartzFunView.h"
```

with

```
#import "BIDGLFunView.h"
```

Next, replace the changeColor: method with the following version:

```
- (IBAction)changeColor:(id)sender {
    UISegmentedControl *control = sender;
    NSInteger index = [control selectedSegmentIndex];

    BIDGLFunView *glView = (BIDGLFunView *)self.view;

    switch (index) {
        case kRedColorTab:
            glView.currentColor = [UIColor redColor];
            glView.useRandomColor = NO;
            break;
        case kBlueColorTab:
            glView.currentColor = [UIColor blueColor];
            glView.useRandomColor = NO;
            break;
        case kYellowColorTab:
            glView.currentColor = [UIColor yellowColor];
            glView.useRandomColor = NO;
            break;
        case kGreenColorTab:
            glView.currentColor = [UIColor greenColor];
            glView.useRandomColor = NO;
            break;
        case kRandomColorTab:
            glView.useRandomColor = YES;
            break;
        default:
            break;
    }
}
```

And, finally, in the changeShape: method, change this line:

```
[(BIDQuartzFunView *)self.view setShapeType:[control
                                selectedSegmentIndex]];
```

to this:

```
[(BIDGLFunView *)self.view setShapeType:[control
                                selectedSegmentIndex]];
```

Updating the Nib

We also need to change the view in our nib. Since we copied the *QuartzFun* project, the view is still configured to use the BIDQuartzFunView for its underlying class, and we need to change that to BIDGLFunView.

Single-click *BIDViewController.xib* to open Interface Builder. Single-click the *Quartz Fun View* in the dock, and then press ⌥⌘3 to bring up the identity inspector. Change the class from *BIDQuartzFunView* to *BIDGLFunView*.

Finishing GLFun

Before you can compile and run this program, you'll need to link in two frameworks to your project. Follow the instructions in Chapter 7 for adding the Audio Toolbox framework (in the "Linking in the Audio Toolbox Framework" section), but instead of selecting *AudioToolbox.framework*, select *OpenGLES.framework* and *QuartzCore.framework*.

Frameworks added? Good. Go run your project. It should look just like the Quartz version. You've now seen enough OpenGL ES to get you started.

> **NOTE:** If you're interested in using OpenGL ES in your iPhone applications, you can find the
> OpenGL ES specification, along with links to books, documentation, and forums where OpenGL
> ES issues are discussed, at http://www.khronos.org/opengles/. Also, visit
> http://www.khronos.org/developers/resources/opengles/, and search for
> "tutorial." And be sure to check out the OpenGL tutorial in Jeff's iPhone blog, at
> http://iphonedevelopment.blogspot.com/2009/05/opengl-es-from-
> ground-up-table-of.html.

Drawing to a Close

In this chapter, we've really just scratched the surface of the iOS drawing ability. You should feel pretty comfortable with Quartz 2D now, and with some occasional references to Apple's documentation, you can probably handle most any drawing requirement that comes your way. You should also have a basic understanding of what OpenGL ES is and how it integrates with iOS view system.

Next up? You're going to learn how to add gestural support to your applications.

Chapter 17

Taps, Touches, and Gestures

The screens of the iPhone, iPod touch, and iPad—with their crisp, bright, touch-sensitive display—are truly things of beauty and masterpieces of engineering. The multitouch screen common to all iOS devices is one of the key factors in the platform's tremendous usability. Because the screen can detect multiple touches at the same time and track them independently, applications are able to detect a wide range of gestures, giving the user power that goes beyond the interface.

Suppose you are in the Mail application staring at a long list of junk e-mail that you want to delete. You can tap each one individually, tap the trash icon to delete it, and then wait for the next message to download, deleting each one in turn. This method is best if you want to read each message before you delete it.

Alternatively, from the list of messages, you can tap the *Edit* button in the upper-right corner, tap each e-mail row to mark it, and then hit the *Delete* button to delete all marked messages. This method is best if you don't need to read each message before deleting it. Another alternative is to swipe a message in the list from side to side. That gesture produces a *Delete* button for that message. Tap the *Delete* button, and the message is deleted.

This example is just one of the countless gestures that are made possible by the multitouch display. You can pinch your fingers together to zoom out while viewing a picture, or reverse-pinch to zoom in. You can long-press an icon to turn on "jiggly mode," which allows you to delete applications from your iOS device.

In this chapter, we're going to look at the underlying architecture that lets you detect gestures. You'll learn how to detect the most common gestures, and how to create and detect a completely new gesture.

Multitouch Terminology

Before we dive into the architecture, let's go over some basic vocabulary. First, a **gesture** is any sequence of events that happens from the time you touch the screen with one or more fingers until you lift your fingers off the screen. No matter how long it takes, as long as one or more fingers remain against the screen, you are still within a gesture (unless a system event, such as an incoming phone call, interrupts it). Note that Cocoa Touch doesn't expose any class or structure that represents a gesture. In some sense, a gesture is a verb, and a running app can watch the user input stream to see if one is happening.

A gesture is passed through the system inside a series of **events**. Events are generated when you interact with the device's multitouch screen. They contain information about the touch or touches that occurred.

The term **touch** refers to a finger being placed on the screen, dragging across the screen, or being lifted from the screen. The number of touches involved in a gesture is equal to the number of fingers on the screen at the same time. You can actually put all five fingers on the screen, and as long as they aren't too close to each other, iOS can recognize and track them all. Now, there aren't many useful five-finger gestures, but it's nice to know the iOS can handle one if necessary. In fact, experimentation has shown that the iPad can handle up to 11 simultaneous touches! This may seem excessive, but could be useful if you're working on a multiplayer game, where several players are interacting with the screen at the same time.

A **tap** happens when you touch the screen with a finger and then immediately lift your finger off the screen without moving it around. The iOS device keeps track of the number of taps and can tell you if the user double-tapped, triple-tapped, or even 20-tapped. It handles all the timing and other work necessary to differentiate between two single-taps and a double-tap, for example.

A **gesture recognizer** is an object that knows how to watch the stream of events generated by a user, and recognize when the user is touching and dragging in a way that matches a predefined gesture. The UIGestureRecognizer class and its various subclasses can help take a lot of work off your hands when you want to watch for common gestures, since it nicely encapsulates the work of looking for a gesture, and can be easily applied to any view in your application.

The Responder Chain

Since gestures are passed through the system inside events, and events are passed through the **responder chain**, you need to have an understanding of how the responder chain works in order to handle gestures properly. If you've worked with Cocoa for Mac OS X, you're probably familiar with the concept of a responder chain, as the same basic mechanism is used in both Cocoa and Cocoa Touch. If this is new material, don't worry; we'll explain how it works.

Responding to Events

Several times in this book, we've mentioned the first responder, which is usually the object with which the user is currently interacting. The first responder is the start of the responder chain. There are other responders as well. Any class that has UIResponder as one of its superclasses is a **responder**. UIView is a subclass of UIResponder, and UIControl is a subclass of UIView, so all views and all controls are responders. UIViewController is also a subclass of UIResponder, meaning that it is a responder, as are all of its subclasses, such as UINavigationController and UITabBarController. Responders, then, are so named because they respond to system-generated events, such as screen touches.

If the first responder doesn't handle a particular event, such as a gesture, it passes that event up the responder chain. If the next object in the chain responds to that particular event, it will usually consume the event, which stops the event's progression through the responder chain. In some cases, if a responder only partially handles an event, that responder will take an action and forward the event to the next responder in the chain. That's not usually what happens, though. Normally, when an object responds to an event, that's the end of the line for the event. If the event goes through the entire responder chain and no object handles the event, the event is then discarded.

Let's take a more specific look at the responder chain. The first responder is almost always a view or control and gets the first shot at responding to an event. If the first responder doesn't handle the event, it passes the event to its view controller. If the view controller doesn't consume the event, the event is then passed to the first responder's parent view. If the parent view doesn't respond, the event will go to the parent view's controller, if it has one.

The event will proceed up the view hierarchy, with each view and then that view's controller getting a chance to handle the event. If the event makes it all the way up through the view hierarchy without being handled by a view or a controller, the event is passed to the application's window. If the window doesn't handle the event, it passes that event to the application's UIApplication object instance.

If UIApplication doesn't respond to the event, there's one more spot where you can build a global catchall as the end of the responder chain: the app delegate. If the app delegate is a subclass of UIResponder (which it normally is if you create your project from one of Apple's application templates), the app will try to pass it any unhandled events. Finally, if the app delegate isn't a subclass of of UIResponder or doesn't handle the event, then the event goes gently into the good night.

This process is important for a number of reasons. First, it controls the way gestures can be handled. Let's say a user is looking at a table and swipes a finger across a row of that table. What object handles that gesture?

If the swipe is within a view or control that's a subview of the table view cell, that view or control will get a chance to respond. If it doesn't respond, the table view cell gets a chance. In an application like Mail, where a swipe can be used to delete a message, the table view cell probably needs to look at that event to see if it contains a swipe gesture. Most table view cells don't respond to gestures, however. If they don't respond, the event proceeds up to the table view, and then up the rest of the responder chain until something responds to that event or it reaches the end of the line.

Forwarding an Event: Keeping the Responder Chain Alive

Let's take a step back to that table view cell in the Mail application. We don't know the internal details of Apple's Mail application, but let's assume, for the nonce, that the table view cell handles the delete swipe and only the delete swipe. That table view cell must implement the methods related to receiving touch events (discussed shortly) so that it can check to see if that event contained a swipe gesture. If the event contains a swipe, then the table view cell takes an action, and that's that; the event goes no further.

If the event doesn't contain a swipe gesture, the table view cell is responsible for forwarding that event manually to the next object in the responder chain. If it doesn't do its forwarding job, the table and other objects up the chain will never get a chance to respond, and the application may not function as the user expects. That table view cell could prevent other views from recognizing a gesture.

Whenever you respond to a touch event, you need to keep in mind that your code doesn't work in a vacuum. If an object intercepts an event that it doesn't handle, it needs to pass it along manually, by calling the same method on the next responder. Here's a bit of fictional code:

```
-(void)respondToFictionalEvent:(UIEvent *)event {
    if (someCondition)
        [self handleEvent:event];
    else
        [self.nextResponder respondToFictionalEvent:event];
}
```

Notice how we call the same method on the next responder. That's how to be a good responder-chain citizen. Fortunately, most of the time, methods that respond to an event also consume the event, but it's important to know that if that's not the case, you need to make sure the event is pushed back into the responder chain.

The Multitouch Architecture

Now that you know a little about the responder chain, let's look at the process of handling gestures. As we've indicated, gestures are passed along the responder chain, embedded in events. This means that the code to handle any kind of interaction with the multitouch screen needs to be contained in an object in the responder chain. Generally, that means we can choose to either embed that code in a subclass of UIView or embed the code in a UIViewController.

So does this code belong in the view or in the view controller?

If the view needs to do something to itself based on the user's touches, the code probably belongs in the class that defines that view. For example, many control classes, such as UISwitch and UISlider, respond to touch-related events. A UISwitch might want to turn itself on or off based on a touch. The folks who created the UISwitch class embedded gesture-handling code in the class so the UISwitch can respond to a touch.

Often, however, when the gesture being processed affects more than the object being touched, the gesture code really belongs in the view's controller class. For example, if the user makes a gesture touching one row that indicates that all rows should be deleted, the gesture should be handled by code in the view controller. The way you respond to touches and gestures in both situations is exactly the same, regardless of the class to which the code belongs.

The Four Touch Notification Methods

Four methods are used to notify a responder about touches. When the user first touches the screen, the system looks for a responder that has a method called touchesBegan:withEvent:. To find out when the user first begins a gesture or taps the screen, implement this method in your view or your view controller. Here's an example of what that method might look like:

```
- (void)touchesBegan:(NSSet *)touches withEvent:(UIEvent *)event {
    NSUInteger numTaps = [[touches anyObject] tapCount];
    NSUInteger numTouches = [touches count];

    // Do something here.
}
```

This method, and each of the touch-related methods, is passed an NSSet instance called touches and an instance of UIEvent. You can determine the number of fingers currently pressed against the screen by getting a count of the objects in touches. Every object in touches is a UITouch event that represents one finger touching the screen. If this touch is part of a series of taps, you can find out the tap count by asking any of the UITouch objects. In the preceding example, a numTaps value of 2 tells you that the screen was tapped twice in quick succession, while a numTouches value of 2 tells you the user tapped the screen with two fingers at once. If both have a value of 2, then the user double-tapped with two fingers.

All of the objects in touches may not be relevant to the view or view controller where you've implemented this method. A table view cell, for example, probably doesn't care about touches that are in other rows or that are in the navigation bar. You can get a subset of touches that has only those touches that fall within a particular view from the event, like so:

```
    NSSet *myTouches = [event touchesForView:self.view];
```

Every UITouch represents a different finger, and each finger is located at a different position on the screen. You can find out the position of a specific finger using the UITouch object. It will even translate the point into the view's local coordinate system if you ask it to, like this:

```
    CGPoint point = [touch locationInView:self];
```

You can get notified while the user is moving fingers across the screen by implementing touchesMoved:withEvent:. This method is called multiple times during a long drag, and each time it is called, you will get another set of touches and another event. In addition to being able to find out each finger's current position from the UITouch objects, you can also discover the previous location of that touch, which is the finger's position the last time either touchesMoved:withEvent: or touchesBegan:withEvent: was called.

When the user's fingers are removed from the screen, another event, touchesEnded:withEvent:, is invoked. When this method is called, you know that the user is finished with a gesture.

There's one final touch-related method that responders might implement. It's called touchesCancelled:withEvent:, and it is called if the user is in the middle of a gesture when something happens to interrupt it, like the phone ringing. This is where you can do any cleanup you might need so you can start fresh with a new gesture. When this method is called, touchesEnded:withEvent: will not be called for the current gesture.

OK, enough theory—let's see some of this in action.

The TouchExplorer Application

We're going to build a little application that will give you a better feel for when the four touch-related responder methods are called. In Xcode, create a new project using the *Single View Application* template. Enter *TouchExplorer* as the *Product Name*, and select *iPhone* for *Device Family*.

TouchExplorer will print messages to the screen, containing the touch and tap count, every time a touch-related method is called (see Figure 17–1).

Figure 17–1. *The TouchExplorer application*

> **NOTE:** Although the applications in this chapter will run on the simulator, you won't be able to see all of the available multitouch functionality unless you run them on a real iOS device. If you've been accepted into the iOS Developer Program, you have the ability to run the programs you write on your device of choice. The Apple web site does a great job of walking you through the process of getting everything you need to prepare to connect Xcode to your device.

We need three labels for this application: one to indicate which method was last called, another to report the current tap count, and a third to report the number of touches. Single-click *BIDViewController.h*, and add three outlets and a method declaration as follows. The method will be used to update the labels from multiple places.

```
#import <UIKit/UIKit.h>

@interface BIDViewController : UIViewController

@property (weak, nonatomic) IBOutlet UILabel *messageLabel;
@property (weak, nonatomic) IBOutlet UILabel *tapsLabel;
@property (weak, nonatomic) IBOutlet UILabel *touchesLabel;
- (void)updateLabelsFromTouches:(NSSet *)touches;
@end
```

Now, select *BIDViewController.xib* to edit the file. Click the *View* icon in the dock to edit the view if the view editor is not already open. Drag a label onto the view, using the blue guidelines to place the label toward the upper-left corner of the view. Use the resize handle to resize the label over to the right-hand blue guideline. Next, use the attribute inspector to set the label alignment to centered. Finally, hold down the option key and drag two more labels out from the original, spacing them one below the other, leaving you with three labels (see Figure 17–1).

Next, control-drag from the *File's Owner* icon to each of the three labels, connecting the top one to the *messageLabel* outlet, the middle one to the *tapsLabel* outlet, and the last one to the *touchesLabel* outlet.

Feel free to play with the fonts and colors if you're feeling a bit Picasso. When you're finished placing the labels, double-click each one, and press the delete key to get rid of the text that's in them.

Next, single-click the *View* icon in the nib dock and bring up the attributes inspector (see Figure 17–2). On the inspector, go to the *View* section and make sure that both *User Interaction Enabled* and *Multiple Touch* are checked. If *Multiple Touch* is not checked, your controller class's touch methods will always receive one and only one touch, no matter how many fingers are actually touching the phone's screen.

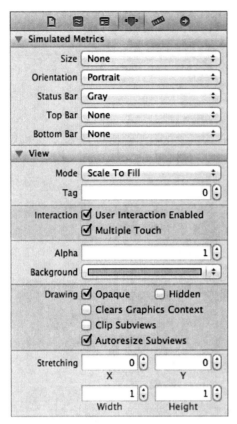

Figure 17–2. *In the View attributes, make sure both User Interaction Enabled and Multiple Touch are checked.*

When you're finished, save the nib. Next, select *BIDViewController.m*, and add the following code at the beginning of the file:

```
#import "BIDViewController.h"

@implementation BIDViewController
@synthesize messageLabel;
@synthesize tapsLabel;
@synthesize touchesLabel;

- (void)updateLabelsFromTouches:(NSSet *)touches {
    NSUInteger numTaps = [[touches anyObject] tapCount];
    NSString *tapsMessage = [[NSString alloc]
        initWithFormat:@"%d taps detected", numTaps];
    tapsLabel.text = tapsMessage;

    NSUInteger numTouches = [touches count];
    NSString *touchMsg = [[NSString alloc] initWithFormat:
        @"%d touches detected", numTouches];
    touchesLabel.text = touchMsg;
}
.
```

.
.

Then insert the following lines of code into the existing `viewDidUnload` methods:

```
- (void)viewDidUnload {
    [super viewDidUnload];
    // Release any retained subviews of the main view.
    // e.g. self.myOutlet = nil;
    self.messageLabel = nil;
    self.tapsLabel = nil;
    self.touchesLabel = nil;
}
```

And add the following new methods at the end of the file:

```
#pragma mark -
- (void)touchesBegan:(NSSet *)touches withEvent:(UIEvent *)event {
    messageLabel.text = @"Touches Began";
    [self updateLabelsFromTouches:touches];
}

- (void)touchesCancelled:(NSSet *)touches withEvent:(UIEvent *)event{
    messageLabel.text = @"Touches Cancelled";
    [self updateLabelsFromTouches:touches];
}

- (void)touchesEnded:(NSSet *)touches withEvent:(UIEvent *)event {
    messageLabel.text = @"Touches Ended.";
    [self updateLabelsFromTouches:touches];
}

- (void)touchesMoved:(NSSet *)touches withEvent:(UIEvent *)event {
    messageLabel.text = @"Drag Detected";
    [self updateLabelsFromTouches:touches];
}
@end
```

In this controller class, we implement all four of the touch-related methods we discussed earlier. Each one sets `messageLabel` so the user can see when each method is called. Next, all four of them call `updateLabelsFromTouches:` to update the other two labels. The `updateLabelsFromTouches:` method gets the tap count from one of the touches, figures out the number of touches by looking at the count of the `touches` set, and updates the labels with that information.

Compile and run the application. If you're running in the simulator, try repeatedly clicking the screen to drive up the tap count, and try clicking and holding down the mouse button while dragging around the view to simulate a touch and drag. Note that a drag is not the same as a tap, so once you start your drag, the app will report zero taps.

You can emulate a two-finger pinch in the iOS simulator by holding down the option key while you click with the mouse and drag. You can also simulate two-finger swipes by first holding down the option key to simulate a pinch, then moving the mouse so the two dots representing virtual fingers are next to each other, and then holding down the shift key (while still holding down the option key). Pressing the shift key will lock the position of the two fingers relative to each other, and you can do swipes and other two-finger gestures. You won't be able to do gestures that require three or more fingers, but you can do most two-finger gestures on the simulator using combinations of the option and shift keys.

If you're able to run this program on your iPhone or iPod touch, see how many touches you can get to register at the same time. Try dragging with one finger, then two fingers, and then three. Try double- and triple-tapping the screen, and see if you can get the tap count to go up by tapping with two fingers.

Play around with the TouchExplorer application until you feel comfortable with what's happening and with the way that the four touch methods work. When you're ready, continue on to see how to detect one of the most common gestures: the swipe.

The Swipes Application

The application we're about to build does nothing more than detect swipes, both horizontal and vertical. If you swipe your finger across the screen from left to right, right to left, top to bottom, or bottom to top, the app will display a message across the top of the screen for a few seconds informing you that a swipe was detected (see Figure 17–3).

Figure 17–3. *The Swipes application will detect both vertical and horizontal swipes.*

Detecting swipes is relatively easy. We're going to define a minimum gesture length in pixels, which is how far the user needs to swipe before the gesture counts as a swipe. We'll also define a variance, which is how far from a straight line our user can veer and still have the gesture count as a horizontal or vertical swipe. A diagonal line generally won't count as a swipe, but one that's just a little off from horizontal or vertical will.

When the user touches the screen, we'll save the location of the first touch in a variable. Then we'll check as the user's finger moves across the screen to see if it reaches a point where it has gone far enough and straight enough to count as a swipe. Let's build it.

Create a new project in Xcode using the *Single View Application* template and a *Device Family* of *iPhone*, and name the project *Swipes*.

Single-click *BIDViewController.h*, and add the following code:

```
#import <UIKit/UIKit.h>

@interface BIDViewController : UIViewController

@property (weak, nonatomic) IBOutlet UILabel *label;
@property CGPoint gestureStartPoint;
@end
```

We start by declaring an outlet for our one label and a variable to hold the first spot the user touches. Then we declare a method that will be used to erase the text after a few seconds.

Select *BIDViewController.xib* to open it for editing. Make sure that the view is set so *User Interaction Enabled* and *Multiple Touch* are both checked using the attributes inspector, and drag a *Label* from the library and drop it on the *View* window. Set up the label so it takes the entire width of the view from blue guideline to blue guideline, and its alignment is centered. Feel free to play with the text attributes to make the label easier to read. Control-drag from the *File's Owner* icon to the label, and connect it to the *label* outlet. Finally, double-click the label and delete its text.

Save your nib. Then return to Xcode, select *BIDViewController.m*, and add the following code at the top:

```
#import "BIDViewController.h"

#define kMinimumGestureLength    25
#define kMaximumVariance          5

@implementation BIDViewController
@synthesize label;
@synthesize gestureStartPoint;

- (void)eraseText {
    label.text = @"";
}
.
.
.
```

We start by defining a minimum gesture length of 25 pixels and a variance of 5. If the user were doing a horizontal swipe, the gesture could end up 5 pixels above or below the starting vertical position and still count as a swipe as long as the user moved 25 pixels horizontally. In a real application, you would probably need to play with these numbers a bit to find what works best for your application.

Insert the following lines of code into the existing `viewDidUnload` method:

```
- (void)viewDidUnload
{
    [super viewDidUnload];
    // Release any retained subviews of the main view.
    // e.g. self.myOutlet = nil;
    self.label = nil;
}
```

And add the following methods at the bottom of the class:

```
#pragma mark -
- (void)touchesBegan:(NSSet *)touches withEvent:(UIEvent *)event {
    UITouch *touch = [touches anyObject];
    gestureStartPoint = [touch locationInView:self.view];
}

- (void)touchesMoved:(NSSet *)touches withEvent:(UIEvent *)event {
    UITouch *touch = [touches anyObject];
    CGPoint currentPosition = [touch locationInView:self.view];

    CGFloat deltaX = fabsf(gestureStartPoint.x - currentPosition.x);
    CGFloat deltaY = fabsf(gestureStartPoint.y - currentPosition.y);

    if (deltaX >= kMinimumGestureLength && deltaY <= kMaximumVariance) {
        label.text = @"Horizontal swipe detected";
        [self performSelector:@selector(eraseText)
                    withObject:nil afterDelay:2];
    } else if (deltaY >= kMinimumGestureLength &&
                deltaX <= kMaximumVariance){
        label.text = @"Vertical swipe detected";
        [self performSelector:@selector(eraseText) withObject:nil
                    afterDelay:2];
    }
}

@end
```

Let's start with the `touchesBegan:withEvent:` method. All we do there is grab any touch from the touches set and store its point. We're primarily interested in single-finger swipes right now, so we don't worry about how many touches there are; we just grab one of them.

```
    UITouch *touch = [touches anyObject];
    gestureStartPoint = [touch locationInView:self.view];
```

In the next method, touchesMoved:withEvent:, we do the real work. First, we get the current position of the user's finger:

```
UITouch *touch = [touches anyObject];
CGPoint currentPosition = [touch locationInView:self.view];
```

After that, we calculate how far the user's finger has moved both horizontally and vertically from its starting position. The function fabsf() is from the standard C math library that returns the absolute value of a float. This allows us to subtract one from the other without needing to worry about which is the higher value:

```
CGFloat deltaX = fabsf(gestureStartPoint.x - currentPosition.x);
CGFloat deltaY = fabsf(gestureStartPoint.y - currentPosition.y);
```

Once we have the two deltas, we check to see if the user has moved far enough in one direction without having moved too far in the other to constitute a swipe. If that's true, we set the label's text to indicate whether a horizontal or vertical swipe was detected. We also use performSelector:withObject:afterDelay: to erase the text after it has been on the screen for 2 seconds. That way, the user can practice multiple swipes without needing to worry if the label is referring to an earlier attempt or the most recent one.

```
if (deltaX >= kMinimumGestureLength && deltaY <= kMaximumVariance) {
    label.text = @"Horizontal swipe detected";
    [self performSelector:@selector(eraseText)
            withObject:nil afterDelay:2];
} else if (deltaY >= kMinimumGestureLength &&
        deltaX <= kMaximumVariance){
    label.text = @"Vertical swipe detected";
    [self performSelector:@selector(eraseText)
            withObject:nil afterDelay:2];
}
```

Go ahead and compile and run the application. If you find yourself clicking and dragging with no visible results, be patient. Click and drag straight down or straight across until you get the hang of swiping.

Automatic Gesture Recognition

The procedure we just used for detecting a swipe wasn't too bad. All the complexity is in the touchesMoved:withEvent: method, and even that wasn't all that complicated. But there's an even easier way to do this. iOS includes a class called UIGestureRecognizer, which eliminates the need for watching all the events to see how fingers are moving. You don't use UIGestureRecognizer directly, but instead create an instance of one of its subclasses, each of which is designed to look for a particular type of gesture, such as a swipe, pinch, double-tap, triple-tap, and so on.

Let's see how to modify the Swipes app to use a gesture recognizer instead of our hand-rolled procedure. As always, you might want to make a copy of your *Swipes* project folder and start from there.

Start off by selecting *BIDViewController.m*, and deleting both the touchesBegan:withEvent: and touchesMoved:withEvent: methods. That's right, you won't be needing them. Then add a couple of new methods in their place:

```
- (void)reportHorizontalSwipe:(UIGestureRecognizer *)recognizer {
    label.text = @"Horizontal swipe detected";
    [self performSelector:@selector(eraseText) withObject:nil afterDelay:2];
}

- (void)reportVerticalSwipe:(UIGestureRecognizer *)recognizer {
    label.text = @"Vertical swipe detected";
    [self performSelector:@selector(eraseText) withObject:nil afterDelay:2];
}
```

These methods implement the actual "functionality" (if you can call it that) that's brought about by the swipe gestures, just as the touchesMoved:withEvent: did previously. Now, add the new code shown here to the viewDidLoad method:

```
- (void)viewDidLoad
{
    [super viewDidLoad];
    // Do any additional setup after loading the view, typically from a nib.

    UISwipeGestureRecognizer *vertical = [[UISwipeGestureRecognizer alloc]
        initWithTarget:self action:@selector(reportVerticalSwipe:)];
    vertical.direction = UISwipeGestureRecognizerDirectionUp|
        UISwipeGestureRecognizerDirectionDown;
    [self.view addGestureRecognizer:vertical];

    UISwipeGestureRecognizer *horizontal = [[UISwipeGestureRecognizer alloc]
        initWithTarget:self action:@selector(reportHorizontalSwipe:)];
    horizontal.direction = UISwipeGestureRecognizerDirectionLeft|
        UISwipeGestureRecognizerDirectionRight;
    [self.view addGestureRecognizer:horizontal];
}
```

There you have it! To sanitize things even further, you can also delete the lines referring to gestureStartPoint from *BIDViewController.h* and *BIDViewController.m* as well (but leaving them there won't harm anything). Thanks to UIGestureRecognizer, all we needed to do here was create and configure some gesture recognizers, and add them to our view. When the user interacts with the screen in a way that one of the recognizers recognizes, the action method we specified is called.

In terms of total lines of code, there's not much difference between these two approaches for a simple case like this. But the code that uses gesture recognizers is undeniably simpler to understand and easier to write. You don't need to give even a moment's thought to the issue of calculating a finger's movement over time, because that's already done for you by the UISwipeGestureRecognizer.

Implementing Multiple Swipes

In the Swipes application, we worried about only single-finger swipes, so we just grabbed any object out of the touches set to figure out where the user's finger was during the swipe. This approach is fine if you're interested in only single-finger swipes, the most common type of swipe used.

But what if you want to handle two- or three-finger swipes? In previous versions of this book, we dedicated about 50 lines of code, and a fair amount of explanation, to achieving this by tracking multiple UITouch instances across multiple touch events. Now that we have gesture recognizers, this is a solved problem. A UISwipeGestureRecognizer can be configured to recognize any number of simultaneous touches. By default, each instance expects a single finger, but you can configure it to look for any number of fingers pressing the screen at once. Each instance responds only to the exact number of touches you specify, so what we'll do is create a whole bunch of gesture recognizers in a loop.

Make a copy of your *Swipes* project folder.

Edit *BIDViewController.m* and modify the viewDidLoad method, replacing it with the one shown here:

```
- (void)viewDidLoad
{
    [super viewDidLoad];
    // Do any additional setup after loading the view, typically from a nib.

    for (NSUInteger touchCount = 1; touchCount <= 5; touchCount++) {
        UISwipeGestureRecognizer *vertical;
        vertical = [[UISwipeGestureRecognizer alloc] initWithTarget:self
            action:@selector(reportVerticalSwipe:)];
        vertical.direction = UISwipeGestureRecognizerDirectionUp
            | UISwipeGestureRecognizerDirectionDown;
        vertical.numberOfTouchesRequired = touchCount;
        [self.view addGestureRecognizer:vertical];

        UISwipeGestureRecognizer *horizontal;
        horizontal = [[UISwipeGestureRecognizer alloc] initWithTarget:self
            action:@selector(reportHorizontalSwipe:)];
        horizontal.direction = UISwipeGestureRecognizerDirectionLeft
            | UISwipeGestureRecognizerDirectionRight;
        horizontal.numberOfTouchesRequired = touchCount;
        [self.view addGestureRecognizer:horizontal];
    }
}
```

Note that in a real application, you might want different numbers of fingers swiping across the screen to trigger different behaviors. You can easily do that using gesture recognizers, simply by having each of them call a different action method.

Now, all we need to do is change the logging, by adding a method that gives us a handy description of the number of touches, and using it in the reporting methods as shown here. Add this method toward the bottom of the BIDViewController class, just above the two swipe-reporting methods:

```
- (NSString *)descriptionForTouchCount:(NSUInteger)touchCount {
    switch (touchCount) {
        case 2:
            return @"Double ";
        case 3:
            return @"Triple ";
        case 4:
            return @"Quadruple ";
        case 5:
            return @"Quintuple ";
        default:
            return @"";
    }
}
```

Next, modify the two swipe-reporting methods as shown:

```
- (void)reportHorizontalSwipe:(UIGestureRecognizer *)recognizer {
    label.text = @"Horizontal swipe detected";
    label.text = [NSString stringWithFormat:@"%@Horizontal swipe detected",
        [self descriptionForTouchCount:[recognizer numberOfTouches]]];
    [self performSelector:@selector(eraseText) withObject:nil afterDelay:2];
}

- (void)reportVerticalSwipe:(UIGestureRecognizer *)recognizer {
    label.text = @"Vertical swipe detected";
    label.text = [NSString stringWithFormat:@"%@Vertical swipe detected",
        [self descriptionForTouchCount:[recognizer numberOfTouches]]];;
    [self performSelector:@selector(eraseText) withObject:nil afterDelay:2];
}
```

Compile and run the app. You should be able to trigger double- and triple-swipes in both directions, and still be able to trigger single-swipes. If you have small fingers, you might even be able to trigger a quadruple- or quintuple-swipe.

> **TIP:** In the simulator, if you hold down the option key, a pair of dots, representing a pair of fingers, will appear. Get them close together, then hold down the shift key, and the dots will stay in the same position relative to each other, allowing you to move the pair of fingers around the screen. Now click and drag down the screen to simulate a double-swipe. Cool!

With a multiple-finger swipe, one thing to be careful of is that your fingers aren't too close to each other. If two fingers are very close to each other, they may register as only a single touch. Because of this, you shouldn't rely on quadruple- or quintuple-swipes for any important gestures, because many people will have fingers that are too big to do those swipes effectively.

Detecting Multiple Taps

In the TouchExplorer application, we printed the tap count to the screen, so you've already seen how easy it is to detect multiple taps. It's not quite as straightforward as it seems, however, because often you will want to take different actions based on the number of taps. If the user triple-taps, you get notified three separate times. You get a single-tap, a double-tap, and finally a triple-tap. If you want to do something on a double-tap but something completely different on a triple-tap, having three separate notifications could cause a problem.

Fortunately, the engineers at Apple anticipated this situation, and they provided a mechanism to let multiple gesture recognizers play nicely together, even when they're faced with ambiguous inputs that could seemingly trigger any of them. The basic idea is that you place a constraint on a gesture recognizer, telling it to not trigger its associated method unless some other gesture recognizer fails to trigger its own method.

That seems a bit abstract, so let's make it real. One commonly used gesture recognizer is represented by the UITapGestureRecognizer class. A tap recognizer can be configured to do its thing when a particular number of taps occur. Imagine we have a view for which we want to define distinct actions that occur when the user taps once or double-taps. You might start off with something like the following:

```
UITapGestureRecognizer *singleTap = [[UITapGestureRecognizer alloc] initWithTarget:
    self action:@selector(doSingleTap)];
singleTap.numberOfTapsRequired = 1;
[self.view addGestureRecognizer:singleTap];

UITapGestureRecognizer *doubleTap = [[UITapGestureRecognizer alloc] initWithTarget:
    self action:@selector(doDoubleTap)];
doubleTap.numberOfTapsRequired = 2;
[self.view addGestureRecognizer:doubleTap];
```

The problem with this piece of code is that the two recognizers are unaware of each other, and they have no way of knowing that the user's actions may be better suited to another recognizer. With the preceding code, if the user double-taps the view, the doDoubleTap method will be called, but the doSingleMethod will also be called—twice!—once for each tap.

The way around this is to create a failure requirement. We tell singleTap that it should trigger its action only if doubleTap doesn't recognize and respond to the user input by adding this single line:

```
[singleTap requireGestureRecognizerToFail:doubleTap];
```

This means that when the user taps once, singleTap doesn't do its work immediately. Instead, singleTap waits until it knows that doubleTap has decided to stop paying attention to the current gesture (that is, the user didn't tap twice). We're going to build on this further with our next project.

In Xcode, create a new project with the *Single View Application* template. Call this new project *TapTaps*, and use the *Device Family* popup to choose *iPhone*.

This application is going to have four labels: one each that informs us when it has detected a single-tap, double-tap, triple-tap, and quadruple-tap (see Figure 17–4).

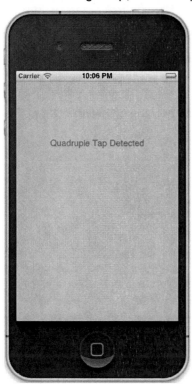

Figure 17–4. *The TapTaps application detects up to four simultaneous taps.*

We need outlets for the four labels, and we also need separate methods for each tap scenario to simulate what you would have in a real application. We'll also include a method for erasing the text fields. Single-click *BIDViewController.h*, and make the following changes:

```
#import <UIKit/UIKit.h>

@interface BIDViewController : UIViewController

@property (weak, nonatomic) IBOutlet UILabel *singleLabel;
@property (weak, nonatomic) IBOutlet UILabel *doubleLabel;
@property (weak, nonatomic) IBOutlet UILabel *tripleLabel;
@property (weak, nonatomic) IBOutlet UILabel *quadrupleLabel;
- (void)tap1;
- (void)tap2;
- (void)tap3;
- (void)tap4;
- (void)eraseMe:(UILabel *)label;
@end
```

Save the file. Then select *BIDViewController.xib* to edit the GUI. Once you're there, add four *Labels* to the view from the library. Make all four labels stretch from blue guideline to blue guideline, set their alignment to centered, and then format them however you see fit. Feel free to make each label a different color, but that is by no means necessary. When you're finished, control-drag from the *File's Owner* icon to each label, and connect each one to *singleLabel*, *doubleLabel*, *tripleLabel*, and *quadrupleLabel*, respectively. Finally, make sure you double-click each label and press the delete key to get rid of any text.

Save your changes. Then select *BIDViewController.m*, and add the following code at the top of the file:

```
#import "BIDViewController.h"

@implementation BIDViewController
@synthesize singleLabel;
@synthesize doubleLabel;
@synthesize tripleLabel;
@synthesize quadrupleLabel;

- (void)tap1 {
    singleLabel.text = @"Single Tap Detected";
    [self performSelector:@selector(eraseMe:)
        withObject:singleLabel afterDelay:1.6f];
}

- (void)tap2 {
    doubleLabel.text = @"Double Tap Detected";
    [self performSelector:@selector(eraseMe:)
        withObject:doubleLabel afterDelay:1.6f];
}

- (void)tap3 {
    tripleLabel.text = @"Triple Tap Detected";
    [self performSelector:@selector(eraseMe:)
        withObject:tripleLabel afterDelay:1.6f];
}

- (void)tap4 {
    quadrupleLabel.text = @"Quadruple Tap Detected";
    [self performSelector:@selector(eraseMe:)
        withObject:quadrupleLabel afterDelay:1.6f];
}

- (void)eraseMe:(UILabel *)label {
    label.text = @"";
}
    .
    .
    .
```

Insert the following lines into the existing `viewDidUnload` method:

```
- (void)viewDidUnload {
```

```
        [super viewDidUnload];
        // Release any retained subviews of the main view.
        // e.g. self.myOutlet = nil;
        self.singleLabel = nil;
        self.doubleLabel = nil;
        self.tripleLabel = nil;
        self.quadrupleLabel = nil;
}
```

Now, add the following code to viewDidLoad:

```
- (void)viewDidLoad {
        [super viewDidLoad];
        // Do any additional setup after loading the view, typically from a nib.
        UITapGestureRecognizer *singleTap =
            [[UITapGestureRecognizer alloc] initWithTarget:self
                                                    action:@selector(tap1)];
        singleTap.numberOfTapsRequired = 1;
        singleTap.numberOfTouchesRequired = 1;
        [self.view addGestureRecognizer:singleTap];

        UITapGestureRecognizer *doubleTap =
            [[UITapGestureRecognizer alloc] initWithTarget:self
                                                    action:@selector(tap2)];
        doubleTap.numberOfTapsRequired = 2;
        doubleTap.numberOfTouchesRequired = 1;
        [self.view addGestureRecognizer:doubleTap];
        [singleTap requireGestureRecognizerToFail:doubleTap];

        UITapGestureRecognizer *tripleTap =
            [[UITapGestureRecognizer alloc] initWithTarget:self
                                                    action:@selector(tap3)];
        tripleTap.numberOfTapsRequired = 3;
        tripleTap.numberOfTouchesRequired = 1;
        [self.view addGestureRecognizer:tripleTap];
        [doubleTap requireGestureRecognizerToFail:tripleTap];

        UITapGestureRecognizer *quadrupleTap =
            [[UITapGestureRecognizer alloc] initWithTarget:self
                                                    action:@selector(tap4)];
        quadrupleTap.numberOfTapsRequired = 4;
        quadrupleTap.numberOfTouchesRequired = 1;
        [self.view addGestureRecognizer:quadrupleTap];
        [tripleTap requireGestureRecognizerToFail:quadrupleTap];
}
```

The four tap methods do nothing more in this application than set one of the four labels
and use performSelector:withObject:afterDelay: to erase that same label after 1.6
seconds. The eraseMe: method erases the text from any label that is passed into it.

The interesting part of this is what occurs in the viewDidLoad method. We start off simply
enough, by setting up a tap gesture recognizer and attaching it to our view.

```
        UITapGestureRecognizer *singleTap =
            [[UITapGestureRecognizer alloc] initWithTarget:self
                                                    action:@selector(tap1)];
```

```
singleTap.numberOfTapsRequired = 1;
singleTap.numberOfTouchesRequired = 1;
[self.view addGestureRecognizer:singleTap];
```

Note that we set both the number of taps (touches in the same position, one after another) required to trigger the action and touches (number of fingers touching the screen at the same time) to 1. After that, we set up another tap gesture recognizer to handle a double-tap.

```
UITapGestureRecognizer *doubleTap =
    [[UITapGestureRecognizer alloc] initWithTarget:self
                                    action:@selector(tap2)];
doubleTap.numberOfTapsRequired = 2;
doubleTap.numberOfTouchesRequired = 1;
[self.view addGestureRecognizer:doubleTap];
[singleTap requireGestureRecognizerToFail:doubleTap];
```

This is pretty similar to the previous code, right up until that last line, in which we give singleTap some additional context. We are effectively telling singleTap that it should trigger its action only in case some other gesture recognizer—in this case doubleTap— decides that the current user input isn't what it's looking for.

Let's think about what this means. With those two tap gesture recognizers in place, a single tap in the view will immediately make singleTap think, "Hey, this looks like it's for me." At the same time, doubleTap will think, "Hey, this looks like it *might* be for me, but I'll need to wait for one more tap." Because singleTap is set up to wait for doubleTap's "failure," it doesn't send its action method right away; instead, it waits to see what happens with doubleTap.

After that first tap, if another tap occurs immediately, then doubleTap says, "Hey, that's mine all right," and fires its action. At that point, singleTap will realize what happened and give up on that gesture. On the other hand, if a particular amount of time goes by (the amount of time that the system considers to be the maximum length of time between taps in a double-tap), doubleTap will give up, and singleTap will see the failure and finally trigger its event.

The rest of the method goes on to define gesture recognizers for three and four taps, and at each point configure one gesture to be dependent on the failure of the next.

```
UITapGestureRecognizer *tripleTap =
    [[UITapGestureRecognizer alloc] initWithTarget:self
                                    action:@selector(tap3)];
tripleTap.numberOfTapsRequired = 3;
tripleTap.numberOfTouchesRequired = 1;
[self.view addGestureRecognizer:tripleTap];
[doubleTap requireGestureRecognizerToFail:tripleTap];

UITapGestureRecognizer *quadrupleTap =
    [[UITapGestureRecognizer alloc] initWithTarget:self
                                    action:@selector(tap4)];
quadrupleTap.numberOfTapsRequired = 4;
quadrupleTap.numberOfTouchesRequired = 1;
[self.view addGestureRecognizer:quadrupleTap];
[tripleTap requireGestureRecognizerToFail:quadrupleTap];
```

Note that we don't need to explicitly configure every gesture to be dependent on the failure of each of the higher-tap-numbered gestures. That multiple dependency comes about naturally as a result of the chain of failure established in our code. Since `singleTap` requires the failure of `doubleTap`, `doubleTap` requires the failure of `tripleTap`, and `tripleTap` requires the failure of `quadrupleTap`, by extension, `singleTap` requires that all of the others fail.

Compile and run the app, and whether you single-, double-, triple-, or quadruple-tap, you should see only one label displayed.

Detecting Pinches

Another common gesture is the two-finger pinch. It's used in a number of applications, including Mobile Safari, Mail, and Photos, to let you zoom in (if you pinch apart) or zoom out (if you pinch together).

Detecting pinches is really easy, thanks to `UIPinchGestureRecognizer`. This one is referred to as a continuous gesture recognizer, because it calls its action method over and over again during the pinch. While the gesture is underway, the recognizer goes through a number of states. The only one we want to watch for is `UIGestureRecognizerStateBegan`, which is the state that the recognizer is in when it first calls the action method after detecting that a pinch is happening. At that moment, the pinch gesture recognizer's `scale` property is always set to `1.0`; for the rest of the gesture, that number goes up and down, relative to how far the user's fingers move from the start. We're going to use the `scale` value to resize the text in a label.

Create a new project in Xcode, again using the *Single View Application* template, and call this one *PinchMe*. The PinchMe application will need only a single outlet for a label, but it also needs a property to hold the size of the label's font at the start of the pinch. Expand the *PinchMe* folder, single-click *BIDViewController.h*, and make the following changes:

```
#import <UIKit/UIKit.h>

@interface BIDViewController : UIViewController

@property (weak, nonatomic) IBOutlet UILabel *label;
@property (assign, nonatomic) CGFloat initialFontSize;
@end
```

Now that we have our outlet, edit *BIDViewController.xib*. In Interface Builder, make sure the view is displayed in its editing window, and drag a single label into the view, aligning it with the upper-left blue guidelines. Grab the lower-right corner of the label and resize it so it hits the lower-right blue guidelines.

Unlike some other examples we've shown, we need to have a bit of text in the label so there's something to see. Double-click the label and change the text to a single capital letter *X*. This letter is what we'll be zooming in and out on. Set the label's alignment to centered. Next, control-drag from the *File's Owner* icon to the label, and connect it to the *label* outlet.

Save the nib. Now, bounce over to *BIDViewController.m*, and add the following code at the top of the file:

```
#import "BIDViewController.h"

@implementation BIDViewController
@synthesize label;
@synthesize initialFontSize;
.
.
.
```

Clean up our outlet in the viewDidUnload method:

```
- (void)viewDidUnload
{
    [super viewDidUnload];
    // Release any retained subviews of the main view.
    // e.g. self.myOutlet = nil;
    self.label = nil;
}
```

Then add the following code to the viewDidLoad method:

```
- (void)viewDidLoad
{
    [super viewDidLoad];
    // Do any additional setup after loading the view, typically from a nib.
    UIPinchGestureRecognizer *pinch = [[UIPinchGestureRecognizer alloc]
        initWithTarget:self action:@selector(doPinch:)];
    [self.view addGestureRecognizer:pinch];
}
```

And add the following method at the end of the file:

```
.
.
.
- (void)doPinch:(UIPinchGestureRecognizer *)pinch {
    if (pinch.state == UIGestureRecognizerStateBegan) {
        initialFontSize = label.font.pointSize;
    } else {
        label.font = [label.font fontWithSize:initialFontSize * pinch.scale];
    }
}
@end
```

In viewDidLoad, we set up a pinch gesture recognizer and tell it to notify us via the doPinch: method when pinching is occurring. Inside doPinch:, we look at the pinch's state to see if it's just starting, and if so, we store the current font size for later use. Otherwise, if the pinch is already in progress, we use the stored initial font size and the current pinch scale to calculate a new font size.

And that's all there is to pinch detection. Compile and run the app to give it a try. As you do some pinching, you'll see the text change size in response (see Figure 17–5). If you're on the simulator, remember that you can simulate a pinch by holding down the option key and clicking and dragging in the simulator window using your mouse.

Figure 17–5. *The PinchMe application detects the pinch guesture, both for zooming in and zooming out.*

Defining Custom Gestures

You've now seen how to detect the most commonly used iPhone gestures. The real fun begins when you start defining your own custom gestures! You've already learned how to use a few of UIGestureRecognizer's subclasses, so now it's time to learn how to create your own gestures, which can be easily attached to any view you like.

Defining a custom gesture is tricky. You've already mastered the basic mechanism, and that wasn't too difficult. The tricky part is being flexible when defining what constitutes a gesture.

Most people are not precise when they use gestures. Remember the variance we used when we implemented the swipe, so that even a swipe that wasn't perfectly horizontal or vertical still counted? That's a perfect example of the subtlety you need to add to your own gesture definitions. If you define your gesture too strictly, it will be useless. If you define it too generically, you'll get too many false positives, which will frustrate the user. In a sense, defining a custom gesture can be hard because you must be precise about a gesture's imprecision. If you try to capture a complex gesture like, say, a figure eight, the math behind detecting the gesture is also going to get quite complex.

The CheckPlease Application

In our sample, we're going to define a gesture shaped like a check mark (see Figure 17–6).

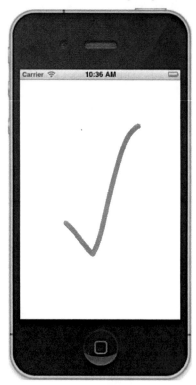

Figure 17–6. *An illustration of our check-mark gesture*

What are the defining properties of this check-mark gesture? Well, the principal one is that sharp change in angle between the two lines. We also want to make sure that the user's finger has traveled a little distance in a straight line before it makes that sharp angle. In Figure 17–6, the legs of the check mark meet at an acute angle, just under 90 degrees. A gesture that required exactly an 85-degree angle would be awfully hard to get right, so we'll define a range of acceptable angles.

Create a new project in Xcode using the *Single View Application* template, and call the project *CheckPlease*. In this project, we're going to need to do some fairly standard analytic geometry to calculate such things as the distance between two points and the angle between two lines. Don't worry if you don't remember much geometry; we've provided you with functions that will do the calculations for you.

Look in the *17 - CheckPlease* folder for two files, named *CGPointUtils.h* and *CGPointUtils.c*. Drag both of these to the *CheckPlease* folder of your project. Feel free to use these utility functions in your own applications.

Control-click in the *CheckPlease* folder, and add a new file to the project. Use the file-creation assistant to create a new Objective-C class called BIDCheckMarkRecognizer. In the *Subclass of* control, type *UIGestureRecognizer*. Then select *BIDCheckMarkRecognizer.h*, and make the following changes:

```
#import <UIKit/UIKit.h>

@interface BIDCheckMarkRecognizer : UIGestureRecognizer

@property (assign, nonatomic) CGPoint lastPreviousPoint;
@property (assign, nonatomic) CGPoint lastCurrentPoint;
@property (assign, nonatomic) CGFloat lineLengthSoFar;
@end
```

Here, we declare three properties: lastPreviousPoint, lastCurrentPoint, and lineLengthSoFar. Each time we're notified of a touch, we're given the previous touch point and the current touch point. Those two points define a line segment. The next touch adds another segment. We store the previous touch's previous and current points in lastPreviousPoint and lastCurrentPoint, which gives us the previous line segment. We can then compare that line segment to the current touch's line segment. Comparing these two line segments can tell us if we're still drawing a single line or if there's a sharp enough angle between the two segments that we're actually drawing a check mark.

Remember that every UITouch object knows its current position in the view, as well as its previous position in the view. In order to compare angles, however, we need to know the line that the previous two points made, so we need to store the current and previous points from the last time the user touched the screen. We'll use these two variables to store those two values each time this method is called, so that we have the ability to compare the current line to the previous line and check the angle.

We also declare a property to keep a running count of how far the user has dragged the finger. If the finger hasn't traveled at least 10 pixels (a value we'll soon define in kMinimumCheckMarkLength), whether the angle falls in the correct range doesn't matter. If we didn't require this distance, we would receive a lot of false positives.

Now select *BIDCheckMarkRecognizer.m*, and make the following changes:

```
#import "BIDCheckMarkRecognizer.h"
#import "CGPointUtils.h"
#import <UIKit/UIGestureRecognizerSubclass.h>

#define kMinimumCheckMarkAngle    50
#define kMaximumCheckMarkAngle     135
#define kMinimumCheckMarkLength    10

@implementation BIDCheckMarkRecognizer
@synthesize lastPreviousPoint;
@synthesize lastCurrentPoint;
@synthesize lineLengthSoFar;

- (void)touchesBegan:(NSSet *)touches withEvent:(UIEvent *)event {
    [super touchesBegan:touches withEvent:event];
    UITouch *touch = [touches anyObject];
```

```
        CGPoint point = [touch locationInView:self.view];
        lastPreviousPoint = point;
        lastCurrentPoint = point;
        lineLengthSoFar = 0.0f;
    }

    - (void)touchesMoved:(NSSet *)touches withEvent:(UIEvent *)event {
        [super touchesMoved:touches withEvent:event];
        UITouch *touch = [touches anyObject];
        CGPoint previousPoint = [touch previousLocationInView:self.view];
        CGPoint currentPoint = [touch locationInView:self.view];
        CGFloat angle = angleBetweenLines(lastPreviousPoint,
                                          lastCurrentPoint,
                                          previousPoint,
                                          currentPoint);
        if (angle >= kMinimumCheckMarkAngle && angle <= kMaximumCheckMarkAngle
            && lineLengthSoFar > kMinimumCheckMarkLength) {
            self.state = UIGestureRecognizerStateEnded;
        }
        lineLengthSoFar += distanceBetweenPoints(previousPoint, currentPoint);
        lastPreviousPoint = previousPoint;
        lastCurrentPoint = currentPoint;
    }
    @end
```

After importing *CGPointUtils.h*, the file we mentioned earlier, we import a special header file called *UIGestureRecognizerSubclass.h*, which contains declarations that are intended for use only by a subclass. The important thing this does is to make the gesture recognizer's state property writable. That's the mechanism our subclass will use to affirm that the gesture we're watching was successfully completed.

Then we define the parameters that we use to decide whether the user's finger-squiggling matches our definition of a check mark. You can see that we've defined a minimum angle of 50 degrees and a maximum angle of 135 degrees. This is a pretty broad range, and depending on your needs, you might decide to restrict the angle. We experimented a bit with this and found that our practice check-mark gestures fell into a fairly broad range, which is why we chose a relatively large tolerance here. We were somewhat sloppy with our check-mark gestures, and so we expect that at least some of our users will be as well. As a wise man once said, "Be rigorous in what you produce and tolerant in what you accept."

The CheckPlease Touch Methods

Let's take a look at the touch methods. You'll notice that each of them first calls the superclass's implementation—something we've never done before. We need to do this in a UIGestureRecognizer subclass so that our superclass can have the same amount of knowledge about the events as we do. Now, on to the code itself.

In touchesBegan:withEvent:, we determine the point that the user is currently touching and store that value in lastPreviousPoint and lastCurrentPoint. Since this method is called when a gesture begins, we know there is no previous point to worry about, so we store the current point in both. We also reset the running line length count to 0.

Then, in touchesMoved:withEvent:, we calculate the angle between the line from the current touch's previous position to its current position, and the line between the two points stored in the lastPreviousPoint and lastCurrentPoint instance variables. Once we have that angle, we check to see if it falls within our range of acceptable angles and check to make sure that the user's finger has traveled far enough before making that sharp turn. If both of those are true, we set the label to show that we've identified a check-mark gesture. Next, we calculate the distance between the touch's position and its previous position, add that to lineLengthSoFar, and replace the values in lastPreviousPoint and lastCurrentPoint with the two points from the current touch so we'll have them next time through this method.

Now that we have a gesture recognizer of our own to try out, it's time to connect it to a view, just as we've done with the others we've used. Select *BIDViewController.h*, and make the following change:

```
#import <UIKit/UIKit.h>

@interface BIDViewController : UIViewController

@property (weak, nonatomic) IBOutlet UILabel *label;
@end
```

Here, we simply define an outlet to a label that we'll use to inform the user when we've detected a check-mark gesture.

Select *BIDViewController.xib* to edit the GUI. Add a *Label* from the library to the upper-left blue guideline, resize it so it spans from left blue guideline to right blue guideline, and set its alignment to centered. Control-drag from the *File's Owner* icon to that label to connect it to the *label* outlet, and double-click the label to delete its text. Save the nib file.

Now, switch over to *BIDViewController.m*, and add the following code to the top of the file:

```
#import "BIDViewController.h"
#import "BIDCheckMarkRecognizer.h"

@implementation BIDViewController
@synthesize label;

- (void)doCheck:(BIDCheckMarkRecognizer *)check {
    label.text = @"Checkmark";
    [self performSelector:@selector(eraseLabel)
            withObject:nil afterDelay:1.6];
}

- (void)eraseLabel {
    label.text = @"";
}
.
.
.
```

This gives us an action method to connect our recognizer to, which in turn triggers the familiar-looking `eraseLabel` method. Next, edit the `viewDidLoad` method, adding the following lines, which connect an instance of our new recognizer to the view:

```
- (void)viewDidLoad
{
    [super viewDidLoad];
    // Do any additional setup after loading the view, typically from a nib.
    BIDCheckMarkRecognizer *check = [[BIDCheckMarkRecognizer alloc] initWithTarget:self
        action:@selector(doCheck:)];
    [self.view addGestureRecognizer:check];
}
```

All that's left now is to add the following code to the existing `viewDidUnload` method:

```
- (void)viewDidUnload
{
    [super viewDidUnload];
    // Release any retained subviews of the main view.
    // e.g. self.myOutlet = nil;
    self.label = nil;
}
```

Compile and run the app, and try out the gesture.

When defining new gestures for your own applications, make sure you test them thoroughly, and if you can, have other people test them for you as well. You want to make sure that your gesture is easy for the user to do, but not so easy that it gets triggered unintentionally. You also need to make sure that you don't conflict with other gestures used in your application. A single gesture should not count, for example, as both a custom gesture and a pinch.

Garçon? Check, Please!

You should now understand the mechanism iOS uses to tell your application about touches, taps, and gestures. You also know how to detect the most commonly used iOS gestures, and even got a taste of how you might go about defining your own custom gestures. The iPhone's interface relies on gestures for much of its ease of use, so you'll want to have these techniques at the ready for most of your iOS development.

When you're ready to move on, turn the page, and we'll tell you how to figure out where in the world you are using Core Location.

Chapter 18

Where Am I? Finding Your Way with Core Location

Every iOS device has the ability to determine where in the world it is using a framework called Core Location. Core Location can actually leverage three technologies to do this: GPS, cell tower triangulation, and Wi-Fi Positioning Service (WPS).

GPS is the most accurate of the three technologies, but it is not available on first-generation iPhones, iPod touches, or Wi-Fi-only iPads. In short, any device with at least a 3G data connection also contains a GPS unit. GPS reads microwave signals from multiple satellites to determine the current location.

> **NOTE:** Technically, Apple uses a version of GPS called Assisted GPS, also known as A-GPS. A-GPS uses network resources to help improve the performance of stand-alone GPS.

Cell tower triangulation determines the current location by doing a calculation based on the locations of the cell towers in the phone's range. Cell tower triangulation can be fairly accurate in cities and other areas with a high cell tower density, but it becomes less accurate in areas where there is a greater distance between towers. Triangulation requires a cell radio connection, so it works only on the iPhone (all models, including the very first) and the iPad with a 3G data connection.

The WPS opton uses the MAC addresses from nearby Wi-Fi access points to make a guess at your location by referencing a large database of known service providers and the areas they service. WPS is imprecise and can be off by many miles.

All three methods put a noticeable drain on the battery, so keep that in mind when using Core Location. Your application shouldn't poll for location any more often than is absolutely necessary. When using Core Location, you have the option of specifying a desired accuracy. By carefully specifying the absolute minimum accuracy level you need, you can prevent unnecessary battery drain.

The technologies that Core Location depends on are hidden from your application. We don't tell Core Location whether to use GPS, triangulation, or WPS. We just tell it how accurate we would like it to be, and it will decide from the technologies available to it which is best for fulfilling your request.

The Location Manager

The Core Location API is actually fairly easy to use. The main class we'll work with is CLLocationManager, usually referred to as the **location manager**. To interact with Core Location, you need to create an instance of the location manager, like this:

```
CLLocationManager *locationManager = [[CLLocationManager alloc] init];
```

This creates an instance of the location manager, but it doesn't actually start polling for your location. You must create an object that conforms to the CLLocationManagerDelegate protocol and assign it as the location manager's delegate. The location manager will call delegate methods when location information becomes available or changes. The process of determining location may take some time—even a few seconds.

Setting the Desired Accuracy

After you set the delegate, you also want to set the requested accuracy. As we mentioned, don't specify a degree of accuracy any greater than you absolutely need. If you're writing an application that just needs to know which state or country the phone is in, don't specify a high level of precision. Remember that the more accuracy you demand of Core Location, the more juice you're likely to use. Also, keep in mind that there is no guarantee that you will get the level of accuracy you have requested.

Here's an example of setting the delegate and requesting a specific level of accuracy:

```
locationManager.delegate = self;
locationManager.desiredAccuracy = kCLLocationAccuracyBest;
```

The accuracy is set using a CLLocationAccuracy value, a type that's defined as a double. The value is in meters, so if you specify a desiredAccuracy of 10, you're telling Core Location that you want it to try to determine the current location within 10 meters, if possible. Specifying kCLLocationAccuracyBest, as we did previously, or specifying kCLLocationAccuracyBestForNavigation (where it uses other sensor data as well) tells Core Location to use the most accurate method that's currently available. In addition, you can also use kCLLocationAccuracyNearestTenMeters, kCLLocationAccuracyHundredMeters, kCLLocationAccuracyKilometer, and kCLLocationAccuracyThreeKilometers.

Setting the Distance Filter

By default, the location manager will notify the delegate of any detected change in the device's location. By specifying a **distance filter**, you are telling the location manager not to notify you of every change, and instead to notify you only when the location changes by more than a certain amount. Setting up a distance filter can reduce the amount of polling your application does.

Distance filters are also set in meters. Specifying a distance filter of 1000 tells the location manager not to notify its delegate until the iPhone has moved at least 1,000 meters from its previously reported position. Here's an example:

```
locationManager.distanceFilter = 1000.0f;
```

If you ever want to return the location manager to the default setting of no filter, you can use the constant kCLDistanceFilterNone, like this:

```
locationManager.distanceFilter = kCLDistanceFilterNone;
```

Just as when specifying the desired accuracy, you should take care to avoid getting updates any more frequently than you really need them, because otherwise you waste battery power. A speedometer app that's calculating the user's velocity based on the user's location will probably want to have updates as fast as possible, but an app that's going to show the nearest fast-food restaurant can get by with a lot less.

Starting the Location Manager

When you're ready to start polling for location, you tell the location manager to start. It will go off and do its thing, and then call a delegate method when it has determined the current location. Until you tell it to stop, it will continue to call your delegate method whenever it senses a change that exceeds the current distance filter.

Here's how you start the location manager:

```
[locationManager startUpdatingLocation];
```

Using the Location Manager Wisely

If you need to determine the current location only and have no need to continuously poll for location, you should have your location delegate stop the location manager as soon as it gets the information your application requires. If you need to continuously poll, make sure you stop polling as soon as you possibly can. Remember that as long as you are getting updates from the location manager, you are putting a strain on the user's battery.

To tell the location manager to stop sending updates to its delegate, call stopUpdatingLocation, like this:

```
[locationManager stopUpdatingLocation];
```

The Location Manager Delegate

The location manager delegate must conform to the CLLocationManagerDelegate protocol, which defines two methods, both of which are optional. One of these methods is called by the location manager when it has determined the current location or when it detects a change in location. The other method is called when the location manager encounters an error.

Getting Location Updates

When the location manager wants to inform its delegate of the current location, it calls the `locationManager:didUpdateToLocation:fromLocation:` method. This method takes three parameters:

- The first parameter is the location manager that called the method.

- The second parameter is a `CLLocation` object that defines the current location of the device.

- The third parameter is a `CLLocation` object that defines the previous location from the last update.

The first time this method is called, the previous location object will be `nil`.

Getting Latitude and Longitude Using CLLocation

Location information is passed from the location manager using instances of the `CLLocation` class. This class has five properties that might be of interest to your application. The latitude and longitude are stored in a property called `coordinate`. To get the latitude and longitude in degrees, do this:

```
CLLocationDegrees latitude = theLocation.coordinate.latitude;
CLLocationDegrees longitude = theLocation.coordinate.longitude;
```

The `CLLocation` object can also tell you how confident the location manager is in its latitude and longitude calculations. The `horizontalAccuracy` property describes the radius of a circle with the `coordinate` as its center. The larger the value in `horizontalAccuracy`, the less certain Core Location is of the location. A very small radius indicates a high level of confidence in the determined location.

You can see a graphic representation of `horizontalAccuracy` in the Maps application (see Figure 18–1). The circle shown in Maps uses `horizontalAccuracy` for its radius when it detects your location. The location manager thinks you are at the center of that circle. If you're not, you're almost certainly somewhere inside the circle. A negative value in `horizontalAccuracy` is an indication that you cannot rely on the values in `coordinate` for some reason.

Figure 18–1. *The Maps application uses Core Location to determine your current location. The outer circle is a visual representation of the horizontal accuracy.*

The CLLocation object also has a property called altitude that can tell you how many meters above or below sea level you are:

```
CLLocationDistance altitude = theLocation.altitude;
```

Each CLLocation object maintains a property called verticalAccuracy that is an indication of how confident Core Location is in its determination of altitude. The value in altitude could be off by as many meters as the value in verticalAccuracy. If the verticalAccuracy value is negative, Core Location is telling you it could not determine a valid altitude.

CLLocation objects also have a timestamp that tells when the location manager made the location determination.

In addition to these properties, CLLocation has a useful instance method that will let you determine the distance between two CLLocation objects. The method is called distanceFromLocation:, and it works like this:

```
CLLocationDistance distance = [fromLocation distanceFromLocation:toLocation];
```

The preceding line of code will return the distance between two CLLocation objects, fromLocation and toLocation. This distance value returned will be the result of a great-circle distance calculation that ignores the altitude property and calculates the distance as if both points were at sea level. For most purposes, a great-circle calculation will be more than sufficient, but if you do want to take altitude into account when calculating distances, you'll need to write your own code to do it.

> **NOTE:** If you're not sure what's meant by *great-circle distance*, you might want to think back to geography class and the notion of a *great-circle route*. The idea is that the shortest distance between any two points on the earth's surface will be found along a route that goes the entire way around the earth: a "great circle." The calculation performed by CLLocation determines the distance between two points along such a route, taking the curvature of the earth into account. Without accounting for that curvature, you would end up with the length of a straight line connecting the two points, which isn't much use, since that line would invariably go straight through some amount of the earth itself!

Error Notifications

If Core Location is not able to determine your current location, it will call a second delegate method named locationManager:didFailWithError:. The most likely cause of an error is that the user denies access. The user must authorize use of the location manager, so the first time your application wants to determine the location, an alert will pop up on the screen asking if it's OK for the current program to access your location (see Figure 18–2).

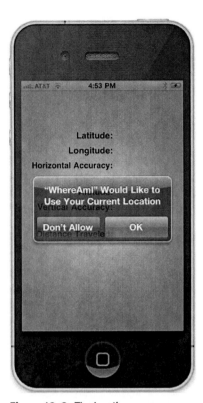

Figure 18–2. *The location manager access must be approved by the user.*

If the user taps the *Don't Allow* button, your delegate will be notified of the fact by the location manager using the `locationManager:didFailWithError:` with an error code of `kCLErrorDenied`. At the time of this writing, the only other error code supported by the location manager is `kCLErrorLocationUnknown`, which indicates that Core Location was unable to determine the location but that it will keep trying. The `kCLErrorDenied` error generally indicates that your application will not be able to access Core Location any time during the remainder of the current session. On the other hand, `kCLErrorLocationUnknown` errors indicate a problem that may be temporary.

> **NOTE:** When working in the simulator, a dialog will appear outside the simulator window, asking to use your current location. In that case, your location will be determined using a super-secret algorithm kept in a locked vault buried deep beneath Apple headquarters in Cupertino.

Trying Out Core Location

Let's build a small application to detect the iPhone's current location and the total distance traveled while the program has been running. You can see what our final application will look like in Figure 18–3.

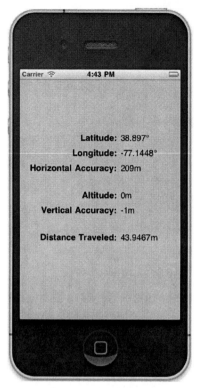

Figure 18–3. *The WhereAmI application in action. This screenshot was taken in the simulator. Notice that the vertical accuracy is a negative number, which tells us it couldn't determine the altitude.*

In Xcode, create a new project using the *Single View Application* template, call the project *WhereAmI*, and set *Device Family* to *iPhone*. Select *BIDViewController.h*, and make the following changes:

```
#import <UIKit/UIKit.h>
#import <CoreLocation/CoreLocation.h>

@interface BIDViewController :
    UIViewController <CLLocationManagerDelegate>

@property (strong, nonatomic) CLLocationManager *locationManager;
@property (strong, nonatomic) CLLocation *startingPoint;
@property (strong, nonatomic) IBOutlet UILabel *latitudeLabel;
@property (strong, nonatomic) IBOutlet UILabel *longitudeLabel;
@property (strong, nonatomic) IBOutlet UILabel *horizontalAccuracyLabel;
@property (strong, nonatomic) IBOutlet UILabel *altitudeLabel;
@property (strong, nonatomic) IBOutlet UILabel *verticalAccuracyLabel;
@property (strong, nonatomic) IBOutlet UILabel *distanceTraveledLabel;
@end
```

First, notice that we've included the Core Location header files. Core Location is not part of either UIKit or Foundation, so we need to include the header files manually. Next, we conform this class to the `CLLocationManagerDelegate` method so that we can receive location information from the location manager.

After that, we declare a `CLLocationManager` pointer, which will be used to hold the instance of the Core Location we create. We also declare a pointer to a `CLLocation`, which we will set to the location we receive in the first update from the location manager. This way, if the user has our program running and moves far enough to trigger updates, we'll be able to calculate how far our user moved. Our delegate will be notified of the previous location with each call, but not the original starting location, which is why we store it.

The remaining properties are all outlets that will be used to update labels on the user interface.

Select *BIDViewController.xib* to create the GUI. Using Figure 18–3 as your guide, drag 12 *Labels* from the library to the *View* window. Six of them should be placed on the left side of the screen, right-justified, and made bold. Give the six bold labels the values *Latitude:*, *Longitude:*, *Horizontal Accuracy:*, *Altitude:*, *Vertical Accuracy:*, and *Distance Traveled:*. Since the *Horizontal Accuracy:* label is the longest, you might place that one first, and then option-drag out copies of that label to create the other five left-side labels. The six right-side labels should be left-justified and placed next to each of the bold labels.

Each of the labels on the right side should be connected to the appropriate outlet we defined in the header file earlier. Once you have all six attached to outlets, double-click each one in turn, and delete the text it holds.

Save your changes. Next, return to Xcode, select *BIDViewController.m,* and make the following changes at the top of the file:

```
#import "BIDViewController.h"

@implementation BIDViewController
@synthesize locationManager;
@synthesize startingPoint;
@synthesize latitudeLabel;
@synthesize longitudeLabel;
@synthesize horizontalAccuracyLabel;
@synthesize altitudeLabel;
@synthesize verticalAccuracyLabel;
@synthesize distanceTraveledLabel;


    .
    .
    .
```

Insert the following lines in `viewDidLoad` to configure the location manager:

```
- (void)viewDidLoad {
    [super viewDidLoad];
    // Do any additional setup after loading the view, typically from a nib.
    self.locationManager = [[CLLocationManager alloc] init];
    locationManager.delegate = self;
    locationManager.desiredAccuracy = kCLLocationAccuracyBest;
    [locationManager startUpdatingLocation];
```

```
}
```

Insert the following lines in `viewDidUnload` to clean up our outlets:

```objc
- (void)viewDidUnload {
    [super viewDidUnload];
    // Release any retained subviews of the main view.
    // e.g. self.myOutlet = nil;
    self.locationManager = nil;
    self.latitudeLabel = nil;
    self.longitudeLabel = nil;
    self.horizontalAccuracyLabel = nil;
    self.altitudeLabel = nil;
    self.verticalAccuracyLabel = nil;
    self.distanceTraveledLabel= nil;
}
```

And insert the following new methods at the end of the file:

```objc
    .
    .
    .
#pragma mark -
#pragma mark CLLocationManagerDelegate Methods
- (void)locationManager:(CLLocationManager *)manager
       didUpdateToLocation:(CLLocation *)newLocation
       fromLocation:(CLLocation *)oldLocation {

    if (startingPoint == nil)
        self.startingPoint = newLocation;

    NSString *latitudeString = [NSString stringWithFormat:@"%g\u00B0",
                                newLocation.coordinate.latitude];
    latitudeLabel.text = latitudeString;

    NSString *longitudeString = [NSString stringWithFormat:@"%g\u00B0",
                                 newLocation.coordinate.longitude];
    longitudeLabel.text = longitudeString;

    NSString *horizontalAccuracyString = [NSString stringWithFormat:@"%gm",
                                          newLocation.horizontalAccuracy];
    horizontalAccuracyLabel.text = horizontalAccuracyString;

    NSString *altitudeString = [NSString stringWithFormat:@"%gm",
                                newLocation.altitude];
    altitudeLabel.text = altitudeString;

    NSString *verticalAccuracyString = [NSString stringWithFormat:@"%gm",
                                        newLocation.verticalAccuracy];
    verticalAccuracyLabel.text = verticalAccuracyString;

    CLLocationDistance distance = [newLocation
                                   distanceFromLocation:startingPoint];
    NSString *distanceString = [NSString stringWithFormat:@"%gm", distance];
    distanceTraveledLabel.text = distanceString;
}
```

```
- (void)locationManager:(CLLocationManager *)manager
      didFailWithError:(NSError *)error {
   NSString *errorType = (error.code == kCLErrorDenied) ?
            @"Access Denied" : @"Unknown Error";
   UIAlertView *alert = [[UIAlertView alloc]
                         initWithTitle:@"Error getting Location"
                         message:errorType
                         delegate:nil
                         cancelButtonTitle:@"Okay"
                         otherButtonTitles:nil];
      [alert show];
}
@end
```

In the viewDidLoad method, we allocate and initialize a CLLocationManager instance, assign our controller class as the delegate, set the desired accuracy to the best available, and then tell our location manager instance to start giving us location updates.

```
- (void)viewDidLoad {
   self.locationManager = [[CLLocationManager alloc] init];
   locationManager.delegate = self;
   locationManager.desiredAccuracy = kCLLocationAccuracyBest;
   [locationManager startUpdatingLocation];
}
```

Updating Location Manager

Since this class designated itself as the location manager's delegate, we know that location updates will come into this class if we implement the delegate method locationmanager:didUpdateToLocation:fromLocation:. Now, let's look at our implementation of that method.

The first thing we do in the delegate method is check whether startingPoint is nil. If it is, then this update is the first one from the location manager, and we assign the current location to our startingPoint property.

```
   if (startingPoint == nil)
      self.startingPoint = newLocation;
```

After that, we update the first six labels with values from the CLLocation object passed in the newLocation argument.

```
   NSString *latitudeString = [NSString stringWithFormat:@"%g\u00B0",
                        newLocation.coordinate.latitude];
   latitudeLabel.text = latitudeString;

   NSString *longitudeString = [NSString stringWithFormat:@"%g\u00B0",
                        newLocation.coordinate.longitude];
   longitudeLabel.text = longitudeString;

   NSString *horizontalAccuracyString = [NSString stringWithFormat:@"%gm",
                        newLocation.horizontalAccuracy];
   horizontalAccuracyLabel.text = horizontalAccuracyString;

   NSString *altitudeString = [NSString stringWithFormat:@"%gm",
```

```
                                   newLocation.altitude];
altitudeLabel.text = altitudeString;

NSString *verticalAccuracyString = [NSString stringWithFormat:@"%gm",
                                    newLocation.verticalAccuracy];
verticalAccuracyLabel.text = verticalAccuracyString;
```

> **NOTE:** Both the longitude and latitude are displayed in formatting strings containing the cryptic-looking \u00B0. This represents the Unicode representation of the degree symbol (°). It's never a good idea to put anything other than ASCII characters directly in a source code file, but including the hex value in a string is just fine, and that's what we've done here.

Determining Distance Traveled

Finally, we determine the distance between the current location and the location stored in startingPoint and display the distance. While this application runs, if the user moves far enough for the location manager to detect the change, the *Distance Traveled:* field will be continually updated with the distance away from where the user was when the application was started.

```
CLLocationDistance distance = [newLocation
                              distanceFromLocation:startingPoint];
NSString *distanceString = [NSString stringWithFormat:@"%gm", distance];
distanceTraveledLabel.text = distanceString;
```

And there you have it. Core Location is fairly straightforward and easy to use.

Before you can compile this program, you need to add *CoreLocation.framework* to your project. You do this the same as you did in Chapter 7 when you added *AudioToolbox.framework*, except you choose *CoreLocation.framework* instead of *AudioToolbox.framework*. Here's a hint: click the first line in the project navigator (the blue *WhereAmI* icon), click the *WhereAmI* icon under *TARGETS*, and then click the *Build Phases* tab. Expand the *Link Binary With Libraries* disclosure triangle, and add your framework.

Compile and run the application, and try it. If you have the ability to run the application on your iPhone or iPad, try going for a drive with the application running and watch the values change as you drive. Um, actually, it's better to have someone else do the driving!

Wherever You Go, There You Are

You've now seen pretty much all there is to Core Location. Although the underlying technology is quite complex, Apple has provided a simple interface that hides most of the complexity, making it quite easy to add location-related features to your applications so you can tell where the users are and identify when they move.

And speaking of moving, when you're ready, proceed directly to the next chapter so we can play with the iPhone's built-in accelerometer.

Chapter 19

Whee! Gyro and Accelerometer!

One of the coolest features of the iPhone, iPad, and iPod touch is the built-in accelerometer—the tiny device that lets iOS know how the device is being held and if it's being moved. iOS uses the accelerometer to handle autorotation, and many games use it as a control mechanism. The accelerometer can also be used to detect shakes and other sudden movement. This capability was extended even further with the introduction of the iPhone 4, which also includes a built-in gyroscope that lets you determine the angle at which the device is positioned around each axis. The gyro and accelerometer are now standard fare on all new iPads and iPod touches. In this chapter, we're going to introduce you to the use of the Core Motion framework to access the gyro and accelerometer values in your application.

Accelerometer Physics

An **accelerometer** measures both acceleration and gravity by sensing the amount of inertial force in a given direction. The accelerometer inside your iOS device is a three-axis accelerometer, meaning that it is capable of detecting either movement or the pull of gravity in three-dimensional space. This means that you can use the accelerometer to discover not only how the device is currently being held (as autorotation does), but also if it's laying on a table, and even whether it's face down or face up.

Accelerometers give measurements in g-forces (*g* for gravity), so a value of 1.0 returned by the accelerometer means that 1 g is sensed in a particular direction, as in these examples:

   ▪ If the device is being held still with no movement, there will be approximately 1 g of force exerted on it by the pull of the earth.

   ▪ If the device is being held perfectly upright, in portrait orientation, it will detect and report about 1 g of force exerted on its y axis.

- If the device is being held at an angle, that 1 g of force will be distributed along different axes depending on how it is being held. When held at a 45-degree angle, the 1 g of force will be split roughly equally between two of the axes.

Sudden movement can be detected by looking for accelerometer values considerably larger than 1 g. In normal usage, the accelerometer does not detect significantly more than 1 g on any axis. If you shake, drop, or throw your device, the accelerometer will detect a greater amount of force on one or more axes. (Please do not drop or throw your own iOS device to test this theory.)

Figure 19–1 shows a graphic representation of the three axes used by the accelerometer. Notice that the accelerometer uses the more standard convention for the y coordinate, with increases in y indicating upward force, which is the opposite of Quartz 2D's coordinate system (discussed in Chapter 16). When you are using the accelerometer as a control mechanism with Quartz 2D, you need to translate the y coordinate. When working with OpenGL ES, which is more likely when you are using the accelerometer to control animation, no translation is required.

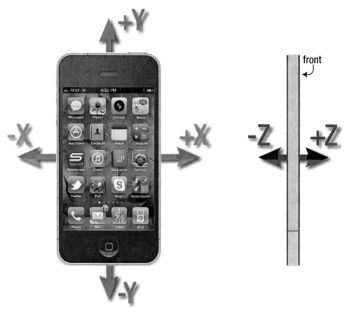

Figure 19–1. *The iPhone accelerometer's axes in three dimensions. The front view of an iPhone 4 on the left shows the x and y axes. The side view of an iPhone 4 on the right shows the z axis.*

Don't Forget Rotation

We mentioned earlier that iPhone 4 also includes a gyroscope sensor that lets you read values describing the device's rotation around its axes.

If the difference between the gyroscope and the accelerometer seems unclear, consider an iPhone lying flat on a table. If you begin to turn the phone around while it's lying flat, the accelerometer values won't change. That's because the forces bent on moving the phone—in this case, just the force of gravity pulling straight down the z axis—aren't changing. (In reality, things are a bit fuzzier than that, and the action of your hand bumping the phone will surely trigger a small amount of accelerometer action.) During that same movement, however, the device's rotation values will change—in particular, the z-axis rotation value. Turning the device clockwise will generate a negative value, and turning it counterclockwise gives a positive value. Stop turning, and the z-axis rotation value will go back to zero.

Rather than registering an absolute rotation value, the gyroscope tells you about changes to the device's rotation as they happen. You'll see how this works in this chapter's first example, coming up shortly.

Core Motion and the Motion Manager

In iOS 4 and up, accelerometer and gyroscope values are accessed using the Core Motion framework. This framework provides, among other things, the CMMotionManager class, which acts as a gateway for all the values describing how the device is being moved by its user. Your application creates an instance of CMMotionManager, and then puts it to use in one of two modes:

- It can execute some code for you whenever motion occurs.
- It can hang on to a perpetually updated structure that lets you access the latest values at any time.

The latter method is ideal for games and other highly interactive applications that need to be able to poll the device's current state during each pass through the game loop. We'll show you how to implement both approaches.

Note that the CMMotionManager class isn't actually a singleton, but your application should treat it like one. You should create only one of these per app, using the normal alloc and init methods. So, if you need to access the motion manager from several places in your app, you should probably create it in your application delegate and provide access to it from there.

Besides the CMMotionManager class, Core Motion also provides a few other classes, such as CMAccelerometerData and CMGyroData, which are simple containers through which your application can access motion data. We'll touch on these classes as we get to them.

Event-Based Motion

We mentioned that the motion manager can operate in a mode where it executes some code for you each time the motion data changes. Most other Cocoa Touch classes offer this sort of functionality by letting you connect to a delegate that gets a message when the time comes, but Core Motion does things a little differently.

Since it's a new framework, available only in iOS 4 and up, Apple decided to let CMMotionManager use another new feature of the iOS 4 SDK: **blocks**. We've already used blocks a couple of times in this book, and now you're going to see another application of this technique.

Use Xcode to create a new *Single View Application* project named *MotionMonitor*, with the *Use Storyboard* option turned off. This will be a simple app that reads both accelerometer data and gyroscope data (if available) and displays the information on the screen.

> **NOTE:** The applications in this chapter do not function on the simulator because the simulator has no accelerometer. Aw, shucks.

First, we need to link Core Motion into our app. This is an optional system framework, so we must add it in. Follow the instructions from Chapter 7 for adding the Audio Toolbox framework (in the "Linking in the Audio Toolbox Framework" section), but instead of selecting *AudioToolbox.framework*, select *CoreMotion.framework*. (In a nutshell, select the project in the project navigator, select the target and the *Build Phases* tab, expand the *Link Binary with Libraries* view, and click the plus button.)

Now, select the *BIDViewController.h* file, and make the following changes:

```
#import <UIKit/UIKit.h>
#import <CoreMotion/CoreMotion.h>

@interface BIDViewController : UIViewController

@property (strong, nonatomic) CMMotionManager *motionManager;
@property (weak, nonatomic) IBOutlet UILabel *accelerometerLabel;
@property (weak, nonatomic) IBOutlet UILabel *gyroscopeLabel;

@end
```

This provides us with a pointer for accessing the motion manager itself, along with outlets to a pair of labels where we'll display the information. Nothing much needs to be explained here, so just go ahead and save your changes.

Next, open *BIDViewController.xib* in Interface Builder. Open the view by selecting its icon in the nib window, and then drag out a *Label* from the library into the view. Resize the label to make it run from the left blue guideline to the right blue guideline, resize it to be about half the height of the entire view, and then align the top of the label to the top blue guideline.

Now, open the attributes inspector and change the *Lines* field from *1* to *0*. The *Lines* attribute is used to specify just how many lines of text may appear in the label, and provides a hard upper limit. If you set it to 0, no limit is applied, and the label can contain as many lines as you like.

Next, option-drag the label to create a copy, and align the copy with the blue guidelines in the bottom half of the view.

Now, control-drag from the *File's Owner* icon to each of the labels, connecting *accelerometerLabel* to the upper one and *gyroscopeLabel* to the lower one.

Finally, double-click each of the labels and delete the existing text.

This simple GUI is complete, so save your work and get ready for some coding.

Next, select *BIDViewController.m*. Here, add the property synthesizers to the top of the implementation block and the memory management calls to the viewDidUnload method:

```
#import "BIDViewController.h"

@implementation BIDViewController
@synthesize motionManager;
@synthesize accelerometerLabel;
@synthesize gyroscopeLabel;
.
.
.
- (void)viewDidUnload
{
    [super viewDidUnload];
    // Release any retained subviews of the main view.
    // e.g. self.myOutlet = nil;
    self.motionManager = nil;
    self.accelerometerLabel = nil;
    self.gyroscopeLabel = nil;
}
.
.
.
```

Now comes the interesting part. Fill in the viewDidLoad method with the following content:

```
- (void)viewDidLoad
{
    [super viewDidLoad];
    // Do any additional setup after loading the view, typically from a nib.
    self.motionManager = [[CMMotionManager alloc] init];
    NSOperationQueue *queue = [[NSOperationQueue alloc] init];
    if (motionManager.accelerometerAvailable) {
        motionManager.accelerometerUpdateInterval = 1.0 / 10.0;
        [motionManager startAccelerometerUpdatesToQueue:queue withHandler:
        ^(CMAccelerometerData *accelerometerData, NSError *error){
            NSString *labelText;
            if (error) {
                [motionManager stopAccelerometerUpdates];
                labelText = [NSString stringWithFormat:
                            @"Accelerometer encountered error: %@", error];
            } else {
                labelText = [NSString stringWithFormat:
                    @"Accelerometer\n-----------\nx: %+.2f\ny: %+.2f\nz: %+.2f",
                    accelerometerData.acceleration.x,
                    accelerometerData.acceleration.y,
                    accelerometerData.acceleration.z];
```

```
            }
            [accelerometerLabel performSelectorOnMainThread:@selector(setText:)
                                        withObject:labelText
                                        waitUntilDone:NO];
        }];
    } else {
        accelerometerLabel.text = @"This device has no accelerometer.";
    }
    if (motionManager.gyroAvailable) {
        motionManager.gyroUpdateInterval = 1.0 / 10.0;
        [motionManager startGyroUpdatesToQueue:queue withHandler:
        ^(CMGyroData *gyroData, NSError *error) {
            NSString *labelText;
            if (error) {
                [motionManager stopGyroUpdates];
                labelText = [NSString stringWithFormat:
                            @"Gyroscope encountered error: %@", error];
            } else {
                labelText = [NSString stringWithFormat:
                            @"Gyroscope\n--------\nx: %+.2f\ny: %+.2f\nz: %+.2f",
                            gyroData.rotationRate.x,
                            gyroData.rotationRate.y,
                            gyroData.rotationRate.z];
            }
            [gyroscopeLabel performSelectorOnMainThread:@selector(setText:)
                                        withObject:labelText
                                        waitUntilDone:NO];
        }];
    } else {
        gyroscopeLabel.text = @"This device has no gyroscope";
    }
}
```

This method contains all the code we need to fire up the sensors, tell them to report to us every 1/10 second, and update the screen when they do so.

Thanks to the power of blocks, it's all really simple and cohesive. Instead of putting parts of the functionality in delegate methods, defining behaviors in blocks lets us see a behavior in the same method where it's being configured. Let's take this apart a bit. We start off with this:

```
self.motionManager = [[CMMotionManager alloc] init];
NSOperationQueue *queue = [[NSOperationQueue alloc] init];
```

This code first creates an instance of CMMotionManager, which we'll use to watch motion events. Then it creates an operation queue, which is simply a container for a pile of work that needs to be done, as you may recall from Chapter 15.

> **CAUTION:** The motion manager wants to have a queue in which it will put the bits of work to be done, as specified by the blocks you will give it, each time an event occurs. It would be tempting to use the system's default queue for this purpose, but the documentation for `CMMotionManager` explicitly warns not to do this! The concern is that the default queue could end up chock-full of these events and have a hard time processing other crucial system events as a result.

Then we go on to configure the accelerometer. We first check to make sure the device actually has an accelerometer. All handheld iOS devices released so far do have one, but it's worth checking in case some future device doesn't. Then we set the time interval we want between updates, specified in seconds. Here, we're asking for 1/10 second. Note that setting this doesn't guarantee that we'll receive updates at precisely that speed. In fact, that setting is really a cap, specifying the best rate the motion manager will be allowed to give us. In reality, it may update less frequently than that.

```
if (motionManager.accelerometerAvailable) {
    motionManager.accelerometerUpdateInterval = 1.0/10.0;
```

Next, we tell the motion manager to start reporting accelerometer updates. We pass in the queue where it will put its work and the block that defines the work that will be done each time an update occurs. Remember that a block always starts off with a caret (^), followed by a parentheses-wrapped list of arguments that the block expects to be populated when it's executed (in this case, the accelerometer data and potentially an error to alert us of trouble), and finishes with a curly brace section containing the code to be executed itself.

```
[motionManager startAccelerometerUpdatesToQueue:queue withHandler:
    ^(CMAccelerometerData *accelerometerData, NSError *error) {
```

Then comes the content of the block. It creates a string based on the current accelerometer values, or it generates an error message if there's a problem. Then it pushes that string value into the `accelerometerLabel`. Here, we can't do that directly, since UIKit classes like `UILabel` usually work well only when accessed from the main thread. Due to the way this code will be executed, from within an `NSOperationQueue`, we simply don't know the specific thread in which we'll be executing. So, we use the `performSelectorOnMainThread:withObject:waitUntilDone:` method to make the main thread handle this.

Note that the accelerometer values are accessed through the `acceleration` property of the `accelerometerData` that was passed into it. The `acceleration` property is of type `CMAcceleration`, which is just a simple `struct` containing three `float` values. `accelerometerData` itself is an instance of the `CMAccelerometerData` class, which is really just a wrapper for `CMAcceleration`! If you think this seems like an unnecessary profusion of classes and types for simply passing three `floats` around, well, you're not alone. Regardless, here's how to use it:

```
NSString *labelText;
if (error) {
```

```
                    [motionManager stopAccelerometerUpdates];
                    labelText = [NSString stringWithFormat:
                                    @"Accelerometer encountered error: %@", error];
                } else {
                    labelText = [NSString stringWithFormat:
                        @"Accelerometer\n-----------\nx: %+.2f\ny: %+.2f\nz: %+.2f",
                        accelerometerData.acceleration.x,
                        accelerometerData.acceleration.y,
                        accelerometerData.acceleration.z];
                }
                [accelerometerLabel performSelectorOnMainThread:@selector(setText:)
                        withObject:labelText
                        waitUntilDone:NO];
```

Then we finish the block, and complete the square-bracketed method call where we were passing that block in the first place. Finally, we provide a different code path entirely, in case the device doesn't have an accelerometer. As mentioned earlier, all iOS devices so far have an accelerometer, but who knows what the future holds in store?

```
            }];
        } else {
            accelerometerLabel.text = @"This device has no accelerometer.";
        }
```

The code for the gyroscope is, as you surely noticed, structurally identical, differing only in the particulars of which methods are called and how reported values are accessed. It's similar enough that there's no need to walk you through it here.

Now, build and run your app on whatever iOS device you have, and try it out (see Figure 19–2). As you tilt your device around in different ways, you'll see how the accelerometer values adjust to each new position, and will hold steady as long as you hold the device steady.

Figure 19–2. *MotionMonitor running on an iPhone 4. Unfortunately, you'll get only a pair of error messages if you run this app in the simulator.*

If you run this on a device with a gyroscope, you'll see how those values change as well. Whenever the device is standing still, no matter which orientation it is in, the gyroscope values will hover around zero. As you rotate the device, you'll see that the gyroscope values change, depending on how you rotate it on its various axes. The values will always move back to zero when you stop moving the device.

Proactive Motion Access

You've seen how to access motion data by passing CMMotionManager blocks to be called as motion occurs. This kind of event-driven motion handling can work well enough for the average Cocoa app, but sometimes it doesn't quite fit an application's particular needs. Interactive games, for example, typically have a perpetually running loop that processes user input, updates the state of the game, and redraws the screen. In such a case, the event-driven approach isn't such a good fit, since you would need to implement an object that waits for motion events, remembers the latest positions from each sensor as they're reported, and is ready to report the data back to the main game loop when necessary.

Fortunately, CMMotionManager has a built-in solution. Instead of passing in blocks, we can just tell it to activate the sensors using the startAccelerometerUpdates and startGyroUpdates methods, after which we simply read the values any time we want, directly from the motion manager!

Let's change our MotionMonitor app to use this approach, just so you can see how it works. Start off by making a copy of your *MotionMonitor* project folder. Next, add a new property to *BIDViewController.h*, a pointer to an NSTimer that will trigger all our display updates:

```
#import <UIKit/UIKit.h>
#import <CoreMotion/CoreMotion.h>

@interface BIDViewController : UIViewController

@property (retain) CMMotionManager *motionManager;
@property (retain) IBOutlet UILabel *accelerometerLabel;
@property (retain) IBOutlet UILabel *gyroscopeLabel;
@property (retain) NSTimer *updateTimer;

@end
```

Now, switch to *BIDViewController.m*, where you'll need to synthesize the new property:

```
@implementation BIDViewController
@synthesize motionManager;
@synthesize accelerometerLabel;
@synthesize gyroscopeLabel;
@synthesize updateTimer;
```

Get rid of the entire viewDidLoad method we had before, and replace it with this simpler version, which just sets up the motion manager and provides informational labels for devices lacking sensors:

```
- (void)viewDidLoad {
    [super viewDidLoad];
    self.motionManager = [[CMMotionManager alloc] init];

    if (motionManager.accelerometerAvailable) {
        motionManager.accelerometerUpdateInterval = 1.0/10.0;
        [motionManager startAccelerometerUpdates];
    } else {
        accelerometerLabel.text = @"This device has no accelerometer.";
    }
    if (motionManager.gyroAvailable) {
        motionManager.gyroUpdateInterval = 1.0/10.0;
        [motionManager startGyroUpdates];
    } else {
        gyroscopeLabel.text = @"This device has no gyroscope.";
    }
}
```

Normally, we use viewDidLoad and viewDidUnload to "bracket" the creation and destruction of properties related to the GUI display. In the case of our new timer, however, we want it to be active only during an even smaller window of time, when the view is actually being displayed. That way, we keep the usage of our main game loop to a bare minimum. We can accomplish this by implementing viewWillAppear: and viewDidDisappear: as follows. Add this code to those two methods:

```
- (void)viewWillAppear:(BOOL)animated {
    [super viewWillAppear:animated];
    self.updateTimer = [NSTimer scheduledTimerWithTimeInterval:1.0/10.0
                                                target:self
                                                selector:@selector(updateDisplay)
                                                userInfo:nil
                                                repeats:YES];
}

- (void)viewDidDisappear:(BOOL)animated {
    [super viewDidDisappear:animated];
    self.updateTimer = nil;
}
```

The code in viewWillAppear: creates a new timer and schedules it to fire once every 1/10 second, calling the updateDisplay method, which we haven't created yet. Add this method just below ViewDidDisappear:

```
- (void)updateDisplay {
    if (motionManager.accelerometerAvailable) {
        CMAccelerometerData *accelerometerData = motionManager.accelerometerData;
        accelerometerLabel.text = [NSString stringWithFormat:
                    @"Accelerometer\n-----------\nx: %+.2f\ny: %+.2f\nz: %+.2f",
                    accelerometerData.acceleration.x,
                    accelerometerData.acceleration.y,
                    accelerometerData.acceleration.z];
    }
    if (motionManager.gyroAvailable) {
        CMGyroData *gyroData = motionManager.gyroData;
        gyroscopeLabel.text = [NSString stringWithFormat:
                    @"Gyroscope\n--------\nx: %+.2f\ny: %+.2f\nz: %+.2f",
                    gyroData.rotationRate.x,
                    gyroData.rotationRate.y,
                    gyroData.rotationRate.z];
    }
}
```

Build and run the app on your device, and you should see that it behaves exactly like the first version. Now you've seen two ways of accessing motion data. Use whichever suits your application best.

Accelerometer Results

We mentioned earlier that the iPhone's accelerometer detects acceleration along three axes, and it provides this information using the CMAcceleration struct. Each CMAcceleration has an x, y, and z field, each of which holds a floating-point value. A value of 0 means that the accelerometer detects no movement on that particular axis. A positive or negative value indicates force in one direction. For example, a negative value for y indicates that downward pull is sensed, which is probably an indication that the phone is being held upright in portrait orientation. A positive value for y indicates some force is being exerted in the opposite direction, which could mean the phone is being held upside down or that the phone is being moved in a downward direction.

Keeping the diagram in Figure 19–1 in mind, let's look at some accelerometer results (see Figure 19–3). Note that, in real life, you will almost never get values this precise, as the accelerometer is sensitive enough to sense even tiny amounts of motion, and you will usually pick up at least some tiny amount of force on all three axes. This is real-world physics, not high-school physics.

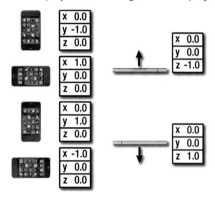

Figure 19–3. *Idealized acceleration values for different device orientations*

Probably the most common usage of the accelerometer in third-party applications is as a controller for games. We'll create a program that uses the accelerometer for input a little later in the chapter, but first, we'll look at another common accelerometer use: detecting shakes.

Detecting Shakes

Like a gesture, a shake can be used as a form of input to your application. For example, the drawing program GLPaint, which is one of the iOS sample code projects, lets users erase drawings by shaking their iOS device, sort of like an Etch A Sketch.

Detecting shakes is relatively trivial. All it requires is checking for an absolute value on one of the axes that is greater than a set threshold. During normal usage, it's not uncommon for one of the three axes to register values up to around 1.3 g, but getting values much higher than that generally requires intentional force. The accelerometer seems to be unable to register values higher than around 2.3 g (at least in our experience), so you don't want to set your threshold any higher than that.

To detect a shake, you could check for an absolute value greater than 1.5 for a slight shake and 2.0 for a strong shake, like this:

```
- (void)accelerometer:(UIAccelerometer *)accelerometer
        didAccelerate:(UIAcceleration *)acceleration {

    if (fabsf(acceleration.x) > 2.0
```

```
        || fabsf(acceleration.y) > 2.0
        || fabsf(acceleration.z) > 2.0) {
        // Do something here...
    }
}
```

This method would detect any movement on any axis that exceeded two g-forces.

You could implement more sophisticated shake detection by requiring the user to shake back and forth a certain number of times to register as a shake, like so:

```
- (void)accelerometer:(UIAccelerometer *)accelerometer
        didAccelerate:(UIAcceleration *)acceleration {

    static NSInteger shakeCount = 0;
    static NSDate *shakeStart;

    NSDate *now = [[NSDate alloc] init];
    NSDate *checkDate = [[NSDate alloc] initWithTimeInterval:1.5f
        sinceDate:shakeStart];
    if ([now compare:checkDate] == NSOrderedDescending
            || shakeStart == nil) {
        shakeCount = 0;
        shakeStart = [[NSDate alloc] init];
    }

    if (fabsf(acceleration.x) > 2.0
        || fabsf(acceleration.y) > 2.0
        || fabsf(acceleration.z) > 2.0) {
        shakeCount++;
        if (shakeCount > 4) {
            // Do something
            shakeCount = 0;
            shakeStart = [[NSDate alloc] init];
        }
    }
}
```

This method keeps track of the number of times the accelerometer reports a value above 2.0, and if it happens four times within a 1.5-second span of time, it registers as a shake.

Baked-In Shaking

There's actually another way to check for shakes—one that's baked right into the responder chain. Remember back in Chapter 17 how we implemented methods like touchesBegan:withEvent: to detect touches? Well, iOS also provides three similar responder methods for detecting motion:

- When motion begins, the motionBegan:withEvent: method is sent to the first responder and then on through the responder chain, as discussed in Chapter 17.

- When the motion ends, the motionEnded:withEvent: method is sent to the first responder.

- If the phone rings, or some other interrupting action happens during the shake, the `motionCancelled:withEvent:` message is sent to the first responder.

This means that you can actually detect a shake without using `CMMotionManager` directly. All you need to do is override the appropriate motion-sensing methods in your view or view controller, and they will be called automatically when the user shakes the phone. Unless you specifically need more control over the shake gesture, you should use the baked-in motion detection rather than the manual method described previously, but we thought we would show you the manual method in case you ever do need more control.

Now that you have the basic idea of how to detect shakes, we're going to break your phone.

Shake and Break

OK, we're not really going to break your phone, but we'll write an application that detects shakes, and then makes your phone look and sound like it broke as a result of the shake.

When you launch the application, the program will display a picture that looks like the iPhone home screen (see Figure 19–4). Shake the phone hard enough, though, and your poor phone will make a sound that you never want to hear coming out of a consumer electronics device. What's more, your screen will look like the one shown in Figure 19–5. Why do we do these evil things? Not to worry. You can reset the iPhone to its previously pristine state by touching the screen.

Figure 19–4. *The ShakeAndBreak application looks innocuous enough...*

Figure 19–5. . . . *but handle it too roughly and—oh no!*

> **NOTE:** Just for completeness, we've included a modified version of ShakeAndBreak in the
> project archives based on the built-in shake-detection method. You'll find it in the project archive
> in a folder named *19 - ShakeAndBreak - Motion Methods*. The magic is in the
> BIDViewController's motionEnded:withEvent: method.

Create a new project in Xcode using the *Single View Application* template. Call the new
project *ShakeAndBreak*. In the *19 - ShakeAndBreak* folder of the project archive, we've
provided the two images and the sound file you need for this application. Drag
home.png, *homebroken.png*, and *glass.wav* to your project. There's also an *icon.png* file
in that folder. Add that to the project as well.

Next, expand the *Supporting Files* folder, and select *ShakeAndBreak-Info.plist* to bring up
the property list editor. We need to add an entry to the property list to tell our application
not to use a status bar. Start off by right-clicking (or control-clicking) anywhere in the
property list editor, and selecting the *Show Raw Keys/Values* option from the context
menu so you can see the real names of the configurations we're setting. Single-click any
row in the property list, and press enter to add a new row. Change the new row's *Key* to
UIStatusBarHidden. The *Value* of this row will default to *NO*. Change it to *YES*. Finally,

expand the array entry named *CFBundleIconFiles*, and press enter to add a new *String* item. Type *icon.png* in the *Value* column (see Figure 19–6).

Key	Type	Value
CFBundleDevelopmentRegion	String	en
CFBundleDisplayName	String	${PRODUCT_NAME}
CFBundleExecutable	String	${EXECUTABLE_NAME}
▼ CFBundleIconFiles	Array	(1 item)
Item 0 ⊕⊖	String	icon.png
CFBundleIdentifier	String	com.Apress.${PRODUCT_NAME:rfc1034identifier}
CFBundleInfoDictionaryVersion	String	6.0
CFBundleName	String	${PRODUCT_NAME}
CFBundlePackageType	String	APPL
CFBundleShortVersionString	String	1.0
CFBundleSignature	String	????
CFBundleVersion	String	1.0
LSRequiresIPhoneOS	Boolean	YES
▶ UISupportedInterfaceOrientations	Array	(3 items)
UIStatusBarHidden	Boolean	YES

Figure 19–6. *Changing the value for CFBundleIconFiles (shown highlighted) and for UIStatusBarHidden (last line in the property list)*

Now, let's start creating our view controller. We're going to need to create an outlet to point to an image view so that we can change the displayed image. We'll also need a couple of UIImage instances to hold the two pictures, a sound ID to refer to the sound, and a Boolean to keep track of whether the screen needs to be reset. Single-click *BIDViewController.h*, and add the following code:

```
#import <UIKit/UIKit.h>
#import <CoreMotion/CoreMotion.h>
#import <AudioToolbox/AudioToolbox.h>

#define kAccelerationThreshold       1.7
#define kUpdateInterval              (1.0f / 10.0f)

@interface BIDViewController : UIViewController
    <UIAccelerometerDelegate>

@property (weak, nonatomic) IBOutlet UIImageView *imageView;
@property (strong, nonatomic) CMMotionManager *motionManager;
@property (assign, nonatomic) BOOL brokenScreenShowing;
@property (assign, nonatomic) SystemSoundID soundID;
@property (strong, nonatomic) UIImage *fixed;
@property (strong, nonatomic) UIImage *broken;
@end
```

Save the header file. Now, select *BIDViewController.xib* to edit the file in Interface Builder. Click the *View* icon to select the view, and then bring up the attributes inspector and change the *Status Bar* popup under *Simulated Metrics* from *Gray* to *None*. Next, drag an *Image View* over from the library to the view in the layout area. The image view should automatically resize to take up the full window, so just place it so that it sits perfectly within the window.

Control-drag from the *File's Owner* icon to the image view, and select the *imageView* outlet. Then save the nib file.

Next, select the *BIDViewController.m* file, and add the following code near the top of the file:

```
#import "BIDViewController.h"

@implementation BIDViewController
@synthesize imageView;
@synthesize motionManager;
@synthesize brokenScreenShowing;
@synthesize soundID;
@synthesize fixed;
@synthesize broken;
.
.
.
```

Give the viewDidLoad method the following implementation:

```
- (void) viewDidLoad {
    [super viewDidLoad];
    // Do any additional setup after loading the view, typically from a nib.

    NSString *path = [[NSBundle mainBundle] pathForResource:@"glass"
                                                     ofType:@"wav"];
    NSURL *url = [NSURL fileURLWithPath:path];
    AudioServicesCreateSystemSoundID((__bridge CFURLRef)url,
                    &soundID);
    self.fixed = [UIImage imageNamed:@"home.png"];
    self.broken = [UIImage imageNamed:@"homebroken.png"];

    imageView.image = fixed;

    self.motionManager = [[CMMotionManager alloc] init];
    motionManager.accelerometerUpdateInterval = kUpdateInterval;
    NSOperationQueue *queue = [[NSOperationQueue alloc] init];
    [motionManager startAccelerometerUpdatesToQueue:queue
                                        withHandler:
      ^(CMAccelerometerData *accelerometerData, NSError *error){
        if (error) {
            [motionManager stopAccelerometerUpdates];
        } else {
            if (!brokenScreenShowing) {
                CMAcceleration acceleration = accelerometerData.acceleration;
                if (acceleration.x > kAccelerationThreshold
                    || acceleration.y > kAccelerationThreshold
                    || acceleration.z > kAccelerationThreshold) {
                    [imageView performSelectorOnMainThread:@selector(setImage:)
                                               withObject:broken
                                            waitUntilDone:NO];
                    AudioServicesPlaySystemSound(soundID);
                    brokenScreenShowing = YES;
                }
```

```
            }
        }
    }];
}
```

Insert the following lines of code into the existing `viewDidUnload` method:

```
- (void)viewDidUnload
{
    [super viewDidUnload];
    // Release any retained subviews of the main view.
    // e.g. self.myOutlet = nil;
    self.imageView = nil;
    self.motionManager = nil;
    self.fixed = nil;
    self.broken = nil;
}
```

And finally, add the following new method at the bottom of the file:

```
    .
    .
    .
#pragma mark -
- (void)touchesBegan:(NSSet *)touches withEvent:(UIEvent *)event {
    imageView.image = fixed;
    brokenScreenShowing = NO;
}
```

```
@end
```

The first method we implement, and the one where most of the interesting things happen, is `viewDidLoad`. First, we create an `NSURL` object pointing to our sound file, load it into memory, and save the assigned identifier in the `soundID` instance variable. In order to satisfy the requirements for ARC, we need to tell the compiler how to manage the memory for the `NSURL` object before passing it off to `AudioServicesCreateSystemSoundID()`, which we do by casting it using the `__bridge` qualifier.

```
    NSString *path = [[NSBundle mainBundle] pathForResource:@"glass"
                                                     ofType:@"wav"];
    NSURL *url = [NSURL fileURLWithPath:path];
    AudioServicesCreateSystemSoundID((__bridge CFURLRef)url,
                                     &soundID);
```

> **NOTE:** New to the `__bridge` qualifier? It's discussed in Chapter 7. In a nutshell, it is used to safely bridge to ARC.

We then load the two images into memory.

```
    self.fixed = [UIImage imageNamed:@"home.png"];
    self.broken = [UIImage imageNamed:@"homebroken.png"];
```

Finally, we set imageView to show the unbroken screenshot and set brokenScreenShowing to NO to indicate that the screen does not currently need to be reset.

```
imageView.image = fixed;
brokenScreenShowing = NO;
```

Then we create a CMMotionManager and an NSOperationQueue (just as we've done before), and start up the accelerometer, sending it a block to be run each time the accelerometer value changes.

```
self.motionManager = [[CMMotionManager alloc] init];
motionManager.accelerometerUpdateInterval = kUpdateInterval;
NSOperationQueue *queue = [[NSOperationQueue alloc] init];
[motionManager startAccelerometerUpdatesToQueue:queue
                                    withHandler:
```

If the block finds accelerometer values high enough to trigger the break, it makes imageView switch over to the broken image and starts playing the breaking sound. Note that imageView is a member of the UIImageView class, which, like most parts of UIKit, is meant to run only in the main thread. Since the block may be run in another thread, we force the imageView update to happen on the main thread.

```
^(CMAccelerometerData *accelerometerData, NSError *error){
    if (error) {
        [motionManager stopAccelerometerUpdates];
    } else {
        if (!brokenScreenShowing) {
            CMAcceleration acceleration = accelerometerData.acceleration;
            if (acceleration.x > kAccelerationThreshold
                || acceleration.y > kAccelerationThreshold
                || acceleration.z > kAccelerationThreshold) {
                [imageView performSelectorOnMainThread:@selector(setImage:)
                                           withObject:broken
                                        waitUntilDone:NO];
                AudioServicesPlaySystemSound(soundID);
                brokenScreenShowing = YES;
            }
        }
    }
}];
```

The last method is one you should be quite familiar with by now. It's called when the screen is touched. All we do in that method is to set the image back to the unbroken screen and set brokenScreenShowing back to NO.

```
imageView.image = fixed;
brokenScreenShowing = NO;
```

Finally, add the *CoreMotion.framework* as well as the *AudioToolbox.framework* so that we can play the sound file. You can link the frameworks into your application by following the instructions from earlier in the chapter.

Compile and run the application, and take it for a test drive. For those of you who don't have the ability to run this application on your iOS device, you might want to give the shake-event-based version a try. The simulator does not simulate the accelerometer hardware, but it does simulate the shake event, so the version of the application in *19 ShakeAndBreak - Motion Method* will work with the simulator.

Go have some fun with it. When you're finished, come on back, and you'll see how to use the accelerometer as a controller for games and other programs.

Accelerometer As Directional Controller

Commonly, instead of using buttons to control the movement of a character or object in a game, the accelerometer is used. In a car-racing game, for example, twisting the iOS device like a steering wheel might steer your car, while tipping it forward might accelerate, and tipping back might brake.

Exactly how you use the accelerometer as a controller will vary greatly depending on the specific mechanics of the game. In the simplest cases, you might just take the value from one of the axes, multiply it by a number, and tack that on to the coordinates of the controlled objects. In more complex games where physics are modeled more realistically, you would need to make adjustments to the velocity of the controlled object based on the values returned from the accelerometer.

The one tricky aspect of using the accelerometer as a controller is that the delegate method is not guaranteed to call back at the interval you specify. If you tell the motion manager to read the accelerometer 60 times a second, all that you can say for sure is that it won't update more than 60 times a second. You're not guaranteed to get 60 evenly spaced updates every second. So, if you're doing animation based on input from the accelerometer, you must keep track of the time that passes between updates and factor that into your equations to determine how far objects have moved.

Rolling Marbles

For our next trick, we're going to let you move a sprite around iPhone's screen by tilting the phone. This is a very simple example of using the accelerometer to receive input. We'll use Quartz 2D to handle our animation.

> **NOTE:** As a general rule, when you're working with games and other programs that need smooth animation, you'll probably want to use OpenGL ES. We're using Quartz 2D in this application for the sake of simplicity and to reduce the amount of code that's unrelated to using the accelerometer. The animation won't be quite as smooth as if we were using OpenGL, but it will be a lot less work.

In this application, as you tilt your iPhone, the marble will roll around as if it were on the surface of a table (see Figure 19–7). Tip it to the left, and the ball will roll to the left. Tip it farther, and it will move faster. Tip it back, and it will slow down and then start going in the other direction.

Figure 19–7. *The Ball application lets you roll a marble around the screen.*

In Xcode, create a new project using the *Single View Application* template, and call this one *Ball*. In the *19 - Ball* folder in the project archive, you'll find an image called *ball.png*. Drag that to your project.

Now, single-click the *Ball* folder, and select **File ➤ New ➤ New File**.... Select *Objective-C class* from the *Cocoa Touch* category, click Next, and name the new class *BIDBallView*. Select *UIView* in the *Subclass of* popup, and click *Create* to save the class files. We'll get back to editing this class a little later.

Select *BIDViewController.xib* to edit the file in Interface Builder. Single-click the *View* icon, and use the identity inspector to change the view's class from *UIView* to *BIDBallView*. Next, switch to the attributes inspector, and change the view's *Background* to *Black Color*. Then save the nib.

Now it's time to edit *BIDViewController.h*. All we need to do here is prepare for Core Motion, so make the following changes:

```
#import <UIKit/UIKit.h>
#import <CoreMotion/CoreMotion.h>

@interface BIDViewController : UIViewController

@property (strong, nonatomic) CMMotionManager *motionManager;
@end
```

Next, switch to *BIDViewController.m*, and add the following lines toward the top of the file:

```
#import "BIDViewController.h"
#import "BIDBallView.h"

#define kUpdateInterval    (1.0f / 60.0f)

@implementation BIDViewController
@synthesize motionManager;
 .
 .
 .
```

Next, populate viewDidLoad with this code:

```
- (void)viewDidLoad {
    [super viewDidLoad];
    // Do any additional setup after loading the view, typically from a nib.

    self.motionManager = [[CMMotionManager alloc] init];
    NSOperationQueue *queue = [[NSOperationQueue alloc] init];
    motionManager.accelerometerUpdateInterval = kUpdateInterval;
    [motionManager startAccelerometerUpdatesToQueue:queue withHandler:
     ^(CMAccelerometerData *accelerometerData, NSError *error) {
         [(BIDBallView *)self.view setAcceleration:accelerometerData.acceleration];
         [self.view performSelectorOnMainThread:@selector(update)
             withObject:nil waitUntilDone:NO];
     }];
}
```

The viewDidLoad method here is similar to some of what we've done elsewhere in this chapter. The main difference is that we are declaring a much higher update interval of 60 times per second. In the block that we tell the motion manager to execute when there are accelerometer updates to report, we pass the acceleration object into our view, and then call a method named update, which updates the position of the ball in the view based on acceleration and the amount of time that has passed since the last update. Since that block can be executed on any thread, and the methods belonging to UIKit objects (including UIView) can be safely used only from the main thread, we once again force the update method to be called in the main thread.

Writing the Ball View

Note that when you entered the code for viewDidLoad in the previous step, you probably saw some error as a result of BIDBallView not being complete. Since we're doing the bulk of our work in the BIDBallView class, we had better write it, huh? Select *BIDBallView.h*, and make the following changes:

```
#import <UIKit/UIKit.h>
#import <CoreMotion/CoreMotion.h>

@interface BIDBallView : UIView
```

```
@property (strong, nonatomic) UIImage *image;
@property CGPoint currentPoint;
@property CGPoint previousPoint;
@property (assign, nonatomic) CMAcceleration acceleration;
@property CGFloat ballXVelocity;
@property CGFloat ballYVelocity;
- (void)update;
@end
```

Let's look at the properties and talk about what we're doing with each of them. The first is a UIImage that will point to the sprite that we'll be moving around the screen.

```
UIImage *image;
```

After that, we keep track of two CGPoint variables. The currentPoint property will hold the current position of the ball. We'll also keep track of the last point where we drew the sprite. That way, we can build an update rectangle that encompasses both the new and old positions of the ball, so that it is drawn at the new spot and erased at the old one.

```
CGPoint     currentPoint;
CGPoint     previousPoint;
```

Next is an acceleration struct, which is how we will get the accelerometer information from our controller.

```
CMAcceleration acceleration;
```

We also have two variables to keep track of the ball's current velocity in two dimensions. Although this isn't going to be a very complex simulation, we do want the ball to move in a manner similar to a real ball. We'll calculate the ball movement in the next section. We'll get acceleration from the accelerometer and keep track of velocity on two axes with these variables.

```
CGFloat ballXVelocity;
CGFloat ballYVelocity;
```

Let's switch over to *BIDBallView.m* and write the code to draw and move the ball around the screen. First, make the following changes at the top of *BIDBallView.m*:

```
#import "BIDBallView.h"

@implementation BIDBallView
@synthesize image;
@synthesize currentPoint;
@synthesize previousPoint;
@synthesize acceleration;
@synthesize ballXVelocity;
@synthesize ballYVelocity;

- (id)initWithCoder:(NSCoder *)coder {
if (self = [super initWithCoder:coder]) {
        self.image = [UIImage imageNamed:@"ball.png"];
        self.currentPoint = CGPointMake((self.bounds.size.width / 2.0f) +
            (image.size.width / 2.0f),
            (self.bounds.size.height / 2.0f) + (image.size.height / 2.0f));
```

```
        ballXVelocity = 0.0f;
        ballYVelocity = 0.0f;
    }
    return self;
}
    .
    .
    .
```

Now, uncomment the commented-out drawRect: method and give it this simple implementation:

```
- (void)drawRect:(CGRect)rect {
    // Drawing code
    [image drawAtPoint:currentPoint];
}
```

Then add these methods to the end of the class:

```
    .
    .
    .
#pragma mark -
- (CGPoint)currentPoint {
    return currentPoint;
}

- (void)setCurrentPoint:(CGPoint)newPoint {
    previousPoint = currentPoint;
    currentPoint = newPoint;

    if (currentPoint.x < 0) {
        currentPoint.x = 0;
        ballXVelocity = 0;
    }
    if (currentPoint.y < 0){
        currentPoint.y = 0;
        ballYVelocity = 0;
    }
    if (currentPoint.x > self.bounds.size.width - image.size.width) {
        currentPoint.x = self.bounds.size.width - image.size.width;
        ballXVelocity = 0;
    }
    if (currentPoint.y > self.bounds.size.height - image.size.height) {
        currentPoint.y = self.bounds.size.height - image.size.height;
        ballYVelocity = 0;
    }

    CGRect currentImageRect = CGRectMake(currentPoint.x, currentPoint.y,
            currentPoint.x + image.size.width,
            currentPoint.y + image.size.height);
    CGRect previousImageRect = CGRectMake(previousPoint.x, previousPoint.y,
            previousPoint.x + image.size.width,
            currentPoint.y + image.size.width);
```

```
    [self setNeedsDisplayInRect:CGRectUnion(currentImageRect,
        previousImageRect)];
}

- (void)update {
    static NSDate *lastUpdateTime;

    if (lastUpdateTime != nil) {
        NSTimeInterval secondsSinceLastDraw =
            -([lastUpdateTime timeIntervalSinceNow]);

        ballYVelocity = ballYVelocity + -(acceleration.y *
            secondsSinceLastDraw);
        ballXVelocity = ballXVelocity + acceleration.x *
            secondsSinceLastDraw;

        CGFloat xAcceleration = secondsSinceLastDraw * ballXVelocity * 500;
        CGFloat yAcceleration = secondsSinceLastDraw * ballYVelocity * 500;

        self.currentPoint = CGPointMake(self.currentPoint.x +
            xAcceleration, self.currentPoint.y + yAcceleration);
    }
    // Update last time with current time
    lastUpdateTime = [[NSDate alloc] init];
}
@end
```

The first thing to notice is that one of our properties is declared as @synthesize, yet we have implemented the mutator method for that property in our code. That's OK. The @synthesize directive will not overwrite accessor or mutator methods that you write; it will just fill in the blanks and provide any methods that you do not write.

Calculating Ball Movement

We are handling the currentPoint property manually, since, when the currentPoint changes, we need to do a bit of housekeeping, such as making sure that the ball has not rolled off the screen. We'll look at that method in a moment. For now, let's look at the first method in the class, initWithCoder:.

Recall that when you load a view from a nib, that class's init or initWithFrame: methods will never be called. Nib files contain archived objects, so any instances loaded from the nib will be initialized using the initWithCoder: method. If we need to do any additional initialization, we must do it in that method.

In this view, we do have some additional initialization, so we've overridden initWithCoder:. First, we load the *ball.png* image. Second, we calculate the middle of the view and set that as our ball's starting point, and we set the velocity on both axes to 0.

```
        self.image = [UIImage imageNamed:@"ball.png"];
        self.currentPoint = CGPointMake((self.bounds.size.width / 2.0f) +
            (image.size.width / 2.0f), (self.bounds.size.height / 2.0f) +
```

```
        (image.size.height / 2.0f));

    ballXVelocity = 0.0f;
    ballYVelocity = 0.0f;
```

Our drawRect: method couldn't be much simpler. We just draw the image we loaded in initWithCoder: at the position stored in currentPoint. The currentPoint accessor is a standard accessor method. The setCurrentPoint: mutator is another story, however.

The first things we do in setCurrentPoint: is to store the old currentPoint value in previousPoint and assign the new value to currentPoint.

```
    previousPoint = currentPoint;
    currentPoint = newPoint;
```

Next, we do a boundary check. If either the x or y position of the ball is less than 0 or greater than the width or height of the screen (accounting for the width and height of the image), then the acceleration in that direction is stopped.

```
    if (currentPoint.x < 0) {
        currentPoint.x = 0;
        ballXVelocity = 0;
    }
    if (currentPoint.y < 0){
        currentPoint.y = 0;
        ballYVelocity = 0;
    }
    if (currentPoint.x > self.bounds.size.width - image.size.width) {
        currentPoint.x = self.bounds.size.width - image.size.width;
        ballXVelocity = 0;
    }
    if (currentPoint.y > self.bounds.size.height - image.size.height) {
        currentPoint.y = self.bounds.size.height - image.size.height;
        ballYVelocity = 0;
    }
```

> **TIP:** Do you want to make the ball bounce off the walls more naturally, instead of just stopping? It's easy enough to do. Just change the two lines in setCurrentPoint: that currently read ballXVelocity = 0; to **ballXVelocity = - (ballXVelocity / 2.0);**. And change the two lines that currently read ballYVelocity = 0; to **ballYVelocity = - (ballYVelocity / 2.0);**. With these changes, instead of killing the ball's velocity, we reduce it in half and set it to the inverse. Now, the ball has half the velocity in the opposite direction.

After that, we calculate two CGRects based on the size of the image. One rectangle encompasses the area where the new image will be drawn, and the other encompasses the area where it was last drawn. We'll use these two rectangles to ensure that the old ball is erased at the same time the new one is drawn.

```
    CGRect currentImageRect = CGRectMake(currentPoint.x, currentPoint.y,
            currentPoint.x + image.size.width,
            currentPoint.y + image.size.height);
```

```
CGRect previousImageRect = CGRectMake(previousPoint.x, previousPoint.y,
        previousPoint.x + image.size.width,
        currentPoint.y + image.size.width);
```

Finally, we create a new rectangle that is the union of the two rectangles we just calculated and feed that to setNeedsDisplayInRect: to indicate the part of our view that needs to be redrawn.

```
[self setNeedsDisplayInRect:CGRectUnion(currentImageRect,
    previousImageRect)];
```

The last substantive method in our class is update, which is used to figure out the correct new location of the ball. This method is called in the accelerometer method of its controller class after it feeds the view the new acceleration object. The first thing this method does is declare a static NSDate variable that will be used to keep track of how long it has been since the last time the update method was called. The first time through this method, when lastUpdateTime is nil, we don't do anything because there's no point of reference. Because the updates are happening about 60 times a second, no one will ever notice a single missing frame.

```
static NSDate *lastUpdateTime;
if (lastUpdateTime != nil) {
```

Every other time through this method, we calculate how long it has been since the last time this method was called. We negate the value returned by timeIntervalSinceNow because lastUpdateTime is in the past, so the value returned will be a negative number representing the number of seconds between the current time and lastUpdateTime.

```
NSTimeInterval secondsSinceLastDraw =
        -([lastUpdateTime timeIntervalSinceNow]);
```

Next, we calculate the new velocity in both directions by adding the current acceleration to the current velocity. We multiply acceleration by secondsSinceLastDraw so that our acceleration is consistent across time. Tipping the phone at the same angle will always cause the same amount of acceleration.

```
ballYVelocity = ballYVelocity + -(acceleration.y *
    secondsSinceLastDraw);
 ballXVelocity = ballXVelocity + acceleration.x *
    secondsSinceLastDraw;
```

After that, we figure out the actual change in pixels since the last time the method was called based on the velocity. The product of velocity and elapsed time is multiplied by 500 to create movement that looks natural. If we didn't multiply it by some value, the acceleration would be extraordinarily slow, as if the ball were stuck in molasses.

```
CGFloat xAcceleration = secondsSinceLastDraw * ballXVelocity * 500;
CGFloat yAcceleration = secondsSinceLastDraw * ballYVelocity * 500;
```

Once we know the change in pixels, we create a new point by adding the current location to the calculated acceleration and assign that to currentPoint. By using self.currentPoint, we use that accessor method we wrote earlier, rather than assigning the value directly to the instance variable.

```
self.currentPoint = CGPointMake(self.currentPoint.x +
```

```
        xAcceleration, self.currentPoint.y + yAcceleration);
```

That ends our calculations, so all that's left is to update `lastUpdateTime` with the current time.

```
    lastUpdateTime = [[NSDate alloc] init];
```

Before you build the app, add the Core Motion framework using the technique mentioned earlier. Once it's added, go ahead and build and run the app.

> **NOTE:** Unfortunately, Ball just will not do much on the simulator. If you want to experience Ball in all its gravity-obeying grooviness, you'll need to join the for-pay iOS Developer Program and install it on your own device.

If all went well, the application will launch, and you should be able to control the movement of the ball by tilting the phone. When the ball gets to an edge of the screen, it should stop. Tip the phone back the other way, and it should start rolling in the other direction. Whee!

Rolling On

Well, we've certainly had some fun in this chapter with physics and the amazing iOS accelerometer and gyro. We created a great April Fools' prank, and you got to see the basics of using the accelerometer as a control device. The possibilities for applications using the accelerometer and gyro are nearly as endless as the universe. So now that you have the basics down, go create something cool and surprise us!

When you feel up to it, we're going to get into using another bit of iOS hardware: the built-in camera.

Chapter 20

The Camera and Photo Library

By now, it should come as no surprise to you that the iPhone, iPad, and iPod touch have a built-in camera and a nifty application called Photos to help you manage all those awesome pictures and videos you've taken. What you may not know is that your programs can use the built-in camera to take pictures. Your applications can also allow the user to select from among the media already stored on the device. We'll look at both of these abilities in this chapter.

Using the Image Picker and UIImagePickerController

Because of the way iOS applications are sandboxed, applications ordinarily can't get to photographs or other data that live outside their own sandboxes. Fortunately, both the camera and the media library are made available to your application by way of an **image picker**.

As the name implies, an image picker is a mechanism that lets you select an image from a specified source. When this class first appeared in iOS, it was used only for images. Nowadays, you can use it to capture video as well.

Typically, an image picker will use a list of images and/or videos as its source (see the left side of Figure 20–1). You can, however, specify that the picker use the camera as its source (see the right side of Figure 20–1).

Figure 20–1. *An image picker in action. Users are presented with a list of images (left), and then once an image is selected, they can move and scale the image (right).*

The image picker interface is implemented by way of a modal controller class called UIImagePickerController. You create an instance of this class, specify a delegate (as if you didn't see that coming), specify its image source and whether you want the user to pick an image or a video, and then launch it modally. The image picker will take control of the device to let the user select a picture or video from the existing media library, or to take a new picture or video with the camera. Once the user makes a selection, you can give the user an opportunity to do some basic editing, such as scaling or cropping an image or trimming away a bit of a video clip. All that behavior is implemented by the UIImagePickerController, so you really don't need to do much heavy lifting here.

Assuming the user doesn't press cancel, the image or video the user takes or selects from the library will be delivered to your delegate. Regardless of whether the user selects a media file or cancels, your delegate has the responsibility to dismiss the UIImagePickerController so that the user can return to your application.

Creating a UIImagePickerController is extremely straightforward. You just allocate and initialize an instance the way you would with most classes. There is one catch, however. Not every device that runs iOS has a camera. Older iPod touches were the first examples of this, and the first-generation iPad is the latest, but more such devices may roll off Apple's assembly lines in the future. Before you create an instance of UIImagePickerController, you need to check to see whether the device your program is currently running on supports the image source you want to use. For example, before letting the user take a picture with the camera, you should make sure the program is running on a device that has a camera. You can check that by using a class method on `UIImagePickerController`, like this:

```
if ([UIImagePickerController isSourceTypeAvailable:
        UIImagePickerControllerSourceTypeCamera]) {
```

In this example, we're passing `UIImagePickerControllerSourceTypeCamera` to indicate that we want to let the user take a picture or shoot a video using the built-in camera. The method `isSourceTypeAvailable:` returns `YES` if the specified source is currently available. You can specify two other values in addition to `UIImagePickerControllerSourceTypeCamera`:

- `UIImagePickerControllerSourceTypePhotoLibrary` specifies that the user should pick an image or video from the existing media library. That image will be returned to your delegate.

- `UIImagePickerControllerSourceTypeSavedPhotosAlbum` specifies that the user will select the image from the library of existing photographs, but that the selection will be limited to the most recent camera roll. This option will run on a device without a camera, but does not do anything useful.

After making sure that the device your program is running on supports the image source you want to use, launching the image picker is relatively easy:

```
UIImagePickerController *picker = [[UIImagePickerController alloc] init];
picker.delegate = self;
picker.sourceType = UIImagePickerControllerSourceTypeCamera;
[self presentModalViewController:picker animated:YES];
```

After we've created and configured the `UIImagePickerController`, we use a method that our class inherited from `UIView` called `presentModalViewController:animated:` to present the image picker to the user.

> **TIP:** The `presentModalViewController:animated:` method is not limited to just presenting image pickers. You can present any view controller to the user, modally, by calling this method on the view controller for a currently visible view.

Implementing the Image Picker Controller Delegate

The object that you want to be notified when the user has finished using the image picker interface needs to conform to the `UIImagePickerControllerDelegate` protocol. This protocol defines two methods: `imagePickerController:didFinishPickingMediaWithInfo:` and `imagePickerControllerDidCancel:`.

The imagePickerController:didFinishPickingMediaWithInfo: method is called when the user has successfully taken a photo or video, or selected an item from the media library. The first argument is a pointer to the UIImagePickerController that you created earlier. The second argument is an NSDictionary instance that will contain the chosen photo or the URL of the chosen video, as well as optional editing information if you enabled editing and the user actually did some editing. That dictionary will also contain the original, unedited image stored under the key UIImagePickerControllerOriginalImage. Here's an example of a delegate method that retrieves the original image:

```
- (void)imagePickerController:(UIImagePickerController *)picker
  didFinishPickingMediaWithInfo:(NSDictionary *)info {

    UIImage *selectedImage = [info objectForKey:UIImagePickerControllerEditedImage];
    UIImage *originalImage = [info objectForKey:UIImagePickerControllerOriginalImage];

    // do something with selectedImage and originalImage

    [picker dismissModalViewControllerAnimated:YES];
}
```

The editingInfo dictionary will also tell you which portion of the entire image was chosen during editing by way of an NSValue object stored under the key UIImagePickerControllerCropRect. You can convert this string into a CGRect like so:

```
NSValue *cropValue = [editingInfo objectForKey:UIImagePickerControllerCropRect];
CGRect cropRect = [cropValue CGRectValue];
```

After this conversion, cropRect will specify the portion of the original image that was selected during the editing process. If you do not need this information, you can just ignore it.

> **CAUTION:** If the image returned to your delegate comes from the camera, that image will not be stored in the photo library. It is your application's responsibility to save the image, if necessary.

The other delegate method, imagePickerControllerDidCancel:, is called if the user decides to cancel the process without capturing or selecting any media. When the image picker calls this delegate method, it's just notifying you that the user is finished with the picker and didn't choose anything.

Both of the methods in the UIImagePickerControllerDelegate protocol are marked as optional, but they really aren't, and here is why: modal views like the image picker must be told to dismiss themselves. As a result, even if you don't need to take any application-specific actions when the user cancels an image picker, you still need to dismiss the picker. At a bare minimum, your imagePickerControllerDidCancel: method will need to look like this in order for your program to function correctly:

```
- (void)imagePickerControllerDidCancel:(UIImagePickerController *)picker {

    [picker dismissModalViewControllerAnimated:YES];
}
```

Road Testing the Camera and Library

In this chapter, we're going to build an application that lets the user take a picture or shoot some video with the camera, or select one from the photo library, and then display the selection on the screen (see Figure 20–2). If the user is on a device without a camera, we will hide the *New Photo or Video* button and allow only selection from the photo library.

Figure 20–2. *The Camera application in action*

Create a new project in Xcode using the *Single View Application* template, naming the application *Camera*. Before working on the code itself, we need to add a couple of frameworks that our application will use. Using the techniques you've used in previous chapters, add the *MediaPlayer* and *MobileCoreServices* frameworks.

We'll add a couple of outlets in this application's view controller. We need one to point to the image view so that we can update it with the image returned from the image picker. We'll also need an outlet to point to the *New Photo or Video* button so we can hide the button if the device doesn't have a camera.

Since we're going to allow users to decide whether to grab a video or an image, we'll use the `MPMoviePlayerController` class to display the chosen video, so we need a property for that. Two more properties keep track of the last selected image and video, along with a string to determine whether a video or image was the last thing chosen. Finally, we'll keep track of the size of the image view, for resizing the captured image to match our display size.

We also need two action methods: one for the *New Photo or Video* button and one for letting the user select an existing picture from the photo library.

Expand the *Camera* folder so that you can get to all the relevant files. Select *BIDViewController.h*, and make the following changes:

```
#import <UIKit/UIKit.h>
#import <MediaPlayer/MediaPlayer.h>

@interface BIDViewController : UIViewController
        <UIImagePickerControllerDelegate, UINavigationControllerDelegate>

@property (weak, nonatomic) IBOutlet UIImageView *imageView;
@property (weak, nonatomic) IBOutlet UIButton *takePictureButton;
@property (strong, nonatomic) MPMoviePlayerController *moviePlayerController;
@property (strong, nonatomic) UIImage *image;
@property (strong, nonatomic) NSURL *movieURL;
@property (copy, nonatomic) NSString *lastChosenMediaType;
@property (assign, nonatomic) CGRect imageFrame;

- (IBAction)shootPictureOrVideo:(id)sender;
- (IBAction)selectExistingPictureOrVideo:(id)sender;
@end
```

The first thing you might notice is that we've actually conformed our class to two different protocols: UIImagePickerControllerDelegate and UINavigationControllerDelegate. Because UIImagePickerController is a subclass of UINavigationController, we must conform our class to both of these protocols. The methods in UINavigationControllerDelegate are optional, and we don't need either of them to use the image picker, but we need to conform to the protocol, or the compiler will give us a warning.

The other thing you might notice is that while we'll be dealing with an instance of UIImageView for displaying a chosen image, we don't have anything similar for displaying a chosen video. UIKit doesn't include any publicly available class like UIImageView that works for showing video content, so instead, we'll use an instance of MPMoviePlayerController, grabbing its view property and inserting it into our view hierarchy. This is a highly unusual way of using any view controller, but it's actually the Apple-approved technique to show video inside a view hierarchy.

Everything else here is pretty straightforward, so save it. Now, select *BIDViewController.xib* to edit the file in Interface Builder.

Designing the Interface

Drag two *Round Rect Buttons* from the library to the window labeled *View*. Place them one above the other, aligning the bottom button with the bottom blue guideline. Double-click the top button, and give it a title of *New Photo or Video*. Then double-click the bottom button, and give it a title of *Pick from Library*. Next, drag an *Image View* from the library, and place it above the buttons. Expand the image view to take up the entire space of the view above the buttons, as shown earlier in Figure 20–2.

Now, control-drag from the *File's Owner* icon to the image view, and select the *imageView* outlet. Drag again from *File's Owner* to the *New Photo or Video* button, and select the *takePictureButton* outlet.

Next, select the *New Photo or Video* button, and bring up the connections inspector. Drag from the *Touch Up Inside* event to *File's Owner*, and select the *shootPictureOrVideo:* action. Now, click the *Pick from Library* button, drag from the *Touch Up Inside* event in the connections inspector to *File's Owner*, and select the *selectExistingPictureOrVideo:* action.

Once you've made these connections, save your changes and return to Xcode.

Implementing the Camera View Controller

Select *BIDViewController.m*, and make the following changes at the beginning of the file:

```
#import "BIDViewController.h"
#import <MobileCoreServices/UTCoreTypes.h>

@interface BIDViewController ()
static UIImage *shrinkImage(UIImage *original, CGSize size);
- (void)updateDisplay;
- (void)getMediaFromSource:(UIImagePickerControllerSourceType)sourceType;
@end

@implementation BIDViewController
  .
  .
  .
```

As we've done a few times earlier in this book, we create a class extension to declare some methods that we don't want to expose in our class's interface, but that will be implemented at some later point in our class definition. The methods we put here are utility methods, meant only for use within the class itself. Note that we also declared a normal C function in this code. A class extension can really contain only methods, so this function doesn't actually belong to the class extension, technically speaking. However, in terms of structuring our code, it "belongs" to our class, so we'll list it here as well.

Now, move along to start defining the class, as follows:

```
.
.
.
@implementation BIDViewController
@synthesize imageView;
@synthesize takePictureButton;
@synthesize moviePlayerController;
@synthesize image;
@synthesize movieURL;
@synthesize lastChosenMediaType;
@synthesize imageFrame;

- (void)viewDidLoad
{
    [super viewDidLoad];
    // Do any additional setup after loading the view, typically from a nib.

    if (![UIImagePickerController isSourceTypeAvailable:
            UIImagePickerControllerSourceTypeCamera]) {
        takePictureButton.hidden = YES;
    }
    imageFrame = imageView.frame;
}
.
.
.
- (void)viewDidAppear:(BOOL)animated
{
    [super viewDidAppear:animated];
    [self updateDisplay];
}
.
.
.
```

The viewDidLoad method will hide the takePictureButton if the device we're running on does not have a camera, and it also grabs the imageView's frame rect, since we're going to need that a little later. We also implement the viewDidAppear: method, having it call the updateDisplay method, which we haven't yet implemented.

It's important to understand the distinction between the viewDidLoad and viewDidAppear: methods. The former is called only when the view has just been loaded into memory. The latter is called every time the view is displayed, which happens both at launch and whenever we return to our controller after showing another full-screen view, such as the image picker.

Next, insert the following lines of code into the existing viewDidUnload method. Normally, viewDidUnload only gets rid of views, but in this case, we're also going to make it get rid of the moviePlayerController. Otherwise, we would have that controller hanging around even when there were no views left to show it in.

.
.
.

```
- (void)viewDidUnload
{
    [super viewDidUnload];
    // Release any retained subviews of the main view.
    // e.g. self.myOutlet = nil;
    self.imageView = nil;
    self.takePictureButton = nil;
    self.moviePlayerController = nil;
}
```

.
.
.

Next, insert the following action methods that we declared in the header:

```
- (IBAction)shootPictureOrVideo:(id)sender {
    [self getMediaFromSource:UIImagePickerControllerSourceTypeCamera];
}
```

```
- (IBAction)selectExistingPictureOrVideo:(id)sender {
    [self getMediaFromSource:UIImagePickerControllerSourceTypePhotoLibrary];
}
```

Each of these simply calls out to one of the utility methods we declared earlier (but still haven't defined), passing in a value defined by UIImagePickerController to specify where the picture or video should come from.

Next, let's implement the delegate methods for the picker view:

```
#pragma mark UIImagePickerController delegate methods
- (void)imagePickerController:(UIImagePickerController *)picker
        didFinishPickingMediaWithInfo:(NSDictionary *)info {
    self.lastChosenMediaType = [info objectForKey:UIImagePickerControllerMediaType];
    if ([lastChosenMediaType isEqual:(NSString *)kUTTypeImage]) {
        UIImage *chosenImage = [info objectForKey:UIImagePickerControllerEditedImage];
        UIImage *shrunkenImage = shrinkImage(chosenImage, imageFrame.size);
        self.image = shrunkenImage;
    } else if ([lastChosenMediaType isEqual:(NSString *)kUTTypeMovie]) {
        self.movieURL = [info objectForKey:UIImagePickerControllerMediaURL];
    }
    [picker dismissModalViewControllerAnimated:YES];
}
```

```
- (void)imagePickerControllerDidCancel:(UIImagePickerController *)picker {
    [picker dismissModalViewControllerAnimated:YES];
}
```

The first delegate method checks to see whether a picture or video was chosen, makes note of the selection (shrinking the chosen image, if any, to precisely fit the display size along the way), and then dismisses the modal image picker. The second one just dismisses the image picker.

Next, let's move on to the function and methods we declared as a class extension early in the file. First up is the shrinkImage() function, which we use to shrink our image down to the size of the view in which we're going to show it. Doing so reduces the size of the UIImage we're dealing with, as well as the amount of memory that imageView needs in order to display it. Add this code toward the end of the file:

```
#pragma mark  -
static UIImage *shrinkImage(UIImage *original, CGSize size) {
    CGFloat scale = [UIScreen mainScreen].scale;
    CGColorSpaceRef colorSpace = CGColorSpaceCreateDeviceRGB();

    CGContextRef context = CGBitmapContextCreate(NULL, size.width * scale,
        size.height * scale, 8, 0, colorSpace, kCGImageAlphaPremultipliedFirst);
    CGContextDrawImage(context,
        CGRectMake(0, 0, size.width * scale, size.height * scale),
        original.CGImage);
    CGImageRef shrunken = CGBitmapContextCreateImage(context);
    UIImage *final = [UIImage imageWithCGImage:shrunken];

    CGContextRelease(context);
    CGImageRelease(shrunken);

    return final;
}
```

Don't worry about the details too much. What you're seeing here is a series of Core Graphics calls that create a new image based on the specified size and render the old image into the new one.

Note that we get a value called scale from the device's main screen and use it as a multiplier when specifying the size of the new image we're creating. The scale is basically the number of physical screen pixels per unit point in all the calls we make. Devices featuring "Retina display", such as the iPhone 4, iPhone 4S, and fourth-generation iPod touch, have a scale of 2.0; all other previous devices, and both iPad versions produced so var, will report 1.0. Using the scale this way lets us make an image that renders at the full resolution of the device on which we're running. Otherwise, the image would end up looking a bit jagged on the iPhone 4 (if you looked really closely, that is).

Next up is the updateDisplay method. Remember that this is being called from the viewDidAppear: method, which is called both when the view is first created and then again after the user picks an image or video and dismisses the image picker. Because of this dual usage, it needs to make a few checks to see what's what and set up the GUI accordingly. The MPMoviePlayerController doesn't let us change the URL it reads from, so each time we want to display a movie, we'll need to make a new controller. All of that is handled here. Add this code toward the bottom of the file:

```
- (void)updateDisplay {
    if ([lastChosenMediaType isEqual:(NSString *)kUTTypeImage]) {
        imageView.image = image;
        imageView.hidden = NO;
        moviePlayerController.view.hidden = YES;
```

```
    } else if ([lastChosenMediaType isEqual:(NSString *)kUTTypeMovie]) {
        [self.moviePlayerController.view removeFromSuperview];
        self.moviePlayerController = [[MPMoviePlayerController alloc]
            initWithContentURL:movieURL];
        moviePlayerController.view.frame = imageFrame;
        moviePlayerController.view.clipsToBounds = YES;
        [self.view addSubview:moviePlayerController.view];
        imageView.hidden = YES;
    }
}
```

The final new method, getMediaFromSource:, is the one that both of our action methods call. This method is pretty simple. It just creates and configures an image picker, using the passed-in sourceType to determine whether to bring up the camera or the media library. Add this code toward the bottom of the file:

```
- (void)getMediaFromSource:(UIImagePickerControllerSourceType)sourceType {
    NSArray *mediaTypes = [UIImagePickerController
        availableMediaTypesForSourceType:sourceType];
    if ([UIImagePickerController isSourceTypeAvailable:
         sourceType] && [mediaTypes count] > 0) {
        NSArray *mediaTypes = [UIImagePickerController
            availableMediaTypesForSourceType:sourceType];
        UIImagePickerController *picker =
        [[UIImagePickerController alloc] init];
        picker.mediaTypes = mediaTypes;
        picker.delegate = self;
        picker.allowsEditing = YES;
        picker.sourceType = sourceType;
        [self presentModalViewController:picker animated:YES];
    } else {
        UIAlertView *alert = [[UIAlertView alloc]
                              initWithTitle:@"Error accessing media"
                              message:@"Device doesn't support that media source."
                              delegate:nil
                              cancelButtonTitle:@"Drat!"
                              otherButtonTitles:nil];
        [alert show];
    }
}
@end
```

That's all we need to do. Compile and run the app. If you're running on the simulator, you won't have the option to take a new picture. If you have the opportunity to run our application on a real device, go ahead and try it. You should be able to take a new picture, and zoom in and out of the picture using the pinch gestures.

If you zoom in before hitting the *Use Photo* button, the cropped image will be the one returned to the application in the delegate method.

It's a Snap!

Believe it or not, that's all there is to letting your users take pictures with the camera so that the pictures can be used by your application. You can even let the user do a small amount of editing on that image if you so choose.

In the next chapter, we're going to look at reaching a larger audience for your iOS applications by making them oh so easy to translate into other languages. *Êtes-vous prêt? Tournez la page et allez directement. Allez, allez!*

Chapter 21

Application Localization

At the time of this writing, the iPhone is available in more than 90 different countries, and that number will continue to increase over time. You can now buy and use an iPhone on every continent except Antarctica. The iPad was held back a bit, as Apple struggled to satisfy demand in what it considered the most important countries first, but now the iPad is also sold all over the world and is nearly as ubiquitous as the iPhone.

If you plan on releasing applications through the App Store, your potential market is considerably larger than just people in your own country who speak your own language. Fortunately, iOS has a robust **localization** architecture that lets you easily translate your application (or have it translated by others) into not only multiple languages, but even into multiple dialects of the same language. Do you want to provide different terminology to English speakers in the United Kingdom than you do to English speakers in the United States? No problem.

That is, localization is no problem if you've written your code correctly. Retrofitting an existing application to support localization is much harder than writing your application that way from the start. In this chapter, we'll show you how to write your code so it is easy to localize, and then we'll go about localizing a sample application.

Localization Architecture

When a nonlocalized application is run, all of the application's text will be presented in the developer's own language, also known as the **development base language**.

When developers decide to localize their application, they create a subdirectory in their application bundle for each supported language. Each language's subdirectory contains a subset of the application's resources that were translated into that language. Each subdirectory is called a **localization project**, or **localization folder**. Localization folder names always end with the extension *.lproj*.

In the iOS Settings application, the user has the ability to set the device's preferred language and region format. For example, if the user's language is English, available regions might be United States, Australia, and Hong Kong—all regions in which English is spoken.

When a localized application needs to load a resource—such as an image, property list, or nib—the application checks the user's language and region and looks for a localization folder that matches that setting. If it finds one, it will load the localized version of the resource instead of the base version.

For users who selected French as their iOS language and France as their region, the application will look first for a localization folder named *fr_FR.lproj*. The first two letters of the folder name are the ISO country code that represents the French language. The two letters following the underscore are the ISO code that represents France.

If the application cannot find a match using the two-letter code, it will look for a match using the language's three-letter ISO code. In our example, if the application is unable to find a folder named *fr_FR.lproj*, it will look for a localization folder named *fre_FR or fra_FR*.

All languages have at least one three-letter code. Some have two three-letter codes: one for the English spelling of the language and another for the native spelling. Only some languages have two-letter codes. When a language has both a two-letter code and a three-letter code, the two-letter code is preferred.

> **NOTE:** You can find a list of the current ISO country codes on the ISO web site (http://www.iso.org/iso/country_codes.htm). Both the two- and three-letter codes are part of the ISO 3166 standard.

If the application cannot find a folder that is an exact match, it will then look for a localization folder in the application bundle that matches just the language code without the region code. So, staying with our French-speaking person from France, the application next looks for a localization project called *fr.lproj*. If it doesn't find a language project with that name, it will look for *fre.lproj*, then *fra.lproj*. If none of those are found, it checks for *French.lproj*. The last construct exists to support legacy Mac OS X applications, and generally speaking, you should avoid it.

If the application doesn't find a language project that matches either the language/region combination or just the language, it will use the resources from the development base language. If it does find an appropriate localization project, it will always look there first for any resources that it needs. If you load a UIImage using imageNamed:, for example, the application will look first for an image with the specified name in the localization project. If it finds one, it will use that image. If it doesn't, it will fall back to the base language resource.

If an application has more than one localization project that matches—for example, a project called *fr_FR.lproj* and one called *fr.lproj*—it will look first in the more specific match, which is *fr_FR.lproj* in this case. If it doesn't find the resource there, it will look in *fr.lproj*. This gives you the ability to provide resources common to all speakers of a language in one language project, localizing only those resources that are impacted by differences in dialect or geographic region.

You should choose to localize only those resources that are affected by language or country. For example, if an image in your application has no words and its meaning is universal, there's no need to localize that image.

Strings Files

What do you do about string literals and string constants in your source code? Consider this source code from the previous chapter:

```
UIAlertView *alert = [[UIAlertView alloc]
    initWithTitle:@"Error accessing photo library"
        message:@"Device does not support a photo library"
        delegate:nil
cancelButtonTitle:@"Drat!"
otherButtonTitles:nil];
[alert show];
```

If you've gone through the effort of localizing your application for a particular audience, you certainly don't want to be presenting alerts written in the development base language. The answer is to store these strings in special text files called **strings files**.

What's in a Strings File?

Strings files are nothing more than Unicode (UTF-16) text files that contain a list of string pairs, each identified by a comment. Here is an example of what a strings file might look like in your application:

```
/* Used to ask the user his/her first name */
"First Name" = "First Name";

/* Used to get the user's last name */
"Last Name" = "Last Name";

/* Used to ask the user's birth date */
"Birthday" = "Birthday";
```

The values between the /* and the */ characters are just comments for the translator. They are not used in the application, and you could skip adding them, though they're a good idea. The comments give context, showing how a particular string is being used in the application.

You'll notice that each line lists the same string twice. The string on the left side of the equal sign acts as a key, and it will always contain the same value, regardless of language. The value on the right side of the equal sign is the one that is translated to the local language. So, the preceding strings file, localized into French, might look like this:

```
/* Used to ask the user his/her first name */
"First Name " = "Prénom";

/* Used to get the user's last name */
"Last Name " = "Nom de famille";
```

```
/* Used to ask the user's birth date */
"Birthday" = "Anniversaire";
```

The Localized String Macro

You won't actually create the strings file by hand. Instead, you'll embed each localizable text string in a special macro in your code. Once your source code is final and ready for localization, you'll run a command-line program, named genstrings, which will search all your code files for occurrences of the macro, pulling out all the unique strings and embedding them in a localizable strings file.

Let's see how the macro works. First, here's a traditional string declaration:

```
NSString *myString = @"First Name";
```

To make this string localizable, do this instead:

```
NSString *myString = NSLocalizedString(@"First Name",
    @"Used to ask the user his/her first name");
```

The NSLocalizedString macro takes two parameters:

- The first parameter is the string value in the base language. If there is no localization, the application will use this string.

- The second parameter is used as a comment in the strings file.

NSLocalizedString looks in the application bundle inside the appropriate localization project for a strings file named *localizable.strings*. If it does not find the file, it returns its first parameter, and the string will appear in the development base language. Strings are typically displayed only in the base language during development, since the application will not yet be localized.

If NSLocalizedString finds the strings file, it searches the file for a line that matches the first parameter. In the preceding example, NSLocalizedString will search the strings file for the string "First Name". If it doesn't find a match in the localization project that matches the user's language settings, it will then look for a strings file in the base language and use the value there. If there is no strings file, it will just use the first parameter you passed to the NSLocalizedString macro.

Now that you have an idea of how the localization architecture and the strings file work, let's take a look at localization in action.

Real-World iOS: Localizing Your Application

We're going to create a small application that displays the user's current **locale**. A locale (an instance of NSLocale) represents both the user's language and region. It is used by the system to determine which language to use when interacting with the user, as well as how to display dates, currency, and time information, among other things. After we create the application, we will then localize it into other languages. You'll learn how to localize nib files, strings files, images, and even your application's display name.

You can see what our application is going to look like in Figure 21–1. The name across the top comes from the user's locale. The words down the left side of the view are static labels that are set in the nib file. The words down the right side are set programmatically using outlets. The flag image at the bottom of the screen is a static UIImageView.

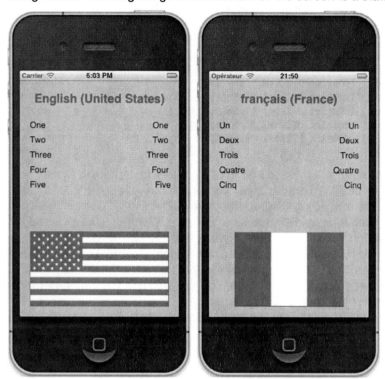

Figure 21–1. *The LocalizeMe application shown with two different language/region settings*

Let's hop right into it.

Setting Up LocalizeMe

Create a new project in Xcode using the *Single View Application* template, and call it *LocalizeMe*.

If you look in the source code archive, within the *21 - LocalizeMe* folder, you'll see a folder named *Images*. Inside that folder, you'll find a pair of folders, one named *English* and one named *French*, each containing a file named *flag.png*. One version of the flag is the US flag, and the other is the French flag.

Start by dragging the English language version of *flag.png* into the project navigator's *LocalizeMe* folder. When prompted, add a copy of that file to the project. We'll get to the French version of that file as we make our way through the chapter.

Now let's add some label outlets to the project. We need to create outlets to a total of six labels: one for the blue title across the top of the view, and five for the words down the right-hand side (see Figure 21–1). Select *BIDViewController.h*, and make the following changes:

```
#import <UIKit/UIKit.h>

@interface BIDViewController : UIViewController

@property (weak, nonatomic) IBOutlet UILabel *localeLabel;
@property (weak, nonatomic) IBOutlet UILabel *label1;
@property (weak, nonatomic) IBOutlet UILabel *label2;
@property (weak, nonatomic) IBOutlet UILabel *label3;
@property (weak, nonatomic) IBOutlet UILabel *label4;
@property (weak, nonatomic) IBOutlet UILabel *label5;
@end
```

Now select the *BIDViewController.xib* file to edit the GUI in Interface Builder. Make sure the *View* window is visible, and then drag a *Label* from the library, dropping it at the top of the view, aligned with the top blue guideline. Resize the label so that it takes the entire width of the view, from blue guideline to blue guideline. With the label selected, open the attributes inspector. Look for the *Font* control, and click the small *T* icon it contains to bring up a small font-selection popup. Click *System Bold* to let this title label stand out a bit from the rest. Then use the attributes inspector to set the text alignment to centered, and set the text color to a bright blue. You can also use the font selector to make the font size larger if you wish. As long as *Autoshrink* is selected in the object attributes inspector, the text will be resized if it gets too long to fit.

With your label in place, control-drag from the *File's Owner* icon to this new label, and select the *localeLabel* outlet.

Next, drag five more *Labels* from the library, and put them against the left margin using the blue guideline, one above the other (see Figure 21–1). Resize the labels so they go about halfway across the view, or a little less. Double-click the top one, and change it from *Label* to *One*. Repeat this procedure with the other four labels, changing the text to the numbers *Two* through *Five*.

Drag five more *Labels* from the library, this time placing them against the right margin. Change the text alignment using the object attributes inspector so that they are right-aligned, and increase the size of the label so that it stretches from the right blue guideline to about the middle of the view. Control-drag from *File's Owner* to each of the five new labels, connecting each one to a different numbered label outlet. Now, double-click each one of the new labels, and delete its text. We will be setting these values programmatically.

Finally, drag an *Image View* from the library over to the bottom part of the view so it touches the bottom and left blue guidelines. In the attributes inspector, select *flag.png* for the view's *Image* attribute, and resize the image to stretch from blue guideline to blue guideline. In the attributes inspector, change the *Mode* attribute from its current value to *Aspect Fit*. Not all flags have the same aspect ratio, and we want to make sure the localized versions of the image look right. Selecting this option will cause the image view to resize any other images put in this image view so they fit, but it will maintain the correct aspect ratio (ratio of height to width). Finally, make the flag bigger, until it hits the right-side blue guideline.

Save your nib. Then switch to *BIDViewController.m*, and insert the following code at the top of the file:

```
#import "BIDViewController.h"

@implementation BIDViewController
@synthesize localeLabel;
@synthesize label1;
@synthesize label2;
@synthesize label3;
@synthesize label4;
@synthesize label5;
.
.
.
```

Now, provide the following implementation for the viewDidLoad method:

```
- (void)viewDidLoad
{
    [super viewDidLoad];
    // Do any additional setup after loading the view, typically from a nib.

    NSLocale *locale = [NSLocale currentLocale];
    NSString *displayNameString = [locale
        displayNameForKey:NSLocaleIdentifier
        value:[locale localeIdentifier]];
    localeLabel.text = displayNameString;

    label1.text = NSLocalizedString(@"One", @"The number 1");
    label2.text = NSLocalizedString(@"Two", @"The number 2");
    label3.text = NSLocalizedString(@"Three", @"The number 3");
    label4.text = NSLocalizedString(@"Four", @"The number 4");
    label5.text = NSLocalizedString(@"Five", @"The number 5");
}
```

Also, add the following code to the existing viewDidUnload method:

```
- (void)viewDidUnload
{
    [super viewDidUnload];
    // Release any retained subviews of the main view.
    // e.g. self.myOutlet = nil;
    self.localeLabel = nil;
    self.label1 = nil;
    self.label2 = nil;
    self.label3 = nil;
    self.label4 = nil;
    self.label5 = nil;
}
```

The only item of note in this class is the viewDidLoad method. The first thing we do there is get an NSLocale instance that represents the user's current locale, which can tell us both the user's language and region preferences, as set in the iPhone's Settings application.

```
NSLocale *locale = [NSLocale currentLocale];
```

The next line of code might need a bit of explanation. NSLocale works somewhat like a dictionary. It can give you a whole bunch of information about the current user's preferences, including the name of the currency and the expected date format. You can find a complete list of the information that you can retrieve in the NSLocale API reference.

In this next line of code, we're retrieving the **locale identifier**, which is the name of the language and/or region that this locale represents. We're using a function called displayNameForKey:value:. The purpose of this method is to return the value of the item we've requested in a specific language.

The display name for the French language, for example, is *Français* in French, but *French* in English. This method gives you the ability to retrieve data about any locale so that it can be displayed appropriately to any users. In this case, we're getting the display name for the locale in the language of that locale, which is why we pass [locale localeIdentifier] in the second argument. The localeIdentifier is a string in the format we used earlier to create our language projects. For an American English speaker, it would be *en_US*, and for a French speaker from France, it would be *fr_FR*.

```
NSString *displayNameString = [locale
            displayNameForKey:NSLocaleIdentifier
            value:[locale localeIdentifier]];
```

Once we have the display name, we use it to set the top label in the view.

```
localeLabel.text = displayNameString;
```

Next, we set the five other labels to the numbers one through five spelled out in our development base language. We also provide a comment telling what each word is. You can just pass an empty string if the words are obvious, as they are here, but any string you pass in the second argument will be turned into a comment in the strings file, so you can use this comment to communicate with the person doing your translations.

```
label1.text = NSLocalizedString(@"One", @"The number 1");
label2.text = NSLocalizedString(@"Two", @"The number 2");
label3.text = NSLocalizedString(@"Three", @"The number 3");
label4.text = NSLocalizedString(@"Four", @"The number 4");
label5.text = NSLocalizedString(@"Five", @"The number 5");
```

Let's run our application now.

Trying Out LocalizeMe

You can use either the simulator or a device to test LocalizeMe. The simulator does seem to cache some language and region settings, so you may want to run the application on the device if you have that option. Once the application launches, it should look like Figure 21–2.

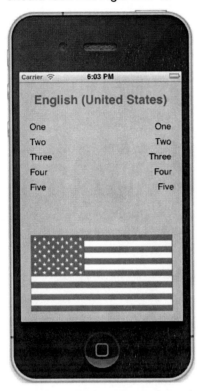

Figure 21–2. *The language running under the authors' base language. Our application is set up for localization but is not yet localized.*

By using the NSLocalizedString macros instead of static strings, we are ready for localization, but we are not localized yet. If you use the Settings application on the simulator or on your iPhone to change to another language or region, the results look essentially the same, except for the label at the top of the view (see Figure 21–3).

Figure 21–3. *The nonlocalized application run on an iPhone set to use the French language in France*

Localizing the Nib

Now, let's localize the nib file. The basic process for localizing any file is the same. In Xcode, single-click *BIDViewController.xib*, and then select **View ➤ Utilities ➤ Show File Inspector** to bring up the file inspector to see detailed information for the nib file.

> **CAUTION:** Xcode will allow you to localize pretty much any file in the navigator. Just because you can, doesn't mean you should. Never localize a source code file. Doing so will cause compile errors, as multiple object files with the same name will be created.

Look for the *Localization* section of the file inspector. You'll see that it shows a single localization: *English*. Click the plus (+) button at the bottom of the *Localization* section, and select *French (fr)* from the popup list that appears (see Figure 21–4).

Figure 21–4. *The file inspector showing localization and other information for BIDViewController.xib*

After adding a localization, take a look at the project navigator. Notice that the *BIDViewController.xib* file now has a disclosure triangle next to it, as if it were a group or folder. Expand it, and take a look (see Figure 21–5).

Figure 21–5. *Localizable files have a disclosure triangle and a child value for each language or region you add.*

In our project, *BIDViewController.xib* is now shown as a group containing two children: one tagged as *English* and one as *French*. The *English* version was created automatically when you created the project, and it represents your development base language.

Each of these files lives in a separate folder, one called *en.lproj* and one called *fr.lproj*. Go to the Finder, and open the *LocalizeMe* folder within your *LocalizeMe* project folder. In addition to all your project files, you should see folders named *en.lproj* and *fr.lproj* (see Figure 21–6).

```
h   BIDAppDelegate.h
m   BIDAppDelegate.m
h   BIDViewController.h
m   BIDViewController.m
    en.lproj              ▶
    flag.png
    fr.lproj              ▶
    LocalizeMe-Info.plist
h   LocalizeMe-Prefix.pch
m   main.m
```

Figure 21–6. *From the outset, our Xcode project included a language project folder for our base language. When we chose to make a file localizable, Xcode created a language project folder for the language we selected as well.*

Note that the *en.lproj* folder was there all along, with its copy of *BIDViewController.xib* inside it all the while. When Xcode finds a resource that has exactly one localized version, it displays it as a single item. As soon as a file has two or more localized versions, they're displayed as a group.

> **TIP:** When dealing with locales, language codes are lowercase, but country codes are uppercase. So, the correct name for the French language project is *fr.lproj*, but the project for Parisian French (French as spoken by people in France) is *fr_FR.lproj*, not *fr_fr.lproj* or *FR_fr.lproj*. The iOS file system is case-sensitive, so it is important to match case correctly.

When you asked Xcode to create the French localization, Xcode created a new localization project in your project folder called *fr.lproj* and placed a copy of *BIDViewController.xib* in that folder. In Xcode's project navigator, *BIDViewController.xib* should now have two children: *English* and *French*. Select *French* to open the nib file that will be shown to French speakers.

The nib file that opens in Interface Builder will look exactly like the one you built earlier, because the nib file you just created is a copy of the earlier one. Any changes you make to this file will be shown to people who speak French. Double-click each of the labels on the left side and change them from *One*, *Two*, *Three*, *Four*, and *Five* to *Un*, *Deux*, *Trois*, *Quatre*, and *Cinq*. Then save the nib.

Your nib is now localized in French. Compile and run the program. After it launches, tap the home button.

If you've already changed your Settings to the French region and language, you should see your translated labels on the left. For those folks who are a bit unsure about how to make those changes, we'll walk you through it.

In the simulator, go to the Settings application, and select the *General* row and then the row labeled *International*. From here, you'll be able to change your language and region preferences (see Figure 21–7).

Figure 21–7. *Changing the language and region—the two settings that affect the user's locale*

You want to change the *Region Format* first, because once you change the language, iOS will reset and return to the home screen. Change the *Region Format* from *United States* to *France* (first select *French*, then select *France* from the new table that appears), and then change *Language* from *English* to *Français*. Click the *Done* button, and the simulator will reset its language. Now, your phone is set to use French.

Run your app again. This time, the words down the left-hand side should show up in French (see Figure 21–8). But the flag and right column of text are still wrong. We'll take care of the flag first.

Figure 21–8. *The application is partially translated into French now.*

Localizing an Image

We could change the flag image directly in the nib by just selecting a different image for the image view in the French localized nib file. Instead of doing that, we'll actually localize the flag image itself.

When an image or other resource used by a nib is localized, the nib will automatically show the correct version for the language (though not for the dialect, at the time of this writing). If we localize the *flag.png* file itself with a French version, the nib will automatically show the correct flag when appropriate.

First, quit the simulator and make sure your application is stopped. Back in Xcode, single-click *flag.png* in the project navigator. Next, bring up the file inspector, and locate the *Localization* section, which you'll see is currently empty. Make sure *flag.png* is still selected, and then press the + button at the bottom of the *Localization* section. Xcode will add an *English* localization for the file and move *flag.png* into the *en.lproj* folder (check the Finder to see this for yourself).

NOTE: In the current version of Xcode 4.2, there seems to be a slight bug in the GUI that shows up here. With *flag.png* selected, clicking the + button as we just described causes the file to be deselected, which makes the inspector suddenly pop into a no selection state. But don't worry; it has done its work properly anyway. Just select *flag.png* in the project navigator again and continue.

You'll see that where *flag.png* is shown in the project navigator, there's still no disclosure triangle to indicate that it's a localized resource. That's because we still have just one version of it, in *en.lproj*, just as *BIDViewController.xib* started out.

Make sure *flag.png* is still selected, and use the *Localization* section of the file inspector to add a new localization. This time when the popup list appears, you'll see both *English* and *French* at the top of the list, since Xcode knows those are already represented in the project. Select *French*, and you'll see that *flag.png* immediately acquires a disclosure icon. Expand that, and you'll see two copies of the flag: one labeled *English* and one labeled *French*.

Switch back to the Finder and your project's directory, and you'll see this situation mirrored in the file system, with *en.lproj* and *fr.lproj* each containing a *flag.png* file. The one in *fr.lproj* is a copy of the original, which is obviously not the correct image. Since Xcode doesn't let you edit image files, the easiest way to get the correct image into the localization project is to just copy that image into the project using the Finder.

Go to the *21 - LocalizeMe* folder, open the *Images* folder, open the *French* folder, and use the *flag.png* file in that folder to replace the *flag.png* you'll find in *LocalizeMe/fr.lproj*.

That's it. You're finished. Back in Xcode, click the image file *flag.png (French)* in the project navigator. You should see the French flag.

Now, try running the app again. We hope you'll see something akin to the image shown in Figure 21–9. If not, do not despair. Chances are the US version of the flag image is being cached, either by the simulator or by the device, if you are running the app that way. We'll go over a couple approaches you can try to get the correct image.

Figure 21–9. *The flag image and application nib are both localized to French now.*

If you are running in the simulator, first quit the app (but *not* the simulator). Click over to Xcode and stop the application. Return to the simulator and select **iOS Simulator ➤ Reset Contents and Settings** to reset the simulator, and then quit the simulator. Return to Xcode and select **Product ➤ Clean** to force a complete rebuild. Now, run the app again. Once you rerun the app, you'll need to reset the region and language to get the French flag to appear. If this still does not work, try doing the simulator reset, quit the simulator, do a **Product ➤ Clean**, quit Xcode, and then start the whole thing again.

If you're running on the device, your iPhone has probably cached the American flag from the last time you ran the application. You can remove the old application from your iPhone using the *Organizer* window in Xcode. Select **Window ➤ Organizer** to bring up the *Organizer* window. Under the *Devices* tab, you'll see a column on the left showing all the iOS devices that Xcode knows about. The currently connected device has a green dot next to its name. Select the *Applications* item belonging to your devices, and you'll see a list of all the applications you've compiled and installed on your own. Find *LocalizeMe* in the list, select it, and click the minus (–) button to remove the old version of that application and the caches associated with it. Now, select **Project ➤ Clean**. When that's complete, build and run the application again. Once the application launches, you'll need to reset the region, and then the language. The French flag should now come up in addition to the French words down the left side (see Figure 21–9).

Generating and Localizing a Strings File

In Figure 21–9, notice that the words on the right side of the view are still in English. In order to translate those, we need to generate our base language strings file and then localize that. To accomplish this, we'll need to leave the comfy confines of Xcode for a few minutes.

Launch *Terminal.app*, which is in */Applications/Utilities/*. When the terminal window opens, type *cd* followed by a space. Don't press return.

Now, go to the Finder, and drag the project folder *21 - LocalizeMe* to the terminal window. As soon as you drop the folder onto the terminal window, the path to the project folder should appear on the command line. Now, press return. The cd command is Unix-speak for "change directory," so what you've just done is steer your terminal session from its default directory over to your project directory.

Our next step is to run the program genstrings and tell it to find all the occurrences of NSLocalizedString in your *.m* files in the *Classes* folder. To do this, type the following command, and then press return:

```
genstrings ./LocalizeMe/*.m
```

When the command is finished executing (it just takes a second on a project this small), you'll be returned to the command line. In the Finder, look in the project folder for a new file called *Localizable.strings*. Drag that to the *LocalizeMe* folder in Xcode's project navigator, but when it prompts you, don't click the *Add* button just yet. First uncheck the box that says *Copy items into destination group's folder (if needed)*, because the file is already in your project folder. Click *Finish* to import the file.

> **CAUTION:** You can rerun genstrings at any time to re-create your base language file, but once you have localized your strings file into another language, it's important that you don't change the text used in any of the NSLocalizedString() macros. That base-language version of the string is used as a key to retrieve the translations, so if you change them, the translated version will no longer be found, and you will need to either update the localized strings file or have it retranslated.

Once the file is imported, single-click *Localizable.strings*, and take a look at it. It contains five entries, because we use NSLocalizableString five times with five distinct values. The values that we passed in as the second argument have become the comments for each of the strings.

The strings were generated in alphabetical order. In this case, since we're dealing with numbers, alphabetical order is not the most intuitive way to present them, but in most cases, having them in alphabetical order will be helpful.

```
/* The number 5 */
"Five" = "Five";
```

```
/* The number 4 */
"Four" = "Four";

/* The number 1 */
"One" = "One";

/* The number 3 */
"Three" = "Three";

/* The number 2 */
"Two" = "Two";
```

Let's localize this sucker.

Make sure *Localizable.strings* is selected, and then repeat the same steps we've performed for the other localizations:

- Open the file inspector if it's not already visible.

- In the *Localizations* section, click the + button once to create an English localization. This will likely unselect the file, but it will create the English localization.

- Reselect *Localizable.strings*. Then click the + button once again to make a French localization.

Back in the project navigator, select the French localization of the file. In the editor, make the following changes:

```
/* The number 5 */
"Five" = "Cinq";

/* The number 4 */
"Four" = "Quatre";

/* The number 1 */
"One" = "Un";

/* The number 3 */
"Three" = "Trois";

/* The number 2 */
"Two" = "Deux";
```

In real life (unless you're multilingual), you would ordinarily send this file out to a translation service to translate the values to the right of the equal signs. In this simple example, armed with knowledge that came from years of watching *Sesame Street*, we can do the translation ourselves.

Now save, compile, and run the app. You should see the relevant numbers translated into French.

Localizing the App Display Name

We want to show you one final piece of localization that is commonly used: localizing the app name that's visible on the home screen and elsewhere. Apple does this for several of the built-in apps, and you might want to do so as well.

The app name used for display is stored in your app's *Info.plist* file, which, in our case, is actually named *LocalizeMe-Info.plist*. You'll find it in the *Supporting Files* folder. Select this file for editing, and you'll see that one of the items it contains, *Bundle display name*, is currently set to *${PRODUCT_NAME}*.

In the syntax used by *Info.plist* files, anything starting with a dollar sign is subject to variable substitution. In this case, this means that when Xcode compiles the app, the value of this item will be replaced with the name of the product in this Xcode project, which is the name of the app itself. This is where we want to do some localization, replacing *${PRODUCT_NAME}* with the localized name for each language. However, as it turns out, this doesn't quite work out as simply as you might expect.

The *Info.plist* file is sort of a special case, and it isn't meant to be localized. Instead, if you want to localize the content of *Info.plist*, you need to make localized versions of a file named *InfoPlist.strings*. Fortunately, that file is already included in every project Xcode creates, so all we need to do is localize it.

Look in the *Supporting Files* folder and find the *InfoPlist.strings* file. Use the file inspector's *Localizations* section to create a French localization using the same steps you did for the previous localizations (it starts off with an English version located in the *en.lproj* folder).

Now, we want to add a line to define the display name for the app. In the *LocalizeMe-Info.plist* file, we were shown the display name associated with a dictionary key called *Bundle display name*, but that's not the real key name! It's merely an Xcode nicety, trying to give a more friendly and readable name. The real name is *CFBundleDisplayName*, which you can verify by selecting *LocalizeMe-Info.plist*, right-clicking anywhere in the view, and selecting **Show Raw Keys/Values**. This shows you the true names of the keys in use.

So, select the English localization of *InfoPlist.strings*, and add the following line:

CFBundleDisplayName = "Localize Me";

Now, select the French localization of the *InfoPlist.strings* file. Edit the file to give the app a proper French name:

CFBundleDisplayName = "Localisez Moi";

If you build and run the app in the simulator right now, you may not see the new name. iOS seems to cache this information when a new app is added, but doesn't necessarily change it when an existing app is replaced by a new version—at least not when Xcode is doing the replacing. So, if you're running the simulator in French but you don't see the new name, don't worry. Just delete the app from the simulator, go back to Xcode, and build and run the app again.

Now, our application is fully localized for the French language.

Auf Wiedersehen

If you want to maximize sales of your iOS application, localize it as much as possible. Fortunately, the iOS localization architecture makes easy work of supporting multiple languages, and even multiple dialects of the same language, within your application. As you saw in this chapter, nearly any type of file that you add to your application can be localized.

Even if you don't plan on localizing your application, get in the habit of using `NSLocalizedString` instead of just using static strings in your code. With Xcode's Code Sense feature, the difference in typing time is negligible, and should you ever want to translate your application, your life will be much, much easier.

At this point, our journey is nearly done. We're almost to the end of our travels together. After the next chapter, we'll be saying *sayonara*, *au revoir*, *auf wiedersehen*, *avtío*, *arrivederci*, *hej då*, and *adiós.* You now have a solid foundation you can use to build your own cool iOS applications. Stick around for the going-away party though, as we still have a few helpful bits of information for you.

Chapter 22

Where to Next?

Well, wow! You're still with us, huh? Great! It sure has been a long journey since that very first iOS application we built together. You've certainly come a long way. We would love to tell you that you now know it all. But when it comes to technology, and especially when it comes to programming, you never know it all.

At its core, programming is about problem solving and figuring things out. It's fun, and it's rewarding. But, at times, you will run up against a puzzle that just seems insurmountable—a problem that appears to have no solution. Sometimes, the answer will come to you if you just take a bit of time away from the problem. A good night's sleep or a few hours of doing something different can often be all that is needed to get you through it. Believe us—you can stare at the same problem for hours, overanalyzing and getting yourself so worked up that you miss an obvious solution. But sometimes, even a change of scenery doesn't help. In those situations, it's good to have friends in high places. This chapter outlines some resources you can turn to when you're in a bind.

Apple's Documentation

Become one with Xcode's documentation browser, grasshopper. The documentation browser is a front end to a wealth of incredibly valuable sample source code, concept guides, API references, video tutorials, and a whole lot more.

There are few areas of iOS that you won't be able to learn more about by making your way through Apple's documentation. And if you get comfortable with Apple's documentation, navigating through uncharted territories and new technologies as Apple rolls them out will be easier.

> **NOTE:** Xcode's documentation browser takes you to the same information you can get to by
> going to Apple's Developer web site at `http://developer.apple.com`.

Mailing Lists

You might want to sign up for these handy mailing lists:

> **Cocoa-dev**: This moderately high-volume list, run by Apple, is primarily about Cocoa for Mac OS X. Because of the common heritage shared by Cocoa and Cocoa Touch, however, many of the people on this list may be able to help you. (Do make sure to search the list archives before asking your question.)
>
> http://lists.apple.com/mailman/listinfo/cocoa-dev

> **Xcode-users**: Another list maintained by Apple, this one is specific to questions and problems related to Xcode.
>
> http://lists.apple.com/mailman/listinfo/xcode-users

> **Quartz-dev**: This is an Apple-maintained mailing list for discussion of the Quartz 2D and Core Graphics technologies.
>
> http://lists.apple.com/mailman/listinfo/quartz-dev

> **Cocoa-unbound**: This list, intended for discussion of both Mac and iOS development, appeared in 2010 in response to the sometimes heavy-handed moderation of some of the Apple-run lists, particularly Cocoa-dev. The posting volume is lower here, and topics can run a bit further afield.
>
> http://groups.google.com/group/cocoa-unbound

> **IPhone SDK Development**: Another third-party list, this one is focused entirely on iOS development. You'll find a medium-sized community here, with a nice cast of regulars.
>
> http://groups.google.com/group/iphonesdkdevelopment

Discussion Forums

These discussion forums allow you to post your questions to a wide range of forum readers:

> **iphonedevbook.com**: As the official forum for this book, this features an active, vibrant community, full of people with the wisdom and sensibility to buy our book, such as yourself.
>
> http://iphonedevbook.com

Apple Developer Forums: This is a web forum set up by Apple specifically for discussing iOS and Mac software development. Many iOS programmers, both new and experienced (including Apple engineers and evangelists), contribute to these forums. It's also the only place you can legally discuss issues with prerelease versions of the SDK that are under nondisclosure agreements. You'll need to sign in with your Apple ID to access this forum.

```
http://devforums.apple.com
```

Apple Discussions, Developer Forums: This link connects you to Apple's community forums for Mac and iOS software developers:

```
http://discussions.apple.com/category.jspa?categoryID=164
```

Apple Discussions, iPhone: This link connects to Apple's community forums for discussing the iPhone:

```
http://discussions.apple.com/category.jspa?categoryID=201
```

Web Sites

Visit these web sites for helpful coding advice:

CocoaHeads: This is the site of a group dedicated to peer support and promotion of Cocoa. It focuses on local groups with regular meetings where Cocoa developers can get together, help each other out, and even socialize a bit. There's nothing better than knowing a real person who can assist you, so if there's a CocoaHeads group in your area, check it out. If there's not, why not start one?

```
http://cocoaheads.org
```

NSCoder Night: NSCoder Nights are weekly, organized meetings where Cocoa programmers get together to code and socialize. Like CocoaHeads meetings, NSCoder Nights are independently organized local events.

```
http://nscodernight.com
```

Stack Overflow: This is a community Q&A site targeted at programmers. Many experienced iOS programmers hang out here and answer questions.

```
http://stackoverflow.com
```

iDeveloper TV: This is a great resource for in-depth video training in iOS and Mac development, for a price. It also contains some nice, free video content, mostly from NSConference (listed in the "Conferences" section of this chapter), which is run by the same people behind iDeveloper TV.

```
http://ideveloper.tv
```

Cocoa Controls: Here, you'll find a huge range of GUI components for both iOS and Mac OS X. Most of them are free and open source. These controls can be useful as is or as examples for further learning.

`http://cocoacontrols.com/`

Blogs

If you still haven't found a solution to your coding dilemma, you might want to read these blogs:

Wil Shipley's blog: Wil is one of the most experienced Objective-C programmers on the planet. His *Pimp My Code* series of blog postings should be required reading for any Objective-C programmer.

`http://www.wilshipley.com/blog`

Wolf Rentzsch's blog: Wolf is an experienced, independent Cocoa programmer and the founder of the (now defunct) C4 independent developers' conference.

`http://rentzsch.tumblr.com`

iDevBlogADay: This is a multiauthor blog, whose authorship rotates daily among several indie developers of iOS and Mac software. Follow this blog, and you'll be exposed to new insights from different developers every day.

`http://idevblogaday.com`

CocoaCast: This has a blog and podcast about various Cocoa programming topics, available in both English and French.

`http://cocoacast.com/`

@ObjectiveC on Twitter: The @objectivec Twitter user posts about new Cocoa-related blog posts. It's worth a follow.

`http://mobile.twitter.com/objectivec`

Mike Ash's blog: Mike is "just this guy, you know?" This RSS feed presents Mike's collection of his ongoing iOS Friday Q&A.

`http://www.mikeash.com/pyblog/`

Conferences

Sometimes, books and web sites aren't enough. Attending an iOS-focused conference can be a great way to get new insights and meet other developers. Fortunately, this is an area that has really boomed over the past few years, and iOS developers have no shortage of interesting conferences to attend. Here are a few:

WWDC: Apple's World Wide Developer Conference is the annual event where Apple typically unleashes the next great new things for its developer community.

`http://developer.apple.com/wwdc`

MacTech: This is a conference for Mac and iOS programmers and IT professionals. It's hosted by the same people who publish *MacTech Magazine*.

`http://www.mactech.com/conference`

NSConference: This multiple-continent event has been held in both the United Kingdom and United States, so far. It's run and promoted by Steve "Scotty" Scott, perhaps the hardest working man in the Mac/iOS conference scene.

`http://nsconference.com`

360 iDev: This approximately once-a-year conference, which is hosted in either San Jose or Denver (flipping between the two year after year), began in 2009.

`http://www.360idev.com`

iPhone/iPad DevCon: This one is a newcomer. At the time of this writing, it has been held only a couple of times, but it's one to keep an eye on.

`http://www.iphonedevcon.com`

Çingleton: So far, there has been just a single instance of the Çingleton Symposium, in October 2011, but plans are in the works for more. Çingleton won't be a singleton.

`http://www.cingleton.com`

Voices That Matter: This series includes conferences on more than just iOS. Some of the conferences have been focused on other mobile platforms and web development. The iOS and iPhone events have been ongoing since 2009.

`http://www.voicesthatmatter.com`

CocoaConf: The second installment of CocoaConf is just weeks away at the time of this writing, so it will have already happened by the time you read this. But don't worry, there's surely more to come.

`http://www.cocoaconf.com`

Follow the Authors

Dave, Jack, and Jeff are all active Twitter users. You can follow them via @davemark, @jacknutting, and @jeff_lamarche, respectively. They have blogs, too:

- Jeff's iOS development blog contains a lot of great technical material. Be sure to check out the comprehensive series on OpenGL ES.

  http://iphonedevelopment.blogspot.com

  http://www.davemark.com

- Jack uses his blog, nuthole.com, to talk about what's going on in his career and his life (technically and otherwise). It's a blog like many others, but this one is Jack's.

  http://www.nuthole.com

> **TIP:** Are you serious about diving more deeply into the iOS SDK, and especially interested in all the great new functionality introduced with the iOS 5 SDK (of which we only scratched the surface in this book)? If so, you should check out *More iOS 5 Development: Further Explorations of the iOS SDK* by Dave Mark, Alex Horovitz, Kevin Kim, and Jeff LaMarche (Apress, 2012).

And if all else fails, drop us an e-mail at begin5errata@iphonedevbook.com. This is the perfect place to send messages about typos in the book or bugs in *our* code. We can't promise to respond to every e-mail message, but we will read all of them. Be sure to read the errata on the Apress site and the forums on http://iphonedevbook.com/forum before clicking *Send*. And please do write and tell us about the cool applications you develop.

Farewell

The programming language and frameworks we've worked with in this book are the end result of more than 20 years of evolution. And Apple engineers are feverishly working round the clock, thinking of that next cool new thing. The iOS platform has just begun to blossom. There is so much more to come.

By making it through this book, you've built yourself a sturdy foundation. You have a solid knowledge of Objective-C, Cocoa Touch, and the tools that bring these technologies together to create incredible new iPhone, iPod touch, and iPad applications. You understand the iOS software architecture—the design patterns that make Cocoa Touch sing. In short, you're ready to chart your own course. We are so proud!

We sure are glad you came along on this journey with us. We wish you the best of luck and hope that you enjoy programming iOS as much as we do.

Index